T0202933

Lecture Notes in Computer Science   13420

More information about this series at https://link.springer.com/bookseries/558

Claudio Di Ciccio · Remco Dijkman ·
Adela del Río Ortega ·
Stefanie Rinderle-Ma (Eds.)

Business Process Management

20th International Conference, BPM 2022
Münster, Germany, September 11–16, 2022
Proceedings

Springer

Editors
Claudio Di Ciccio ⓘD
Sapienza University of Rome
Rome, Italy

Remco Dijkman ⓘD
Eindhoven University of Technology
Eindhoven, The Netherlands

Adela del Río Ortega ⓘD
Universidad de Sevilla
Seville, Spain

Stefanie Rinderle-Ma ⓘD
Technical University of Munich
Garching, Germany

ISSN 0302-9743   ISSN 1611-3349 (electronic)
Lecture Notes in Computer Science
ISBN 978-3-031-16102-5   ISBN 978-3-031-16103-2 (eBook)
https://doi.org/10.1007/978-3-031-16103-2

Preface

This volume contains the papers presented at the 20th International Conference on Business Process Management (BPM 2022) held during September 11–16, 2022, in Münster, Germany. As with the previous two editions, the BPM 2022 conference still had to deal with the impact of the COVID-19 pandemic, which caused uncertainties and additional workload. Despite all struggles, the BPM community has continued as a determined and flexible community. The fruits of its research contributed to BPM 2022 can be found in the following excellent conference program.

BPM 2021 was held in a hybrid setting, thus giving the attendees the chance to enjoy the conference remotely and to connect again with one another while being physically in the same place whenever possible. BPM 2022 strived for a full in-person celebratory 20th anniversary, flanked by a multitude of events, such as the Blockchain, CEE, and RPA fora, workshops, tutorials, and wonderful social events, that provided the opportunity for exchange of the latest BPM ideas and networking. The program also included three invited keynote talks about seminal topics in business process management.

BPM 2022 followed the history and philosophy of previous editions with respect to the three main research tracks "Foundations" (Track I), "Engineering" (Track II), and "Management" (Track III), reflecting the different communities of the conference series. Track I (chaired by Claudio Di Ciccio) addressed computer science research methods for researching the underlying principles of BPM, computational theories, algorithms, semantics, novel languages, and architectures. Track II (chaired by Remco Dijkman) dealt with engineering aspects of information systems research, including business process intelligence, process mining, process modeling, and process enactment, and employed rigorous and repeatable empirical evaluations. Track III (chaired by Adela del Río Ortega) aimed at advancing the understanding of how BPM can deliver business value, by improving and transforming organizations through process-oriented capabilities, and examining process thinking based on empirical observations. Stefanie Rinderle-Ma served as the Consolidation Chair.

Overall, we received 114 abstract submissions, out of which 97 went into review as full papers. Out of these 97 full paper submissions, 32 accounted for Track I, 35 for Track II, and 30 for Track III. The review process followed the high-quality standards of the BPM conference series. Each paper was reviewed by at least three Program Committee (PC) members of the respective track. Then, an extensive discussion phase between the reviewers and a Senior PC member followed. Finally, the discussion was summarized by the Senior PC member in a meta-review. This thorough review process resulted in seven accepted papers for Track I, six accepted papers for Track II, and nine accepted papers for Track III, totaling 22 contributions included in the main research track (19.3% acceptance rate). Moreover, the review process resulted in the inclusion of 13 papers in the BPM Forum program, published in a separate volume of the Springer LNBIP series; these papers aim at presenting highly innovative research and ideas.

Figure 1 illustrates in the form of a word cloud the variety of topics that are covered in the proceedings. It highlights some subjects that have been of interest to the community for some years already, such as process mining and analysis. However, the figure also evidences that robotic process automation (RPA) has attracted significant interest this year. A number of terms refer to the research methods, core concepts, and data structures to which the community usually resorts (some of those may arouse your curiosity, such as 'workflow graph' and 'execution context'). The main topics of this volume are reflected in the session themes, including analytics, design methods, process mining practice, task mining, systems, and process mining.

Fig. 1. A word cloud illustrating the topics covered in the proceedings.

"Open Science" is a major principle for the BPM community aiming at reproducibility and replicability of the research results. Following the tradition started in 2020, the authors were explicitly requested to link one or more repositories with additional artifacts such as data sets, prototypes, and interview protocols alongside implemented prototypes to their papers.

The BPM 2022 program opened with an exciting keynote on each day of the conference. On the first day, Gero Decker from SAP/Signavio talked about "BPM products for the next 20 years" and shed light on the question of how academia and the BPM community can become an essential part of this development. On the second day, Jan Mendling from Humboldt-Universität zu Berlin, in his talk on "Taking the next steps towards Business Process Science", gave exciting insights into the development of the business process science discipline with a particular focus on the development and strengths of the business process community. On the third day, Chiara Ghidini from the Fondazione Bruno Kessler (FBK) discussed "Perspectives on BPM: an ideal ground for Integrative AI?" and outlined new challenges that arise at the interface of integrative AI and the BPM field. In addition to the keynote abstracts and papers, this volume features papers accompanying five tutorials.

We would like to express our gratitude for the exceptional support of the different BPM conference committees, especially the tracks' Program Committees and Senior Program Committee members. They made the rigorous and extensive review procedure possible and, in turn, enabled the high-quality research output reflected by the papers in this volume. In addition to the committees of the BPM 2022 main track and BPM 2022 Forum, committees for the workshops, the tutorials, the RPA Forum, the Blockchain Forum, the Industry Forum, the Central and Eastern European (CEE) Forum, the Demonstration and Resources Track, the Doctoral Consortium, the BPM Dissertation Award, and the Journal First Track did a tremendous job in reviewing and selecting high-quality contributions to the different tracks and fora.

We also acknowledge our sponsors for their support in making this conference happen: we are very grateful for the platinum sponsorship by celonis, SAP Signavio, and MR.KNOW. Through their financial support, the conference could take place as it did. We are also thankful for our bronze (cronos, Provinzial, and viadee) and academic sponsors (Deutsche Forschungsgemeinschaft, German Research Foundation, DFG), Springer, the University of Münster, and the European Research Center for Information Systems (ERCIS). Finally, we also appreciate the use of EasyChair for streamlining an intensive reviewing process.

Our special thanks go to Jörg Becker as the General Chair of BPM 2022, together with the Organizing Committee Chairs Katrin Bergener and Armin Stein, and their group. The Münster team did an invaluable job in planning and organizing an unforgettable conference, especially in light of the still challenging times we are living in, with the high degree of uncertainty this also adds to the organizational tasks.

Last but not least, we thank you as the readers of this volume and wish you a great experience in diving into the latest BPM research.

September 2022

Claudio Di Ciccio
Remco Dijkman
Adela del Río Ortega
Stefanie Rinderle-Ma

Organization

The 20th International Conference on Business Process Management (BPM 2022) was organized by the University of Münster, and took place in Münster, Germany.

Steering Committee

Mathias Weske (Chair)	HPI, University of Potsdam, Germany
Wil van der Aalst	RWTH Aachen University, Germany
Boualem Benatallah	University of New South Wales, Australia
Jörg Desel	Fernuniversität Hagen, Germany
Marlon Dumas	University of Tartu, Estonia
Jan Mendling	Humboldt Universität zu Berlin, Germany
Manfred Reichert	University of Ulm, Germany
Stefanie Rinderle-Ma	Technical University of Munich, Germany
Hajo Reijers	Utrecht University, The Netherlands
Michael Rosemann	Queensland University of Technology, Australia
Shazia Sadiq	University of Queensland, Australia
Barbara Weber	University of St. Gallen, Switzerland

Executive Committee

General Chair

Jörg Becker	University of Münster, Germany

Main Conference Program Committee Chairs

Claudio Di Ciccio (Track I Chair)	Sapienza University of Rome, Italy
Remco Dijkman (Track II Chair)	Eindhoven University of Technology, The Netherlands
Adela del Río Ortega (Track III Chair)	University of Seville, Spain
Stefanie Rinderle-Ma (Consolidation Chair)	Technical University of Munich, Germany

Workshop Chairs

Cristina Cabanillas	University of Seville, Spain
Agnes Koschmider	Kiel University, Germany
Niels F. Garmann-Johnsen	University of Agder, Norway

Demonstration and Resources Chairs

Christian Janiesch	TU Dortmund, Germany
Chiara Di Francescomarino	Fondazione Bruno Kessler, Italy
Thomas Grisold	University of Liechtenstein, Liechtenstein

Tutorial Chairs

Bettina Distel	University of Münster, Germany
Minseok Song	POSTECH, South Korea

Industry Forum Chairs

Jan vom Brocke	University of Liechtenstein, Liechtenstein
Jan Mendling	Humboldt-Universität zu Berlin, Germany
Michael Rosemann	Queensland University of Technology, Australia

Blockchain Forum Chairs

Raimundas Matulevicius	University of Tartu, Estonia
Qinghua Lu	CSIRO Data61, Australia
Walid Gaaloul	Télécom SudParis, France

RPA Forum Chairs

Andrea Marrella	Sapienza University of Rome, Italy
Bernhard Axmann	Technical University of Ingolstadt, Germany

Central and Eastern European Forum Chairs

Vesna Bosilj Vukšić	University of Zagreb, Croatia
Renata Gabryelczyk	University of Warsaw, Poland
Mojca Indihar Štemberger	University of Ljubljana, Slovenia
Andrea Kő	Corvinus University of Budapest, Hungary

Doctoral Consortium Chairs

Hajo Reijers	Utrecht University, The Netherlands
Robert Winter	University of St. Gallen, Switzerland

BPM Dissertation Award Chair

Jan Mendling	Humboldt Universität zu Berlin, Germany

Journal First Track Chairs

Matthias Weidlich	Humboldt Universität zu Berlin, Germany
Amy Van Looy	Ghent University, Belgium

Publicity Chairs

Flavia Maria Santoro   University of the State of Rio de Janeiro, Brazil
Abel Armas Cervantes   University of Melbourne, Australia
Manuel Resinas   University of Seville, Spain

Organizing Committee Chairs

Katrin Bergener   University of Münster, Germany
Armin Stein   University of Münster, Germany

Track I: Foundations

Senior Program Committee

Chiara Di Francescomarino   Fondazione Bruno Kessler, Italy
Dirk Fahland   Eindhoven University of Technology,
                                     The Netherlands
Chiara Ghidini   Fondazione Bruno Kessler, Italy
Thomas Hildebrandt   University of Copenhagen, Denmark
Richard Hull   New York University, USA
Sander J. J. Leemans   Queensland University of Technology, Australia
Andrea Marrella   Sapienza University of Rome, Italy
Fabrizio Maria Maggi   Free University of Bozen-Bolzano, Italy
Marco Montali   Free University of Bolzano-Bozen, Italy
Oscar Pastor Lopez   Universitat Poliècnica de València, Spain
Artem Polyvyanyy   University of Melbourne, Australia
Manfred Reichert   University of Ulm, Germany
Arthur ter Hofstede   Queensland University of Technology, Australia
Wil van der Aalst   RWTH Aachen University, Germany
Jan Martijn van der Werf   Utrecht University, The Netherlands
Hagen Voelzer   IBM Research Europe, Germany
Matthias Weidlich   Humboldt-Universität zu Berlin, Germany
Mathias Weske   HPI, University of Potsdam, Germany

Program Committee

Lars Ackermann   University of Bayreuth, Germany
Adriano Augusto   University of Melbourne, Australia
Ahmed Awad   University of Tartu, Estonia
Patrick Delfmann   University of Koblenz-Landau, Germany
Rik Eshuis   Eindhoven University of Technology,
                                     The Netherlands
Peter Fettke   DFKI and Saarland University, Germany

Valeria Fionda	University of Calabria, Italy
Ulrich Frank	University of Duisburg-Essen, Germany
María Teresa Gómez-López	University of Seville, Spain
Guido Governatori	CSIRO Data61, Australia
Gianluigi Greco	University of Calabria, Italy
Giancarlo Guizzardi	Free University of Bozen-Bolzano, Italy
Antonella Guzzo	University of Calabria, Italy
Akhil Kumar	Pennsylvania State University, USA
Irina Lomazova	National Research University Higher School of Economics, Russia
Qinghua Lu	CSIRO Data61, Australia
Xixi Lu	Utrecht University, The Netherlands
Felix Mannhardt	Eindhoven University of Technology, The Netherlands
Werner Nutt	Free University of Bozen-Bolzano, Italy
Chun Ouyang	Queensland University of Technology, Australia
Luigi Pontieri	National Research Council of Italy (CNR), Italy
Daniel Ritter	SAP, Germany
Andrey Rivkin	Free University of Bozen-Bolzano, Italy
Arik Senderovich	University of Toronto, Canada
Tijs Slaats	University of Copenhagen, Denmark
Monique Snoeck	KU Leuven, Belgium
Ernest Teniente	Universitat Politècnica de Catalunya, Spain
Eric Verbeek	Eindhoven University of Technology, The Netherlands
Karsten Wolf	University of Rostock, Germany
Francesca Zerbato	University of St. Gallen, Switzerland

Track II: Engineering

Senior Program Committee

Boualem Benatallah	University of New South Wales, Australia
Andrea Burattin	Technical University of Denmark, Denmark
Josep Carmona	Universitat Politècnica de Catalunya, Spain
Jochen De Weerdt	Katholieke Universiteit Leuven, Belgium
Marlon Dumas	University of Tartu, Estonia
Avigdor Gal	Technion, Israel
Massimo Mecella	Sapienza University of Rome, Italy
Jorge Munoz-Gama	Pontificia Universidad Católica de Chile, Chile
Luise Pufahl	TU Berlin, Germany
Hajo A. Reijers	Utrecht University, The Netherlands
Shazia Sadiq	University of Queensland, Australia

Pnina Soffer	University of Haifa, Israel
Boudewijn van Dongen	Eindhoven University of Technology, The Netherlands
Barbara Weber	University of St. Gallen, Switzerland
Ingo Weber	TU Berlin, Germany
Moe Thandar Wynn	Queensland University of Technology, Australia

Program Committee

Marco Aiello	University of Stuttgart, Germany
Robert Andrews	Queensland University of Technology, Australia
Abel Armas Cervantes	University of Melbourne, Australia
Cristina Cabanillas	University of Seville, Spain
Fabio Casati	University of Trento, Italy
Massimiliano de Leoni	University of Padua, Italy
Johannes De Smedt	KU Leuven, Belgium
Benoît Depaire	Hasselt University, Belgium
Joerg Evermann	Memorial University of Newfoundland, Canada
Walid Gaaloul	Télécom SudParis, France
Luciano García-Bañuelos	Tecnológico de Monterrey, Mexico
Laura Genga	Eindhoven University of Technology, The Netherlands
Daniela Grigori	Université Paris Dauphine-PSL, France
Georg Grossmann	University of South Australia, Australia
Mieke Jans	Hasselt University, Belgium
Anna Kalenkova	University of Melbourne, Australia
Dimka Karastoyanova	University of Groningen, The Netherlands
Agnes Koschmider	Kiel University, Germany
Henrik Leopold	Kühne Logistics University, Germany
Francesco Leotta	Sapienza Università di Roma, Italy
Elisa Marengo	Free University of Bozen-Bolzano, Italy
Rabeb Mizouni	Khalifa University, United Arab Emirates
Timo Nolle	Technical University of Darmstadt, Germany
Helen Paik	University of New South Wales, Australia
Cesare Pautasso	University of Lugano, Switzerland
Pierluigi Plebani	Politecnico di Milano, Italy
Pascal Poizat	Université Paris Nanterre, France
Simon Poon	University of Sydney, Australia
Barbara Re	Università di Camerino, Italy
Manuel Resinas	University of Seville, Spain
Stefan Schönig	Universität Regensburg, Germany
Marcos Sepúlveda	Pontificia Universidad Católica de Chile, Chile

Natalia Sidorova	Eindhoven University of Technology, The Netherlands
Renuka Sindhgatta	IBM, India
Minseok Song	POSTECH, South Korea
Niek Tax	Meta, UK
Irene Teinemaa	Bloomberg, The Netherlands
Nick van Beest	CSIRO Data61, Australia
Han van der Aa	University of Mannheim, Germany
Sebastiaan J. van Zelst	RWTH Aachen University, Germany
Seppe Vanden Broucke	Katholieke Universiteit Leuven, Belgium
Karolin Winter	Technical University of Munich, Germany
Nicola Zannone	Eindhoven University of Technology, The Netherlands

Track III: Management

Senior Program Committee

Wasana Bandara	Queensland University of Technology, Australia
Daniel Beverungen	Paderborn University, Germany
Paul Grefen	Eindhoven University of Technology, The Netherlands
Marta Indulska	University of Queensland, Australia
Peter Loos	Saarland University, Germany
Jan Mendling	Humboldt University, Germany
Jan Recker	University of Hamburg, Germany
Maximilian Röglinger	University of Bayreuth, Germany
Michael Rosemann	Queensland University of Technology, Australia
Flavia Maria Santoro	University of the State of Rio de Janeiro, Brazil
Mojca Indihar Štemberger	University of Ljubljana, Slovenia
Peter Trkman	University of Ljubljana, Slovenia
Amy van Looy	Ghent University, Belgium
Jan vom Brocke	University of Liechtenstein, Liechtenstein

Program Committee

Yvonne Lederer Antonucci	Widener University, USA
Marco Comuzzi	Ulsan National Institute of Science and Technology, South Korea
Barbara Dinter	Chemnitz University of Technology, Germany
Bedilia Estrada-Torres	Universidad de Sevilla, Spain
Renata Gabryelczyk	University of Warsaw, Poland
Thomas Grisold	University of Liechtenstein, Liechtenstein

Kanika Goel	Queensland University of Technology, Australia
Tomislav Hernaus	University of Zagreb, Croatia
Christian Janiesch	TU Dresden, Germany
Andrea Kő	Corvinus University of Budapest, Hungary
John Krogstie	Norwegian University of Science and Technology, Norway
Michael Leyer	University of Rostock, Germany
Alexander Mädche	Karlsruhe Institute of Technology, Germany
Martin Matzner	FAU Erlangen-Nürnberg, Germany
Ralf Plattfaut	Fachhochschule Südwestfalen, Germany
Geert Poels	Ghent University, Belgium
Gregor Polancic	University of Maribor, Slovenia
Pascal Ravesteijn	HU University of Applied Sciences Utrecht, The Netherlands
Kate Revoredo	Wirtschaftsuniversität Wien, Austria
Dennis Riehle	University of Muenster, Germany
Stefan Sackmann	University of Halle-Wittenberg, Germany
Estefanía Serral	KU Leuven, Belgium
Oktay Turetken	Eindhoven University of Technology, The Netherlands
Inge van de Weerd	Utrecht University, The Netherlands
Irene Vanderfeesten	Open University of the Netherlands, The Netherlands
Axel Winkelmann	University of Wuerzburg, Germany
Bastian Wurm	Wirtschaftsuniversität Wien, Austria
Michael Zur Muehlen	Stevens Institute of Technology, USA

Additional Reviewers

Faria Khandaker	Lorenzo Rossi	Iris Beerepoot
Vinicius Stein Dani	Christoph Tomitza	Víctor Gálvez
Vladimir Bashkin	Manuel Weber	Laura Lohoff
Christoph Drodt	Hendrik Wache	
Boming Xia	Wouter van der Waal	Gregor Kipping
Yue Liu	Lukas-Valentin Herm	Florian Kragulj
Monika Kaczmarek-Heß	Sandra Zilker	Sebastian Dunzer
Nour Assy	Sven Weinzierl	Johannes Damarowsky
Ebaa Alnazer	Mario Nadj	Peyman Toreiniz
Dominik Janssen	Leonard Nake	
Robin Pesl	Ulrich Gnewuch	Martin Böhmer
Brian Setz	Carolin Vollenberg	Arjen Wierikx

Data, Conceptual Knowledge, and AI: What Can They Do Together? (Abstract of Invited Talk)

Chiara Ghidini

Fondazione Bruno Kessler, Trento, Italy
ghidini@fbk.eu

In the last few years several disciplines have called for approaches that go beyond vertical and separated areas and instead push for integrated approaches. One notable field in which this integrated, or better *integrative*, approach is considered absolutely necessary is Artificial Intelligence (AI), in particular with the integration of symbolic (or knowledge based) and sub-symbolic (or data driven) techniques and representations, or the integration of different vertical areas (e.g., Natural Language Processing and Knowledge Representation). Somehow differently from AI, where areas separated into almost different disciplines along the years, BPM has an integrative nature "by definition". Data are crucial, but mainly to produce explicit and human readable knowledge and models. Similarly, models with no ground on data remain somehow detached from reality. Also, data and knowledge are multi-dimensional, and many interesting results have been obtained by reconciling (or integrating) the object- and process-centric views on data. The explosion of AI, and its increased usage in BPM, should reinforce the integrative nature of BPM research, rather than push for data driven black box solutions. In this talk I will discuss how data, conceptual knowledge, and AI can work together when dealing with specific challenges in process discovery, predictive and proactive process monitoring, and explainability.

Contents

Keynotes

Advancing Business Process Science via the Co-evolution of Substantive and Methodological Knowledge

Jan Mendling[1,2,3]($\boxtimes$) (iD)

[1] Humboldt-Universität zu Berlin, Unter den Linden 6, 10099 Berlin, Germany
jan.mendling@hu-berlin.de
[2] Weizenbaum-Institut e. V., Hardenbergstraße 32, 10623 Berlin, Germany
[3] Wirtschaftsuniversität Wien, Welthandelsplatz 1, 1020 Vienna, Austria

Abstract. The International Conference on Business Process Management (BPM) is a conference series with some remarkable successes over the last 20 years. In this paper, we discuss how neighboring fields have made progress. A key observation is the co-evolution of the problem and solution spaces: methodological innovations yield substantive advancements and, in turn, substantive findings help to improve methods. We discuss implications of this observation for business process science.

Keywords: Business process classification · Substantive knowledge · Methodological knowledge · Types of business processes · Co-evolution of problem and solution spaces

1 Introduction

The International Conference on Business Process Management (BPM) celebrates its 20th edition in 2022. The conference looks back at some major achievements in fostering research on business processes and developing a vital research community around this topic. Key conference papers are highly cited and exemplify the impact of research on business process management [74]. Also the number of participants has constantly grown with a peak of almost 500 at the last pre-Covid conference in Vienna. All these are indications of success.

Success in the past is, however, only a weak predictor of future impact, while innovation is a key factor of change and progress [80]. We know from research in our own field that there are at least two sources inspiring innovation: problems and opportunities [34]. Previous keynotes to the BPM conference have discussed problems. For instance at the 10th anniversary BPM conference, Van der Aalst called for increasing the "concern for real-life use cases" to address the relevance problem of BPM research [3]. By reviewing BPM conference papers from 2003 to 2014, Recker and Mendling identify the need to further develop the rigor of "methodologically strong empirical and theoretical research" as an integral part of the BPM conference [74].

© Springer Nature Switzerland AG 2022
C. Di Ciccio et al. (Eds.): BPM 2022, LNCS 13420, pp. 3–18, 2022.
https://doi.org/10.1007/978-3-031-16103-2_1

These and other problem-oriented reviews of BPM research are essentially inward-looking. They focus on what is done within the BPM research community. In this paper, we aim to complement this inward-looking with an outward-looking perspective. An analysis that is outward-looking is driven by an outsider's perspective, looks at developments in other research fields, and foregrounds opportunities [30]. Such an outsider's perspective of a research domain can be taken regarding two different areas of knowledge: substantive knowledge (what is researched) and methodological knowledge (how is researched) [21, p. 257]. We refer to the field established by both substantive and methodological knowledge on business processes as *business process science* [18,54]. First, we review substantive concepts at the heart of research on business processes and identify key classification concepts. Second, we reflect upon methodological advancements of research areas that have had a substantial impact in recent years. From these reflections, we draw conclusions for the BPM conference and how a science of business processes [18,54] can be further developed.

This paper is structured as follows. Section 2 discusses the research area covering substantive knowledge on business processes. Section 3 describes methodological knowledge on how business processes can be researched. Section 4 reviews recent advancements in fields such as network analysis, image processing, and text processing. Section 5 presents directions for business process science and the BPM conference.

2 Substantive Knowledge on Business Processes

This section makes an attempt at sketching the boundaries of substantive knowledge on business processes. To this end, we first define what a business process is and how it relates to similar notions. Second, we discuss which different classification schemes have been proposed for business processes.

2.1 What is a Business Process?

The notion of a business process refers to a specific plan of action. It is related to a plethora of other notions emphasizing a *plan* (plan of action, modus operandi, strategy, tactic, procedure, approach, method, technique, algorithm), a *performance* (work, routine, operation, practice), or a *potential* (capability, ability, capacity, competence), often without making their mutual distinction explicit. Also at the task level, plenty of terms are used (activity, action, function, step, job, skill) with largely overlapping semantics.

The notion of a business process has some specific characteristics that are distinctive. These are emphasized by several similar definitions of what a business process is. Dumas et al. [30, p. 6] define a

> "business process as a collection of inter-related events, activities, and decision points that involve a number of actors and objects, which collectively lead to an outcome that is of value to at least one customer."

This definition points to two salient characteristics of a business process. First, a business process is a plan of action towards a desired outcome. As such, it shares characteristics with types of change that Van de Ven and Poole call *teleological*: it is driven by goal formulation, followed by implementation, evaluation, and modification of goals based on what was learned by the involved actors [89]. Second, a business process involves a number of actors for executing different activities. The efficiencies of organizing work processes by *division of labour* were first described by Adam Smith in 1776 [83]. However, increasing division of labour does not only provide benefits in terms of specialization and efficiency, but also entails an increasing cost of communication and coordination. The trade-off between benefits provided by division of labour and costs of coordination implies an optimum. This optimum depends on technology [17], which makes business process science a research area that strongly builds on information systems for process coordination and task automation. Due to the involvement of different actors and different pieces of technology, business process science is concerned with social phenomena, technical phenomena, and socio-technical phenomena, as exemplified by the five-level research framework for process mining [19].

2.2 Types of Business Processes

Classification plays an important role for summarizing and ordering knowledge [21, p. 330]. Many classification schemes are relevant for business processes, including those being used for tasks. Research on business processes builds on the distinction between the process and its subordinate tasks. This distinction is conceptual and relative, but not ontological. The process is foregrounded as a white box, while the subordinate tasks are black-boxed and conceptually pushed to the background. Once we foreground one of these tasks, they can be considered as processes on their own. On the one hand, any tasks, as small as it might be, can be made a process by breaking it up into at least two steps to be assigned to two different persons. This principle is generally applicable, either because a task has work components that can be separated or by the option to add a control step to check whether the original task yielded the desired outcome. On the other hand, we can look at business processes at a more abstract level, making them appear as tasks. The name of this abstract task can be derived from, a.o., the main activity, the trigger, or the desired outcome of the process [45].

Principle of Mutability: *Any business process* can be abstracted to appear as a complex task and *any task* can be organized as a business process.

It is a consequence of this observation that classification schemes for tasks are also applicable for business processes. We identify such schemes in four different fields. We discuss them proceeding from coarse-granular to fine-granular.

Research in *economics* has been interested in business processes for a long time. Mind that Adam Smith, mentioned above, is considered as one of the founding fathers of this discipline. In essence, its prime interest has been on the question which impact technological progress has on business processes at the macro-economic level. In recent decades, progress of information technology has been a specific focus. It was found that this progress leads to increasing computerization of not only routine cognitive tasks, but also non-routine analytical tasks and routine manual tasks [11]. The consequence of this technological change is a change of the optimal trade-off between task specialization and coordination [17]. Recent advancements of machine learning are expected to further accelerate business process redesign and impact the workforce at the macro-economic level [20]. Key classification categories in this economics discussion are the routine vs. non-routine and cognitive vs. manual task dichotomy, as well as the degree of how well defined inputs and outputs are. A fine-granular classification scheme is the International Standard Industrial Classification of All Economic Activities (ISIC), available as Revision 4 published by the United Nations Statistics Division [27]. This classification includes more than 750 hierarchically organized productive activities.

Research in *management* has looked at business processes, a.o., from the perspective of organizational learning [47]. The term *organizational routine* is more prominent in this discourse than the term business process [13,31], though this concept largely overlaps with the concept of a business process [94]. Lillrank defines routine processes as those that are repetitively executed in a similar, habituated way, while he considers non-routine processes as not repetitive [48]. He also defines the notion of a standard process that is executed identically according to formal rules or algorithms. This spectrum from standard to non-routine is connected with the notion of complexity. Task complexity [22], later extended to routine complexity [39], is associated with uncertainty and variation in inputs, paths, and outcomes. Also interdependence, knowledge intensity, and differentiation are relevant in this context [98]. Various measures have been defined to calculate complexity from event logs of a business process [10]. Survey-based research found that complexity is negative associated with process performance [64,93].

Research on *work psychology* focuses on the notion of "job", which is closely related to tasks "that employees complete for their organizations on a daily basis" [67]. An important contribution in this domain is the *generalized work activities* taxonomy [70]. This taxonomy organizes 42 work activities in four major categories. First, work in the category *Information Input* is concerned with gathering and evaluating information. Second, *Mental Processes* includes activities of data processing, reasoning, and decision making. Third, *Work Output* covers activities of physical, manual work, and performing complex technical activities. Fourth, *Interacting with Others* refers to activities of communicating and interacting, coordinating and managing, and administering. The business processes of an organization determine which work activities are performed and by use of which technology [86], and the generalized work activities taxonomy helps to analyze them at a fine-granular level.

Research on *linguistics* has focused on the development of a systematic inventory of actions. Most prominent is the work on WordNet [32,63]. WordNet has been used for defining and organizing the MIT Process Handbook [52]. Its hierarchy builds on inheritance relations between verbs and the eight generic verbs defined by WordNet. These eight verbs are *to create, to modify, to preserve, to destroy, to combine, to separate, to decide, to manage*. Other resources have explicitly focused on verbs, such as Levin's 49 verb classes [46], Verbnet [79], FrameNet [12], or more recently UBY [38], which integrates many of the other resources. The application of these works for managing business processes is discussed in [59].

The described classifications are important examples from different fields of research that are interested in tasks and business processes. These classifications differ in terms of their granularity and their focus. Some are developed bottom-up in an enumerative fashion, while others define key criteria top-down. Research on business processes also differs in terms of ends-in-view [56]. For this reason, a key consideration for any research on business processes is to analyze which classification might turn out to be fruitful for particular research objectives. It is one of the strengths of research presented at the BPM conference that it is most often generic. On the downside, there are also opportunities to take differences between business processes more explicitly into account. The mentioned classification schemes can be helpful in this regard.

3   Methodological Knowledge for Studying Business Processes

Various applied research fields are concerned with technological questions such as how can a specific task be done or how can artifacts be designed capable of performing that task [21]. Research on BPM has developed various generic algorithms, methods, and technologies for managing business processes. These have been used in other applied fields for gaining insights into how specific tasks can be done efficiently. Such applied works building on BPM concepts can be found in various industries including construction [68], railways [72], healthcare [53], agriculture [33], or manufacturing [87]. From this perspective, research on BPM provides methodological knowledge for studying business processes.

3.1   How to Research Business Processes?

Methods are central for doing research. They describe how knowledge can be obtained systematically. According to Bunge, a method can be considered scientific if it is *intersubjective* (it does not depend on the person who performs it), *controllable* (it can be checked by other methods), and *explainable* (theories explain how it works) [21, p. 251]. If a method is scientific, it is also called a *research method*. Bunge states that the "task of methodology is to find . . . optimal inquiry strategies" [21]. This means that methodological knowledge is concerned with how we can research, specifically business processes in our case here.

A specific method is tied to a specific objective (greek: *telos*). The teleological nature is something that methods and business processes share. The desired outcome of conducting research by using methods is to obtain novel insights.

If successful, we speak of a *contribution*. The starting point is a *research problem*, that is something that a research community does not yet fully understand. Bunge describes *empirical problems* (measure what is the phenomenon), *theoretical problems* (explain why the phenomenon occurs), and *technological problems* (design how an artifact can achieve a goal efficiently) [21, p. 255]. The nature of the research problem and the aspired contribution point to appropriate methods.

No matter which research problem and aspired contribution, there are some general capabilities that methods for researching business processes have to be able to uncover. These include the observation of events and activities, their ordering, their reference to time, and the outcomes they bring forth [24], the surrounding causal conditions, action strategies, context conditions, and consequences [85], and in particular characteristics of the work system in which business processes are embedded [8]. The focus on desired outcomes, actors and activities, as well as supporting technology are of specific importance due to their constituting role for business processes.

3.2   Types of Process Research Methods

There are various methods that can be used for researching business processes. They can be classified in different ways. We organize methods according to the research paradigm in which they are grounded, the purposes they are used for, the data they use as input, and the outputs they produce. Again, we proceed from coarse-granular to fine-granular.

Research methods can be classified according to the *research paradigm* they relate to. Research on business processes has used methods of formal science, behavioural science, and engineering science [74]. Recommendations for *formal science* are presented in [5, 41]. Methods of *behavioural science* cover the classical spectrum of methods used by the empirical social sciences [73]. *Engineering science* (also called design science) include methods that support the design of technological artifacts [57, 69, 84]. Also the track structure of the BPM conference reflects these three categories.

Research on business process management distinguishes methods along the spectrum of *BPM activities* they support. Various classifications for BPM activities have been proposed [51]. Kettinger et al. identify six stages for structuring business process improvement methods including envision, initiate, diagnose, redesign, restructure, and evaluate [42]. Within the BPM lifecycle, Dumas et al. describe 16 different methods only for supporting process redesign [30]. Cua et al. compare total quality management, just-in-time, and total productive maintenance and find many methods that are also related to business processes [26]. All the methods in this category focus to a large extent on industrial applications.

Research methods can be classified according to which *input data* they use. In *information visualization*, time-oriented data is classified in terms of scale (qualitative or quantitative), reference (abstract or spatial), kind (event or state), and the number of variables (univariate or multivariate) [7, 61]. The *empirical social sciences* distinguish quantitative methods such as survey research or experimental research, qualitative methods including case studies, action research or grounded theory [73]. Recently, computational data processing using large-scale digital trace data has been added as a separate category [73]. A large class of such digital trace data relevant for business processes is event sequence data [95], often stored in event logs [1]. A sequence in this context is an ordered list of events, in which events can be represented as a simple or complex symbolic representation, a numerical value, a vector of real values, or as a complex data type [95]. An example of a method that works with numerical sequences is statistical time series analysis. Examples of methods using symbolic sequences are social sequence analysis [6] and process mining [88]. Support vector machines are an example that can be used for sequence classification using numerical as well as symbolic data. The type of input data determines which kind of algorithmic processing is possible.

Research methods can be classified according to the *output* they produce. Many of the classical deductive research methods yield measures and statistics, while inductive methods help to construct hypotheses. Research on business processes often benefits from visual representations such as models and charts. Methods that produce such representations build on algorithms that are implemented by analysis tools [14]. The visual representations that they produce afford a precision that textual accounts cannot always provide [62]. Analysis tools facilitate the interactive manipulation of these representations for supporting sense-making [29]. A classification of such methods is defined by the ESeVis framework. It distinguishes instance-based and model-based representations as the first axis and the visual arrangement as matrix, timeline, hierarchy, sankey, and chart as the second axis [97]. The strong emphasis on algorithms that produce visual representations is a strength of the BPM conference.

Many methods developed by the BPM community have already been used as a research method. For instance, a study in software engineering uses process mining to analyze the consistency of reporting styles (which are essentially sequences) of experimental studies [75]. Also in BPM, there is a study that compares the sequences proposed by methods using process mining [50]. These examples highlight the potential to apply methods developed in BPM as research methods in other fields [35].

4   Substantive and Methodological Advancements

This section drills down into one specific area of business process science: the development of new and better algorithms for analyzing processes. This area is classically covered by the BPM conference where many process mining algorithms have been presented for the first time. At a first glance, the focus of this area looks like a matter of methodological knowledge. We will discuss this in more detail in the following. First, we revisit insights from research on design and innovation. Then, we look at other fields of algorithm research that made impressive advancements in recent years. From that, we discuss implications for business process science.

4.1   Co-evolution of Problems and Solutions

An algorithmic design provides a solution to an algorithmic problem [57]. It is, however, a fallacy to assume that a solution emerges straight from a fixed and given problem. The research field of design studies emphasizes that design is concerned with *ill-defined problems*, which require a designer to adopt solution-focused strategies and abductive thinking [25]. This means that if the problem is ill-defined, it needs equal exploration as the potential solution. There are important observations supporting this point in software engineering, design studies, and organization science.

Much of the classical research in *software engineering* is driven by the aim to abstract from concrete experiences of success and failure towards prescriptive models for designing systems [76]. One of the early works deviating from this path is the ethnographic study by Guindon on how software developers actually work on a project [36]. She observes two things. First, there is a general tendency that the focus of work shifts from requirements towards high-level design towards implemented solution. Second, however, at all stages there are jumps back to requirements and to high-level design. As a consequence of this pattern, work on requirements and on design only comes to an end when the solution is completed.

In the area of *design studies*, similar observations are made by Dorst and Cross [28]. Their study follows designers who work on a solution to the problem of designing a bin for public trains in the Netherlands. In essence, what they find is the co-evolution of problem and solution. A better understanding of the problem informs the solution, while also progress on the solution side provides a better understanding of the actual problem. This co-evolution implies a step by step refinement of both the problem and solution spaces.

Research in *organization science* has often worked with the assumption that an organizational problem is given, such that problem solving can be viewed as a search for a satisfactory or optimal solution. Von Hippel and von Krogh propose that such a problem formulation is not required. They suggest a search process that is targeted towards the identification of viable problem-solution pairs [91]. They support their proposal with a conceptual argument based on Simon's observations on ill-structured problems [82].

These findings from other fields lead to propositions for research on algorithms supporting process analysis. In this context, algorithmic problems can be matched with algorithm designs as solutions. Research on the co-evolution of problems and solution spaces provides support for the following propositions:

P1: better algorithms leads to a better understanding of algorithmic problems;
P2: better understanding of algorithmic problems leads to better algorithms.

In the context of business process science, this implies that methodological knowledge on how to research business processes should lead to better substantive knowledge, and vice versa. Evidence for the implications in both directions can be found in the history of science and technology [40,44].

4.2   Advancements in Neighboring Fields

Next, we review work in neighboring fields with the aim to check if there is support for our proposition and, if yes, how work of the research community can be organized accordingly. We first turn to graph algorithms, then to natural language processing, and finally to image processing.

In a reflection on the field of algorithm engineering, specifically *graph algorithms*, Sanders highlights the benefits of designing and evaluating algorithms based on a research cycle with falsifiable hypotheses in a Popperian sense and a feedback loop [78]. He describes some "astonishing breakthroughs" thanks to the availability of realistic problem instances. Specifically, he points to the realistic road network data that became available in 2005 and informed algorithmic improvements that provided a speedup of up to six orders of magnitude as compared to the classical Dijkstra algorithm. Key to these methodological advances has been a substantive understanding of real-world data sets and their characteristics to which new algorithms could be tailored.

Word-sense disambiguation is an important problem of *natural language processing*. It takes as input a word mentioned in a specific context and aims to return as an output its explicit semantics [15]. In this way, it addresses the problem of polysemy that words such as "application" have different meanings like "applying for a benefit" or "software system". Solutions for word-sense disambiguation have classically built on knowledge resources such as WordNet [32,63] or BabelNet [65], and more recently also on supervised and hybrid models [15]. Key for the evaluation of new techniques for word-sense disambiguation are goldstandard datasets, in which words are already mapped to their correct semantics. These datasets have to be tagged by humans who need to agree on annotations. Recent techniques yield an F1-score of up to 80% [15]. Notably, the availability of gold standards has not only inspired the development of new techniques, but also informed the definition of new algorithmic tasks such as word-in-context, lexical substitution, or definition modeling [15].

Object recognition is an important problem of *image processing*. As input, it takes an image and provides as output an annotation of objects that are visible in the image. A central resource for research in this area is the ImageNet dataset, which organizes more than 14 million images using the concept hierarchy of WordNet [77]. A separate stream of research on object recognition is concerned with the accurate construction of large-scale datasets. Candidate images are collected using search engine queries. Labels of these images are then determined my humans using crowdsourcing platforms like Amazon Mechanical Turk. Quality assurance builds on gold standard images and worker training [77]. For certain tasks, this yields an accuracy of 97% to 99%. Measures such as precision and recall are used for evaluation of object recognition techniques. Today, techniques are dominated by those that build on convolutional neural networks, strongly informed by available datasets [77]. Also in this field, different novel task types have emerged including image classification, single-object localization, or object detection [77].

In summary, we observe with reference to these research fields that both propositions P1 and P2 have supporting evidence.

5   Implications for Business Process Science

The observations on neighboring fields have implications for business process science. They offer us a base for comparison and a source of inspiration. More specifically, we reflect upon the state of BPM research in relation to substantive knowledge on tasks and datasets as well as on methodological knowledge on algorithms and their evaluation.

Regarding *tasks*, progress has been made in various streams of BPM research informing the definition of new types of tasks and problems. Examples are the challenges described by the process mining manifesto [2], the process mining related use cases [4], the challenges of smart BPM [55], the 25 challenges of semantic process modeling [58], or classes of prediction targets [66]. These are indications that progress on solutions informed the identification of new problems. Some imbalances have to be noted. Some tasks are very popular (35 process discovery algorithms are listed in [9]) while others are hardly considered. The availability of BPI Challenge datasets might explain the preference of researchers for working on process discovery. Another observation relates to the coverage of the described tasks and challenges. The focus on event logs and process models is strong while challenges closer to the tasks of process analysts like finding bottlenecks or potential for automation receive too little attention.

Regarding *datasets*, progress has also been made thanks to available datasets. Examples are the SAP Reference Model [60], the BPM Academic Initiative [43, 92], the Process Model Matching Contest [23], and the annual BPI Challenges since 2011 [49]. The number of datasets has grown, mostly thanks to the BPI Challenges. What has received less attention is the systematic construction of gold standards for certain tasks. Often, the sharing of datasets emerges from an opportunity; less attention is given to the construction of desired datasets. Such desirable datasets could be the full dataset of an ERP system of a large company, a full email record of different actors involved in a specific process, or screen recordings of actors conducting their work on a desktop computer. There is some awareness of the importance of datasets, e.g. by now also accepting resource papers together with the BPM Demos. However, the BPM community does not have anything that compares to WordNet or ImageNet, and nothing that relates datasets to the classification schemes we discussed above. Building such datasets could imspire fully new streams of research.

Regarding *algorithms*, there is on the one hand progress in terms of the number of algorithms that are proposed. Alone for automatic process discovery, Augusto et al. list 35 algorithms [9]. On the other hand, many algorithms focus on popular tasks and build on similar ideas. More diversity of approaches than in process mining can be observe in the field of visual analytics, at least in terms of the visual representations [37,97]. Some initiatives have been launched to increase exchange between both fields, such as the Dagstuhl seminar on "Human in the (Process) Mines". This has the potential to import concepts that are new to the BPM community.

Regarding *evaluations*, there is a strong emphasis on accuracy measures. Precision and recall are highly relevant in this context [71]. Strong emphasis in many papers is on the demonstration of improvements over the most recent best accuracy measures. Such improvement competition rewards overfitting and there is suspicion that recent gains of evaluations using ImageNet could be subject to this problem [16]. Research designs that are hardly used are design studies involving users of analytical systems. Such studies are specifically suited for tasks that are not fully crisp, but expose some fuzziness [81]. Identifying an improvement potential for a business process belongs to this category. Examples of design studies on process mining techniques are [90,96]. Finally, there is the opportunity to study how algorithm performance depends on characteristics of the process and the corresponding dataset. One example in this category is [10].

In summary, there are opportunities for learning from neighboring fields. The experiences of these fields are a solid basis for taking the next steps in advancing business process science by driving the co-evolution of substantive knowledge and methodological knowledge.

Acknowledgement. The research by Jan Mendling was supported by the Einstein Foundation Berlin under Grant No. EPP-2019-524.

References

1. van der Aalst, W.M.P.: Process mining: a 360 degree overview. In: van der Aalst, W.M.P., Carmona, J. (eds.) Process Mining Handbook. LNBIP, vol. 448, pp. 3–34. Springer, Cham (2022). https://doi.org/10.1007/978-3-031-08848-3_1
2. van der Aalst, W., et al.: Process mining Manifesto. In: Daniel, F., Barkaoui, K., Dustdar, S. (eds.) BPM 2011, Part I. LNBIP, vol. 99, pp. 169–194. Springer, Heidelberg (2012). https://doi.org/10.1007/978-3-642-28108-2_19
3. Aalst, W.M.P.: A decade of business process management conferences: personal reflections on a developing discipline. In: Barros, A., Gal, A., Kindler, E. (eds.) BPM 2012. LNCS, vol. 7481, pp. 1–16. Springer, Heidelberg (2012). https://doi.org/10.1007/978-3-642-32885-5_1
4. Van der Aalst, W.M.: Business process management: a comprehensive survey. International Scholarly Research Notices 2013 (2013)
5. van der Aalst, W.M.: How to write beautiful process-and-data-science papers? arXiv preprint arXiv:2203.09286 (2022)
6. Abbott, A.: A primer on sequence methods. Organ. Sci. **1**(4), 375–392 (1990)
7. Aigner, W., Miksch, S., Schumann, H., Tominski, C.: Visualization of Time-Oriented Data, vol. 4. Springer, London (2011). https://doi.org/10.1007/978-0-85729-079-3
8. Alter, S.: Work system theory: overview of core concepts, extensions, and challenges for the future. J. Assoc. Inf. Syst. **14**, 72 (2013)
9. Augusto, A., et al.: Automated discovery of process models from event logs: review and benchmark. IEEE Trans. Knowl. Data Eng. **31**(4), 686–705 (2019). https://doi.org/10.1109/TKDE.2018.2841877

10. Augusto, A., Mendling, J., Vidgof, M., Wurm, B.: The connection between process complexity of event sequences and models discovered by process mining. Inf. Sci. **598**, 196–215 (2022)
11. Autor, D.H., Levy, F., Murnane, R.J.: The skill content of recent technological change: an empirical exploration. Q. J. Econ. **118**(4), 1279–1333 (2003)
12. Baker, C.F., Fillmore, C.J., Lowe, J.B.: The Berkeley FrameNet project. In: COLING 1998 Volume 1: The 17th International Conference on Computational Linguistics (1998)
13. Becker, M.C.: Organizational routines: a review of the literature. Ind. Corp. Change **13**(4), 643–678 (2004)
14. Berente, N., Seidel, S., Safadi, H.: Research commentary-data-driven computationally intensive theory development. Inf. Syst. Res. **30**(1), 50–64 (2019)
15. Bevilacqua, M., Pasini, T., Raganato, A., Navigli, R.: Recent trends in word sense disambiguation: a survey. In: International Joint Conference on Artificial Intelligence, pp. 4330–4338. International Joint Conference on Artificial Intelligence, Inc. (2021)
16. Beyer, L., Hénaff, O.J., Kolesnikov, A., Zhai, X., Oord, A.V.D.: Are we done with ImageNet? arXiv preprint arXiv:2006.07159 (2020)
17. Borghans, L., Ter Weel, B.: The division of labour, worker organisation, and technological change. Econ. J. **116**(509), F45–F72 (2006)
18. vom Brocke, J., et al.: Process science: the interdisciplinary study of continuous change. Available at SSRN 3916817 (2021)
19. vom Brocke, J., Jans, M., Mendling, J., Reijers, H.A.: A five-level framework for research on process mining. Bus. Inf. Syst. Eng. **63**(5), 483–490 (2021). https://doi.org/10.1007/s12599-021-00718-8
20. Brynjolfsson, E., Mitchell, T.: What can machine learning do? Workforce implications. Science **358**(6370), 1530–1534 (2017)
21. Bunge, M.: Treatise on Basic Philosophy Volume 5 Epistemology and Methodology I: Exploring the World. D. Reidel Publishing Company, Netherlands (1983)
22. Campbell, D.J.: Task complexity: a review and analysis. Acad. Manage. Rev. **13**(1), 40–52 (1988)
23. Cayoglu, U., et al.: Report: the process model matching contest 2013. In: Lohmann, N., Song, M., Wohed, P. (eds.) BPM 2013. LNBIP, vol. 171, pp. 442–463. Springer, Cham (2014). https://doi.org/10.1007/978-3-319-06257-0_35
24. Cloutier, C., Langley, A.: What makes a process theoretical contribution? Organ. Theory **1**(1), 2631787720902473 (2020)
25. Cross, N.: The nature and nurture of design ability. Des. Stud. **11**(3), 127–140 (1990)
26. Cua, K.O., McKone, K.E., Schroeder, R.G.: Relationships between implementation of TQM, JIT, and TPM and manufacturing performance. J. Oper. Manage. **19**(6), 675–694 (2001)
27. Division, U.N.S.: International Standard industrial classification of all economic activities (ISIC). No. 4, United Nations Publications (2008)
28. Dorst, K., Cross, N.: Creativity in the design process: co-evolution of problem-solution. Des. Stud. **22**(5), 425–437 (2001)
29. Du, F., Shneiderman, B., Plaisant, C., Malik, S., Perer, A.: Coping with volume and variety in temporal event sequences: strategies for sharpening analytic focus. IEEE Trans. Visual. Comput. Graph. **23**(6), 1636–1649 (2016)
30. Dumas, M., Rosa, M.L., Mendling, J., Reijers, H.A.: Fundamentals of Business Process Management, 2nd edn. Springer, Heidelberg (2018). https://doi.org/10.1007/978-3-662-56509-4

31. Feldman, M.S., Pentland, B.T.: Reconceptualizing organizational routines' as a source of flexibility and change. Adm. Sci. Q. **48**(1), 94–118 (2003)
32. Fellbaum, C. (2010). WordNet. In: Poli, R., Healy, M., Kameas, A. (eds.) Theory and Applications of Ontology: Computer Applications, pp. 231–243. Springer, Dordrecht. https://doi.org/10.1007/978-90-481-8847-5_10
33. Fritz, M., Schiefer, G.: Tracking, tracing, and business process interests in food commodities: a multi-level decision complexity. Int. J. Prod. Econ. **117**(2), 317–329 (2009)
34. Grisold, T., Groß, S., Stelzl, K., vom Brocke, J., Mendling, J., Röglinger, M., Rosemann, M.: The five diamond method for explorative business process management. Bus. Inf. Syst. Eng. **64**(2), 149–166 (2022). https://doi.org/10.1007/s12599-021-00703-1
35. Grisold, T., Wurm, B., Mendling, J., vom Brocke, J.: Using process mining to support theorizing about change in organizations. In: 53rd Hawaii International Conference on System Sciences, HICSS 2020, Maui, Hawaii, USA, 7–10 January 2020, pp. 1–10. ScholarSpace (2020). https://hdl.handle.net/10125/64417
36. Guindon, R.: Designing the design process: exploiting opportunistic thoughts. Hum. Comput. Interact. **5**(2–3), 305–344 (1990)
37. Guo, Y., Guo, S., Jin, Z., Kaul, S., Gotz, D., Cao, N.: A survey on visual analysis of event sequence data. IEEE Trans. Vis. Comput. Graph. (2021). https://doi.org/10.1109/TVCG.2021.3100413
38. Gurevych, I., Eckle-Kohler, J., Hartmann, S., Matuschek, M., Meyer, C.M., Wirth, C.: UBY-a large-scale unified lexical-semantic resource based on LMF. In: Proceedings of the 13th Conference of the European Chapter of the Association for Computational Linguistics, pp. 580–590 (2012)
39. Hærem, T., Pentland, B.T., Miller, K.D.: Task complexity: extending a core concept. Acad. Manage. Rev. **40**(3), 446–460 (2015)
40. Hansson, S.O. (ed.): The Role of Technology in Science: Philosophical Perspectives. PET, vol. 18. Springer, Dordrecht (2015). https://doi.org/10.1007/978-94-017-9762-7
41. ter Hofstede, A.H., Proper, H.A.: How to formalize it? Formalization principles for information system development methods. Inf. Softw. Technol. **40**(10), 519–540 (1998)
42. Kettinger, W.J., Teng, J.T., Guha, S.: Business process change: a study of methodologies, techniques, and tools. MIS Q. **21**, 55–80 (1997)
43. Kunze, M., Luebbe, A., Weidlich, M., Weske, M.: Towards understanding process modeling – the case of the bpm academic initiative. In: Dijkman, R., Hofstetter, J., Koehler, J. (eds.) BPMN 2011. LNBIP, vol. 95, pp. 44–58. Springer, Heidelberg (2011). https://doi.org/10.1007/978-3-642-25160-3_4
44. Latour, B., Woolgar, S.: Laboratory Life: The Construction of Scientific Facts. Princeton University Press, Princeton (2013)
45. Leopold, H., Mendling, J., Reijers, H.A., La Rosa, M.: Simplifying process model abstraction: techniques for generating model names. Inf. Syst. **39**, 134–151 (2014)
46. Levin, B.: English Verb Classes and Alternations: A Preliminary Investigation. University of Chicago press, Chicago (1993)
47. Levitt, B., March, J.G.: Organizational learning. Annu. Rev. Sociol. **14**, 319–340 (1988)
48. Lillrank, P.: The quality of standard, routine and nonroutine processes. Organ. Stud. **24**(2), 215–233 (2003)

49. Lopes, I.F., Ferreira, D.R.: A survey of process mining competitions: the BPI challenges 2011–2018. In: Di Francescomarino, C., Dijkman, R., Zdun, U. (eds.) BPM 2019. LNBIP, vol. 362, pp. 263–274. Springer, Cham (2019). https://doi.org/10.1007/978-3-030-37453-2_22
50. Malinova, M., Gross, S., Mendling, J.: A study into the contingencies of process improvement methods. Inf. Syst. **104**, 101880 (2022)
51. Malinova, M., Mendling, J.: Identifying do's and don'ts using the integrated business process management framework. Bus. Process Manage. J. **24**, 882–899 (2018)
52. Malone, T.W., Crowston, K., Herman, G.A.: Organizing Business Knowledge: The MIT Process Handbook. MIT Press, Cambridge (2003)
53. McNulty, T., Ferlie, E.: Reengineering Health Care: the Complexities of Organizational Transformation. OUP, Oxford (2002)
54. Mendling, J.: From scientific process management to process science: towards an empirical research agenda for business process management. In: ZEUS, pp. 1–4 (2016)
55. Mendling, J., Baesens, B., Bernstein, A., Fellmann, M.: Challenges of smart business process management: an introduction to the special issue. Decis. Support Syst. **100**, 1–5 (2017). https://doi.org/10.1016/j.dss.2017.06.009
56. Mendling, J., Berente, N., Seidel, S., Grisold, T.: The philosopher's corner: pluralism and pragmatism in the information systems field: the case of research on business processes and organizational routine. ACM SIGMIS Database DATABASE Adv. Inf. Syst. **52**(2), 127–140 (2021)
57. Mendling, J., Depaire, B., Leopold, H.: Theory and practice of algorithm engineering. arXiv preprint arXiv:2107.10675 (2021)
58. Mendling, J., Leopold, H., Pittke, F.: 25 challenges of semantic process modeling. Int. J. Inf. Syst. Softw. Eng. Big Co. **1**(1), 78–94 (2015)
59. Mendling, J., Recker, J., Reijers, H.A.: On the usage of labels and icons in business process modeling. Int. J. Inf. Syst. Model. Des. (IJISMD) **1**(2), 40–58 (2010)
60. Mendling, J., Verbeek, H., van Dongen, B.F., van der Aalst, W.M., Neumann, G.: Detection and prediction of errors in EPCs of the SAP reference model. Data Knowl. Eng. **64**(1), 312–329 (2008)
61. Mennis, J.L., Peuquet, D.J., Qian, L.: A conceptual framework for incorporating cognitive principles into geographical database representation. Int. J. Geogr. Inf. Sci. **14**(6), 501–520 (2000)
62. Meyer, R.E., Höllerer, M.A., Jancsary, D., Van Leeuwen, T.: The visual dimension in organizing, organization, and organization research: core ideas, current developments, and promising avenues. Acad. Manage. Ann. **7**(1), 489–555 (2013)
63. Miller, G.A.: WordNet: An Electronic Lexical Database. MIT Press, Cambridge (1998)
64. Münstermann, B., Eckhardt, A., Weitzel, T.: The performance impact of business process standardization. Bus. Process Manage. J. **30**, 125–134 (2010)
65. Navigli, R., Bevilacqua, M., Conia, S., Montagnini, D., Cecconi, F.: Ten years of BabelNet: a survey. In: IJCAI, pp. 4559–4567 (2021)
66. Neu, D.A., Lahann, J., Fettke, P.: A systematic literature review on state-of-the-art deep learning methods for process prediction. Artif. Intell. Rev. **5**, 801–827 (2022)
67. Oldham, G.R., Fried, Y.: Job design research and theory: past, present and future. Organ. Behav. Hum. Decis. Process. **136**, 20–35 (2016)
68. Pan, Y., Zhang, L.: A BIM-data mining integrated digital twin framework for advanced project management. Autom. Constr. **124**, 103564 (2021)

69. Peffers, K., Tuunanen, T., Rothenberger, M.A., Chatterjee, S.: A design science research methodology for information systems research. J. Manage. Inf. Syst. **24**(3), 45–77 (2007)
70. Peterson, N.G., et al.: Understanding work using the occupational information network (O*NET): implications for practice and research. Pers. Psychol. **54**(2), 451–492 (2001)
71. Polyvyanyy, A., Solti, A., Weidlich, M., Ciccio, C.D., Mendling, J.: Monotone precision and recall measures for comparing executions and specifications of dynamic systems. ACM Trans. Softw. Eng. Methodol. (TOSEM) **29**(3), 1–41 (2020)
72. Rau, I., Rabener, I., Neumann, J., Bloching, S.: Managing environmental protection processes via BPM at Deutsche Bahn. In: vom Brocke, J., Mendling, J. (eds.) Business Process Management Cases. MP, pp. 381–396. Springer, Cham (2018). https://doi.org/10.1007/978-3-319-58307-5_20
73. Recker, J.: Scientific Research in Information Systems: A Beginner's Guide, 2nd edn. Springer, Heidelberg (2021). https://doi.org/10.1007/978-3-642-30048-6
74. Recker, J., Mendling, J.: The state of the art of business process management research as published in the bpm conference. Bus. Inf. Syst. Eng. **58**(1), 55–72 (2016). https://doi.org/10.1007/s12599-015-0411-3
75. Revoredo, K., Djurica, D., Mendling, J.: A study into the practice of reporting software engineering experiments. Empir. Softw. Eng. **26**(5), 113 (2021). https://doi.org/10.1007/s10664-021-10007-3
76. Royce, W.W.: Managing the development of large software systems-concepts and techniques. In: Proceedings of IEEE WESCON, August 1970, pp. 1–9 (1970)
77. Russakovsky, O., et al.: ImageNet large scale visual recognition challenge. Int. J. Comput. Vis. **115**(3), 211–252 (2015)
78. Sanders, P.: Algorithm engineering – an attempt at a definition. In: Albers, S., Alt, H., Näher, S. (eds.) Efficient Algorithms. LNCS, vol. 5760, pp. 321–340. Springer, Heidelberg (2009). https://doi.org/10.1007/978-3-642-03456-5_22
79. Schuler, K.K.: VerbNet: A Broad-Coverage, Comprehensive Verb Lexicon. University of Pennsylvania, Philadelphia (2005)
80. Schumpeter, J.A.: The analysis of economic change. Rev. Econ. Stat. **17**(4), 2–10 (1935)
81. Sedlmair, M., Meyer, M., Munzner, T.: Design study methodology: reflections from the trenches and the stacks. IEEE Trans. Vis. Comput. Graph. **18**(12), 2431–2440 (2012)
82. Simon, H.A.: The structure of ill structured problems. Artif. Intell. **4**(3–4), 181–201 (1973)
83. Smith, A.: The Wealth of Nations: An inquiry into the nature and causes of the Wealth of Nations. Harriman House Limited (2010/1776)
84. Staples, M.: Critical rationalism and engineering: methodology. Synthese **192**(1), 337–362 (2014). https://doi.org/10.1007/s11229-014-0571-6
85. Strauss, A., Corbin, J.: Basics of Qualitative Research. Sage publications, Thousand Oaks (1990)
86. Tijdens, K.G., De Ruijter, E., De Ruijter, J.: Comparing tasks of 160 occupations across eight european countries. Employee Relations **36**, 110–127 (2014)
87. Tonelli, F., Demartini, M., Loleo, A., Testa, C.: A novel methodology for manufacturing firms value modeling and mapping to improve operational performance in the industry 4.0 era. Procedia CIRP **57**, 122–127 (2016)
88. Van Der Aalst, W.: Process Mining: Data Science in Action, 2nd edn. Springer, Heidelberg (2016). https://doi.org/10.1007/978-3-662-49851-4

89. Van de Ven, A.H., Poole, M.S.: Explaining development and change in organizations. Acad. Manage. Rev. **20**(3), 510–540 (1995)
90. Vidgof, M., Djurica, D., Bala, S., Mendling, J.: Interactive log-delta analysis using multi-range filtering. Softw. Syst. Model. **21**(3), 847–868 (2022). https://doi.org/10.1007/s10270-021-00902-0
91. Von Hippel, E., Von Krogh, G.: Crossroads-identifying viable "need-solution pairs": problem solving without problem formulation. Organ. Sci. **27**(1), 207–221 (2016)
92. Weske, M., Decker, G., Dumas, M., La Rosa, M., Mendling, J., Reijers, H.A.: Model collection of the business process management academic initiative. Zenodo (2020)
93. Wüllenweber, K., Beimborn, D., Weitzel, T., König, W.: The impact of process standardization on business process outsourcing success. Inf. Syst. Front. **10**(2), 211–224 (2008)
94. Wurm, B., Grisold, T., Mendling, J., vom Brocke, J.: Business process management and routine dynamics. In: Cambridge Handbook of Routine Dynamics, pp. 513–524. Cambridge University Press (2021)
95. Xing, Z., Pei, J., Keogh, E.: A brief survey on sequence classification. ACM SIGKDD Explor. Newsl. **12**(1), 40–48 (2010)
96. Yeshchenko, A., Ciccio, C.D., Mendling, J., Polyvyanyy, A.: Visual drift detection for event sequence data of business processes. IEEE Trans. Vis. Comput. Graph. **28**(8), 3050–3068 (2022). https://doi.org/10.1109/TVCG.2021.3050071
97. Yeshchenko, A., Mendling, J.: A survey of approaches for event sequence analysis and visualization using the esevis framework. CoRR abs/2202.07941 (2022). https://arxiv.org/abs/2202.07941
98. Zelt, S., Recker, J., Schmiedel, T., Vom Brocke, J.: Development and validation of an instrument to measure and manage organizational process variety. PloS One **13**(10), e0206198 (2018)

Tutorials

BPM in Digital Transformation: New Tools and Productivity Challenges

Joaquín Peña[✉][iD], Alfonso Bravo[iD], and Manuel Resinas[iD]

Universidad de Sevilla, Sevilla, Spain
joaquinp@lsi.us.es

Abstract. Digital transformation (DT) has brought an unprecedented pace of change. At the same time, it has also created an environment where knowledge workers have to deal with an increasingly Volatile, Uncertain, Complex, and Ambiguous (VUCA) workplace. In this scenario, the design, development, implementation, execution, and evolution of business processes have changed in the last years. In this tutorial, we cover two consequences of these changes that deserve special attention for the impact they can have in the near future: (i) the new tools being used to support the execution of processes, and (ii) the human aspect of process execution since in this new context -multiple changing processes executed in parallel- productivity challenges appear that affects directly process performance. We illustrate how these new tools are used to manage processes, and the challenges to be addressed for research and practice using real case studies extracted from empirical studies (+1500 participants) and transfer projects with +14000 direct users affected from SMEs and international companies in different sectors (commodities, engineering, manufacturing, banking, retail, etc.) using the productivity methodology we have developed for addressing those projects: The FAST Productivity Methodology.

Keywords: Business process management · Collaborative work · Productivity · Digital transformation · Work stream collaboration tools · Board-based tools

1 Introduction and Motivation

Digital transformation is changing the way we work and do business. Although it presents an opportunity to use technology for improving productivity, many new challenges have appeared in the last years [18] mainly derived from the Volatile, Uncertain, Complex, and Ambiguous (VUCA) environments in which businesses are immersed today [3]. Those changes affect how processes are managed and implemented, from their design to their monitoring and optimization, challenging traditional approaches to BPM [2].

Among others, there are two factors affecting processes that are especially relevant: (i) the popularization and generalization of a new category of tools

© Springer Nature Switzerland AG 2022
C. Di Ciccio et al. (Eds.): BPM 2022, LNCS 13420, pp. 21–26, 2022.
https://doi.org/10.1007/978-3-031-16103-2_2

that can be used for implementing processes: Work Stream Collaboration Tools (WSCT) such as Microsoft Teams or Slack, and Board-Based Tools (BBT), such as Trello or MS Planner [10,12]; and (ii) the human aspect of process execution since in this context - with multiple changing processes executed in parallel with unstructured work - new productivity challenges appear affecting directly process performance.

WSCTs are highly flexible and configurable tools that allow a team of knowledge workers to perform online conversations, file sharing, and collaborative task management, amongst others. This flexibility has proven to be beneficial to executing the growing number of unstructured and continuously changing work processes performed by organizations nowadays [5].

Regarding the human perspective of process execution, process participants execute multiple processes in parallel using the aforementioned tools, email, or other information systems. They are constantly making decisions about which tasks or processes to prioritize, they have to deal with information overload, and they are affected by interruptions and their motivation [18]. This situation creates severe challenges for the productivity of knowledge workers that directly affect process performance and hence, need to be addressed like other aspects of process performance.

In this tutorial, we first detail WSCTs and BBTs and how they can be used to execute business processes (cf. Sect. 2). Then, we discuss the human perspective of process execution and which productivity challenges can negatively affect the productivity of BPs (cf. Sect. 3). Later, we report on some solutions for the main challenges and real case studies derived from our experience applying the FAST Productivity Methodology in many projects with SME and big companies reaching more than 14,000 users (cf. Sect. 4). Finally, we discuss the implications for future research and practice (cf. Sect. 5).

2   New Tools for Executing Processes: WSCT and BBT

Traditional BPM software systems (BPMS) for executing and managing processes usually require a software team that elicits the requirements and implement software to manage the processes. This is a long process that makes the evolution of processes difficult. However, in a VUCA environment, processes change very rapidly to adapt to the changes in the business. This problem, which has also been identified in the context of digital transformation, causes software cannot keep the pace of changes and becomes obsolete before being useful [2]. In contrast, WSCTs and BBTs provide flexible tools for collaborative work that are intensively used to manage formal and informal processes. The processes managed with these tools are usually subject to frequent changes or are not yet implemented with traditional BPMSs.

Thus, when implementing processes, we must consider two new categories of tools: WSCT and BBT. On the one hand, WSC is a concept coined in 2018 that refers to "products that deliver a persistent conversational workspace for group collaboration and can be arranged into public or private channels (often organized by topic/project)" [9]. These tools are designed to improve team coordination, performance, communications, and productivity [12]. This emergent class of collaboration technology can combine a diverse number of features, including instant messaging, calls, optimised search, (shared) calendars and notifications, real-time document collaboration, task managers, and cloud storage with version control, amongst others. They also typically integrate with other enterprise applications and bots and can be accessed on mobile or desktop devices. Thus, WSC tools are very powerful and flexible but need rules and methodologies to be correctly used [7]. However, there is neither previous experience nor a strong research body that offer guidelines to design good solutions based on WSCTs [18].

On the other hand, BBTs such as Trello or Planner are structured around boards that contain cards organized in lists. This structure allows users to organize a wide variety of formal or informal information and work processes flexibly. For instance, cards can represent process instances, and their evolution can be implemented by moving cards between lists. In addition, these tools can be integrated into WSCTs, complementing informal communication with a more structured definition of work processes. The flexibility of BBTs means that in every situation, the user is required to design new boards from scratch, which is not a straightforward task, especially for non-technical users. To alleviate this problem, in [16], the authors developed 8 BBT design patterns, from which four are devoted to processes helping users design the board for different purposes.

3   The Human Perspective of Process Execution: Productivity Challenges

Although these new tools allow a new flexible and rapid way of implementing processes, the human perspective of its execution also presents many challenges. Today, knowledge workers perform not only structured work defined by processes but also structured work in the form of projects, unstructured or uncertain projects, and unplanned work. The work in this last category usually reaches the worker through unexpected interruptions, emails, or instant messages. Thus, structured work of processes must compete with interruptions for workers' attention. This means that a very well-designed process can have inferior performance if the participants have, for example, a high number of interruptions.

Consequently, for process improvement, it is necessary to monitor, analyze, and improve the situation not only from the process perspective, but also from the worker's perspective. There is already some work headed in that direction. For instance, Pika et al. [17] describes a framework for analyzing and evaluating resource behavior like utilization, preferences, productivity, or collaboration patterns through mining process event logs. Palvalin [15] introduces a conceptual model of knowledge work productivity, which consists of two significant elements: work environment, which includes physical, virtual, and social environment, and knowledge worker, which includes individual work practices and well-being at work. Finally, in del-Río-Ortega et al. [18], the authors perform an empirical study with 365 knowledge workers that were using WSCTs from three companies. The result is a set of 14 productivity challenges, namely: interruptions, prioritization/goals, organizational coordination, planning/task management, work overload, lack of knowledge, email management, volatility, lack of focus, bureaucracy, meetings, software, motivation, and information overload.

4   Solutions to Focus, Achieve, Sustain and Target

As a result of analyzing the solutions to those 14 challenges and our industry experience applying them, we organize the solutions we must take into account for improving productivity and managing processes in four main principles corresponding to the letters F.A.S.T. as follows:

- **F**: *Focus*: In an environment with rapid changes we need to define, review, and rapidly update priorities. To achieve this, agile methods using BBTs to implement a *business compass* can be used. A *business compass* provides a dashboard where the priorities between processes/projects or individual goals/responsibilities are represented and periodically reviewed [5,8].
- **A**: *Achieve*: In a context with many tasks from different sources, unplanned tasks, steps of a process, or tasks from collaborative projects, workers need a reliable storage that collects all types of individual and collective tasks. This storage is usually called an *external brain*. Workers also need methodologies to coordinate work (planning, synchronization, and retrospectives), mainly agile methods. Both storage and method help systematize and give order to the communication between workers using WSCT [4–6].
- **S**: *Sustain*: The amount of information received nowadays is very high. In this context, for maintaining the path towards our goals, it is mandatory to filter inputs to focus on what is important for the business, automatically (email rules, or AI filters) and manually, deciding and registering the work to be done in the *external brain* [1]. In addition, humans are not machines, for sustaining motivation and productive energy, workers must nurture their lifestyle (sleep, exercise, eating, etc.) and implement tools to control work stress such as creating task lists or systematizing decision-making [13].
- **T**: *Target*: In the context of an elevated number of interruptions, we need tools to improve concentration. For that, we can use a tool called the *concentration bubble* that consists of time and space boxing (blocking time in the calendar and switching off communication) to create organizational spaces free of interruptions for doing focused work [11,14].

Figure 1 shows the relation between the 14 productivity challenges described in [18] and these four principles. Green cells with an "x" indicate that the principle at hand includes tools that significantly address a challenge. Grey cells with an "~" indicate that the principle includes tools that partially contribute to solving the challenge. As seen in the table, all challenges are covered by at least one principle.

Challenges /Principles	Interruptions	Prioritisation/goals	Organisational coordination	Planning/task management	Work overload	Lack of knowledge or training	Email management	Volatility	Concentration/focus	Bureaucracy	Meetings	Software	Motivation	Information overload
Focus	~	x	~	~	~			~	~			~	x	~
Achieve		~	x	x	~		~	x	~	x	x	~	x	
Sustain: Filter	x	~			x	x	x	~	~			~	~	x
Sustain: Energy/ lifestyle		~		~	~	x			~			~	x	
Target	x	~		~					x				~	x

Fig. 1. Relation between challenges and the FAST principles

Based on these principles, we have developed *The FAST Productivity Methodology*[1] to apply them. This methodology has been successfully used in many projects of digital transformation of SME and international companies, reaching more than 14,000 users.

5   Conclusions and Future Research Challenges

Knowledge workers in VUCA digitized environments have developed new ways of working and executing processes. In addition, current digital transformation environments and tools bring new productivity challenges directly affecting process performance. In this new context, there are two main conclusions for future research:

- The use of new tools to run processes. These new tools, although flexible, can be structured to execute processes. Further research is necessary to design, monitor, and manage the evolution of processes in those tools.
- The importance of the context of process participants. Users are usually executing multiple processes in parallel together with other unstructured work. This forces them to continuously decide which tasks or processes to prioritize and creates constant interruptions that affect their workflow. Dealing with these productivity challenges has an impact on process performance. Therefore, further research is necessary to detect, analyze and resolve these productivity challenges. Furthermore, the context of process participants should be considered for process (re-)design, monitoring, and execution management.

---

[1] www.fastproductivity.com.

Acknowledgements. This work has been funded by RTI2018-101204-B-C22 funded by MCIN/AEI/10.13039/501100011033/ and ERDF A way of making Europe; grant P18-FR-2895 (Junta de Andalucia/FEDER, UE); and US-1381595 (US/JUNTA/FEDER, UE).

References

1. Allen, D.: Getting Things Done: the Art of Stress-Free Productivity. Penguin, New York (2015)
2. Baiyere, A., Salmela, H., Tapanainen, T.: Digital transformation and the new logics of business process management. Eur. J. Inf. Syst. **29**(3), 238–259 (2020)
3. Bennett, N., Lemoine, J.: What VUCA really means for you. Harv. Bus. Rev. **92**(1/2) (2014)
4. Bettoni, M., Bernhard, W., Bittel, N., Mirata, V.: The art of new collaboration: three secrets. In: ECKM 2018, pp. 1133–1141, July 2018
5. Busse, R., Weidner, G.: A qualitative investigation on combined effects of distant leadership, organisational agility and digital collaboration on perceived employee engagement. Leadersh. Org. Dev. J. **41**(4), 535–550 (2020)
6. Chasanidou, D., Elvesaeter, B., Berre, A.J.: Enabling team collaboration with task management tools. In: Proceedings of the 12th International Symposium on Open Collaboration, pp. 1–9 (2016)
7. Dawson, R., Hough, J., Hill, J., Winterford, B., Alexandrov, D.: Implementing Enterprise 2.0. Advanced Human Technologies, Sydney (2009)
8. Denning, S.: The age of agile. Strategy Leadersh. **45**(1), 3–10 (2017)
9. Gotta, M., Dewnarain, G., Preset, A.: Market Guide for Workstream Collaboration. Gartner Research, Stamford (2018)
10. Gotta, M., Preset, A., Elliot, B.: Embrace Workstream Collaboration to Transform Team Coordination and Performance. Gartner Research, Stamford (2017)
11. Gupta, A., Li, H., Sharda, R.: Should I send this message? Understanding the impact of interruptions, social hierarchy and perceived task complexity on user performance and perceived workload. Decis. Support Syst. **55**(1), 135–145 (2013)
12. Kerravala, Z., Michels, D.: Business Agility Drives the Need for Workstream Communications and Collaboration. ZK Research, Westminster (2015)
13. McKeown, G.: Essentialism: the Disciplined Pursuit of Less. Currency, Redfern (2020)
14. Newport, C.: Deep Work: Rules for Focused Success in a Distracted World. Hachette, Paris (2016)
15. Palvalin, M.: What matters for knowledge work productivity? Empl. Relat. **41**(1), 209–227 (2019)
16. Peña, Joaquín, Bravo, Alfonso, del-Río-Ortega, Adela, Resinas, Manuel, Ruiz-Cortés, Antonio: Design patterns for board-based collaborative work management tools. In: La Rosa, Marcello, Sadiq, Shazia, Teniente, Ernest (eds.) CAiSE 2021. LNCS, vol. 12751, pp. 177–192. Springer, Cham (2021). https://doi.org/10.1007/978-3-030-79382-1_11
17. Pika, A., Leyer, M., Wynn, M.T., Fidge, C.J., Hofstede, A.H.M.T., Aalst, W.M.P.V.D.: Mining resource profiles from event logs. ACM Trans. Manage. Inf. Syst. (TMIS) **8**(1), 11–130 (2017)
18. del Río-Ortega, A., Peña, J., Resinas, M., Ruiz-Cortés, A.: Productivity challenges in digital transformation and its implications for workstream collaboration tools. In: HICSS 2021, pp. 1–10. ScholarSpace/AIS Electronic Library (AISeL) (2021)

Multi-dimensional Process Analysis

Dirk Fahland(✉)(iD)

Eindhoven University of Technology, Eindhoven, The Netherlands
d.fahland@tue.nl

Abstract. Processes are complex phenomena that emerge from the interplay of human actors, materials, data, and machines. Process science develops effective methods and techniques for studying and improving processes. The BPM field has developed mature methods and techniques for studying and improving process executions from the control-flow perspective, and the limitations of control-flow focused thinking are well-known. Current research explores concepts from related disciplines to study behavioral phenomena "beyond" control-flow. However, it remains challenging to relate models and concepts of other behavioral phenomena to the dominant control-flow oriented paradigm.

This tutorial introduces several recently developed simple models that naturally describe behavior beyond control-flow, but are inherently compatible with control-flow oriented thinking. We discuss the *Performance Spectrum* to study performance patterns and their propagation over time, *Event Knowledge Graphs* to study networks of behavior over data objects and actors, and *Proclets* as a formal model for reasoning over control-flow, data object, queue and actor behavior. For each model, we discuss which phenomena can be studied, which insights can be gained, which tools are available, and to which other fields they relate.

Keywords: Process management · Process thinking · Process mining

1 Introduction

Processes are complex phenomena that emerge when human actors process materials and data using various tools (digital and physical). *Process science* is the discipline of studying, managing, and improving processes (and how they change) by developing effective methods and techniques to do so [3]. The fields of *Business Process Management* (BPM) and *Process Mining* study *process behavior*, and have developed mature theories, models, and solutions for studying and improving process executions, primarily, from the *control-flow* perspective [1,7] which describes the order in which activities are performed on a case, i.e., in a process execution.

The limitations of control-flow focused thinking about process behavior are well-known within BPM [16], leading to active research on how to incorporate ideas from other disciplines to study behavioral phenomena "beyond" control-flow such as queuing theory [17], organizations and routines research [2], and

© Springer Nature Switzerland AG 2022
C. Di Ciccio et al. (Eds.): BPM 2022, LNCS 13420, pp. 27–33, 2022.
https://doi.org/10.1007/978-3-031-16103-2_3

databases [4]. However, it remains challenging to relate models of other behavioral phenomena to the dominant control-flow oriented paradigm. A significant challenge is to make pragmatically related concepts (e.g., "we execute the process by creating and updating a number of related documents") also formally compatible so that different phenomena can be studied naturally.

This tutorial gently *introduces the audience to a different mindset of thinking about and analyzing process behavior beyond control-flow.* Section 2 maps out two fundamental dimensions in thinking about process behavior. Section 3 details the concepts and ready-to-use tools presented in the tutorial to let participants make first steps on their own on publicly available data.

2   Dimensions in Process Thinking

Studying processes comes with a specific way of thinking. *Process thinking* essentially considers that "everything flows" which inherently requires to study emergence within dynamics [18].

For processes owned by or involving by humans (in contrast to natural processes) [3], process thinking focuses on understanding, designing, and implementing goal-oriented behaviors in social and technical systems and organizations of all kinds and sizes. Here, process thinking essentially structures the flow of information and material between various actors and resources in terms of processes: several coherent steps designed to achieve common and individual goals together [7]. Throughout a process, multiple actors, resources, physical objects and information entities interact and synchronize with each other.

The *scope of process thinking* varies depending on "how much dynamics" to consider. A most basic classification covers the following questions:

- "How many entities describe a dynamic?" – the *inner scope* of process thinking.
- "How many dynamics to consider?" – the *outer scope* of process thinking.

The most basic answers to each are "one" and "multiple", giving rise to the four quadrants of multi-dimensional process thinking shown in Fig. 1.

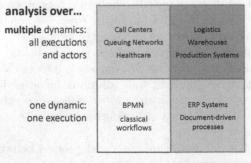

Fig. 1. Four quadrants of multi-dimensional process thinking

Focusing on Process Executions

Processes are primarily studied through the dynamics of their executions. Thereby each execution is understood as a single dynamic, i.e., a separate object of study.

One Execution - Single Entity. Standard industrial process modeling languages, such as BPMN [7], and classical process mining [1] focus primarily on describing and analyzing information handling dynamics as they are found in many administrative procedures, for instance in insurance companies or universities.

From this angle, each process is scoped in terms of individual cases or individual documents, e.g., a student application or an insurance claim. The information in a case is processed independently of other cases along a single process description, often in a workflow system. In terms of Fig. 1, process thinking here encompass a *single-dimensional inner scope* (information flow within a case) structured into a *single-dimensional outer scope* (study flow per case).

One Execution - Multiple Entities. Most organizations, for example in manufacturing and retail, operate multiple processes over shared data and materials, such as Order-to-Cash or Purchase-to-Pay processes. These processes are often supported by complex systems such as Enterprise Resource Planning (ERP) systems.

Processes here are centered around updating and managing a collection of related documents [15]. Process thinking in this scenario requires to consider the interlinked dynamics of multiple processes and objects together (*multi-dimensional inner scope*) – with the aim of reasoning about each specific dynamic, i.e., one customer order, individually (*single-dimensional outer scope*).

Taking the System and Organization into the Picture

The dynamics of each case relies on the system that moves it forward, for example, the actors and machines doing the actual work. There are more dynamics to consider:

- How does the involvement of actors and machines influence the dynamics of a case, e.g., by their availability, workload, and capabilities?
- How do changes to physical properties of materials and machines involved in the process influence the dynamics of a case?
- How do the underlying systems influence the dynamics, for instance through queueing, prioritizing, assigning of work, or (reliability of) automation of steps?

However, the availability of actors, materials, and systems, in turn, is influenced by the amount, nature, and progress of all cases. Studying these interdependencies requires to consider all cases in the process together (and not in isolation).

Multiple Executions - One Entity. Processes for manufacturing and logistics, such as baggage handling at airports [19] combine information handling with material flows.

Each individual physical item (a bag) is processed along a logical process flow (*single-dimensional inner scope*); all bags move within a shared physical

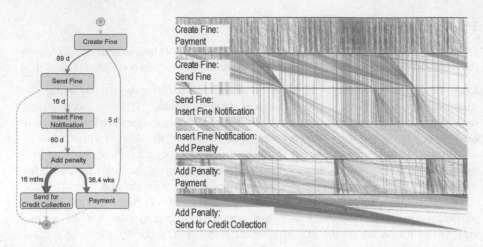

Fig. 2. Performance spectrum (right) visualizes all cases over time [6].

environment of conveyor belts, carts, machines, and workers. The processing of one material item depends not only on the logical process but also on all other items that surround it: all cases together define whether work accumulates at a particular machine, work cannot be completed at the desired quality, or target deadlines are not met (*multi-dimensional outer scope*). These dynamics also arise in call centers and hospitals [17]; they cannot be observed, analyzed, and improved when studying each case in isolation.

Multiple Executions - Multiple Entities. More advanced logistics operations, such as warehouse automation and manufacturing systems, also consider material flows that are being merged together, through batch processing and manufacturing steps.

Analyzing and improving processes in such systems requires both a *multidimensional inner scope* and a *multi-dimensional outer scope* of process thinking.

3   Analyzing Processes in Multiple Dimensions

The tutorial covers techniques to analyze processes in the three multi-dimensional quadrants of Fig. 1. The tutorial's first two parts show how to reveal multi-dimensional dynamics in event data; the third part reflects on how to express these dynamics in formal models.

We first revisit the shortcomings of classical process performance analysis, where process models (discovered or modeled) are annotated with performance information. We then introduce the *Performance Spectrum* [5,6] as a simple data structure and visual analytics technique that allows to study the performance of all executions of a process over time, see Fig. 2. We show how to identify and classify performance patterns and their interpretation in terms of other process management concepts such as batching, actor behavior, and queues (top-left

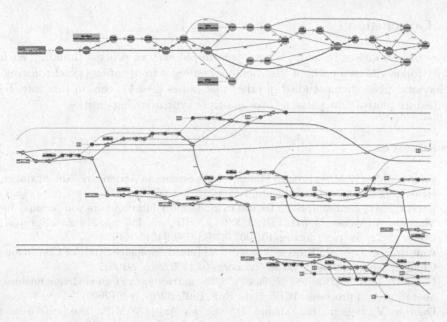

Fig. 3. Event knowledge graph of a single loan application with multiple offers (top) and actors processing multiple loan applications together (bottom) [8].

quadrant of Fig. 1). We discuss how to automatically detect performance patterns [13] to obtain models of how performance problems propagate to other parts of the process [19]. These techniques enable thinking about "multiple process executions at once" (Fig. 3).

We then turn to the well-known problem of analyzing process behavior over multiple objects [4], i.e., the bottom-right quadrant of Fig. 1. Sequential event logs are unable to correctly represent such behavior [8]. *Event knowledge graphs* [8,11] avoid shortcomings of sequential event logs. They can be constructed from classical event data and allow deep insights into behavior of a multi-dimensional inner scope using a few basic queries in standard graph databases [10]. Interestingly, treating actors as "objects" results in event knowledge graphs describing actor behavior [14] allowing to study and actor routines and habits over processes [2]. These techniques enable "network thinking" for processes as a whole, not just process executions.

The final part discusses how to model the phenomena described in the first two parts, with a focus of integrating various perspectives. The compositional paradigm of *synchronous proclets* [9] allows to dynamically compose behavioral models of object behavior, actor behavior, queuing dynamics, and information flow. Proclets model multi-object processes [9] as well as material handling processes [12], i.e., multi-dimensional inner and outer scope. The compositionality of proclets extends to their behavior, allowing to reason across all three aspects to infer missing information [12].

4   Conclusion

This tutorial invites to adopt a new, yet simple, way of process thinking: each entity follows its own path; process behavior emerges from entities synchronizing. Behavioral phenomena studied in other disciplines [2–4,17] can, in this way, be studied as control-flow patterns over multiple synchronizing entities.

References

1. van der Aalst, W.M.P.: Process Mining - Data Science in Action, 2nd edn. Springer, Heidelberg (2016). https://doi.org/10.1007/978-3-662-49851-4
2. Becker, M.C., Pentland, B.T.: Digital twin of an organization: are you serious? In: Marrella, A., Weber, B. (eds.) BPM 2021. LNBIP, vol. 436, pp. 243–254. Springer, Cham (2022). https://doi.org/10.1007/978-3-030-94343-1_19
3. vom Brocke, J., et al.: Process Science: The Interdisciplinary Study of Continuous Change, September 2021. https://doi.org/10.2139/ssrn.3916817
4. Cohn, D., Hull, R.: Business artifacts: a data-centric approach to modeling business operations and processes. IEEE Data Eng. Bull. **32**(3), 3–9 (2009)
5. Denisov, V., Belkina, E., Fahland, D., van der Aalst, W.M.P.: The performance spectrum miner: visual analytics for fine-grained performance analysis of processes. In: BPM 2018 Demos. CEUR Workshop Proceedings, vol. 2196, pp. 96–100. CEUR-WS.org (2018)
6. Denisov, V., Fahland, D., van der Aalst, W.M.P.: Unbiased, fine-grained description of processes performance from event data. In: Weske, M., Montali, M., Weber, I., vom Brocke, J. (eds.) BPM 2018. LNCS, vol. 11080, pp. 139–157. Springer, Cham (2018). https://doi.org/10.1007/978-3-319-98648-7_9
7. Dumas, M., Rosa, M.L., Mendling, J., Reijers, H.A.: Fundamentals of Business Process Management, 2nd edn. Springer, Heidelberg (2018). https://doi.org/10.1007/978-3-662-56509-4
8. Esser, S., Fahland, D.: Multi-dimensional event data in graph databases. J. Data Semant. **10**(1–2), 109–141 (2021). https://doi.org/10.1007/s13740-021-00122-1
9. Fahland, D.: Describing behavior of processes with many-to-many interactions. In: Donatelli, S., Haar, S. (eds.) PETRI NETS 2019. LNCS, vol. 11522, pp. 3–24. Springer, Cham (2019). https://doi.org/10.1007/978-3-030-21571-2_1
10. Fahland, D.: Multi-dimensional-process- mining/eventgraph_tutorial, April 2022. https://doi.org/10.5281/zenodo.6478615
11. Fahland, D.: Process mining over multiple behavioral dimensions with event knowledge graphs. In: van der Aalst, W.M.P., Carmona, J. (eds.) Process Mining Handbook. LNBIP, vol. 448, pp. 274–319. Springer, Cham (2022). https://doi.org/10.1007/978-3-031-08848-3_9
12. Fahland, D., Denisov, V., van der Aalst, W.M.P.: Inferring unobserved events in systems with shared resources and queues. Fundam. Informaticae **183**(3–4), 203–242 (2021)
13. Klijn, E.L., Fahland, D.: Performance mining for batch processing using the performance spectrum. In: Di Francescomarino, C., Dijkman, R., Zdun, U. (eds.) BPM 2019. LNBIP, vol. 362, pp. 172–185. Springer, Cham (2019). https://doi.org/10.1007/978-3-030-37453-2_15

14. Klijn, E.L., Mannhardt, F., Fahland, D.: Classifying and detecting task executions and routines in processes using event graphs. In: Polyvyanyy, A., Wynn, M.T., Van Looy, A., Reichert, M. (eds.) BPM 2021. LNBIP, vol. 427, pp. 212–229. Springer, Cham (2021). https://doi.org/10.1007/978-3-030-85440-9_13
15. Lu, X., Nagelkerke, M., van de Wiel, D., Fahland, D.: Discovering interacting artifacts from ERP systems. IEEE Trans. Serv. Comput. 8(6), 861–873 (2015)
16. Recker, J., Mendling, J.: The state of the art of business process management research as published in the BPM conference - recommendations for progressing the field. Bus. Inf. Syst. Eng. 58(1), 55–72 (2016)
17. Senderovich, A.: Queue mining: service perspectives in process mining. In: BPM 2017 Demo Track and Dissertation Award. CEUR Workshop Proceedings, vol. 1920. CEUR-WS.org (2017). https://ceur-ws.org/Vol-1920/paper6.pdf
18. Shannon, N., Frischherz, B.: Process thinking. In: Metathinking. MP, pp. 29–33. Springer, Cham (2020). https://doi.org/10.1007/978-3-030-41064-3_4
19. Toosinezhad, Z., Fahland, D., Köroglu, Ö., van der Aalst, W.M.P.: Detecting system-level behavior leading to dynamic bottlenecks. In: ICPM 2020, pp. 17–24. IEEE (2020)

Theory and Practice - What, With What and How is Business Process Management Taught at German Universities?

Julian Koch[✉], Jannis Koch, Maximilian Sträßner[✉], and André Coners

University of Applied Sciences Southwestfalia, Hagen, Germany
{koch.julian,koch.jannis,Straessner.maximilian,
coners.andre}@fh-swf.de

Abstract. Process management has been around for about 100 years. However, it only became visible in German university teaching in the late 90s of the 20th century. Business process management (BPM), as offered in the curricula of business informatics and business administration, usually includes the design, implementation, and optimization of processes. In this context, various instruments, competences, procedures, and methods can usually be taught and conveyed. In this tutorial we present a methodology and an analysis of these teaching contents on BPM at German universities.

Keywords: Curriculum mining · Curriculum BPM · Text mining · Platform curriculum

1 Introduction

Business Process Management (BPM) is a cross-organizational management discipline based on the analysis, design and implementation of processes [1]. BPM provides a variety of tools and methods to define and implement sequences of activities that add value to create efficiency and continuity in organizational work [2]. Due to its great relevance in industry, it is also strongly represented in university teaching of business administration and hybrid courses such as information systems. However, information about individual course contents is difficult to obtain. The information needed to study BPM in teaching is available, but unfortunately in the form of module handbooks, most of which are only accessible in PDF format. As a result, it is not possible to track what methods are taught and the depth of teaching about BPM. By combining various crawl and text mining methodologies, we have managed to build a database that contains up to 75% of all German courses (and their module handbooks). The goal of this tutorial is to discuss a teaching strategy and content analysis for the subject BPM in higher education using this database. In this tutorial, we will use our new method to analyze module handbook data and demonstrate the benefits that this particular database can provide. The following research questions could be answered exemplarily through our work:

© Springer Nature Switzerland AG 2022
C. Di Ciccio et al. (Eds.): BPM 2022, LNCS 13420, pp. 34–39, 2022.
https://doi.org/10.1007/978-3-031-16103-2_4

1) Which BPM concepts, technologies, frameworks, and paradigms are most taught?
2) In which teaching forms or didactic concepts is BPM taught?
3) Are there noticeable differences between the Business Administration and Business Information Systems degree programmes regarding the teaching of BPM?
4) Are there noticeable differences between universities and universities of applied sciences about the teaching of BPM?

The aim of this tutorial is to present, discuss and exemplify how the textual content of module handbooks of universities and universities of applied sciences (hereafter referred to as "U" and "FH") can be collected, structured, analyzed and measured. This should enable statements to be made about the extent to which and the form in which BPM is included in German university teaching and, in further analyses, how it is taught. To be able to carry out the intended investigation in a meaningful way, a list of search operators is first created using the example of the lecture hall in the field of BPM. This then yields a list of suitable word combinations to identify BPM within the module description.

2 Curriculum Mining for BPM

As previously noted, the text from module handbooks from universities and universities of applied sciences is analyzed and measured in this work (hereafter referred to as "U" and "FH"). This enables generalizations to be made regarding the volume and style of BPM instruction in German higher education. A list of search operators has to be created in order to conduct the targeted research in a meaningful manner [3]. Each search term must be thoroughly evaluated and verified. If necessary, adjustments to the wording, language or other factors must also be taken into account. [4]. A list of appropriate word combinations for locating BPM in the module handbook data was produced as a result.

The first step in our tutorial is to explain and present the compilation of the data sources of the U- and FHs in Germany (cf. Fig. 1). For this purpose, we select the mass study programs Business Administration and Business Informatics as examples. In the next step, we show how the identified websites of the study programs are searched for relevant module handbooks using an automated crawler, downloaded, converted, extracted, and stored in a database. Once the crawling process is complete, the necessary conversion process begins by converting the database into a structured text format. This is a required step that always occurs before the text pre-processing [4]. Then, an appropriate text mining application is used to reduce noise in the database, e.g., tokenization, lowercase, stopword-removal, stemming, and lemmatization [5]. The database then contains semi-structured data, e.g., the name of the university or college, the program of study, the module name and the complete module description. The presentation and discussion of this structuring of all module handbooks subsequently enables us to perform a variety of measurements and analyses, e.g., for comparison between institution types (U's and FH's), study programs or module offerings, etc. Text mining algorithms are used in the tutorial to search the module descriptions for the given search operators of the exemplary defined teaching area BPM. To answer our exemplary guiding questions together with the auditorium, a comparative analysis follows, both on the level of higher education institutions – U's and FH's - and on the level of study programs - Business

Informatics and Business Administration. To this end, the two types of Institutions can be compared, for example, at the program level according to the average number of BPM-related content per teaching module, or the proportion of modules with BPM content in the chronological semester sequence of the programs. Our tutorial will then measure the number of degree programs with a stand-alone BPM teaching module and testify and discuss implications with the auditorium accordingly. This will include presenting, evaluating, and discussing the results at both the university level (U's and FH's) and the degree program level. For example, it could be shown what the exact distribution of the proportion of modules with BPM content looks like over the period of the (usually six-semester) degree programs.

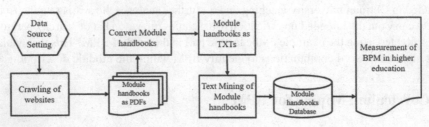

Fig. 1. Research process

To prepare our tutorial and as a test run of our database, we first examined mentions of "business process management" within our sample. As shown in Table 1 our sample consists of 814 programs, divided into 351 Business Informatics and 463 Business Administration programs (and their associated module handbooks). The first finding reveals that not all module handbooks include "Business Process Management". 34% of Business Informatics and 35% of Business Administration module handbooks were identified with relevant hits. What is striking here is that there seems to be no difference between the two, after all, very different specializations, when it comes to the curriculum inclusion of Business Process Management. If a dividing line is also drawn here between universities and HEIs, it is noticeable that the proportion of mentions of BPM diverges greatly here. While only 24% of universities in Informatics and Business and 25% in Business Administration have included BPM in their curricula, the figure for universities of applied sciences is significantly higher. Here, the proportion of BPM in the curriculum is around 50% for both fields of study, so there really do seem to be significant differences between the two types of institution. BPM, which is highly applied in practice, is also taught more likely at FHs with a higher practical orientation.

In order to examine our question postulated in the introduction: "Which BPM concepts, technologies, frameworks, and paradigms are most taught?" more closely, we first concentrated on process modeling in preparation for the tutorial. According to the literature, there are three main modeling methods that are used in practice Event-driven process chains, BPMN and UML diagrams [6]. After we had identified the module handbooks with BPM content in the first step, we could now further search these module handbooks for specific content. As shown in Fig. 2, the event-driven process chain was found in 73% of the Business Informatics manuals and in 70% of the Business

Table 1. BPM distribution in study programs Business informatics and Business Administration

Search term	Study program	Crawl hits	Module handbooks	Ratio
Business Process Management	Business Informatics	119	351	34%
	Business Administration	162	463	35%
Event-driven process chain	Business Informatics	87	351	25%
	Business Administration	113	463	24%
BPMN	Business Informatics	66	351	19%
	Business Administration	71	463	15%
UML	Business Informatics	3	351	1%
	Business Administration	0	463	0%

Administration manuals with BPM content. Thus, it was clearly ahead of BPMN (Business Process Model and Notation) with 55% and 44%, respectively. Surprisingly, UML was listed in only very few module manuals. Here, the rate for both fields of study was less than one percent. This leads to the conclusion that in the German curriculum, BPMN and EPC are much more important than UML. This development could be examined more closely over time in connection with the updating of the data, which is carried out every semester.

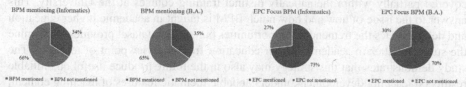

Fig. 2. Proportions of BPM mentions in the study programs Informatics and Business Administration

In addition, we show the distribution of the proportion of modules with "Business Process Management"-related content over the period of the (mostly six-semester) degree programs, differentiated by Informatics and Business and Business Administration (cf. Fig. 3). Here it becomes clear that BPM and the Modules related to it are generally taught at later stages of the Curriculum. In our sample (n = 814) not a single study program has BPM located in the first Semester. The practical approach of Business Process Management and the complexity it entails mean, that students only gain access to BPM content after they have already learned the basics. This has become prevalent in the implementation of BPM in curricula. In summary, it can be stated that the differences between the two specializations are also rather marginal here. Mostly, BPM is taught between the 4th and the 6th semester, this may also be due to the fact that in our evaluation so far, no distinction was made between compulsory and elective modules. Initial research from our side, suggests that BPM plays a major role in the elective sector.

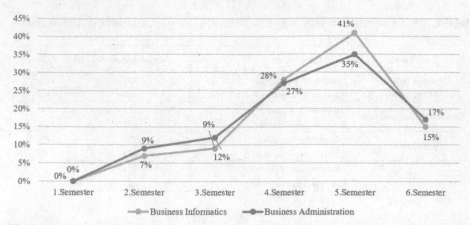

Fig. 3. Distribution of methods in the study programs Business Administration and Computer Science

3 Conclusion

Concluding, our findings demonstrate that at FHs, the emphasis is on teaching business process management (BPM) in standalone modules and topics related to process modeling, particularly in the later semesters of the degree programs. In contrast, BPM is only covered partially within thematically distinct training courses at the University. This answer to the issue of how and how much BPM is taught in academic higher education and demonstrates the tremendous opportunities our special dataset provides to examine the subject matter in academic higher education from the viewpoint of teaching. The study demonstrates that this strategy may also in the future produce useful quantifiable results on teaching development, understandable thematic features of teaching content, and measurable study trajectories. The proposed tutorial addresses practitioners as well as PhD students and researchers to answer data-driven questions in the context of higher education in Germany. This is especially true because of the broad thematic coverage, which is possible by considering all modularized courses of study without thematic restrictions. Especially for young researchers, the connectivity of the methodology presented here in the tutorial and the resulting possibilities of content-based data analysis with large data sets could be stimulation and inspiration for their own research.

References

1. Rosemann, M., vom Brocke, J.: The six core elements of business process management. In: vom Brocke, J., Rosemann, M. (eds.) Handbook on Business Process Management 1. IHIS, pp. 105–122. Springer, Heidelberg (2015). https://doi.org/10.1007/978-3-642-45100-3_5
2. vom Brocke, J., Rosemann, M. (eds.): Handbook on Business Process Management 2. IHIS, Springer, Heidelberg (2015). https://doi.org/10.1007/978-3-642-45103-4
3. Han, J., Kamber, M., Pei, J.: Data Mining. Concepts and Techniques, 3rd edn. The Morgan Kaufmann Series in Data Management Systems (2012)
4. Feldman, R., Sanger, J.: The text mining handbook. Advanced Approaches in Analyzing Unstructured Data. Cambridge University Press, Cambridge, New York (2007)

5. Allahyari, M., et al.: A Brief Survey of Text Mining: Classification, Clustering and Extraction Techniques (2017)
6. Dumas, M., La Rosa, M., Mendling, J., Reijers, H.A.: Fundamentals of Business Process Management. Springer, Heidelberg (2018). https://doi.org/10.1007/978-3-662-56509-4

How to Leverage Process Mining in Organizations - Towards Process Mining Capabilities

Gregor Kipping[1]([⊠]) [iD], Djordje Djurica[2] [iD], Sandro Franzoi[1] [iD], Thomas Grisold[1] [iD], Laura Marcus[4,5,6,7] [iD], Sebastian Schmid[4,6,7] [iD], Jan vom Brocke[1] [iD], Jan Mendling[2,3] [iD], and Maximilian Röglinger[4,6,7] [iD]

[1] University of Liechtenstein, Vaduz, Liechtenstein
{gregor.kipping,sandro.franzoi,thomas.grisold,
jan.vom.brocke}@uni.li
[2] Vienna University of Economics and Business (WU), Vienna, Austria
djordje.djurica@wu.ac.at
[3] Humboldt University of Berlin, Berlin, Germany
jan.mendling@hu-berlin.de
[4] University of Bayreuth, Bayreuth, Germany
[5] University of Applied Sciences Augsburg, Augsburg, Germany
[6] FIM Research Center, Augsburg and Bayreuth, Germany
{laura.marcus,sebastian.schmid,maximilian.roeglinger}@fim-rc.de
[7] Branch Business & Information Systems Engineering of the Fraunhofer FIT, Augsburg and Bayreuth, Germany

Abstract. Process mining is a fast-growing technology concerned with managing and improving business processes. While the technology itself has been thoroughly scrutinized by prior research, we are only beginning to understand the managerial and organizational implications of process mining. Creating such knowledge is essential for a successful adoption and use of process mining in organizations. We conduct a qualitative-inductive interview study to explore how process mining can be leveraged in organizations. To this end, we systematically examine the needs and experiences of practitioners with process mining at different levels, including heads of process mining, process analysts, and data engineers. Complementing our tutorial, this article provides a theoretical background, outlines our research approach, and presents preliminary findings.

Keywords: Process mining · Organizational implications · Process mining capabilities

1 Introduction

Process mining draws on process event log data to visualize, analyze, and improve business process work [8]. It is associated with a range of economic benefits that are, for example, tied to significantly increased customer satisfaction or cost reduction [3].

© Springer Nature Switzerland AG 2022
C. Di Ciccio et al. (Eds.): BPM 2022, LNCS 13420, pp. 40–46, 2022.
https://doi.org/10.1007/978-3-031-16103-2_5

While research in this field has been mainly concerned with technical matters, several recent works called for research around managerial and organizational aspects of process mining [e.g., 10], in order to leverage the full potential of process mining [11]. Understanding the organizational perspective involved in process mining is crucial to capitalize on the possible benefits of the technology [4, 7].

In this short paper, we explore how process mining can be leveraged in organizations successfully. We consider individual stakeholders and the respective capabilities that they need in order to capitalize on the benefits of process mining. To this end, we conducted an interview study with practitioners to assess and analyze their expectations, needs, and capabilities concerning process mining in organizations. We focused on various organizational roles that deal with process mining, namely heads of process mining, process analysts, and data engineers. Key to our findings is that each role is linked to specific tasks that, in turn, translate into different expectations, use cases, and required capabilities.

2   Research Background

Process mining, a technology at the interface of data mining and Business Process Management (BPM), is a relatively new and high-in-demand technology that uses actual process data stored in information systems to display, analyze, and monitor the performances of business processes. Along with process visualization, it can be used for conformance checking, process analysis, and process enhancement [8]. A recent study by Deloitte highlights the practical relevance of process mining; 95% of the companies surveyed stated that they had either already implemented process mining or were planning to do so [2].

Extensive research has been conducted with the primary focus of improving existing or developing new algorithms [e.g., 1, 8]. However, in addition to the development and improvement of algorithms, there are also non-technical aspects that are crucial for the adoption and management of process mining [10]. To this end, recent research has focused on the practical implications of process mining, including project success factors [6], methodologies to conduct projects [9], case studies, or Delphi studies [7]. Despite these works, we are only beginning to understand how process mining is adopted, used, and managed in practice [3, 10], and what potentials it bears for identifying, understanding, and intervening into processes [5, 11]. Specifically, there is a lack of research around the capabilities and competencies that are required to successfully implement and scale process mining.

3   Research Method

We conducted an interview study to examine capabilities associated with process mining. We interviewed participants across different industries to eliminate bias from our results, which allows us to create a more nuanced view on how process mining can be leveraged in organizations successfully. To this extent, our research focused on three different roles. One of them relates to the strategic matters, such as selecting processes to be mined or defining use cases (head of process mining). The other two roles are concerned with operative tasks, such as analyzing results or improving accuracy (process analyst and data engineer).

Our interviews were semi-structured, and the interview protocols were divided into several parts; they included general information about the aim of the project, general questions about the implementation of process mining in the participants' organization, questions about the participants' tasks, necessary skills for completing those tasks, and technologies that are used. The interviews were conducted both in person and online. Each interview lasted for around one hour. So far, we interviewed six participants, from five different companies and four different industries. We only selected participants, who work for companies that already have a process mining team in place. As a result, the participants were three process analysts, two heads of process mining, and one data engineer. We provide more information in Table 1.

Table 1. Information about interviewees

Role	Responsibility	Industry	Company size
Head of process mining	Translating company-wide strategic goals into tangible targets and driving the adoption of process mining in the organization	1. Manufacturing 2. Energy	1. 31 000 employees 2. 20 000 employees
Process analyst	Building and developing actionable insights	1. Automotive 2. Manufacturing 3. Insurance	1. 100 000 employees 2. 15 000 employees 3. 13 000 employees
Data engineer	Driving the technical implementation of process mining and providing ongoing technical support	1. Automotive	1. 100 000 employees

4   Preliminary Results

4.1   Role-Related Tasks, Technologies, and Skills

In the following section, we will present our preliminary results. We derive the capabilities required for process mining by taking a closer look at the tasks, technologies, and skills for each role as shown in Table 2.

We can see some commonalities across the different roles. For example, each role indicates that communication skills to talk with a multitude of stakeholders are an important part of their daily job. Furthermore, it is important to understand the department's needs and align them with the capabilities needed for process mining. As such, the ability to quickly understand a domain problem and translate business requirements into technical requirements is key. Additionally, we found that data analysis is not only performed

Table 2. Core tasks, technologies, and skills of process mining practitioners

Role	Tasks	Technologies	Skills
Head of process mining	• Communication with departments • Coordination and team leadership • Enabling of continuous process mining usage • Implementation of KPIs • Data preparation and analysis	• Process mining tool	• Communication skills • Data science and statistics skills • Programming • Project management • Translation of business requirements into technical requirements
Process analyst	• Communication with departments • Data preparation and analysis • Internal sales • Presentation of results	• Cloud • Process mining tool • Process modeling tool • Python • SQL	• Communication skills • Data preprocessing, analysis, and visualization • Domain and business knowledge • Programming • Project management
Data engineer	• Communication with departments • ETL operations • Data preparation and analysis • Process and data understanding	• Cloud • Process mining tool • Python • Spark • SQL	• Communication skills • Domain and business knowledge • Problem solving mindset • Programming

by process analysts. All roles reported to engage in some form of data preparation and analysis, albeit to varying degrees. Similarly, both process analysts and heads of process mining perform tasks related to project management. This overlap in tasks of the respective roles is also reflected in the technologies used. All roles use a process mining tool (e.g., Celonis) or a process modeling tool (e.g., Adonis) for their core process mining activities. There is a strong overlap between data engineers and process analysts as both roles use programming languages such as Python or SQL as well as some cloud technology (e.g., AWS).

There are, however, also clear differences between the roles. Heads of process mining reported, for example, that they are hardly involved in the technical realization of process mining projects and their work revolves more around managerial tasks. On the contrary, data engineers primarily engage in the technical implementation of process mining projects. Accordingly, the skillset is the most technology-oriented out of all roles. Process analysts can be located between these two roles and, thus, have the broadest requirement profile. They must continuously perform a balancing act between technical and business matters related to process mining. This is also reflected in the skillset, which includes technical skills (such as programming) as well as business-related skills (such as project management). When asked for their backgrounds, process analysts had diverse prior experiences ranging from business administration to statistics studies.

4.2 General Observations

In addition to role-specific insights, we report on general observations that we made across all interviewees. These are summarized in Table 3.

Table 3. General organizational implications of process mining

Reported benefits	Main goals	Challenges	Future use cases
• Easier and faster process improvement • Increased process transparency	• Cost minimization • Resource minimization • Risk minimization • Transparency maximization	• Commitment • Data quality • Discrepancy between model and reality • Internal resistance	• External data incorporation • Data streaming • Real-time data availability • Stronger specialization

All interviewees reported that prior to the use of process mining, knowledge about processes was only implicitly represented. Process deviations and their underlying causes were often unknown. While process improvement initiatives were potentially successful, the interviewees reported that the procedure was tedious and lacked standardization. To this end, there was agreement across all interviewees that process mining enabled the companies to improve processes in easier and faster ways. Also, it was reported that process transparency had been increased. However, we observed that these advantages are accompanied by several challenges. For instance, interviewees stated that some processes cannot be optimized by applying process mining techniques, which is mainly due to a multitude of unavoidable process deviations caused by human behavior. One interviewee, for example, indicated that process interruptions are caused by customers being able to contact them, which leads to a complicated process model. Another common challenge of process mining that we identified is a lack of commitment in companies. We found that this manifests itself, for example, in a lack of resources. Another challenge that was reported concerns internal resistance. According to the interviews, employees often feel monitored by the presence of process mining and refuse to support the technology because they are afraid that their domain knowledge will become obsolete. Lastly, process mining success is perceived to be hindered by a lack of data quality. However, the interviewees remain optimistic about the potential of process mining within their respective company. Most notably, a lot of potential is seen in the further development of data sources. One interviewee stated that external data sources such as weather forecasts should be incorporated. Also, it was suggested that companies should move from batch processing to stream processing whereby data is available in real-time. Lastly, we found that process analysts are required to be generalists as their portfolio of tasks is versatile. As a result, the interviewees expect a stronger specialization of the roles as the technology matures.

5 Conclusion

In this paper, we investigated how process mining can be leveraged in organizations. We conducted a qualitative-inductive interview study to examine practitioners' needs and experiences at different levels. We identified core competencies, tasks, and skills of practitioners who have different roles with regards to process mining, as well as general organizational implications that were reported across all roles. Our preliminary results show that process mining is advancing as an essential part of modern management in order to cope with the ever-increasing dynamics in contemporary organizational work. Overall, it is crucial that those who deal with process mining have the necessary skills and competences to make process mining successful. In future research, we aim to expand these preliminary findings by creating a deeper understanding about necessary capabilities of process mining stakeholders.

Acknowledgements. This proposal has been funded by the ERASMUS+ program of the European Union (EU Funding 2021-1-LI01-KA220-HED-000027575 "Developing Process Mining Capabilities at the Enterprise Level"). We would like to express our gratitude to the European Union and AIBA Liechtenstein for their support.

References

1. Augusto, A., et al.: Automated discovery of process models from event logs: review and benchmark. IEEE Trans. Knowl. Data Eng. **31**(4), 686–705 (2019)
2. Galic, G., Wolf, M.: Global Process Mining Survey 2021: Delivering Value with Process Analytics - Adoption and Success Factors of Process Mining. Deloitte (2021). https://www2.deloitte.com/de/de/pages/finance/articles/global-process-mining-survey-2021.html. Accessed 6 June 2022
3. Grisold, T., Mendling, J., Otto, M., vom Brocke, J.: Adoption, use and management of process mining in practice. Bus. Process. Manag. J. **27**(2), 369–387 (2021)
4. Grisold, T., et al.: Digital innovation and business process management: opportunities and challenges as perceived by practitioners. Commun. Assoc. Inf. Syst. **49**, 556–571 (2021)
5. Grisold, T., Wurm, B., Mendling, J., vom Brocke, J.: Using process mining to support theorizing about change in organizations. In: Proceedings of the 53rd Hawaii International Conference on System Sciences (HICSS). Maui, Hawaii (2020)
6. Mans, R.R., et al.: Business process mining success. In: Proceedings of the 21st European Conference on Information Systems (ECIS). Utrecht, Netherlands (2013)
7. Martin, N., et al.: Opportunities and challenges for process mining in organizations: results of a delphi study. Bus. Inf. Syst. Eng. **63**(5), 511–527 (2021). https://doi.org/10.1007/s12599-021-00720-0
8. van der Aalst, W.M.P.: Process Mining - Data Science in Action, 2nd edn. Springer, Heidelberg (2016). https://doi.org/10.1007/978-3-662-49851-4

9. van Eck, M.L., Lu, X., Leemans, S.J.J., van der Aalst, W.M.P.: PM2: a process mining project methodology. In: Proceedings of the 27th International Conference on Advanced Information Systems Engineering (CAiSE). Stockholm, Sweden (2015)
10. vom Brocke, J., Jans, M., Mendling, J., Reijers, H.A.: A five-level framework for research on process mining. Bus. Inf. Syst. Eng. **63**(5), 483–490 (2021). https://doi.org/10.1007/s12599-021-00718-8
11. vom Brocke, J., et al.: Process science. the interdisciplinary study of continuous change. SSRN Electron. J. 1–9 (2021)

Mastering Robotic Process Automation with Process Mining

Simone Agostinelli[1], Andrea Marrella[1(✉)], Luka Abb[2],
and Jana-Rebecca Rehse[2]

[1] Sapienza Universitá di Roma, Rome, Italy
{agostinelli,marrella}@diag.uniroma1.it
[2] Universität Mannheim, Mannheim, Germany
{luka.abb,rehse}@uni-mannheim.de

Abstract. Robotic Process Automation (RPA) is an emerging automation technology that creates software (SW) robots to partially or fully automate rule-based and repetitive tasks (aka routines) previously performed by human users in their applications' user interfaces (UIs). Successful usage of RPA requires strong support by skilled human experts, from the detection of the routines to be automated to the development of the executable scripts required to enact SW robots. In this paper, we discuss how process mining can be leveraged to minimize the manual and time-consuming steps required for the creation of SW robots, enabling new levels of automation and support for RPA. We first present a reference data model that can be used for a standardized specification of UI logs recording the interactions between workers and SW applications to enable interoperability among different tools. Then, we introduce a pipeline of processing steps that enable us to (1) semi-automatically discover the anatomy of a routine directly from the UI logs, and (2) automatically develop executable scripts for performing SW robots at run-time. We show how this pipeline can be effectively enacted by researchers/practitioners through the SmartRPA tool.

Keywords: Robotic Process Automation · Process mining · User Interface (UI) logs · Reference data model for UI logs · Segmentation · Automated generation of SW robots from UI Logs · SmartRPA

1 Introduction

Robotic Process Automation (RPA) is an emerging automation technology in the Business Process Management (BPM) domain that creates software (SW) robots to partially or fully automate rule-based and repetitive tasks (or simply *routines*) performed by human users in their applications' user interfaces (UIs) [1]. Despite the growing attention around RPA, when considering state-of-the-art RPA technology, it becomes apparent that the current generation of RPA tools is driven by predefined rules and manual configurations made by expert users rather than automated techniques [7,8,10].

© Springer Nature Switzerland AG 2022
C. Di Ciccio et al. (Eds.): BPM 2022, LNCS 13420, pp. 47–53, 2022.
https://doi.org/10.1007/978-3-031-16103-2_6

The traditional life-cycle of an RPA project can be summarized as follows [14]: (1) `determine` which process steps are good candidates to be automated in the form of routines; (2) `model` the selected routines trough *flowchart diagrams*, which involve the specification of the actions, routing constructs, data flow, etc. that define the behaviour of a SW robot; (3) `develop` each modeled routine by generating the SW code required to concretely enact the associated SW robot on a target computer system; (4) `deploy` the SW robots in their environment to perform their actions; (5) `monitor` the performance of SW robots to detect bottlenecks and exceptions; and (6) `maintain` routines over time, updating the SW robots when needed. The majority of the previous steps, particularly the early ones, require the support of skilled human experts, which need to understand the anatomy of the candidate routines to automate through interviews, walk-troughs, and detailed observation of workers conducting their daily work, cf. step (1), and manually define the flowchart diagrams representing the structure of such routines, cf. step (2). These diagrams will drive the development of the executable scripts (also called RPA scripts), allowing for the concrete enactment of SW robots at run-time, cf. steps (3) and (4). The problem is that this high degree of human involvement contradicts the underlying objective of RPA, i.e., an increased level of automation.

In this paper, we discuss how process mining can be leveraged to address this problem, enabling new levels of automation and support for RPA. Building on the RPM (Robotic Process Mining) framework [16], we show that the generation of SW robots can be achieved in a semi-automated way directly from the UI logs recording the interactions between workers and SW applications during one or more routine(s) executions, thus eliminating the manual and time-consuming steps (1) and (2) required for modeling the details of the routine structure.

Specifically, in Sect. 2, we first present a reference data model that enables a standardized specification of UI logs. Then, in Sect. 3, we show how the RPM framework can be effectively enacted by researchers/practitioners through the SmartRPA approach [4,6] and its implemented tool [5], which enables to interpret the UI logs keeping track of many routine executions, and to generate SW robots that emulate the most suitable routine variant for any specific intermediate user input that is required during the routine execution. Finally, in Sect. 4 we conclude the paper by tracing future work.

2 Specifying and Collecting UI Logs

The main source of data for RPA are UI logs, which are a particular kind of event log that record low-level manual activities during the execution of a task in an information system. Examples of events in a UI log include clicking a button, entering a string into a text field, ticking a checkbox, or selecting a value from a dropdown. The specific scope of a UI log, including the definition of relevant activities and attributes to cover, depends on the context in which the log is collected and the purpose for which it is used. Hence, the first challenges when collecting UI logs are often (1) to determine what kind of data is available and (2) to design the data collection process so that the logs are comprehensive enough to cover the desired automation use cases.

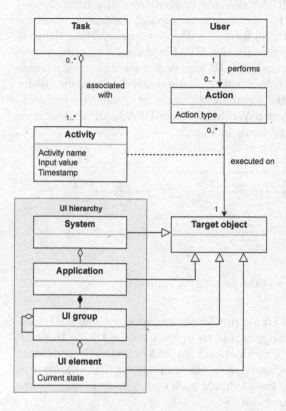

Fig. 1. Reference data model for UI logs [2]

To specify a UI log for RPA, one needs to determine which attributes can and should be recorded and how they relate to each other. The UI log should be as standardized as possible to allow for interoperability between different tools, but they also need to be adapted to the individual scenario. To achieve this, they can refer to the reference data model for process-related UI logs, shown in Fig. 1. This reference model defines the core attributes of UI logs but remains flexible with regard to the scope, level of abstraction, and case notion [2]. It defines the activity of a UI log as a combination of an action (e.g., *click* or *input*) and a target object in the user interface. It further specifies the possible instances of target objects and their hierarchical relation, as well as task and user components that provide additional (business) context.

After specifying the UI log structure, the actual data needs to be recorded. Generally, there are three ways to achieve this: application-independent logging with screen capture and OCR technology [14,18], application-specific logging with plug-ins [4,17], and application-internal logging within the an application's source code. Not all options are feasible in each application context and they each have certain assets and drawbacks. For example, application-internal logging will typically produce the highest data quality, but it is only possible if we have access to the application's source code. Application-independent and application-specific logging have to externally reconstruct the events that happen within the application, but can be applied to any tool independent of its origin.

3   SmartRPA: From UI Logs to SW Robots

The approach underlying SmartRPA takes inspiration from the RPM framework presented by Leno et al. in [16]. RPM aims to support analysts to produce executable specifications of routines, in form of SW robots, interpreting the routine executions stored in a UI log. Specifically, RPM envisions a pipeline of two main stages that consist of: *(i)* interpreting UI logs corresponding to executions of one or more routine executions, by identifying the candidate routines to be automated with RPA tools (i.e., the *segmentation* issue [9]); and *(ii)* synthesizing executable RPA scripts to enact SW robots. SmartRPA incorporates these stages within a larger approach, as shown in Fig. 2.

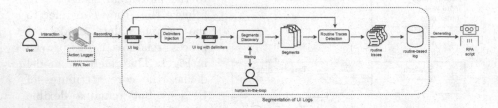

Fig. 2. Overview of the SmartRPA approach

Starting from an unsegmented UI log previously recorded with an RPA tool, the first stage of the SmartRPA approach is to inject into the UI log the *end-delimiters* of the routines under examination. An end-delimiter is a dummy action added to the UI log immediately after the user action that is known to complete a routine execution. The knowledge of such end-delimiters is crucial to make the approach work, as discussed in [3].

For tackling the segmentation issue, we rely on three main steps: *(i)* a *frequent-pattern identification technique* [11] to automatically derive the routine segments from a UI log (i.e., *routine segments* describe the different behaviours of the routine(s) under analysis, in terms of repeated patterns of performed user actions), *(ii)* a *human-in-the-loop interaction* to filter out those segments not allowed (i.e., wrongly discovered from the UI log) by any real-world routine execution by means of *declarative constraints* [13], and *(iii)* a routine traces detection component that leverages *trace alignment* in process mining [12] to cluster all user actions belonging to a specific routine segment into well-bounded routine traces (i.e., a *routine trace* represents an execution instance of a routine within a UI log). Such traces are finally stored in a dedicated *routine-based log*, which captures exactly all the user actions happened during many different executions of the routine.

Commercial RPA tools can eventually employ routine-based logs to synthesize executable scripts in the form of SW robots that will emulate the routine behaviour on the UI without the manual modeling of the routines. To this end, the SmartRPA tool[1] is able to automatically synthesize executable

---

[1] https://github.com/bpm-diag/smartRPA

scripts for enacting SW robots at run-time. Notably, the SW robots generated by SmartRPA are obtained to handle the intermediate user inputs that are required during the routine execution, thus enabling to emulate the most suitable routine variant for any specific combination of user inputs as observed in the UI log. This makes the synthesis of SW robots performed by SmartRPA *reactive* to any user decision found during a routine execution, thus allowing the potential run-time generation of as many SW robots as the routine variants to be emulated [6].

4 Concluding Remarks

The goal of RPA is to automate routines and high-volume tasks, but it currently requires substantial manual intervention of expert users. In this paper, we offer a twofold contribution towards an intelligent and fully automated generation of SW robots from the users' observed behavior as recorded in UI logs. First, we introduce a reference data model for a standardized specification of UI logs, which enforces interoperability among different RPM-based tools. Second, we present a pipeline of processing steps, implemented trough the SmartRPA approach, to develop executable RPA scripts by solely interpreting the UI logs at hand.

The reference model provides a common, application-independent conceptual framework for user interactions. However, it still has to prove its utility in practice. We therefore want to encourage readers to adopt the model for capturing UI logs in their projects. Compared with the literature approaches to automated RPA script generation from UI logs [15,18], which enable to automate only the most frequent routine variant among the ones discovered in the UI log, SmartRPA provides a reactive approach that emulates the most suitable routine variant for any specific intermediate user input that is required during the routine execution. As a consequence, this makes the working of SW robots generated by SmartRPA flexible and adaptable to several real-world situations.

The main weakness of SmartRPA relates to the quality of information recorded in real-world UI logs. Since a UI log is fine-grained, routines executed with many different strategies may potentially affect the identification of the routine segments. In addition, SmartRPA is based on a semi-supervised assumption, since the end-delimiters required to untangle the segmentation issue are known a-priori. Conversely, on the positive side, the employed segmentation technique is able to outperform existing literature approaches in terms of supported segmentation variants, in particular when there are many interleaved routine executions recorded in the UI log [3]. For this reason, we consider this contribution as an important step towards the development of an unsupervised approach that employs machine learning techniques to automatically identify the end-delimiters.

Acknowledgments. This work has been partially supported by the H2020 project DataCloud and the Sapienza grant BPbots.

References

1. van der Aalst, W.M.P., Bichler, M., Heinzl, A.: Robotic process automation. Bus. Inf. Syst. Eng. **60**(4), 269–272 (2018). https://doi.org/10.1007/s12599-018-0542-4
2. Abb, L., Rehse, J.R.: A reference data model for process-related user interaction logs. In: Di Ciccio, C., et al. (eds.) BPM 2022, LNCS 13420, pp. 57–74. Springer, Cham (2022)
3. Agostinelli, S., Leotta, F., Marrella, A.: Interactive segmentation of user interface logs. In: Hacid, H., Kao, O., Mecella, M., Moha, N., Paik, H. (eds.) ICSOC 2021. LNCS, vol. 13121, pp. 65–80. Springer, Cham (2021). https://doi.org/10.1007/978-3-030-91431-8_5
4. Agostinelli, S., Lupia, M., Marrella, A., Mecella, M.: Automated generation of executable RPA scripts from user interface logs. In: Asatiani, A., et al. (eds.) BPM 2020. LNBIP, vol. 393, pp. 116–131. Springer, Cham (2020). https://doi.org/10.1007/978-3-030-58779-6_8
5. Agostinelli, S., Lupia, M., Marrella, A., Mecella, M.: SmartRPA: a tool to reactively synthesize software robots from user interface logs. In: Nurcan, S., Korthaus, A. (eds.) CAiSE 2021. LNBIP, vol. 424, pp. 137–145. Springer, Cham (2021). https://doi.org/10.1007/978-3-030-79108-7_16
6. Agostinelli, S., Lupia, M., Marrella, A., Mecella, M.: Reactive Synthesis of Software Robots in RPA from User Interface Logs. Computers in Industry (2022)
7. Agostinelli, S., Marrella, A., Mecella, M.: Research challenges for intelligent robotic process automation. In: Di Francescomarino, C., Dijkman, R., Zdun, U. (eds.) BPM 2019. LNBIP, vol. 362, pp. 12–18. Springer, Cham (2019). https://doi.org/10.1007/978-3-030-37453-2_2
8. Agostinelli, S., Marrella, A., Mecella, M.: Towards Intelligent Robotic Process Automation for BPMers (2020). https://arxiv.org/abs/2001.00804
9. Agostinelli, S., Marrella, A., Mecella, M.: Exploring the challenge of automated segmentation in robotic process automation. In: Cherfi, S., Perini, A., Nurcan, S. (eds.) RCIS 2021. LNBIP, vol. 415, pp. 38–54. Springer, Cham (2021). https://doi.org/10.1007/978-3-030-75018-3_3
10. Chakraborti, T., et al.: From robotic process automation to intelligent process automation. In: Asatiani, A., et al. (eds.) BPM 2020. LNBIP, vol. 393, pp. 215–228. Springer, Cham (2020). https://doi.org/10.1007/978-3-030-58779-6_15
11. Cook, D.J., Krishnan, N.C., Rashidi, P.: Activity discovery and activity recognition: a new partnership. IEEE Trans. Cybern. **43**(3), 820–828 (2013). https://doi.org/10.1109/TSMCB.2012.2216873
12. de Leoni, M., Lanciano, G., Marrella, A.: Aligning partially-ordered process-execution traces and models using automated planning. In: Twenty-Eight International Conference on Automated Planning and Scheduling (ICAPS 2018), pp. 321–329 (2018). https://aaai.org/ocs/index.php/ICAPS/ICAPS18/paper/view/17739
13. van Der Aalst, W.M., Pesic, M., Schonenberg, H.: declarative workflows: balancing between flexibility and support. Comp. Sc.-Res. Dev. **23**(2) (2009). https://doi.org/10.1007/s00450-009-0057-9
14. Jimenez-Ramirez, A., Reijers, H.A., Barba, I., Del Valle, C.: A method to improve the early stages of the robotic process automation lifecycle. In: Giorgini, P., Weber, B. (eds.) CAiSE 2019. LNCS, vol. 11483, pp. 446–461. Springer, Cham (2019). https://doi.org/10.1007/978-3-030-21290-2_28

15. Leno, V., Deviatykh, S., Polyvyanyy, A., Rosa, M.L., Dumas, M., Maggi, F.M.: Robidium: automated synthesis of robotic process automation scripts from UI logs. In: BPM Demonstration and Resources (2020). https://ceur-ws.org/Vol-2673/paperDR08.pdf

16. Leno, V., Polyvyanyy, A., Dumas, M., La Rosa, M., Maggi, F.M.: Robotic process mining: vision and challenges. Bus. Inf. Syst. Eng. **63**(3), 301–314 (2020). https://doi.org/10.1007/s12599-020-00641-4

17. Leno, V., Polyvyanyy, A., Rosa, M.L., Dumas, M., Maggi, F.M.: Action logger: enabling process mining for robotic process automation. In: BPM Demonstration and Resources (2019). https://ceur-ws.org/Vol-2420/paperDT2.pdf

18. Linn, C., Zimmermann, P., Werth, D.: Activity mining - a new level of detail in mining business processes. In: Workshops der INFORMATIK, pp. 245–258 (2018). https://dl.gi.de/20.500.12116/17225

Task Mining

A Reference Data Model
for Process-Related User Interaction Logs

Luka Abb[✉][iD] and Jana-Rebecca Rehse[iD]

University of Mannheim, Mannheim, Germany
{luka.abb,rehse}@uni-mannheim.de

Abstract. User interaction (UI) logs are high-resolution event logs that record low-level activities performed by a user during the execution of a task in an information system. Each event in a UI log corresponds to a single interaction between the user and the interface, such as clicking a button or entering a string into a text field. UI logs are used for purposes like task mining or robotic process automation (RPA), but each study and tool relies on a different conceptualization and implementation of the elements and attributes that constitute user interactions. This lack of standardization makes it difficult to integrate UI logs from different sources and to combine tools for UI data collection with downstream analytics or automation solutions. To address this, we propose a universally applicable reference data model for process-related UI logs. Based on a review of scientific literature and industry solutions, this model includes the core attributes of UI logs, but remains flexible with regard to the scope, level of abstraction, and case notion. We provide an implementation of the model as an extension to the XES interchange standard for event logs and demonstrate its practical applicability in a real-life RPA scenario.

Keywords: User behavior mining · UI Log · Data model · Robotic process automation · Task mining

1 Introduction

User interaction (UI) logs are high-resolution event logs that record low-level, manual activities performed by a user during the execution of a task in an information system (IS) [1]. Each event in a UI log corresponds to a single interaction between the user and the graphical user interface (GUI) of a software application. Examples include clicking a button, entering a string into a text field, ticking a checkbox, or selecting an item from a dropdown [26]. Multiple recent research streams use this type of data, for example to analyze usage patterns in software applications [6,12,28], to identify candidate routines for robotic process automation (RPA) [7,26,35], or to derive RPA automation and test scripts [3,8]. In addition, companies like Celonis and UiPath offer tools that record and process UI data for inspecting and automating task executions [5].

© Springer Nature Switzerland AG 2022
C. Di Ciccio et al. (Eds.): BPM 2022, LNCS 13420, pp. 57–74, 2022.
https://doi.org/10.1007/978-3-031-16103-2_7

The UI logs currently used in research differ substantially. The data collected in a specific research context is usually limited in scope and tailored to the proposed analysis technique or automation approach. This results in considerable variation regarding the number, type, and granularity of recorded events and corresponding attributes. Even when researchers record the same attributes at a similar level of detail, there is no common definition of UI log attributes to which they can adhere. Instead, they often rely on ad-hoc conceptualizations of elementary notions like activities and UI components. The situation is similar in industry, where each vendor has developed their own UI log format tailored to the capabilities of their recording software [27].

This lack of standardization makes it difficult to integrate UI logs from different sources [26,30]. It also poses a challenge for the interoperability of data collection and downstream processing tools: logs recorded by one tool are usually only compatible with the associated analytics or automation approach. Combining data collection and processing tools requires considerable preprocessing effort or is entirely infeasible if the necessary attributes cannot be recorded [27].

In this paper, we address these challenges by proposing a reference data model for process-related UI logs. This model provides a data structure and an accompanying interchange format that others can reuse to conceptualize and capture UI logs in a process context. To ensure widespread applicability, the model is designed such that it subsumes and integrates the commonalities of existing process-related UI logs, but remains flexible with regard to the their differences. To identify those commonalities and differences, we conduct a literature review in Sect. 3 and a review of industry solutions in Sect. 4. The reference data model, along with its underlying design principles and an accompanying interchange format, is presented in Sect. 5. In Sect. 6, we demonstrate how the data model can be instantiated in practice by applying it in a real-life RPA scenario. Finally, we conclude the paper with a discussion in Sect. 7.

2   Background and Related Work

Event Logs. Process mining extracts information from *event logs*, i.e., collections of *events* recorded in an IS [32]. An event log consists of *cases* that each correspond to one process instance. Each case contains a trace of events that occurred during the execution of the process instance and can have additional attributes, for example, the size of an order in an order-to-cash process. Events are related to a particular step in a process with an *activity* label (e.g., create invoice) and can also have additional attributes.

Data Formats. To enable the exchange of event data between different ISs, the business process management (BPM) community has developed interchange formats that define the structure and general contents of event logs. The current main format is XES (eXtensible Event Stream), which was introduced in 2010 to replace the older MXML format and was accepted as the official IEEE standard for event data in 2016 [38]. In XES, an event log consists of a three-level hierarchy of log, trace, and event objects. The format is designed to be highly generic, with a minimal set of explicitly defined attributes on each of the three levels. Additional attributes, with a commonly understood semantic meaning, can be introduced by XES extensions. For example, the concept extension introduces the "name" attribute, which stores names for event logs, traces, and events. Although researchers have recently pointed out shortcomings of XES and proposed more flexible, object-centric alternatives such as OCEL [17], XES remains the most common event log format and is supported by many process mining tools.

UI Logs. UI logs are a particular type of event log in which events correspond to low-level interactions of a user with a GUI. They can be recorded either internally by adding logging capabilities to an application, or externally by dedicated logging tools. These tools record screen coordinates for each action and map them to parts of the GUI using optical character recognition technology.

User Behavior Mining. UI logs essentially record how users behave while they are engaged with an application. They can be analyzed by means of data or process mining techniques to gain data-driven insights into user behavior. We refer to this analysis of UI logs as *user behavior mining (UBM)* [1]. UBM can serve different purposes, including the analysis of software usage patterns, the design of new user assistance components, or the automation of tasks.

Task Mining. One application of UBM is to enhance traditional process mining by providing a more detailed view of execution steps. Event logs gathered from ERP systems like SAP or Oracle capture the main tasks in a process, like creating an order, but they do not provide insights into how employees actually perform these tasks. Recording and analyzing detailed task executions is referred to as *task mining* [32] or *desktop activity mining* [29]. These techniques can give companies deeper insights into their processes than traditional process mining alone, and they can also help software vendors to optimize their products, for example, by identifying common usability issues.

Robotic Process Automation. UI logs can be used to automate tasks and entire processes by having bots emulate the recorded user interactions. This approach to automation is called *robotic process automation (RPA)* [22] and has lately received considerable attention in research and practice. Within RPA, UBM techniques can be used to derive automation scripts, but also for *robotic process mining* [26], which for example encompasses the identification of suitable tasks for automation from UI logs [24].

Web Usage Mining. Another field that is concerned with the analysis of user behavior is *web usage mining* [34], i.e., the analysis of clickstream user data recorded during interactions with websites. Web usage mining is often process-agnostic; its main purpose is to optimize websites, for example, by adapting their content and structure to users' browsing behavior [13,19]. The primary data source for web usage mining are server UI logs that are generated in a standardized logging format like the Extended Log Format [37] and have a fixed set of attributes. These include the URL of the current and previous page request, the resource accessed, timestamps, identifying data like the user's IP address, and technical data about the user's web browser and operating system.

UBM in Other Domains. In addition to the research areas mentioned above, interaction logs have been used as a source of data-driven insights into user behavior in several other domains, such as human-computer-interaction [14,15], information retrieval [21], and visualization [18]. The logs in these domains can take various forms, but they generally record user interactions at a much lower level of detail than the process-related UI logs that we focus on in this paper.

3   Literature Review

This paper's goal is to develop a reference data model that subsumes and integrates the commonalities of current approaches for capturing process-related UI logs, but stays flexible with regard to their differences. In this section, we review UI logs from scientific literature to identify those commonalities and differences.

3.1   Research Method

We conducted a structured literature review [23] in SpringerLink, IEEE Xplore, and ACM Digital Library. As search terms, we used "log" combined with (1) "user interact*" and "user interface" (2) "task mining" and "desktop activity mining" as common terms for high-resolution process mining, and (3) "robotic process automation" and "robotic process mining" as important applications of UI logs. We limited our search to papers written in English and published after 2015 because we focus on the current state of the art. The relevance of the initial search results was assessed based on their title and abstract. This yielded a set of potentially relevant papers, on which we performed a forward-backward-search to also cover papers that our search terms might have missed.

To ascertain the relevance of the identified papers, we scanned their full text for passages on UI logs or recording approaches for them. Papers were considered as relevant if (1) they contained a concrete UI log or (2) they described the UI log collection process in enough detail to infer the captured attributes.

Table 1. The papers found in the database search

Search Term	IEEE Xplore	ACM DL	SpringerLink
log AND "user interact*"	[24]	[12]	[2,3,7,35]
log AND "user interface"			
log AND "task mining"			[26]
log AND "desktop activity mining"			
log AND "robotic process automation"			[5,22]
log AND "robotic process mining"			

As listed in Table 1, we found 9 relevant publications in the initial search. Several papers appeared in more than one query, but are only listed under the search term that we first found them with. The forward-backward search returned another 10 publications. Although we did not explicitly search for web usage mining logs, our search returned papers about server-side and also client-side web usage mining (recording web activities by adding tracking software to a browser), but none of these met the above-listed criteria. In our review, we therefore only included one exemplary clickstream log from a process mining context: the BPI Challenge 2016 [11], in which the Dutch Employee Insurance Agency recorded eight months of user activities on their website. Our final result was hence a set of 20 relevant publications.

3.2 Results

Some of the 20 relevant publications covered the same use case and data collection approach and were therefore treated as duplicates, resulting in 12 unique approaches. The majority of papers cover RPA [2–5, 7–9, 20, 22, 24–27, 35]. Four publications [6, 10, 12, 28] focus on software process mining [33]. The remaining two are general approaches to analyzing low-level user interactions with broader applications [11, 29].

Commonalities. Although the reviewed UI logs were fairly heterogeneous, we found a set of six core attributes that are recorded in more than half of them. Table 2 indicates which of the 12 approaches include which attributes (•).

Table 2. UI log attributes as found in the literature review

Source	Action type	Target element	UI Hierarchy	Application	Input value	Timestamp
[2–5]	•	•	•	•	•	•
[6]	•	•		Single	•	•
[9]	•	•	•	•	•	•
[10]	•	•		Single		•
[11]		•	•	Single	•	•
[12]	•			Single		•
[20]	•	•	•	•		•
[8,22]	•			•	•	•
[7,24–27]	•	•	•	•	•	•
[28]	•	•	•	Single		•
[29]	•	•		•	•	•
[35]		•	•	•	•	•

1. An *action type*, which describes the action a user takes. Actions are most often divided into mouse and keyboard inputs, but some logs further distinguish between different mouse buttons, string inputs, and hotkeys. Only two logs do not record action types: the BPIC 2016 clickstream log [11] and Urabe et al. [35], who only record that an interaction has taken place.

2. The atomic *target UI element*, on which the user action is executed. This attribute is recorded in most UI logs, except for two: Dev et al. [12] only record the usage of specific functions, such as crop in a graphics editor, and Jimenez-Ramirez et al. [8,22] record click coordinates and screenshots, but only use them to match similar user actions and do not map them to target elements.
3. The *software application* that the user interacts with. This could be a web browser, an ERP system, or an office application. This attribute is always recorded when researchers track user actions across multiple applications, but is not captured when the tracking is limited to a single application.
4. One or multiple attributes that specify the location of the target element in the application's *UI hierarchy*. For example, an Excel cell is located in a worksheet (hierarchy level 1), which belongs to a workbook (hierarchy level 2) [27]. UI hierarchy attributes are included in about half the reviewed logs.
5. The *input value* that the user writes into a text field. Input values are included in about half of the reviewed logs.
6. A *timestamp*, which records the exact date and time at which the action occurred. This is recorded in all logs.

Differences. Most authors characterize user interactions through an action type, i.e., *what* the user does, and a target element, i.e., *where* they do it. However, the set of possible values for the action type, and hence the level of detail at which actions are recorded, differs considerably. For example, Agostinelli et al. [3] record aggregated action types abstracted from raw hardware input (e.g., clickButton and clickTextField), whereas Jimenez-Ramirez et al. [22] make the low-level differentiation between left, right, and middle mouse clicks. Which other attributes are included in a UI log differs between approaches: whereas timestamps and the application in focus (where applicable) are recorded in all logs, input values and information on the location of a target element within the application's UI hierarchy are included in about half of them. Examples for other, less common attributes that are only recorded in few approaches include the current value of a text field [5,9,27], user IDs [5,10,27,35], other resources involved [6,11], and associations to higher-level process steps [20,29].

Another interesting finding was that most of the reviewed UI logs are initially unlabeled, i.e., they do not have a concrete case notion [16]. In some publications, events in unlabelled logs are later grouped into cases based on different attributes. These attributes include external session IDs created automatically by a system [11] or manually by users [3,27,28], user IDs [10], or case IDs from associated higher-level event logs [20].

4   Review of Industry Solutions

To ensure broad applicability of our reference data model, we also review industry approaches for conceptualizing and capturing UI logs. Because those approaches are core to the industry solutions' functionality and business secrets, the available material for this review may be less specific than scientific papers. Therefore, we conduct the industry review in this section as an addition to the literature review, meant to confirm and complement the established findings.

4.1   Research Method

Selection Strategy. An initial analysis indicated that RPA tools are presently the only industry solutions that collect UI logs on a large scale. Some vendors also advertise task mining capabilities, but their primary focus is on recording UI logs for the automation of routines. Because the RPA market is highly fractured and fast-moving, we could not conduct a complete review. Instead, we opted to analyze a sample of companies that can be seen as representative for the market. Therefore, we selected the companies that the 2021 Gartner Magic Quadrant RPA report[1] attributes with a "high ability to execute" and/or a "high completeness of vision": UiPath, Automation Anywhere, Microsoft Power Automate, Blue Prism, NICE, WorkFusion, Pegasystems, Appian, EdgeVerve Systems, and Servicetrace. We also included Celonis Task Mining, which is the only major product that uses UI logs primarily for low-level process mining.

Review Approach. In analyzing those eleven tools, we focused on finding the commonalities and differences between the industry logs and the scientific logs. Specifically, we wanted to know whether the industry logs capture the same set of six core attributes found in the scientific logs (commonalities) and whether the industry logs capture any other attributes that could be relevant for a widely applicable reference data model (differences). To answer those questions, we collected freely available material about the tools.[2] This included trial or demo versions, documentations, and promotional material, such as videos showcasing the recording process. After collecting the material, we had to exclude two companies from our list, Pegasystems and EdgeVerve Systems, because we could not obtain sufficient information on the functionalities of their recording software.

4.2   Results

Commonalities. For each industry solution, we analyzed whether it also records the six core attributes found in the literature review. The results are summarized in Table 3. All reviewed tools record action types, target elements, input values, applications, and timestamps. Similar to what we found in the literature review, the recordable action types differ considerably between tools.

Differences. We also examined whether the industry solutions systematically record any other attributes, but we did not find any. However, we did find a significant difference between industry logs and scientific logs in how they capture

---

[1] https://www.gartner.com/en/documents/3988021.

[2] A full list of the material that we analyzed can be found at https://gitlab.uni-mannheim.de/jpmac/ui-log-data-model/-/blob/46b363dc75b992a43398f501e3a4cb0e755107d0/industry_review_sources.pdf.

Table 3. UI log attributes as found in the industry review

Company	Action type	Target element	UI Hierarchy	Application	Input value	Time-stamp
UiPath	•	•	Screenshots	•	•	•
MS power automate	•	•	•	•	•	•
Automation anywhere	•	•	•	•	•	•
Celonis	•	•	Screenshots	•	•	•
Blue prism	•	•	Screenshots	•	•	•
Workfusion	•	•	Screenshots	•	•	•
NICE	•	•	Screenshots	•	•	•
Appian	•	•	Screenshots	•	•	•
ServiceTrace	•	•	Screenshots	•	•	•

information on the location of elements within the UI hierarchy. In research, this information is explicitly recorded in UI log attributes, but most industry tools instead store it as screenshots outside of the log. Some tools also use the UI hierarchy to construct selectors that uniquely identify an element within an application's GUI, similar to file paths. Another difference between industry logs and scientific logs concerns the case notion. In the industry solutions, the case ID is always a task or process label that is manually added to the log. Additional business context attributes also need to be added by users and are not recorded by the tool.

5   Reference Data Model

In this section, we introduce our reference data model for user interactions. We consider a reference model to be a conceptual model that serves to be reused for the design of other conceptual models [31]. Under such a reuse-oriented conceptualization, (universal) applicability of the model is not a defining property. However, maximizing the model's application scope increases its reuse potential and therefore its value to the community. Therefore, we designed the model in an inductive or bottom-up fashion [31]: based on the commonalities and differences between existing UI logs that we found in the literature and industry reviews, we constructed a model that subsumes those commonalities, but remains flexible with regard to their differences. In the following, we first elaborate on the principles that guided our design process in Subsect. 5.1. The reference model is presented in detail in Subsect. 5.2. In Subsect. 5.3, we provide a data interchange format for UI log data as a supplement to the reference model.

5.1   Design Principles

In the literature and industry reviews, we found that the main commonality between existing UI logs are the six core attributes. The main differences between them concerned the scope, the level of abstraction, and the case notion. Based on these findings, our data model follows four fundamental design principles:

1. **Minimal set of core components**: The essential characteristics of user interactions, as found in the reviews, are modeled as the components and standard attributes of the data model. Because the model is intended to be non-specific and universally applicable, we include no other elements, thus keeping the number of components and standard attributes to a minimum.
2. **Flexible scope**: To ensure flexibility in scope, the data model can be extended with any number of additional components and all components can have an arbitrary number of attributes. Also, nearly all components and standard attributes are optional. The only non-optional component and attribute that ensure the existence of a UI log are the activity and its name.
3. **Flexible level of abstraction**: To enable user interactions to be modeled in various application contexts and at various levels of abstraction, the domain of the standard attributes in the data model, such as the action type, is left unspecified and can be determined at the point of instantiation. Furthermore, all components are modeled as classes and can be subclassed. Explicit subclasses are only defined for the target object, because they are inherent to the structure of user interfaces and the way they are embedded in ISs.
4. **No explicit case notion**: Whereas the case notion of a business processes is tied to its instances, UI logs are not inherently structured along any data dimension. The reviews have shown that they can have many possible case identifiers. The data model therefore does not include an explicit case notion. Instead, the case notion needs to be defined at the point of instantiation.

5.2   Reference Model Components

The reference data model is depicted as a UML diagram in Fig. 1. It consists of nine components, modeled as classes, and their interrelations, modeled as associations. Each class has an ID and can have any number of attributes. Some components have standard attributes that have a particular significance for user interactions. In the following, we define and explain the individual components.

Components that Define the User Interaction. In our model, user interactions have two parts. First, the *action* component with its *action type* standard attribute that describes what the user does. Common action types, as observed in the reviews, correspond to the functionalities of standard peripheral input devices, such as left or right mouse clicks, single keystrokes, or keystroke combinations for shortcuts. Higher-level distinctions are also possible. For example, when collecting data in an ERP system, actions can be divided into input actions, which make changes to a business object, and navigation actions, which only serve to navigate the GUI.

The second part of an interaction is the *target object* that the action is executed on. It is instantiated as one of four object types in the UI hierarchy, as explained below. The action type and target object together determine the central model component: the *activity*. It is uniquely defined as a combination of an action and a target object and acts as the event label, like in a traditional event log. An activity has three standard attributes: the *activity name*, an optional *input value* that denotes, e.g., the string that is entered into a text field, and a *timestamp* to indicate its execution time. The activity name is determined as a function of the action type of the corresponding action and the identifier of the corresponding target object, for example, a concatenation. The timestamp is a very common attribute in traditional event logs as well as UI logs. It is, however, not a strictly required attribute in the data model, since there are alternative ways to introduce a notion of order into an event log [36].

Components that Define the UI Hierarchy. The UI hierarchy integrates the various types of UI element context data into a general structure. It consists of four components, which form a tree-shaped composition hierarchy: UI element, UI group, application, and system. The UI element and UI group levels mirror the hierarchical structure of virtually all GUIs (e.g., the document object model of a website). The application and system levels go beyond the actual GUI and position it within an IS, which makes it possible to record application- and system-level user interactions and allows the UI log to be compatible with cross-application and even cross-system UI tracking.

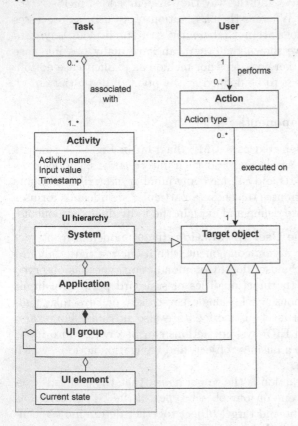

Fig. 1. User interaction data model

Most actions are executed on the atomic *UI elements*, which form the lowest level. Examples include buttons, text boxes, dropdowns, checkboxes, or sliders. Elements can be stateful, such as a non-empty text box or a greyed-out button. Capturing this state is necessary, for example, to track the effects of copy/paste actions or to differentiate between activity outcomes. The state of a UI element is therefore recorded in its *current state* standard attribute.

UI elements are combined into *UI groups*, which can be nested within other UI groups. In many cases, these UI groups are explicit design elements of the user interface, but our model does not impose grouping criteria and allows UI groups to be formed from arbitrary sets of UI elements. A simple example that we saw in the literature review is an Excel cell (UI element), which is part of a worksheet (UI group), which is again part of a workbook (UI group). Modeling UI groups has two main advantages. First, it allows to uniquely identify functionally identical UI elements. For the example above, recording information about UI groups allows us to distinguish between the cell A1 in separate Excel worksheets. This idea is used in many industry solutions to generate element selectors from screen captures. Second, UI groups can be useful for event abstraction, i.e., mapping user interactions to higher-level conceptual tasks, if these tasks are closely tied to particular UI groups. For example, all interactions with elements in a login mask (enter username, enter password, click login) can directly be abstracted to the "login" task.

UI elements and UI groups belong to an *application*, i.e., a single program instance. Some actions are directly executed on the application and are not tied to lower-level elements, such as "undo" or application-specific hotkeys.

The root node of the UI hierarchy is the *system*, on which the applications run and actions are recorded. Similar to application-level actions, it is also possible to capture system-level actions, such as the Ctrl-Alt-Del key combination to open the Task Manager on a Windows system.

Components that Define the Context. Finally, the data model includes two components that put UIs in a conceptual context: *user* and *task*. These exist in some form for all UI logs, which is why they are included in the model. In contrast, other potential context components, such as organizational or resource attributes, are use-case-specific and can be considered by extending the model.

The user is the entity that initiates any interaction. Each action is associated with a single user. Because user IDs and attributes depend on the data collection environment (e.g., device IDs in mobile applications or IP addresses on websites), the model does not specify any attributes for users. This also means that the user component is not necessarily restricted to humans and can model computer-initiated interactions, for example when recording partially automated processes.

The task component associates the recorded user interactions with conceptual tasks or routines, which makes it possible to map low-level GUI interactions to higher-level user activities. This abstraction is an essential prerequisite for being able to perform meaningful analysis on UI logs or to use them for automation.

5.3   Exchange Format

To further increase the applicability and reuse potential of our data model, we implemented it as an extension to the XES standard for event logs.[3] This *UIlog* extension provides a standardized exchange format for UI logs as a supplement to the data model. The implementation considers the activity equivalent to the event label and does not include the activity name or timestamp standard attributes because those are already provided by the concept and time extensions. The other components and standard attributes are defined at event level,

---

[3] The XML specification for the *UILog* extension is available at https://gitlab.uni-mannheim.de/jpmac/ui-log-data-model/-/raw/main/UILog_extension.

i.e., as attributes of an activity instance. The generic target object is not directly implemented, but can instead be specified through attributes that correspond to its four UI hierarchy subclasses: the target object is the lowest-level UI hierarchy component that exists for this event. For example, for an event with a UI element attribute, the target object is always this UI element, whereas for an event with no UI element or UI group attributes, the target object is the application.

6   Working Example

To demonstrate the practical utility of the reference data model, we describe in this section how it can be instantiated in a real-life scenario. The scenario is based on an RPA project, which we are currently conducting in cooperation with an ERP system vendor. The project is set in the medical technology industry, where companies are required to regularly validate their ISs to ensure that they are in compliance with external quality regulations. The validation of an IS involves manually executing a number of predefined workflows step-by-step according to a rigid execution plan, checking the result of each step against a set of acceptance criteria, and documenting the result. Manually executing a well-defined validation workflow is a repetitive and time-consuming task. The goal of our project is to automate this task using RPA. We want to record how process experts interact with the UI of the ERP system during validation and then train bots to emulate their actions.

In the following, we use the example of a keyword creation workflow to show how an artificial UI log that captures one execution of this workflow may instantiate the data model. The keyword creation workflow consists of five consecutive steps, which are executed on the GUI parts shown in Fig. 2. The user (1) logs in (a), (2) selects the right client and profile (b), (3) navigates through the dashboard (c) to reach the explorer tree (d, left), (4) creates a new keyword (d, right), and (5) logs out. The main acceptance criterion is that the newly created keyword shows up in the explorer tree after refreshing.

Table 4 shows a UI log for one case, i.e., one execution of the keyword creation workflow. It includes the action type, target UI element, and one level of UI groups, plus input value and current state where applicable. The captured action types are left and right clicks, text input, selected keyboard shortcuts, and none. The activity label of an event (in most cases) consists of the concatenated action type and target object identifier.

The first two events in the log do not correspond to single user interactions, but instead take advantage of the UI group concept to directly abstract to higher-level tasks. Instead of recording each event in the login and client selection masks separately, the task is tracked only at completion and the content of the text fields is read out when the user presses the "Login" and "Set Profile" buttons. For these abstracted activities (marked with an "A_" prefix), the action type is "none"; they are defined only through the target UI group, independent of the performed actions. This approach can be used for simple tasks with the same execution pattern in all workflows. Its main upside is that it reduces noise,

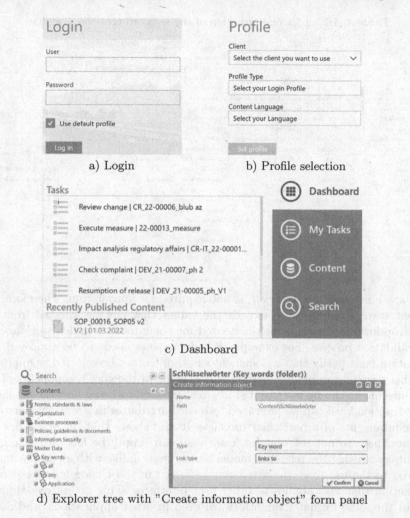

a) Login   b) Profile selection

c) Dashboard

d) Explorer tree with "Create information object" form panel

Fig. 2. The user interface of the application described in the example

which is a common problem in UI logs [26]. In our scenario, activities like initially entering a wrong password do not affect the outcome of the workflow and are therefore not relevant for automating it. By abstracting during data collection, those activities are automatically disregarded. Other advantages of abstraction are reduced implementation effort and smaller UI logs.

For effective automation, various user inputs need to be tracked. Therefore, the instantiation of the input value attribute in the log is flexible and depends on the action type and target object: When a user writes into a textbox, the input value is the entered string. When an item is selected from a dropdown, the input value records the label of that item. For abstracted activities, the input value captures the string values of all relevant UI group elements as a map.

Table 4. UI log for one execution of the keyword creation workflow

Activity	Action type	UI element	UI group	Input value	Current state
A_Login	none		login mask	{username: pren, password: dts123}	
A_Profile Selection	none		user select client	{client: base, profile: author}	
click content	left click	content	dashboard ov		
click masterdata	left click	masterdata	explorer tree		
click masterdata node expand	left click	masterdata node expand	explorer tree		
click keywords node expand	left click	keywords node expand	explorer tree		
rclick keywords	right click	keywords	explorer tree		
click ppanel new	left click	ppanel new	explorer tree		
click new information object	left click	new information object	explorer tree		
click name	left click	name	fpanel keyword		
input name	input	name	fpanel keyword	MyKeyword	
click dd type	left click	dd type	fpanel keyword		[keyword, keywords folder]
click dd type	left click	dd type	fpanel keyword	keyword	[keyword, keywords folder]
click dd linksto	left click	dd linksto	fpanel keyword		[linksto]
click dd linksto	left click	dd linksto	fpanel keyword	linksto	[linksto]
click confirm	left click	confirm	fpanel keyword		
click keywords node expand	left click	keywords node expand	explorer tree		
KEY_F5 explorer tree	KEY_F5		explorer tree		
click logout	left click	logout	explorer tree		
click confirm	left click	confirm	dialog logout		

Most state information, however, is not required for automation. Therefore, the current state attribute only records the values that can be selected from list and dropdown elements, which is needed for some more complex workflows in the validation process. For example, if a documents needs to be approved, the validation must verify that a document's author cannot be selected as approver.

This simple example demonstrates how some of the core components of the reference model can be instantiated in a real-life scenario, and how the flexibility in abstraction level can be leveraged to record attributes in a way that matches the requirements of a particular use case. It also shows that, in practice, components that are not relevant for a use case can simply be left out. The main advantage of using the reference model here is that, unlike with an ad-hoc model tailored to the use case, the attributes captured in the UI log follow a general convention that also applies to other user interfaces. This makes recording UI logs in the same format straightforward even in other applications, and makes it possible to develop automation or task mining solutions that are independent of the recording approach used.

7   Discussion and Conclusion

In this paper, we propose a reference data model for UI logs. Based on reviews of scientific literature and industry solutions, it has a set of core components to capture essential characteristics of user interactions and is flexible with regard to scope, abstraction level, and case notion. We implement the model as an XES extension and exemplarily show how it can be instantiated in a real-life RPA scenario.

Contribution. Our main objective is to address the issues that arise from the lack of standardization of UI logs. Therefore, we derive the reference from existing UI logs. Most of the components are directly adopted from the core attributes identified in our reviews. Our contribution is their integration into a unified framework with well-defined relations. For example, we propose a rigid interpretation of an activity by defining it as a combination of an action and a target element. We also expand on the location context of UI elements and explicitly define four distinct types of target objects in an unambiguous hierarchy.

To this unified framework, we add additional, less frequently collected components and standard attributes that are particularly relevant for a complete model of user interactions. For instance, the current state is an important property of stateful UI elements when the log is intended to be used for automation. In the UI hierarchy, we add the system on top of the commonly recorded application to model system-level user interactions. We also introduce the user and task context components to add (optional) generic business context to UI logs.

Limitations. One limitation of our work concerns its grounding in existing UI logs. Despite following a methodical approach, we do not claim that our reviews or the model are complete or exhaustive. There could be unidentified UI logs or future UI logs in different use cases, which are not well represented by the model. For instance, our data model is only intended to model user interactions with graphical user interfaces, and we did not consider alternative input types, for example from voice commands or eye-tracking devices. The model may also be somewhat biased towards automation use cases because RPA solutions are overrepresented in the two reviews that it is based on.

Another limitation is that the XES standard is not particularly well-suited for UI logs. It does not support explicitly defining the relations between attributes, so all components of the UI hierarchy have to be implemented at event level. Therefore, even if many events involve the same target object, UI group, application, system and their attributes need to be included each time, leading to considerable redundancy. XES also assumes a single case notion, contrary to the flexible case notion that we intend for the data model.

Conceptually, implementing the model as an extension to the Object-Centric Event Log format OCEL [17] would be more appealing, because users, tasks, and UI hierarchy elements could be modeled as objects, reducing the redundancy. However, OCEL does not support extensions and has two main limitations with regard to UI logs. First, it does not support dynamic object attributes that can differ between events, such as the current value of a textbox that may change between interactions. Second, object attributes cannot be tied to certain object types, so for example the current state attribute cannot be limited to UI elements only. Therefore, we decided to implement the model as an XES extension.

Future Work. Our reference model can contribute to the field by providing a common, application-independent conceptual framework for user interactions. However, like any reference model, it needs to prove its utility in practice. We therefore want to encourage researchers and practitioners to adopt the model for capturing UI logs in their projects, and to extend it both with regard to new use cases and with regard to conceptual aspects, such as user privacy.

References

1. Abb, L., Bormann, C., van der Aa, H., Rehse, J.R.: Trace clustering for user behavior mining. In: European Conference on Information Systems, AIS (2022)
2. Agostinelli, S., Leotta, F., Marrella, A.: Interactive segmentation of user interface logs. In: Hacid, H., Kao, O., Mecella, M., Moha, N., Paik, H. (eds.) ICSOC 2021. LNCS, vol. 13121, pp. 65–80. Springer, Cham (2021). https://doi.org/10.1007/978-3-030-91431-8_5
3. Agostinelli, S., Leotta, F., Marrella, A.: Interactive segmentation of user interface logs. In: Hacid, H., Kao, O., Mecella, M., Moha, N., Paik, H. (eds.) ICSOC 2021. LNCS, vol. 13121, pp. 65–80. Springer, Cham (2021). https://doi.org/10.1007/978-3-030-91431-8_5
4. Agostinelli, S., Marrella, A., Mecella, M.: Automated segmentation of user interface logs. In: Robotic Process Automation, pp. 201–222. De Gruyter Oldenbourg (2021)
5. Agostinelli, S., Marrella, A., Mecella, M.: Exploring the challenge of automated segmentation in robotic process automation. In: Cherfi, S., Perini, A., Nurcan, S. (eds.) RCIS 2021. LNBIP, vol. 415, pp. 38–54. Springer, Cham (2021). https://doi.org/10.1007/978-3-030-75018-3_3
6. Ardimento, P., Bernardi, M.L., Cimitile, M., Ruvo, G.D.: Learning analytics to improve coding abilities: a fuzzy-based process mining approach. In: International Conference on Fuzzy Systems, pp. 1–7. IEEE (2019)
7. Bosco, A., Augusto, A., Dumas, M., La Rosa, M., Fortino, G.: Discovering automatable routines from user interaction logs. In: Hildebrandt, T., van Dongen, B.F., Röglinger, M., Mendling, J. (eds.) BPM 2019. LNBIP, vol. 360, pp. 144–162. Springer, Cham (2019). https://doi.org/10.1007/978-3-030-26643-1_9
8. Chacón Montero, J., Jimenez-Ramirez, A., Gonzalez Enríquez, J.: Towards a method for automated testing in robotic process automation projects. In: International Workshop on Automation of Software Test, pp. 42–47 (2019)
9. Choi, D., R'bigui, H., Cho, C.: Candidate digital tasks selection methodology for automation with robotic process automation. Sustainability 13(16), 8980 (2021)
10. Damevski, K., Shepherd, D.C., Schneider, J., Pollock, L.: Mining sequences of developer interactions in visual studio for usage smells. IEEE Trans. Softw. Eng. 43(4), 359–371 (2017)
11. Dees, M., van Dongen, B.: BPI challenge 2016 (2016). https://data.4tu.nl/articles/dataset/BPI_Challenge_2016_Clicks_Logged_In/12674816/1
12. Dev, H., Liu, Z.: Identifying frequent user tasks from application logs. In: International Conference on Intelligent User Interfaces, pp. 263–273. ACM (2017)
13. Ding, A., Li, S., Chatterjee, P.: Learning user real-time intent for optimal dynamic web page transformation. Inf. Syst. Res. 26(2), 339–359 (2015)
14. Dumais, S., Jeffries, R., Russell, D.M., Tang, D., Teevan, J.: Understanding user behavior through log data and analysis. In: Olson, J.S., Kellogg, W.A. (eds.) Ways of Knowing in HCI, pp. 349–372. Springer, New York (2014). https://doi.org/10.1007/978-1-4939-0378-8_14
15. Fern, X., Komireddy, C., Grigoreanu, V., Burnett, M.: Mining problem-solving strategies from HCI data. ACM Trans. Comput. Human Interact. 17(1), 1–7 (2010)
16. Ferreira, D.R., Gillblad, D.: Discovering process models from unlabelled event logs. In: Dayal, U., Eder, J., Koehler, J., Reijers, H.A. (eds.) BPM 2009. LNCS, vol. 5701, pp. 143–158. Springer, Heidelberg (2009). https://doi.org/10.1007/978-3-642-03848-8_11

17. Ghahfarokhi, A.F., Park, G., Berti, A., van der Aalst, W.M.P.: OCEL: a standard for object-centric event logs. In: Bellatreche, L., et al. (eds.) ADBIS 2021. CCIS, vol. 1450, pp. 169–175. Springer, Cham (2021). https://doi.org/10.1007/978-3-030-85082-1_16
18. Guo, H., Gomez, S., Ziemkiewicz, C., Laidlaw, D.: A case study using visualization interaction logs and insight metrics to understand how analysts arrive at insights. IEEE Trans. Visual. Comput. Graph. 22(1), 51–60 (2016)
19. Ho, S., Bodoff, D., Tam, K.: Timing of adaptive web personalization and its effects on online consumer behavior. Inf. Syst. Res. 22(3), 660–679 (2010)
20. Hofmann, A., Prätori, T., Seubert, F., Wanner, J., Fischer, M., Winkelmann, A.: Process selection for RPA projects: a holistic approach. In: Robotic Process Automation, pp. 77–90. De Gruyter Oldenbourg (2021)
21. Islamaj Dogan, R., Murray, G., Névéol, A., Lu, Z.: Understanding PubMed® user search behavior through log analysis. Database 2009 (2009)
22. Jimenez-Ramirez, A., Reijers, H.A., Barba, I., Del Valle, C.: A method to improve the early stages of the robotic process automation lifecycle. In: Giorgini, P., Weber, B. (eds.) CAiSE 2019. LNCS, vol. 11483, pp. 446–461. Springer, Cham (2019). https://doi.org/10.1007/978-3-030-21290-2_28
23. Kitchenham, B.: Procedures for performing systematic reviews. Keele University, Technical report (2004)
24. Leno, V., Augusto, A., Dumas, M., La Rosa, M., Maggi, F.M., Polyvyanyy, A.: Identifying candidate routines for robotic process automation from unsegmented UI logs. In: International Conference on Process Mining, pp. 153–160. IEEE (2020)
25. Leno, V., Augusto, A., Dumas, M., La Rosa, M., Maggi, F.M., Polyvyanyy, A.: Discovering data transfer routines from user interaction logs. Inf. Syst. 107, 101916 (2021)
26. Leno, V., Polyvyanyy, A., Dumas, M., La Rosa, M., Maggi, F.M.: Robotic process mining: vision and challenges. Bus. Inf. Syst. Eng. 63(3), 301–314 (2021)
27. Leno, V., Polyvyanyy, A., La Rosa, M., Dumas, M., Maggi, F.: Action logger: Enabling process mining for robotic process automation. In: BPM Demos. Springer (2019)
28. Linares-Vásquez, M., White, M., Bernal-Cárdenas, C., Moran, K., Poshyvanyk, D.: Mining android app usages for generating actionable GUI-based execution scenarios. In: Working Conference on Mining Software Repositories, pp. 111–122. IEEE (2015)
29. Linn, C., Zimmermann, P., Werth, D.: Desktop activity mining - a new level of detail in mining business processes. In: INFORMATIK, pp. 245–258. Köllen (2018)
30. López-Carnicer, J.M., del Valle, C., Enríquez, J.G.: Towards an OpenSource logger for the analysis of RPA projects. In: Asatiani, A., et al. (eds.) BPM 2020. LNBIP, vol. 393, pp. 176–184. Springer, Cham (2020). https://doi.org/10.1007/978-3-030-58779-6_12
31. Rehse, J.R., Fettke, P.: A procedure model for situational reference model mining. Enterprise Model. Inf. Syst. Architect. 14(3), 1–9 (2019)
32. Reinkemeyer, L.: Process Mining in Action: Principles. Use Cases and Outlook. Springer, Switzerland (2020). https://doi.org/10.1007/978-3-030-40172-6
33. Rubin, V.A., Mitsyuk, A.A., Lomazova, I.A., van der Aalst, W.: Process mining can be applied to software too! In: International Symposium on Empirical Software Engineering and Measurement. ACM (2014)
34. Srivastava, J., Cooley, R., Deshpande, M., Tan, P.N.: Web usage mining: Discovery and applications of usage patterns from web data. In: SIGKDD Explorations, vol. 1, pp. 12–23. ACM (2000)

35. Urabe, Y., Yagi, S., Tsuchikawa, K., Oishi, H.: Task clustering method using user interaction logs to plan RPA introduction. In: Polyvyanyy, A., Wynn, M.T., Van Looy, A., Reichert, M. (eds.) BPM 2021. LNCS, vol. 12875, pp. 273–288. Springer, Cham (2021). https://doi.org/10.1007/978-3-030-85469-0_18

36. van der Aalst, W.M.P., Santos, L.: May I take your order? In: Marrella, A., Weber, B. (eds.) BPM 2021. LNBIP, vol. 436, pp. 99–110. Springer, Cham (2022). https://doi.org/10.1007/978-3-030-94343-1_8

37. WWW Consortium: Extended log file format (1995). https://www.w3.org/TR/WD-logfile.html

38. XES Working Group: IEEE standard for eXtensible Event Stream (XES) for achieving interoperability in event logs and event streams. IEEE Std 1849 (2016)

Analyzing Variable Human Actions
for Robotic Process Automation

A. Martínez-Rojas[1]([✉]) [iD], A. Jiménez-Ramírez[1] [iD], J. G. Enríquez[1] [iD],
and H. A. Reijers[2] [iD]

[1] Departamento de Lenguajes y Sistemas Informáticos, Escuela Técnica Superior de
Ingeniería Informática, Avenida Reina Mercedes, s/n, 410121 Sevilla, Spain
`{amrojas,ajramirez,jgenriquez}@us.es`
[2] Department of Information and Computing Sciences, Utrecht University,
Princetonplein 5, Utrecht 3584 CC, The Netherlands
`h.a.reijers@uu.nl`

Abstract. Robotic Process Automation (RPA) provides a means to
automate mundane and repetitive human tasks. Task Mining approaches
can be used to discover the actions that humans take to carry out a
particular task. A weakness of such approaches, however, is that they
cannot deal well with humans who carry out the same task differently
for different cases according to some hidden rule. The logs that are used
for Task Mining generally do not contain sufficient data to distinguish
the exact drivers behind this variability. In this paper, we propose a new
Task Mining framework that has been designed to support engineers who
wish to apply RPA to a task that is subject to variable human actions.
This framework extracts features from User Interface (UI) Logs that
are extended with a new source of data, namely screen captures. The
framework invokes Supervised Machine Learning algorithms to generate
decision models, which characterize the decisions behind variable human
actions in a machine-and-human-readable form. We evaluated the proposed Task Mining framework with a set of synthetic UI Logs. Despite
the use of only relatively small logs, our results demonstrate that a high
accuracy is generally achieved.

Keywords: Robotic Process Automation · Process discovery · Task
mining · Decision model discovery

1 Introduction

Robotic Process Automation (RPA) is a software technology that facilitates the
automation of human tasks, especially when they are structured and repetitive.

This research has been supported by the Spanish Ministry of Science, Innovation and
Universities under the NICO project (PID2019-105455GB-C31) and the Centro para el
Desarrollo Tecnológico Industrial (CDTI) of Spain under the CODICE project (EXP
00130458/IDI-20210319 - P018-20/E09) and by the FPU scholarship program, granted
by the Spanish Ministry of Education and Vocational Training (FPU20/05984).

C. Di Ciccio et al. (Eds.): BPM 2022, LNCS 13420, pp. 75–90, 2022.
https://doi.org/10.1007/978-3-031-16103-2_8

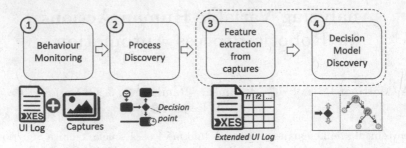

Fig. 1. Proposed framework for variability analysis through interpretable decisions from UI Logs. The current paper focuses on steps 3 and 4.

In contrast to other automation approaches (i.e., API-based), RPA works by closely mimicking the way humans interact with computer applications [2].

Some of the typical benefits are that the technology helps to save costs, increases agility, and improves quality [9,18] while its level of intrusiveness is low [33]. Due to these wide range of benefits, industry has adopted this technology on a wide scale in recent years [10].

Most RPA projects start out by observing how human workers perform work that is to be automated. To support this initial RPA analysis, approaches such as Task Mining [1,31] and Robotic Process Mining [22] are highly suitable. What all these techniques have in common is that they operate on UI Logs, i.e., a series of timestamped events (e.g., mouse clicks and keystrokes), obtained by monitoring and recording user interfaces.

An open issue concerning discovering the process model behind the UI Log is to disclose the drivers behind the variations that are shown in a process model. These variations indicate that human operators take different decisions for different cases, but it is generally not possible to find rules which explain how these decisions were made. That, however, is crucial knowledge for the engineer who aims to develop the RPA bot. A clear example occurs in the development of a business process outsourcing operation, where it is necessary to work with different systems that, in their turn, are virtualized. In this type of scenario, it is difficult to understand certain operator decisions that depend specifically on the context of the problem being addressed. Although efforts have been made to overcome this problem, they rely on what is observable in the log [3,12,19,32]. This approach has inherent limitations since human work is sometimes based on information that is simply not captured in the log, i.e., it only 'appears on the screen. Therefore, human decision-making remains *hidden* in this respect and, as a result, difficult to automate.

Against this backdrop, this paper proposes a framework to analyze the variability in human actions automatically. The framework leverages information on the screen to detect factors that influence human decisions – an angle that existing approaches have neglected so far.

In previous work, we proposed (1) a tool to monitor the user behavior which generates a UI Log, including one screen capture for each event (cf. step 1 in Fig. 1) [25], and (2) a method to analyze such logs to discover the underlying process model (cf. step 2 in Fig. 1) [15]. This paper significantly extends these contributions by: (1) proposing a novel approach to systematically analyze the screen captures to extract information which is, then, incorporated into the UI Logs (cf. step 3 in Fig. 1) using image-processing techniques [26] (e.g., Optical Character Recognition), (2) presenting a method to discover decision models which explain the variability that is found in the UI Log (cf. step 4 in Fig. 1) using a Machine Learning (ML) approach, and (3) evaluating the approach with synthetics problems of different complexity to demonstrate the efficacy and efficiency of the framework.

It should be noted that industrial RPA platforms often do incorporate sophisticated task mining techniques, as well as features for image processing. The point is that these capabilities are not integrated in approaches to analyze task variability.

The rest of the paper is organized as follows. Section 2 provides the background in topics like behavior monitoring, ML, and image processing. Section 3 introduces a synthetic business case to motivate the proposal. Section 4 elaborates on the novel method to discover the decision models. Section 5 reports the empirical evaluation performed to validate the method. Sect. 6 presents a critical discussion. Section 7 reviews similar approaches in the literature. Finally, Sect. 8 summarizes the work and describes future research lines.

2 Background

The approach that is presented in this paper builds on behavior monitoring techniques, process discovery, Graphical User Interface (GUI) analysis, and ML.

For *behavioral monitoring*, there are several industrial solutions for keylogging[1] that capture the interaction of a human interacting with a system. In addition, other approaches have been proposed in academia, taking a further step in how to automate certain stages of robotization [3,11,20,24,25]. It should be noted that there are different formats proposed for capturing events, although the most representative for this work is the *UI Log* from [15] which defines it as an extension of the XES format—standard for event logs in Process Mining—which incorporate attributes like the *app_name* (i.e., the name of the app), *event_type* (i.e., mouse click or keystroke), *click_type* (i.e., left, right, or middle), *click_coords* (i.e., position of the mouse on the screen), the *keystroke* (i.e., the keys that are typed), and the *screenshot* (i.e., the screen capture associated to this event path).

Using a UI Log, many proposals exist for *process discovery*, i.e., to automatically or semi-automatically discover the underlying process model that is associated with human behavior [5,6,13,15,19,23]. Moreover, these proposals include functionalities to clean the UI Log from irrelevant information, so that noise in the resulting process model is filtered; and to select variants/cases/activities

---

[1] Availabe at: www.spyrix.com and bestxsoftware.com/es/.

according to the frequency, length, and other criteria that are useful to identify process candidates to robotize. The resulting process model may contain decision points and separate branches for different process variants.

In the field of *GUI analysis*, approaches exist than can identify the GUI components within an image [28,34]. GUI components are atomic graphical elements with predefined functionality, displayed within a GUI of a software application [28]. Besides locating the element on the screen (i.e., calculating the bounding boxes), they can classify them by the type of element, e.g., image, button, or text. Nevertheless, when dealing with texts, Optical Character Recognition (OCR) techniques are more appropriate. For instance, KerasOCR [17] allows extracting words and their bounding boxes from screen captures. Such an image-based technique allows for extracting the content and structure of an image in a computer-readable way (cf. Fig. 2).

In the *ML* domain, supervised algorithms exist that focus on finding rules to explain a given dataset, e.g., extracting a decision tree [27]. Datasets are commonly represented in tabular form: each column is an input variable or a label (i.e., what needs to be predicted or classified), and each row is a member of the dataset. In a classification problem, the algorithm tries to find patterns in the input variables that help to explain the labels. Decision trees are an example of classification algorithms that, besides just providing a classification, do so in a human-interpretable way [14].

3   Running Example

This section describes an artificial business case to explain and motivate the problem addressed in this paper. Figure 3 depicts an excerpt of the elementary user interfaces for registering a customer in the context of a telecom company.

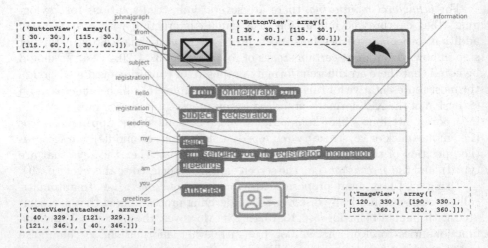

Fig. 2. Example of GUI analysis applied to a sample screen capture. Extracted buttons are in blue, images are in green, and texts are in red. To the sake of readability, some extracted coordinate are not shown. (Color figure online)

In this registration process, the human operator checks her email inbox and reviews the pending emails regarding the registration tasks. After opening the email, the operator has to validate that all provided information related to the new customer is correct. More precisely, the customer *ID card* is expected to be included as an attachment. If it is indeed included (cf. Fig. 3a), all the customer data has to be registered into a CRM system (cf. Fig. 3b). Otherwise, in case the *ID card* is missing (cf. Fig. 3c), an email has to be sent to the customer requesting such data (cf. Fig. 3d). Regardless which of these two situations occurred (i.e., Variant 1 or Variant 2 of Fig. 3), the operator returns to their inbox to process the next mail. This process must be repeated several times during the day to process the entire queue of emails in the operator's inbox.

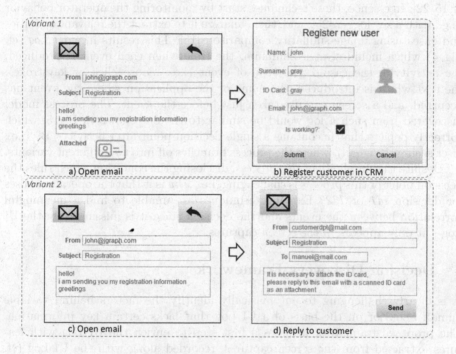

Fig. 3. Mockups of motivating example

Timestamp	CaseId	ActivityId	MorKeyb	Coordinates	TextInput	NameApp	Screenshot
12312313	1	A	MOUSE	123,32	""	Mail client	image0001.png
12312314	1	B	MOUSE	32,43	""	Mail client	image0003.png
12312315	1	C	MOUSE	44,12	""	Mail client	image0004.png
12312316	1	D	KEYBOARD	234,367	"28362233J"	CRM	image0005.png
12312317	1	D	MOUSE	23,55	""	CRM	image0007.png
12312318	2	A	MOUSE	123,32	""	Mail client	image0008.png
12312319	2	B	MOUSE	32,43	""	Mail client	image0010.png
12312320	2	C	MOUSE	44,12	""	Mail client	image0011.png
...	...	...	...	...	...	...	...

Fig. 4. Excerpt of a UI Log obtained form a keylogger.

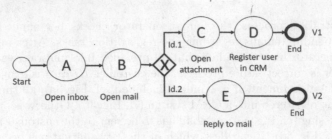

Fig. 5. Process model discovered from the UI Log.

Task Mining techniques can be applied to discover the operator process model [3,15,22]. In essence, these techniques start by monitoring the operator behavior (e.g., with a key logger [25]) and, then, analyze it to extract the relevant activities and cases using image-similarity comparison [15]. This results in a *UI Log* (cf. Fig. 4) which includes, at a minimum, the time when each event is produced, the activityId, the caseId, the type of event (i.e., mouse click or keystroke), the text which is introduced, the name of the application where the event has occurred, and a screen capture taken just before the event. The process model discovered from such a log would be similar to the one shown in Fig. 5, which correctly depicts that it contains a single decision point after activity "B" (i.e., after seeing the email), where the process branches off into two different variants.

Despite the simplicity of this process, disclosing the condition which rules the decision point of this process is challenging, i.e., *why* is it that the operator choses for decision *Id1* or *Id2*? Existing techniques are unable to find a meaningful correlation between the events and the decision since it is missing from the UI Log – it only appears in the screen captures.

4   Decision Discovery Framework

It is clearly challenging to automatically identify the factors behind variable human behavior on the basis of a UI Log that lacks certain key information. The proposed framework includes a first step to enrich the UI Log with features extracted from the screen captures, recorded along with the UI Log (cf. Sect. 4.1). The second step that is carried out by the framework leverages this new information, which is derived from the screen captures, to deliver a decision model. In our opinion, it is crucial that such a model can be interpreted by a machine and also be understood by a human (cf Sect. 4.2).

4.1   Feature Extraction

This step transforms the screen captures that are taken for each event in the UI Log into structured information (i.e., features), which can be incorporated back into the log. Since a variety of features can be extracted from a screen capture, the current framework offers a common interface for these extractors (cf. Definition 1) which can be implemented accordingly to the project necessities.

Definition 1. *A **Feature Extractor** is a tuple <Name, Function> that represents a software component named Name, with a Function that receives an image and returns a list of pairs <key, value>, where the key is the name of a feature that presents a value in the given image.*

Once implemented, the framework applies these feature extractors to each event in the UI Log, i.e., the *Function* of each extractor is applied to the screen capture of each event. If the feature extracted is new in the UI Log (i.e., none of the columns of the UI Log have the same name as the *key*), a new column is appended to the log, and the *value* is assigned to this event. The rest of the log events have an empty value for this new column. Otherwise, if the feature already exists, the *value* is assigned to this event at the existing column *key*.

To illustrate this process, we will describe the **UI Element Occurrence** extractor in more detail. It extracts the occurrences of UI elements in the sense that the output of its *Function* contains as many *keys* as different UI elements are found in the screen capture. The *value* associated with each *key* is a number greater than 0, which expresses the number of occurrences of this *key* in the screen capture. The extractor includes the detection and classification of each component to determine which type of UI Element they belong to. For this purpose, each screen capture is processed in three phases by the *Function*:

1. To detect the UI elements, image-processing techniques are applied to find the elements within the image. In this phase, edge detection algorithms can be used, such as Canny's algorithm [7], which we applied. Each detected UI element is then cropped to deal with these separately in the next phase.
2. To classify the detected UI elements according to the type of GUI component they belong to, an ML model—previously trained for conducting such a task— is used.[2] Specifically, we adapted the convolutional neural network proposed in Moran's work [28], which is able to detect 14 different types of UI elements.
3. To return the number of occurrences for each type of UI element, the detected UI elements in the first phase are first grouped, according to the class determined during the second phase, and then summed.

Consequently, after applying the UI Element Occurrence extractor, 14 columns—one for each type of UI element—are appended to the resulting UI Log. They are then available, along with the additional columns from other extractors.

Running Example. When considering the screen capture in Fig. 3.a, which relates to our running example, 2 image buttons can be observed (i.e., the 'envelope' for returning to the email home page, and the 'arrow' for reply), as well as 3 texts. Therefore, the returning list of the *Function* of the UI Element Occurrence extractor is: $\{<\#\_ImageButton, 2>, <\#\_TextView, 3>\}$. This results in an UI Log that includes the $\#\_ImageButton$ column, which gets value 2, and the $\#\_TextView$ column, which gets value 3. The values are shown in Fig. 6 under the header 'Occurrence Extractor info'.

---

[2] We trained the model with this dataset: https://doi.org/10.5281/zenodo.2530277.

				Occurrence Extractor info			Other extractors	
Timestamp	MorKeyb	Coordenates	TextInput	#_ImageButton	#_TextView	#_Button	Subject	Attached
12312313	MOUSE	123,32	""	2	15 ...	1	56,67 ...	
12312314	MOUSE	32,43	""	2	3 ...	0	74,68 ...	72,91
12312315	MOUSE	44,12	""	2	5 ...	0	74,68 ...	72,76
12312316	KEYBOARD	234,367	"28362233J"	0	10 ...	2	...	
12312317	MOUSE	23,55	""	0	14 ...	2	...	
12312318	MOUSE	123,32	""	2	15 ...	1	56,67 ...	
...	...	...	...	...	...	...	...	...

Fig. 6. A sample UI Log including features (columns) extracted from the screen captures.

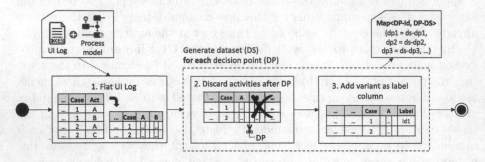

Fig. 7. Proposed algorithm for generating labeled datasets

4.2 Decision Model Discovery

Once the UI Log is enriched with the extracted features, the decision model discovery takes place. To enable this, first a labeled dataset is generated for each decision point in the process model. Secondly, a classification algorithm is applied to disclose the rules existing in the dataset, which will help to explain the decision in the process model. In the remainder of this section, each of these steps will be explained.

Generating the Labeled Dataset. This step processes the enriched UI Log to convert it into a labeled dataset usable by supervised ML algorithms (e.g., classification trees) in the next step. Specifically, one dataset is created for each decision point that appears in the process model.

Given a decision point, the objective is to determine which branch is chosen, providing a detailed explanation. Therefore, the *label* of the resulting dataset is the branch of the decision point. With this aim, Fig. 7 illustrates, in an activity diagram, the way of creating an appropriate dataset for each decision point that appears in a process model.

First, it receives both the UI Log and the discovered process model and returns a map where the keys are the *IDs* of the decision point in the process model, and the values are the *dataset* extracted to each one. The algorithm starts by flattening the UI Log (cf. step 1 in Fig. 7). This operation is done by putting all UI Log events of each case in the same row of the dataset (i.e., the dataset will contain one row for each different case in the UI Log). In this way, the dataset will include columns for an *ID* (i.e., auto-incremental number starting in 0), the *caseID*, a *TimeStampEnd* (i.e., corresponding to the timestamp of the last event of this case), and a *TimeStampStart* (i.e., corresponding to the timestamp of the first event of this case), and the rest of the UI Log attributes for each activity of this case (i.e., event type, click coordinates, each of the extracted features, etc.). Regarding these latter attributes, the column names are the attribute names prefixing its *activityID*. For instance, the EventType attribute of activity A will be stored as: *EventType_A*.

Then, for each decision point in the process model, the columns of the dataset are filtered in such a way that only the columns related to the activities that precede that decision point are kept (cf. step 2 in Fig. 7). Note that the columns of the events *after* the decision point relate to events in the *future*, so they should not be considered when discovering the decision model of that particular decision. Aferwards, the *label* column is added to the dataset. For each row (i.e., each process case), its value is the branch which is taken for this decision point, (cf. step 3 in Fig. 7).

Discovering the Decision Model with Classification Algorithms. As we explained, the labeled dataset generated at this point will be used to train a supervised ML model. This model will classify the *label* column based on the rest of the dataset columns. There is a wide range of algorithms that can be used for this purpose. For our framework, we use decision trees since both humans and machines can easily interpret them. This kind of model expresses the discovered rules of the classification in the form of a tree: the tree nodes are a column of the dataset (i.e., UI Log attributes) while the tree edges are non-overlapping conditions as evaluated over the node. Our framework implements four common used algorithms to construct decision trees: CART [8], ID3 [29], C4.5 [30] and CHAID [16]. Although these techniques are known to be similar, they are implemented to explore their behavior in this context.

To make the tree even more understable for humans, the framework takes the discovered features and rules to **highlight** them into the associated screen captures, in this way linking them with their visual information.

ID	Tstamp_Start	Tstamp_End	Case	EventType_A	EventType_B	Coor_A	Coor_B	TextInput_A	TextInput_B	NameApp_A	...
0	12312313	12312317	1	MOUSE	MOUSE	32,45	123,42	-	-	Mail client	...
1	12312318	12312322	2	MOUSE	MOUSE	32,45	123,42	-	-	Mail client	...
2	12312323	12312326	3	MOUSE	MOUSE	32,45	123,42	-	-	Mail client	...
3	12312327	12312330	4	MOUSE	MOUSE	32,45	123,42	-	-	Mail client	...

...	#_Button_A	#_ImageView_B	...	Attached_A	Attached_B	from_A ... subject_B	Label
...	1	1	...	-	74,91	54,67 ...	Id1
...	1	1	...	-	74,91	54,67 ... 74,68	Id1
...	1	0	...	-	-	54,67 ... 74,68	Id2
...	1	0	...	-	-	54,67 ... 74,68	Id2

Fig. 8. Dataset extracted from the UI Log of Fig. 4

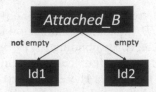

Fig. 9. Decision Tree for the decision point in the process model of Fig. 5

Running Example. In our running example, only one decision point is discovered from the UI Log (cf. Fig. 5). Therefore, only one dataset is generated (cf. Fig. 8). The dataset contains a series of columns that describe the attributes for each case and a *label* column that indicates the decision which is made at the decision point. The decision tree that is obtained when trained on this dataset is shown in Fig. 9. As can be inferred from the framed columns in Fig. 8, the decision *Id2* is made when there is nothing in column *Attached_B*; otherwise, the decision *Id1* is taken. When looking at the screen capture of activity *B* (cf. Fig. 3a and 3c), we can see that these actions correspond to the absence or presence of the word "Attachment" in the email, respectively. Note that the classification algorithm provides one tree, although other alternatives exist. For instance, the number of ImageView UI elements in activity *B* explains the behavior too.

5   Empirical Evaluation

Purpose: This empirical evaluation aims to analyze our proposal to discover the conditions that drive decisions from a UI Log. Particularly, it focuses on situations where the conditions depend on information that does appear on the screen but not in the log itself.

Objects: The evaluation is based on a set of synthetic problems, which resemble realistic use cases in the administrative domain. More precisely, 3 different processes (P) are created, each of them with a different level of complexity. Complexity is measured in terms of the number of activities, the number of variants to execute the process, and the number of visual features that affect the decision to choose between variants. The processes are:

P1   Client creation. A process with 5 activities and 2 variants. The single decision in this process is made based on the existence of an attachment in the reception email.
P2   Client validation. A process with 7 activities and 2 variants. The decision is made based on the user's response to a query.
P3   Client deletion. A process with 7 activities and 4 variants. The decisions are made based on two conditions: (1) the existence of pending invoices and (2) the existence of an attachment to justify the payment of the invoices.

These processes all contain a single decision point, although the one in P3 is rather complex. All processes include (1) synthetic screen captures for their activities and (2) a sample event log with a single instance for each variant. To generate the objects for the evaluation, we generate event logs of different sizes ($|L|$) for each of these processes by deriving events from the sample event log. We consider log sizes in the range of $\{10, 25, 50, 100\}$ events. Note that we consider complete instances in the log and, thus, we remove the last instance if it goes beyond $|L|$. Some of these logs are generated with a balanced number of instances, while others are unbalanced ($B?$) in the sense that more than a 20% frequency difference exists between the most frequent and less frequent variants. To average the result over a collection of problems, 30 instances are randomly generated for each tuple $<P, |L|, B?>$.[3]

Independent Variables: The independent variables of this empirical evaluation are (1) the process (i.e., P), (2) the log size (i.e., $|L|$), and (3) whether the log is balanced or not (i.e., $B?$).

Response Variables: The efficiency and efficacy of the approach are evaluated in terms of: (1) the average time spent on each of the framework phases, i.e., feature extraction (tFE) and decision model discovery (tDD), (2) the number of columns included in the log after the feature extraction phase ($\#CL$), (3) the number of columns included in the dataset after the flattening phase ($\#CD$), and (4) the average accuracy of the discovered model (Ac).

Evaluation Design: For each of the 3 processes, 30 instances are randomly generated by varying the graphical look&feel of the screen captures, including/excluding some UI elements that do not affect the process, as well as replacing texts. For each of these 90 instances, 8 different logs are generated considering the different values of $|L|$ and $B?$. To do this, the instances of the sample event log of the process are used as templates to create similar ones by applying changes in the events while keeping their logic (e.g., a mouse click at a random place inside the same button). The framework is then executed for each of the 720 objects and the response values are calculated considering the average values for the 30 instances. To measure accuracy, we record 100% if the model correctly identifies the condition and 0% otherwise.

Execution Environment: The evaluation was run on a machine with Windows 10, an Intel i9-7900X processor at 3.30 GHz, 64 Gb of RAM, and 10 cores.

Evaluation Results and Data Analysis: Table 1 shows the experiment results. For each problem, identified by P, $B?$, and $|L|$, the average values for the 30 scenarios are shown for each response variable. Some of them are calculated for the feature extraction phase (i.e., tFE and $\#CL$,) and the others are calculated for the decision model discovery (i.e., tDD and Ac for each algorithm).

Regarding the feature extraction phase, we can observe that it is a time-consuming task (i.e., tFE), which takes longer as the number of activities of the process increase (i.e., P) and the size of the log (i.e., $|L|$) grows. This behavior is expected since the extraction algorithms need to be applied to each screen capture that exists in the log before the decision point. Nonetheless, the number of

---

[3] The set of problems are available at: https://doi.org/10.5281/zenodo.5734323.

Table 1. Experiment results

P	B?	$|L|$	tFE^a	#CL	#CD	tDD_{CART}^b	tDD_{ID3}^b	$tDD_{C4.5}^b$	tDD_{CHAID}^b	Acc_{CART}^c	Acc_{ID3}^c	$Acc_{C4.5}^c$	Acc_{CHAID}^c
P1	Yes	10	108	25	39	323	339	330	335	12.9	12.9	12.9	12.9
		25	258	25	39	332	361	356	350	83.9	83.9	83.9	83.9
		50	497	25	39	352	394	393	379	100.0	100.0	100.0	100.0
		100	1,029	25	39	334	360	358	352	100.0	100.0	100.0	100.0
	No	10	112	25	39	324	340	329	334	12.9	12.9	12.9	12.9
		25	273	25	39	330	354	351	344	83.9	83.9	83.9	83.9
		50	501	25	39	350	390	388	376	96.8	96.8	96.8	96.8
		100	977	25	39	331	359	358	349	100.0	100.0	100.0	100.0
P2	Yes	10	79	25	79	704	716	708	701	0.0	0.0	0.0	0.0
		25	137	25	79	399	421	404	819	31.0	31.0	31.0	31.0
		50	319	25	79	1,323	1,572	1,405	2,102	93.1	93.1	93.1	93.1
		100	588	25	79	3,495	3,781	3,593	3,531	100.0	100.0	100.0	100.0
	No	10	70	25	79	308	321	307	316	0.0	0.0	0.0	0.0
		25	158	25	79	563	597	570	583	6.9	6.9	6.9	6.9
		50	363	25	79	1,787	1,853	2,175	2,463	75.9	75.9	75.9	75.9
		100	659	25	79	4,627	4,584	3,737	4,764	100.0	100.0	100.0	100.0
P3	Yes	10	51	25	79	1,086	1,090	1,096	1,090	0.0	0.0	0.0	0.0
		25	189	25	79	1,155	1,101	1,058	1,014	0.0	0.0	0.0	0.0
		50	333	25	79	1,818	2,200	1,981	2,748	32.3	48.4	32.3	48.4
		100	724	25	79	2,848	3,039′	3,008	5,950	100.0	100.0	96.8	100.0
	No	10	51	25	79	1,085	1,102	1,088	1,087	0.0	0.0	0.0	0.0
		25	219	25	79	1,378	1,094	1,489	1,082	0.0	0.0	0.0	0.0
		50	394	25	79	1,735	2,359	1,893	2,284	19.4	45.2	19.4	45.2
		100	737	25	79	5,473	5,626	3,762	5,908	77.4	100.0	93.6	100.0

[a]Expressed in seconds; [b]Expressed in milliseconds; [c]Expressed in %.

features that are extracted and included in the UI Log (i.e., #CL) only depends on the extractor itself that, for this experiment, the *UI Element Occurrence* extractor (cf. Sect. 4.1) obtains a fixed number of features for each screenshot, i.e., 14—one for each UI element. The other 11 columns are the standard ones defined for the UI Log.

Regarding the decision model discovery phase, the number of columns included in the dataset for training the classification algorithm (i.e., #CD) depends on the number of activities before the decision point. More precisely, P1 has 2 activities, while P2 and P3 both have 4 activities. In this phase, it can be observed that the time for decision model discovery, which is expressed in *ms* (i.e., tDD), is negligible in comparison with tFE, which is expressed in *s*. Moreover, tDD depends on both $|L|$, since the flattener algorithm needs to run over all the events, and #CD, since the tree's training runs over the entries in the dataset. In turns, tDD seems not to be influenced by the algorithm. When analyzing the accuracy of the classification tree (i.e., Ac), it is clear that the framework has a better performance with higher values of $|L|$ since there are more entries in the dataset. However, as expected, the accuracy decreases when the process becomes more complex (i.e., P): more columns in the dataset are not relevant for the decision, i.e., can be considered noise. In addition, for P3, we observe more differences between the performance of the algorithms while in P2 and P3 all the algorithms present the same behavior. This situation may be caused since P3 decision depends on more than one feature. Furthermore, having more possible variants (i.e., 4) may affect the performance of the algorithms because fewer rows are obtained for each variant to train the decision models. Unlike the previous response variables, Ac is influenced by whether the log is balanced or not (i.e., $B?$). This behavior can be expected since an unbalanced dataset offers fewer opportunities to distinguish between data and noise. It is important to note that the accuracy is 100% for most cases with $|L| = 100$, which is a reasonably small number for this kind of logs. This provides the insight that the framework *can generally explain the variability within these processes*. This result is very encouraging, particularly when considering the small log sizes included in this experimental set-up.

6 Discussion

Discovering why decisions are made in a process is of utmost importance when the aim is to analyze the variability or even to automate such decisions. The framework proposed in this paper was motivated by the fact that existing Task Mining approaches fail to leverage screen captures when mining UI Logs. In general, alternatives exist to make a transparent analysis of the UI, e.g., navigating the DOM tree or accessing Windows GUI API. Yet, we discovered that considering screen captures is necessary for specific situations like in Business Process Outsourcing scenarios [15], where access to the front-end of the information systems is usually secured or virtualized by systems like Citrix. Although combining both sources of information—transparent analysis of UIs and screen captures—could bring benefits when they are available, this proposal focuses on these situations based exclusively on screenshots proposing a framework that can support a process analyst to (1) accelerate the analysis phase since the automatically-generated decision models include rules linked to the screen captures, which are readable by the analyst, (2) accelerate the development phase since the rules are in a computer-readable format too, and (3) unleash candidates to automate that would be discarded otherwise because no rule could be found to describe the human variability in UI Log. Our experimental evaluation, as conducted with synthetic problems of a variety of complexities, showed positive results. It is noteworthy that these problems use logs that includes one event for each activity while, in practice, there might be more than one events for a single activity, i.e., several events performed over the same screen. However, the approach has not been tested on real UI Logs nor real screen captures; these are clearly limitations to our work.

There are further limitations that can be observed. Specifically, the feature extraction phase highly relies on the image-processing techniques (i.e., the OCR and GUI analysis), which may occasionally present wrong classifications or detections. Moreover, several alternatives exist for these techniques, and they evolve fast due to highly active communities. Nonetheless, the proposed framework is not limited to any of these techniques. After all, it defines a high-level interface to incorporate improved techniques easily. In addition, it is noteworthy that the number of extracted features is directly related to the number of columns in the dataset to be used to train the classifier. Although the aim is to not overlook any relevant feature from the screen captures, non-relevant features may just produce noise that would negatively affect its performance unless more entries are included in the dataset (i.e., longer UI Logs). Fortunately, if the number of dataset columns increases, the evaluation has demonstrated that the framework keeps a reasonably good performance if the size of the log increases as well.

Finally, the validation has intentionally considered problems that rely on the screen captures' content to make the decision. This was done to ensure that the evaluation covers the two phases of the framework. However, the classification algorithm may take into account *all* the information contained in the UI Log, which makes it compatible with existing related approaches, e.g., SmartRPA [4], which is very suitable when screen captures are not required.

7   Related Work

There exist other proposals in the literature that can be applied for decision model discovery. Rozinat and Van der Aalst [32] use decision trees to analyze choices made based on data dependencies that affect the routing of a case. However, this approach does not consider information on the screen nor provides the possibility of showing graphically to a non-expert user why a decision was made.

Agostinelli et al. [3] and Leno et al. [22] cover the complete RPA lifecycle, from event capture to the automatic generation of scripts. Data capture is based on an *Action Logger* that captures the information through plugins [23] or separately within the system [3]. Furthermore, although this capture is mainly focused on the keyboard and mouse events, they capture the DOM tree in those events which interact with the web browser. In contrast to these approaches, our work focuses on screen captures as the primary information source.

Leno et al. [21] present an algorithm that generates a kind of "association rules" between events and results. The information gathering is based on tailor-made plugins. Their approach proposes similar solutions as is the case for our framework. However, in their approach it is not possible to capture the information that the user generates outside the context of these plugins. Finally, Gao et al. [12] propose a solution based on the implementation of decision trees for the algorithmic deduction of RPA rules, based on the captured user behavior. Similar to [21], screen captures and their features are left out from this analysis.

8   Conclusion and Future Work

This paper proposes a framework for discovering decision models from event logs that are extended with screen captures. The framework includes a phase to extract features from such screen captures, as well as a phase to discover decision models using these features. Furthermore, we illustrated our proposal by means of a running example and provided an extensive empirical evaluation. Our proposal advances the state-of-the-art on Task Mining for its application to RPA by providing insights into the drivers for variability in human behavior that is targeted to be automated.

For future work, we plan: (1) to validate the current approach in a real-life context with people with different profiles. So that both the feature extraction algorithm and the decision discovery algorithm can be tested with industrial data and scenarios; (2) to investigate the robustness of the proposal against noise injected at the event level; (3) to investigate mechanisms to reduce noise by filtering/selecting the appropriate features, e.g., by using eye-tracking technologies, which could focus the attention of the feature extraction algorithms on those screen regions where human attention is devoted to; (4) to analyze further classification algorithms to provide more alternatives to express the observed variability; in this way, they can be compared in terms of usability and understandability; and (5) analyze other UI elements detection and classification techniques to compare them with the accuracy of Canny's algorithm and CNN.

References

1. van der Aalst, W.M.P.: Process Mining: Data Science in Action. Springer, Heidelberg (2016). https://doi.org/10.1007/978-3-662-49851-4
2. Scheppler, B., Weber, C.: Robotic process automation. Informatik Spektrum **43**(2), 152–156 (2020). https://doi.org/10.1007/s00287-020-01263-6
3. Agostinelli, S., Lupia, M., Marrella, A., Mecella, M.: Automated generation of executable RPA scripts from user interface logs. In: BPM, pp. 116–131 (2020)
4. Agostinelli, S., Lupia, M., Marrella, A., Mecella, M.: SmartRPA: a tool to reactively synthesize software robots from user interface logs. In: CAiSE, pp. 137–145 (2021)
5. Augusto, A., et al.: Automated discovery of process models from event logs: review and benchmark. IEEE Trans. Knowl. Data Eng. **31**(4), 686–705 (2018)
6. Bazhenova, E., Bülow, S., Weske, M.: Discovering decision models from event logs. In: BIS, pp. 237–251 (2016)
7. Brahmbhatt, S.: Shapes in Images, pp. 67–93. Apress, Berkeley, CA (2013)
8. Breiman, L., Friedman, J.H., Olshen, R.A., Stone, C.J.: Classification and Regression Trees. Wadsworth and Brooks, Monterey, CA (1984)
9. Capgemini, C.: Robotic Process Automation - Robots conquer business processes in back offices (2017)
10. Denagama Vitharanage, I.M., Bandara, W., Syed, R., Toman, D.: An empirically supported conceptualisation of robotic process automation (RPA) benefits. In: ECIS (2020)
11. Egger, A., ter Hofstede, A.H., Kratsch, W., Leemans, S.J., Röglinger, M., Wynn, M.T.: Bot log mining: using logs from robotic process automation for process mining. In: ER, pp. 51–61 (2020)
12. Gao, J., van Zelst, S.J., Lu, X., van der Aalst, W.M.: Automated robotic process automation: a self-learning approach. In: OTM, pp. 95–112 (2019)
13. Geyer-Klingeberg, J., Nakladal, J., Baldauf, F., Veit, F.: Process mining and robotic process automation: a perfect match. In: BPM, pp. 124–131 (2018)
14. Gilpin, L.H., Bau, D., Yuan, B.Z., Bajwa, A., Specter, M., Kagal, L.: Explaining explanations: an overview of interpretability of machine learning. In: 2018 IEEE 5th DSAA, pp. 80–89 (2018)

15. Jimenez-Ramirez, A., Reijers, H.A., Barba, I., Del Valle, C.: A method to improve the early stages of the robotic process automation lifecycle. In: CAiSE, pp. 446–461 (2019)
16. Kass, G.V.: An exploratory technique for investigating large quantities of categorical data. J. R. Statist. Soc. Ser. C (Appl. Statist.) **29**(2), 119–127 (1980)
17. Keras OCR. https://github.com/faustomorales/keras-ocr. Accessed Mar 2022
18. Lacity, M., Willcocks, L.: What knowledge workers stand to gain from automation. Harv. Bus. Rev. **19**, 1–6 (2015)
19. Leno, V., Armas-Cervantes, A., Dumas, M., La Rosa, M., Maggi, F.M.: Discovering process maps from event streams. In: ICSSP, pp. 86–95 (2018)
20. Leno, V., Augusto, A., Dumas, M., La Rosa, M., Maggi, F.M., Polyvyanyy, A.: Identifying candidate routines for robotic process automation from unsegmented UI logs. In: ICPM, pp. 153–160 (2020)
21. Leno, V., Dumas, M., La Rosa, M., Maggi, F.M., Polyvyanyy, A.: Automated discovery of data transformations for robotic process automation. arXiv preprint arXiv:2001.01007 (2020)
22. Leno, V., Polyvyanyy, A., Dumas, M., La Rosa, M., Maggi, F.M.: Robotic process mining: vision and challenges. Bus. Inf. Syst. Eng. **63**, 1–14 (2020)
23. Leno, V., Polyvyanyy, A., La Rosa, M., Dumas, M., Maggi, F.M.: Action logger: enabling process mining for robotic process automation. In: CEUR Workshop Proceedings (2019)
24. Leshob, A., Bourgouin, A., Renard, L.: Towards a process analysis approach to adopt robotic process automation. In: ICEBE, pp. 46–53 (2018)
25. López-Carnicer, J.M., del Valle, C., Enríquez, J.G.: Towards an opensource logger for the analysis of RPA projects. In: BPM, pp. 176–184 (2020)
26. Majumder, B.P., Potti, N., Tata, S., Wendt, J.B., Zhao, Q., Najork, M.: Representation learning for information extraction from form-like documents. In: ACL, pp. 6495–6504 (2020)
27. Mitchell, T.: Machine Learning. McGraw Hill, New York (1997)
28. Moran, K., Bernal-Cárdenas, C., Curcio, M., Bonett, R., Poshyvanyk, D.: Machine learning-based prototyping of graphical user interfaces for mobile apps. IEEE Trans. Softw. Eng. **46**(2), 196–221 (2018)
29. Quinlan, J.R.: Induction of decision trees. Mach. Learn. **1**(1), 81–106 (1986)
30. Quinlan, J.R.: C4. 5: Programs For Machine Learning. Elsevier, Amsterdam (2014)
31. Reinkemeyer, L.: Process Mining in Action: Principles, Use Cases and Outlook. Springer, Switzerland (2020). https://doi.org/10.1007/978-3-030-40172-6
32. Rozinat, A., van der Aalst, W.M.P.: Decision mining in ProM. In: Dustdar, S., Fiadeiro, J.L., Sheth, A.P. (eds.) BPM 2006. LNCS, vol. 4102, pp. 420–425. Springer, Heidelberg (2006). https://doi.org/10.1007/11841760_33
33. Willcocks, L., Lacity, M.: A new approach to automating services. MIT Sloan Manage. Rev. **58**(1), 40–49 (2016)
34. Xu, Z., Baojie, X., Guoxin, W.: Canny edge detection based on open CV. In: 2017 13th ICEMI, pp. 53–56 (2017)

The SWORD is Mightier Than the Interview: A Framework for Semi-automatic WORkaround Detection

Wouter van der Waal(✉) , Iris Beerepoot , Inge van de Weerd ,
and Hajo A. Reijers

Utrecht University, Utrecht, The Netherlands
{w.g.vanderwaal,i.m.beerepoot,g.c.vandeweerd,h.a.reijers}@uu.nl

Abstract. Workarounds can give valuable insights into the work processes that are carried out within organizations. To date, workarounds are usually identified using qualitative methods, such as interviews. We propose the semi-automated WORkaround Detection (SWORD) framework, which takes event logs as input. This extensible framework uses twenty-two patterns to semi-automatically detect workarounds. The value of the SWORD framework is that it can help to identify workarounds more efficiently and more thoroughly than is possible by the use of a more traditional, qualitative approach.

Through the use of real hospital data, we demonstrate the applicability and effectiveness of the SWORD framework in practice. We focused on the use of three patterns, which all turned out to be applicable to the characteristics of the data set. The use of two of these patterns also led to the identification of actual workarounds. Future work is geared to the extension of the patterns within the framework and the enhancement of techniques that can help to identify these in real-world data.

Keywords: Workarounds · Automated detection · Event data · Healthcare · Process mining · Business process analysis

1 Introduction

Many organizations use standard operating procedures to streamline their work. When procedures are clear, people know what to do. Still, it often happens that work is performed in a way that is different from the prescribed procedure. When confronted with unexpected situations, limited time, or a lack of resources, workers may be unable to follow a procedure and may feel compelled to perform a workaround to solve a problem [13].

Some workarounds are beneficial and can be leveraged to improve organizational procedures [2,6]. In other cases, not following a procedure may be harmful or outright dangerous. Workarounds can result in noncompliance, privacy issues,

© Springer Nature Switzerland AG 2022
C. Di Ciccio et al. (Eds.): BPM 2022, LNCS 13420, pp. 91–106, 2022.
https://doi.org/10.1007/978-3-031-16103-2_9

or negative effects in the process downstream [15,16]. Whatever the effect, it is important that process owners are provided with insights into the occurrence of workarounds. These insights can help to prevent workarounds from happening again or to improve the concerned procedure [6]. In addition, being able to structurally and comprehensively identify workarounds would allow for the monitoring of their emergence, diffusion, and evolution, potentially enabling process analysts to detect and respond to new workarounds faster than with current techniques. This motivates our focus on workaround detection.

To date, most studies in which workarounds were identified and analyzed relied on qualitative methods, primarily through interviewing and observing users during their work [6]. This approach has led to valuable insights related to the mechanics and effects of workarounds, as well as the motivations of the people using them [3,15,32]. However, the use of qualitative methods is labor-intensive; furthermore, users may not disclose their normal behavior when they are aware of being observed [33]. Similar to how process mining is used to solve an otherwise time-consuming problem [1], our focus is on the use of event logs and other quantitative analysis techniques.

In this paper, we introduce the Semi-automated WORkaround Detection (SWORD) framework. The automated part of the SWORD framework uses 22 patterns to identify potential workarounds from an event log. Of these patterns, 16 are based on existing literature, while the remaining six patterns are new. Whether any pattern can be used in a particular situation is dependent on the characteristics of the data in the event log at hand. While the detection of potential workarounds can be performed in a highly automated fashion, the actual confirmation of the occurrence of workarounds still needs to be done by domain experts. This proposed approach can be expected to partly mitigate the change in behavior that people might exhibit when they are aware of being observed because the data is collected in a non-obtrusive way. Also, since the SWORD framework automates the analysis of event data, it is less labor-intensive to use than finding workarounds through interviews.

In earlier work, we already established that event logs from a Health Information System (HIS) can be used to automatically detect and monitor known workarounds [5]. We stay in line with our earlier focus on the healthcare domain with this present work. Studies have shown that in hospital settings specifically workarounds are a widespread phenomenon [28]: Nurses share each other's passwords to save time, physicians send each other X-rays via WhatsApp to get quick second opinions, and secretaries use shadow systems on paper to track the department's occupation. The principles behind the SWORD framework are nonetheless transparent and we expect that many of the patterns can be transferred to other domains.

To show the potential of the SWORD framework, we apply three of the patterns to real hospital data in the setting of an illustrative case study. The data was obtained through our cooperation with the University Hospital Utrecht (UMCU). Our evaluation shows that the proposed approach is feasible in the sense that the patterns allow for an automated analysis of the data. Furthermore, the use of the SWORD framework was helpful to identify actual workarounds in medical practice.

The structure of this paper is as follows. In Sect. 2, we will first clarify what workarounds are and to what extent they can be detected, according to related work. We will explain in more detail how we analyzed the literature as well as the set-up of our case study in Sect. 3. The results of these steps are reported upon in Sect. 4. Finally, we will discuss the implications of our work in Sect. 5.

2 Related Work

Workarounds occur when users intend to reach a goal but perceive a block to do so using the official procedure [13]. Similar to the conformance-checking field of process mining, we can look for differences between process variants to detect them [23]. However, we see two important differences. First of all, a workaround requires an *intent* to reach a business goal. So, fraud, deception, and errors are not in scope. Secondly, the user is unable to achieve this goal by using the intended procedure [13]. So, accidentally following a different route is not a workaround.

In addition, while non-conformance usually supposes strict rules [18,22] or a known process model to conform to [29], workarounds can occur without these. Deviance mining also uses process mining but looks at smaller deviations between processes [24], which may also be useful to mine for workarounds. Quite different from a deviance, a workaround can be very common. If a workaround is sufficiently effective, it may be shared throughout the organization, potentially becoming more common than the official procedure [17].

One approach specifically used to automatically detect workarounds is to use deep learning [33]. Neural networks are trained to recognize different workaround types from event logs. These methods can be difficult to use in practice because they require a large amount of labeled training and testing data. In addition, even if neural networks reach a high classification accuracy, it is difficult to explain why this is happening [26], which is often required if you want to use the results in a healthcare environment.

Outside of the control-flow, the time, resource, and data perspectives are valuable for conformance checking [23]. By investigating workarounds discovered using qualitative methods, such as interviews and observations, previous studies show that these perspectives can also be used to recognize different types of workarounds [5]. We will continue this multi-perspective approach to investigate if, in addition to recognizing workarounds, we can discover new workarounds using event logs.

3 Research Method

Our research method consists of two phases: a phase in which we define a list of workaround detection patterns and a phase in which we test them. The two phases and their underlying activities are depicted in Fig. 1. We will describe the phases in more detail in this section.

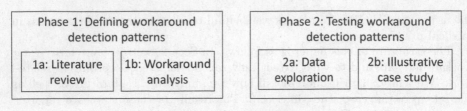

Fig. 1. The phases of our research method

3.1   Phase 1: Defining the Workaround Detection Patterns

As described in the previous section, detecting workarounds has similarities with several approaches in process mining, such as conformance checking. Both focus on differences between the intended and actual process. To investigate these approaches, we carried out a literature review to collect an overview of process mining approaches that can be applied to workaround detection.

After an initial literature search, we selected five (systematic) literature reviews as the starting point of our own study. The reviews focused on several (sometimes overlapping) process mining topics, namely conformance checking [12], process variant analysis [31], predictive process monitoring [11], deviance mining [24], and process mining in healthcare [4]. In the next step, we selected relevant papers from these reviews. We used the following inclusion criteria: (1) the described approach focuses on the differences between process variants, from the control-flow, data, resource, or time perspective, and (2) the described approach uses event data for their analysis.

For example, we did not include supervised learning methods because it is not feasible to label all traces with a workaround or normative label. Note, also, that we keep to the main pattern in situations where slightly differing variants exist. For example, there are multiple version of trace alignment; we do not distinguish between these here. Discovered papers were also searched for new references using reverse snowballing until no new detection patterns came up. Overall, we analyzed 37 papers in detail and included 12 papers in our final analysis, covering 16 detection patterns. The result of this activity is presented in Table 2.

Second, we carried out an analysis of 81 workarounds. These were gathered in previous studies that have been carried out in five different healthcare organizations [5]; a general hospital, two district hospitals, and two specialized care centers. The authors were able to detect multiple discovered workarounds using quantitative methods in a completely different top clinical hospital. This shows that we can expect similar behavior in other healthcare organizations. They are documented as 'workaround snapshots' and include a description of (1) the setting in which the workaround was found, (2) the workaround compared to the normative process, (3) a motivation, and (4) the expected effect of the workaround on cost, time, quality, and flexibility. For an example, see Table 1.

Table 1. Example snapshot

Setting	At a urology clinic. Before a nurse administers medication to a patient, she checks with another nurse to see it is the correct medicine/dosage
Workaround	The nurses do not register the verification check in the HIS
Motivation	Registering the check takes a lot of time, so the nurses only do this check vocally
Effect	This workaround saves time and thus costs, but the data quality is lower. Since the information is not registered, it may lead to errors at a later time

Using the comparison between the workaround and the normative process, we determined if and how each workaround could be monitored using (event log) data. Similar detection patterns were grouped together. This workaround analysis yielded another six patterns, which have been added to the list in Table 2.

3.2 Phase 2: Testing the Workaround Detection Patterns

In the second phase of our study, we evaluated the list of workaround detection patterns. We followed a two-step approach for this. First, we wanted to establish whether the data necessary to detect a pattern was stored in the HISs. Therefore, we analyzed the data structure of the tables in which the event data of the relevant processes and workarounds are stored. For each pattern, we searched for a table containing the required data to identify it. At this point, we considered if we could use that data to discover new workarounds, or if the pattern relied on specific knowledge and could only be used to monitor known workarounds.

Second, we conducted an illustrative case study in which we took three of the identified workaround detection patterns and tried to find them in real data. We selected patterns that could be applied to the available data and with which we expected to find meaningful differences. The goal of this step was to confirm whether our approach would work on real data. At the same time, the IT department of the University Medical Center (UMC) Utrecht was exploring to implement process mining techniques to better facilitate quality improvement projects and consulted us for our expertise. We then performed a technical proof-of-concept on deploying process mining techniques to detect workarounds using the data from the UMC Utrecht. This academic hospital cares for more than 200,000 patients and has around 12,000 employees. SQL was used to capture the event data that we used in our analyses from the HIS used by the UMC Utrecht (HiX, ChipSoft, Amsterdam). We pseudonymized the data as early as possible by assigning random unique values to all patient, resource, and hospitalization identifiers. After data extraction, we used R for further analysis and visualization.

Time Between Activities Out of Bounds. The first pattern covers one of the most common workarounds: batching. When taking patient measurements, hospital staff should register the results directly in the HIS, before doing anything else. Instead, they often take measurements from multiple patients in a row, write the results down on a note and register all patient data afterward. This behavior can be detected using the "Time between activities" pattern.

All manually entered patient measurements in the hospital, except for data from the intensive care unit and operating room, are stored in a single table. We extracted the patient ID, resource ID, and registration timestamp of all measurements between two defined moments. To ensure the right events are captured, directly subsequent measurement registrations of the same patient by one resource are removed. Not doing so could result in counting multiple consecutive registrations for the same patient, which would obviously not be batching.

We compared the three shifts of an average, regular Tuesday, which had some overlap. The day shift ran from 7:00 h until 16:00 h, the evening shift from 15:00 h until 0:00 h, and the night shift from 23:00 h until 8:00 h. There were 1403 measurement events, covering 942 unique patients and 460 resources during the day shift, 785 measurement events, 488 unique patients and 320 resources during the evening shift, and 371 measurement events, 245 unique patients and 132 resources during the night shift.

Long Duration Between Time of Event and Time of Logging. The second workaround covers the registration delay by comparing the difference between the time of activity and the time of registration. This data is registered in the same measurements table we used during the batching workaround. We again used the patient ID and timestamps of the registration. We also extracted the time of activity and the type of measurement, to see if there are differences in delays between types. Note that the time of activity is generally registered manually, so it is unlikely these are exact values.

We have checked this over the same day we used for the first pattern. This time, we did not separate the three shifts, so we included all measurements that occurred between 7:00 h on day one and 8:00 h on day two. A single measurement registration could cover multiple measurement types. For example, a registration could cover both a length and weight measurement at the same time. In those cases, we counted these as a separate registration for each measurement type, all using the same timestamps. In total, there were 10118 measurements by 638 resources, covering 1032 unique patients and 204 different measurement types.

Activities Executed by a Single Resource. The final workaround we investigated concerns resources that stay the same, while they should have differed over the trace. During the triage process at the emergency department (ED), both a nurse and a physician should see the patient. The main ED table in the HIS logs both the "seen by nurse" and "seen by physician" activities. We also extracted the resource logging the activities and ED registration identifiers, which is unique for every ED registration and makes for a good case ID.

We have used the data covering one year. This contained 5993 ED registrations. Both the "seen by nurse" and the "seen by physician" activities were logged 5508 times (92%). In 5460 cases (91%) both had seen the patient.

4   Results

4.1   Workaround Detection Patterns - Literature Review

Researchers in process mining fields other than workaround mining, such as conformance checking, deviance mining, predictive process monitoring, and process performance analysis, have developed patterns to detect differences between similar process variants. Since we can recognize different workarounds using varying perspectives, we have structured our review the same way. We start with control-flow and follow this with the data, resource, and time perspective.

The control-flow perspective is used in most fields, often by comparing traces to process models [5,8,27]. Alternatively, some activities should never co-occur [8] or should occur close to specific others [21]. Some events on their own can already be interesting to monitor [7,9,24]. Repeated behavior can also be an indication something is going wrong. We can look at how often activities repeat [8,9,24,30] or if there are loops in a trace [9,20,24,30,34].

We find the most use of the data perspective in the conformance checking field, where we can simply look at the values of data objects [5,8–10,19,31]. If they deviate too much, this can show unintended behavior in the trace. Alternatively, the exact value might not be important, but the value should not change during the trace [8,10,19].

The resource perspective shows similar patterns. Some events should always be executed by a specific resource or it needs to stay the same during (part of) the trace [5,19]. For example, certain medications should only be prescribed by a physician. While not a detection pattern in itself, earlier mentioned patterns can also be used using a resource as case ID. In this way, we can investigate the behavior of resources. For example, if we do so and notice a resource is repeating the same activity often, something might be going wrong [30].

We can find deviating patterns using the time perspective in multiple fields. From a conformance checking viewpoint, some activities may need to be executed at a specific time [19]. Process performance analysis naturally takes time into account too. We can use the time of activity since the start of the trace to predict the performance of the entire trace [7]. Multiple fields distinguish between process variants by looking at the time between activities [10,19,30,34], the duration of a single activity [30,31,34], or the total time of a trace [31].

Table 2 shows an overview and description of all 22 detection patterns.

Table 2. Workaround detection patterns

	Detection pattern	Explanation	Reference
Control-flow	Occurrence of an activity	A specific activity occurs	[7, 9, 24]
	Occurrence of recurrent activity sequence	A recurrent activity sequence occurs within a trace	[9, 20, 24, 30, 34]
	Frequent occurrence of activity	An activity frequently occurs within a trace	[8, 9, 24, 30]
	Occurrence of activities in an order different from process model	The order of activities in a trace is other than in a predefined process model	[8, 27]
	Occurrence of mutually exclusive activities	Specific activities occur that are mutually exclusive within a trace	[8]
	Occurrence of unusual neighboring activities	An activity is directly followed by an activity other than usual	[21]
	Occurrence of directly repeating activity	An activity is immediately repeated within a trace	
	Missing occurrence of activity	A specific activity is missing in the trace	
Data	Data object with value outside boundary	The value of a data object deviates from the usual values	[8–10, 19, 31]
	Change in value between events	Data values change unexpectedly between events	[8, 10, 19]
	Specific information in free-text fields	Information is logged in free-text fields instead of dedicated fields	
Resource	Activity executed by unauthorized resource	An activity is executed by a resource other than those authorized	[19]
	Activities executed by multiple resources	Activities within the same trace are executed by multiple resources	[19]
	Activities executed by a single resource	Activities within the same trace are all executed by the same resource	[30]
	Frequent occurrence of activity for a resource	An activity occurs more frequently for one resource compared to other resources	
	Frequent occurrence of value for a resource	A data value occurs more frequently for one resource compared to other resources	
Time	Occurrence of activity outside of time period	An activity occurs outside of the usual time period	[19]
	Delay between start of trace and activity is out of bounds	There is a deviation in the delay between the start of the trace and the time of an activity	[7]
	Time between activities out of bounds	There is a deviation in the time between activities	[10, 19, 30, 34]
	Duration of activity out of bounds	There is deviation in the duration of an activity	[30, 31, 34]
	Duration of trace out of bounds	There is a deviation in the duration of a trace	[31]
	Delay between event and logging is out of bounds	There is a deviation in the delay between time of event and time of logging	

4.2   Detection Pattern Analysis

To investigate what data is required to detect each workaround detection pattern, we have used the data structure of HiX. Using this specific HIS, we explored which columns in the data we would need to use to find each pattern.

To apply these patterns, it should always be clear to which trace an event belongs. Depending on the available data, we can use timestamps (e.g., by grouping events that are temporally close), specific activities (e.g., a trace starts with logging in and ends with logging out), or dedicated case IDs.

Table 3 shows an overview of the required data. We distinguish between four data types that may be required: activity, time, data, and resource. Each detection pattern can require a different level of quality for these types.

– Some patterns need *specific* data. This data must be known beforehand. E.g., a data field may require a certain value. Because of the huge number of data fields, it is not feasible to find these values automatically. We cannot find new workarounds with these patterns, only monitor discovered ones.
– We generally require *high-quality* data. Activity names should be distinguishable from each other, timestamps need to determine when an event happened, data needs to be complete, or we need to know which resource executed the event. The exact requirements differ per process. E.g., for one case, "register measurement" is a good activity name, but for another, we need the measurement type.
– For time, *low-quality* data may be sufficient if high-quality is not available. In that case, timestamps only need to be precise enough to determine a correct event order.
– Some data types are *not be needed* to find a pattern. For example, if we investigate if the right resource performed an activity, we do not require timestamps.

Note that these patterns do not require a specific case ID focus. Patient IDs can be useful to check if people do not repeat work that has already been done by someone else. On the other hand, using resource IDs would show more information about how a single person is working.

4.3   Illustrative Case Study

We selected three patterns to test if we could apply the SWORD framework: "Time between activities out of bounds", "Delay between event and logging out of bounds", and "Activities executed by a single resource".

Time Between Activities Out of Bounds. We tested if we could find the batching workaround in real data. While we are confident that this workaround is likely to be used when measurements in a hospital setting are manually being logged, we do not know where and when it is used in the UMC Utrecht. We can detect this workaround by analyzing the time between events. Since we are

Table 3. Workaround detection patterns with the required data. Those marked with
* can only be used to monitor known workarounds.

	Detection patterns	Activity	Time	Data	Resource
Control-flow	*Occurrence of activity	Specific	–	–	–
	Occurrence of recurrent activity sequence	High quality	Low quality	–	–
	Frequent occurrence of activity	High quality	–	–	–
	*Occurrence of activities in an order different from process model	Specific+ Process Model	Low quality	–	–
	*Occurrence of mutually exclusive activities	Specific	–	–	–
	*Occurrence of unusual neighboring activities	Specific	Low quality	–	–
	Occurrence of directly repeating activity	High quality	Low quality	–	–
	*Missing occurrence of activity	Specific	–	–	–
Data	*Data object with value outside boundary	–	–	Specific	–
	Change in value between events	–	Low quality	High quality	–
	*Specific information in free-text fields	–	–	Specific	–
Resource	Activity executed by unauthorized resource	–	–	–	High quality
	Activities executed by multiple resources	High quality	Low quality	–	High quality
	Activities executed by a single resource	High quality	Low quality	–	High quality
	*Frequent occurrence of activity for a resource	High quality	–	–	Specific
	*Frequent occurrence of value for a resource	–	–	High quality	Specific
Time	Occurrence of activity outside of time period	High quality	High quality	–	–
	Delay between start of trace and activity is out of bounds	High quality	High quality	–	–
	Time between activities out of bounds	High quality	High quality	–	–
	Duration of activity out of bounds	High quality	High quality	–	–
	Duration of trace out of bounds	–	High quality	–	–
	Delay between event and logging out of bounds	–	High quality	High quality	–

interested in resource behavior, we use resources as the case ID. We only need to look at measurement registration events. If the time between the events is very short, there cannot have been enough time to measure a patient, so the employee is most likely practicing batching. This workaround is relatively easy to recognize and does not require a field expert to do so, allowing us to test the SWORD framework without requiring interviews with them.

Figure 2 shows graphs containing only manual measurement registration events for the three different shifts in a single day. Every row contains the measurements of a single resource. Since we removed directly subsequent measurements of the same patient, horizontally close events show registrations where there cannot have been enough time to do a new measurement and thus can be considered batching, these are marked in red. We can see that batching occurs more often at the start and end of shifts, which is especially clear during the day shift in Fig. 2a.

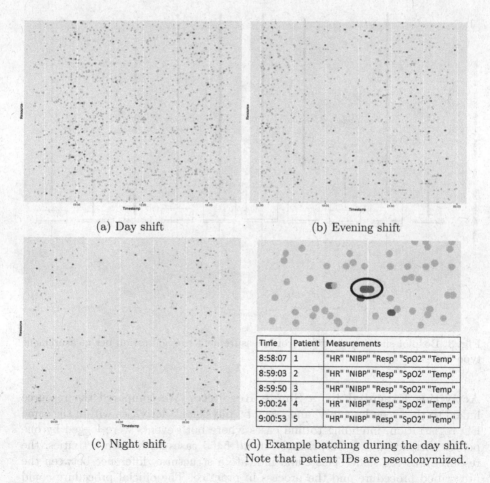

(a) Day shift

(b) Evening shift

(c) Night shift

(d) Example batching during the day shift. Note that patient IDs are pseudonymized.

Time	Patient	Measurements
8:58:07	1	"HR" "NIBP" "Resp" "SpO2" "Temp"
8:59:03	2	"HR" "NIBP" "Resp" "SpO2" "Temp"
8:59:50	3	"HR" "NIBP" "Resp" "SpO2" "Temp"
9:00:24	4	"HR" "NIBP" "Resp" "SpO2" "Temp"
9:00:53	5	"HR" "NIBP" "Resp" "SpO2" "Temp"

Fig. 2. Measurement events per resource for three consecutive shifts. **Grey dots** are normal measurements. **Red dots** are part of batching. (Color figure online)

Delay Between Event and Logging Out of Bounds. We compared the time of activity and the time of registration to determine the registration delay. Since there were 204 different measurement types, we limit our results to the most common ten: early warning score (EWS), heart rate (HR), length, non-invasive blood pressure (NIBP), resting pulse (Resp), oxygen saturation (SpO2), temperature (Temp), visual analog scales (VAS), numeric VAS during activity (VASNRSact), and weight. To aid visibility, we filtered out delays of over an hour.

Figure 3 shows the results. Every measurement type has a boxplot showing the delays linked to it. Every type has its own distribution and thus every type has its own time delays that are considered outliers. All outliers can be considered to be different from the common process, but without an expert, we cannot conclude these are workarounds, mistakes, or something else.

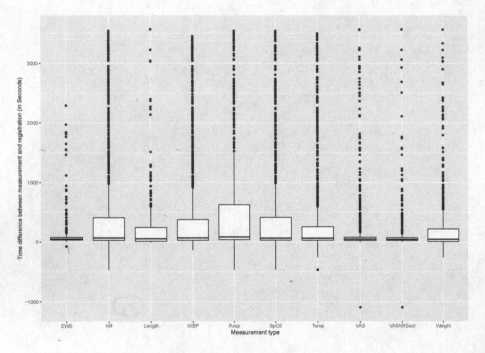

Fig. 3. Boxplot showing the delay from measurement to registration per measurement type

Activities Executed by a Single Resource. We compared the resource linked to the "seen by nurse" and "seen by physician" activities within the same ED registration, intending to find cases where both events were logged by one person instead of two. Surprisingly, in *all* 5460 cases with both activities, the resource was the same. This could indicate a structural difference between the prescribed procedure and the process in practice. The official procedure could also be different than what we expect, so this would require expert input.

Conclusion. We have tested three of the detection patterns in our SWORD framework to see if we could use them with real data. We were able to apply all three patterns and found meaningful results in that they either clearly point to workarounds or warrant further investigation.

5 Discussion and Conclusion

In this paper, we have introduced the SWORD framework. With our review, we have discovered twenty-two patterns that are incorporated in this framework. Sixteen of them are based on literature. We determined the remaining six patterns based on previously discovered workarounds.

Our illustrative case study shows that we can detect workarounds using simple data. We successfully used the time between measurements to find batching and we found clear differences in delays between measurements and their registration for different measurement types.

Our framework is based on the mixed methods approach that is used to recognize various known workarounds in data [5]. We use the same perspectives for detection; control-flow, data, resource, and time. The framework points to specific patterns that can be used for each of these perspectives.

Compared to a neural network approach [33], the SWORD framework is more focused. Instead of using a full event log, the patterns have simpler data requirements. They also do not require data to be labeled as a workaround or normative process beforehand. This saves time and effort from experts. Also, we can find workarounds that are not similar to those in the data.

Note that we have only tested three of the twenty-two patterns with real data. While the requirements for remaining patterns are determined using the HIS structure, in practice, the actual data may not fit completely to it. Multiple snapshots from [5] describe this behavior. While some data should be logged in a certain field, users find it easier to log it in free-text fields instead. This can make it difficult to find these patterns. In the future, we will investigate to what extent these workarounds patterns can be detected with real data.

The SWORD framework uses the patterns as singular options to detect differences, but some workarounds can be detected with multiple patterns [5]. To improve detection, we could use machine learning methods, such as classification [25] or clustering [14]. These can combine detection patterns, allowing us to effectively consider processes from multiple angles at the same time.

Acknowledgment

 This publication is part of the WorkAround Mining (WAM!) project (with project number 18490) which is (partly) financed by the Dutch Research Council (NWO).

References

1. van der Aalst, W.M.P., van Dongen, B.F., Herbst, J., Maruster, L., Schimm, G., Weijters, A.J.M.M.: Workflow mining: a survey of issues and approaches. Data Knowl. Eng. **47**(2), 237–267 (2003). https://doi.org/10.1016/s0169-023x(03)00066-1

2. Alter, S.: Theory of workarounds. Commun. Assoc. Inf. Syst. **34** (2014). https://doi.org/10.17705/1cais.03455

3. Azad, B., King, N.: Enacting computer workaround practices within a medication dispensing system. Eur. J. Inf. Syst. **17**(3), 264–278 (2008). https://doi.org/10.1057/ejis.2008.14

4. Batista, E., Solanas, A.: Process mining in healthcare: a systematic review. In: 2018 9th International Conference on Information, Intelligence, Systems and Applications. IEEE, July 2018. https://doi.org/10.1109/iisa.2018.8633608

5. Beerepoot, I., Lu, X., van de Weerd, I., Reijers, H.A.: Seeing the signs of workarounds: a mixed-methods approach to the detection of nurses' process deviations. In: Proceedings of the Annual Hawaii International Conference on System Sciences. Hawaii International Conference on System Sciences (2021). https://doi.org/10.24251/hicss.2021.456

6. Beerepoot, I., van de Weerd, I.: Prevent, redesign, adopt or ignore: improving healthcare using knowledge of workarounds. In: Twenty-Sixth European Conference on Information Systems (2018)

7. Bolt, A., de Leoni, M., van der Aalst, W.M.P.: Process variant comparison: using event logs to detect differences in behavior and business rules. Inf. Syst. **74**, 53–66 (2018). https://doi.org/10.1016/j.is.2017.12.006

8. Borrego, D., Barba, I.: Conformance checking and diagnosis for declarative business process models in data-aware scenarios. Expert Syst. Appl. **41**(11), 5340–5352 (2014). https://doi.org/10.1016/j.eswa.2014.03.010

9. Bose, R.P.J.C., van der Aalst, W.M.P.: Discovering signature patterns from event logs. In: 2013 IEEE Symposium on Computational Intelligence and Data Mining. IEEE, April 2013. https://doi.org/10.1109/cidm.2013.6597225

10. Burattin, A., Maggi, F.M., Sperduti, A.: Conformance checking based on multi-perspective declarative process models. Expert Syst. Appl. **65**, 194–211 (2016). https://doi.org/10.1016/j.eswa.2016.08.040

11. Di Francescomarino, C., Ghidini, C., Maggi, F.M., Milani, F.: Predictive process monitoring methods: which one suits me best? In: Weske, M., Montali, M., Weber, I., vom Brocke, J. (eds.) BPM 2018. LNCS, vol. 11080, pp. 462–479. Springer, Cham (2018). https://doi.org/10.1007/978-3-319-98648-7_27

12. Dunzer, S., Stierle, M., Matzner, M., Baier, S.: Conformance checking: a state-of-the-art literature review. In: Proceedings of the 11th International Conference on Subject-Oriented Business Process Management. ACM Press (2019). https://doi.org/10.1145/3329007.3329014

13. Ejnefjäll, T., Ågerfalk, P.J.: Conceptualizing workarounds: meanings and manifestations in information systems research. Commun. Assoc. Inf. Syst. **45**, 340–363 (2019). https://doi.org/10.17705/1cais.04520

14. Ester, M., Kriegel, H.P., Sander, J., Xu, X.: A density-based algorithm for discovering clusters in large spatial databases with noise. In: Proceedings of the 2nd International Conference on Knowledge Discovery and Data Mining (1996)

15. Ignatiadis, I., Nandhakumar, J.: The effect of ERP system workarounds on organizational control: an interpretivist case study. Scand. J. Inf. Syst. **21**(2), 3 (2009)

16. Ilie, V.: Psychological reactance and user workarounds. a study in the context of electronic medical records implementations. In: ECIS (2013)

17. Lauer, T., Rajagopalan, B.: Examining the relationship between acceptance and resistance in system implementation. In: AMCIS 2002 Proceedings, p. 179 (2002)

18. Lazovik, A., Aiello, M., Papazoglou, M.: Associating assertions with business processes and monitoring their execution. In: Proceedings of the 2nd International Conference on Service Oriented Computing. ACM Press (2004). https://doi.org/10.1145/1035167.1035182

19. de Leoni, M., van der Aalst, W.M.P.: Aligning event logs and process models for multi-perspective conformance checking: an approach based on integer linear programming. In: Daniel, F., Wang, J., Weber, B. (eds.) BPM 2013. LNCS, vol. 8094, pp. 113–129. Springer, Heidelberg (2013). https://doi.org/10.1007/978-3-642-40176-3_10

20. Lo, D., Cheng, H., Han, J., Khoo, S.C., Sun, C.: Classification of software behaviors for failure detection. In: Proceedings of the 15th ACM SIGKDD International Conference on Knowledge Discovery and Data Mining. ACM Press (2009). https://doi.org/10.1145/1557019.1557083

21. Lu, X., Fahland, D., van den Biggelaar, F.J.H.M., van der Aalst, W.M.P.: Detecting deviating behaviors without models. In: Reichert, M., Reijers, H.A. (eds.) BPM 2015. LNBIP, vol. 256, pp. 126–139. Springer, Cham (2016). https://doi.org/10.1007/978-3-319-42887-1_11

22. Mahbub, K., Spanoudakis, G.: A framework for requirents monitoring of service based systems. In: Proceedings of the 2nd International Conference on Service Oriented Computing, pp. 84–93. ACM Press (2004). https://doi.org/10.1145/1035167.1035181

23. Mannhardt, F., de Leoni, M., Reijers, H.A., van der Aalst, W.M.P.: Balanced multi-perspective checking of process conformance. Computing **98**(4), 407–437 (2015). https://doi.org/10.1007/s00607-015-0441-1

24. Nguyen, H., Dumas, M., La Rosa, M., Maggi, F.M., Suriadi, S.: Business process deviance mining: review and evaluation. arXiv preprint arXiv:1608.08252, August 2016

25. Osuna, E., Freund, R., Girosi, F.: An improved training algorithm for support vector machines. In: Neural Networks for Signal Processing VII. Proceedings of the 1997 IEEE Signal Processing Society Workshop, pp. 276–285. IEEE (1997). https://doi.org/10.1109/nnsp.1997.622408

26. Ras, G., van Gerven, M., Haselager, P.: Explanation methods in deep learning: users, values, concerns and challenges. In: Escalante, H.J., et al. (eds.) Explainable and Interpretable Models in Computer Vision and Machine Learning. TSSCML, pp. 19–36. Springer, Cham (2018). https://doi.org/10.1007/978-3-319-98131-4_2

27. Rebuge, Á., Ferreira, D.R.: Business process analysis in healthcare environments: a methodology based on process mining. Inf. Syst. **37**(2), 99–116 (2012). https://doi.org/10.1016/j.is.2011.01.003

28. Röder, N., Wiesche, M., Schermann, M.: A situational perspective on workarounds in IT-enabled business processes: a multiple case study. In: ECIS (2014)

29. Rozinat, A., van der Aalst, W.M.P.: Conformance checking of processes based on monitoring real behavior. Inf. Syst. **33**(1), 64–95 (2008). https://doi.org/10.1016/j.is.2007.07.001

30. Swennen, M., et al.: Capturing resource behaviour from event logs. In: SIMPDA, pp. 130–134 (2016)

31. Taymouri, F., La Rosa, M., Dumas, M., Maggi, F.M.: Business process variant analysis: survey and classification. Knowl. Based Syst. **211**, 106557 (2021). https://doi.org/10.1016/j.knosys.2020.106557
32. Tucker, A.L.: The impact of workaround difficulty on frontline employees' response to operational failures: a laboratory experiment on medication administration. Manag. Sci. **62**(4), 1124–1144 (2016). https://doi.org/10.1287/mnsc.2015.2170
33. Weinzierl, S., Wolf, V., Pauli, T., Beverungen, D., Matzner, M.: Detecting workarounds in business processes-a deep learning method for analyzing event logs. In: Proceedings of the 28th European Conference on Information Systems, May 2020
34. Wynn, M.T., et al.: ProcessProfiler3D: a visualisation framework for log-based process performance comparison. Decis. Support Syst. **100**, 93–108 (2017). https://doi.org/10.1016/j.dss.2017.04.004

Design Methods

Back to the Roots – Investigating the Theoretical Foundations of Business Process Maturity Models

Vanessa Felch[(✉)] [iD] and Björn Asdecker [iD]

University of Bamberg, Feldkirchenstr. 21, 96050 Bamberg, Germany
{vanessa.felch,bjoern.asdecker}@uni-bamberg.de

Abstract. For years, doubts have been raised about the usefulness of business process maturity models (BPMMs). In addition to methodological shortcomings and limited applicability of the models, another frequently voiced critique is a weak theoretical foundation. This conceptual paper analyzes previously released BPMMs and the related literature. It shows that the vast majority of articles do not refer to any theory to clarify the general underlying assumptions of the models. Instead, they resort to other existing models. In addition, the suitability of the few theoretical approaches to which some authors have referred is highly questionable. A further comparison of the theories' suitability issues with some of the fundamental criticisms of BPMMs reveals remarkable parallels. Against this background, the article at hand creates awareness of the need to consciously select and document the theoretical foundations of future BPMMs. In addition, it contributes to the epistemological discussion on BPMMs, how to evolve and improve the development of maturity models.

Keywords: Business process · Business process management · Maturity model · Theoretical foundation · Theory

1 Introduction

In recent years, numerous business process maturity models (BPMMs) have been developed and published in response to ever-expanding practical interest [54, 63]. Likewise, doubts about the quality and usefulness of specific models have increased. The first documented concern dates to 2007, when de Bruin and Rosemann [13, p. 644] critically commented that "[...] a number of available models appear to be 'power-point deep' in that they are proprietary in nature, have not been rigorously developed and tested, and are not supported by tools that enable them to be applied within a wide range of organizations". Subsequently, other authors, e.g., Pöppelbuß and Röglinger [54] and van Looy [81], joined this critique and repeatedly pointed out the varying quality of specific BPMMs. Tarhan et al. [76] conducted a comprehensive literature review and, in line with previous observations, expressed doubts about the usefulness of several BPMMs.

Supplementary Information The online version contains supplementary material available at https://doi.org/10.1007/978-3-031-16103-2_10.

Felch and Asdecker [19] took note of the growing critique and examined whether this perception can actually be substantiated. Their analysis followed the premise that, in academia, the quality of the published model would be reflected in the quality of the publication outlet. They considered BPMMs published between 1990 and 2019 and used journal impact factors as a quality indicator. The results showed that "[…] articles about BPMMs are published in less-recognized journals, which are of minor relevance in the scientific community" [19, p. 373]. This finding was attributed to methodological short-comings and the models' limited usefulness and applicability. In addition, the authors stressed that "[…] there may be other reasons for not publishing BPMMs in higher quality journals" [19, p. 379].

This article focuses on the theoretical foundation of these BPMMs, another essential criterion for high-quality publications [8, 71, 73]. We understand a theoretical foun-dation as the perspective that establishes the common ground for the investigation and provides the lens through which researchers contribute to their research questions. Previ-ous research has occasionally criticized the poor theoretical grounding of BPMMs (e.g., [44, 81]). However, a systematic analysis has not yet taken place. To (1) investigate whether a lack of theoretical grounding corresponds to individual cases or is rather the common rule and (2) propose ways of improving the current situation, we address the following two research questions:

RQ1: Are the existing BPMMs theoretically grounded, and if so, how?
RQ2: Are the currently used theoretical approaches suitable?

Methodologically, this study systematically reviews the literature to show that the vast majority of BPMM articles do not refer to any theory. Based on this surprising observation, abductive reasoning leads to the conclusion that some of the fundamental criticisms of those models can be attributed to the weaknesses of the few theoretical lenses that have been employed. Over a decade ago, Becker et al. [4, p. 9] called for more work that takes a "critical perspective on maturation". Responding to this call for research, the contribution of this paper is threefold: First, it creates awareness of the need to deliberately choose and document the theoretical foundation of future BPMMs. Second, it highlights the shortcomings of previous theoretical foundations and argues that these may explain some of the fundamental criticisms of BPMMs. Third, it provides ideas for alternative theoretical foundations and proposes potential avenues on how to evolve BPMMs. Thus, this conceptual paper adds to the epistemological discussion on how to evolve and improve BPMM design in particular and maturity model design in general.

2   Why a Solid Theoretical Foundation is Necessary and Useful – Even in Practitioner-Oriented Domains

While a theoretical foundation is widely considered necessary for empirical work, this is less obvious for other studies – particularly if they have strong roots in business practice. Therefore, it is necessary to demonstrate why a solid theoretical foundation is needed even in practitioner-oriented domains such as BPMM research.

In general, a theory is an evidence-based "[…] system composed of two core con-stituents: (1) constructs or concepts and (2) propositions as relationships between those constructs" [41, p. 4]. Theories either are the result of original research or form the basis for new research [42]. In the latter case, this is usually referred to as the theoretical foundation or theoretical grounding. The theoretical foundation serves multiple purposes. Most notably, it provides scholars with (1) the explicit assumptions and boundaries of the research and (2) the key variables and their interrelationships that describe the phenomenon of interest [23, 42]. In metaphorical terms, the role of the theoretical foundation is similar to the building's bedrock or the body's skeleton.

By combining the main goals of theories, Gregor [23] distinguished five types: theories for (1) analysis, (2) explanation, (3) prediction, (4) explanation & prediction, and (5) design & action. Types 1–4 provide well-grounded insights into a phenomenon under investigation. In most cases, maturity models assume that organizations' capabilities mature in predefined stages and that progressing toward higher stages is better. However, such assumptions are merely proposed hypotheses that require empirical evaluations to be considered as theory (types 1–4). Type 5, in contrast, relates to method development by providing explicit specifications for the construction of an artifact. For BPMMs, which are considered an artifact, various procedure models and design guidelines exist (e.g., [3, 14, 54]). These models and guidelines can be categorized as type 5. However, the study at hand focuses on types 1–4, which provide the basic assumptions about how organizations and processes evolve and how BPMMs actually work.

Counterintuitive to the term, theories always contain a practical side, suggesting how something should be done. Gregor [23, p. 613] summarized that theories "[...] are practical because they allow knowledge to be accumulated in a systematic manner and this accumulated knowledge enlightens professional practice". The application of theories is helpful to researchers and practitioners alike [23, 42]. Through theory, researchers can better describe, explain, or predict the phenomena under study [42]. Especially in practitioner-oriented domains, theories serve as a guide to determine which assumptions and variables should be considered when structuring and designing management tools, such as BPMMs, to reach more informed and efficient decisions [11, 12, 22, 36, 42]. Consequently, top-tier outlets require a theoretical foundation in articles [8, 11, 66, 71]. Straub [72, p. viii] strongly supported this approach and therefore recommended to practice: "Rather than prescribing 'snake oil,' and 'untested management miracle-cures' (Pfeffer and Sutton 2006a, p. 1), practitioners should judiciously adopt only evidence-based management prescriptions derived from scientifically based evidence culled from carefully conducted social science and organizational research".

Regarding BPMMs, Niehaves et al. [44] and van Looy [81] have noted a weak theoretical foundation. This is surprising since the common basis of all maturity models is the assumption of predictable patterns in terms of organizational change, which in all circumstances requires some kind of theory. The question of if and how this has been done is the core of a systematic literature review presented in the following section.

3   Current Theoretical Foundations of BPMMs

Several systematic reviews have analyzed the research on BPMMs, of which the most comprehensive one is by Tarhan et al. [76]. Their review focused on academic journals, conference proceedings, and books published between 1990 and 2014. This study was recently expanded by Felch and Asdecker [19], using the same search terms and databases to address the latest BPMM developments (2015–2019). Together, both reviews identified 130 relevant BPMM references, of which 25 were released models. While investigating the theoretical foundation of those BPMMs (Sect. 3.1), we found only a few that referred to an actual theory, which motivated a broader systematic literature review (Sect. 3.2). Before presenting the literature search results, it is necessary to point out that few scholars have considered BPMMs as theory. For instance, Röglinger et al. [58, p. 330] noted that such models "[…] typically represent theories about how an organization's capabilities evolve in a stage-by-stage manner along an anticipated, desired, or logical path". Referring to another BPMM could thus be considered an appropriate theoretical foundation. While acknowledging this perspective, we still disagree. Most notably, such an implicit approach would require a backward search that would contradict the basic scientific principles of clarity, transparency, and reproducibility. Moreover, many maturity models lack sufficient evidence, especially when newly designed. Thus, an essential component of a scientific theory is missing.

3.1 Review of Theories Serving as Base for Released BPMMs

To investigate the theoretical grounding, we analyzed 25 BPMMs classified as 'release' according to the literature reviews by Tarhan et al. [76] and Felch and Asdecker [19] (see supplementary material A). Both initial model developments and refinements were among the 25 articles. In this case, each paper was analyzed separately (cf. [25, 27, 64, 65]). In the case of almost complete textual consistency (cf. [59–62]), only the most detailed article ([62] incl. supplementary electronic material) was included in the analysis. We examined the sections of the remaining contributions preceding the BPMM design. The analysis shows that previous articles paid little attention to the theoretical foundation prior to model development (see supplementary material B). Instead, various papers compared outstanding or thematically appropriate models (e.g., [10, 34]), described the model purpose (e.g., [40, 86]), and/or presented definitions for the terms 'maturity' or 'maturity model' (e.g., [26, 27]). Only one article, Chaghooshi et al. [10], referred to Nolan's stage theory [47].

Overall, the results provide a strong indication that existing BPMMs are rarely embedded in theories. Instead, the vast majority of scholars simply reviewed the literature to identify thematically related models while highlighting their weaknesses, which – in turn – were then used to justify the newly developed model (cf. [13, 44]). While this step is explicitly required by various procedure models (cf. [3, 14]), it does not constitute a theoretical foundation as such. The analysis confirms the views by Niehaves et al. [44], van Looy [81], and Pöppelbuß et al. [55, p. 511], who stated that "[…] the design of maturity models has been too often informed by existing models (e.g., the CMM and CMMI) instead of applying these meaningful theoretical approaches". This preliminary finding motivated a more comprehensive analysis of the existing BPMM literature presented in the following section.

3.2 Search of Potential Theories for BPMMs

To identify relevant articles, the search string ("BPMM" OR ("business process" AND "maturity model")) AND ("theory" OR "theories" OR "theoretical" OR "foundation") was used in the title, abstract, and keywords within four scientific databases: Business Source Ultimate via EBSCO, ScienceDirect, Scopus, and Web of Science. The initial search results were screened by applying the following selection criteria. First, only English publications were considered. Second, the results were limited to articles in journals, conference proceedings, and books. A total of 160 references were retrieved (see Table 1).

Table 1. Search string results and selection process

Digital library	EBSCO	Science Direct	Scopus	Web of Science
Initially retrieved	8	5	61	86
Initially selected	7	4	30	34
Duplicates removed	49			
Finally selected	5			

After reviewing the title, abstract, and keywords, 85 studies were excluded. These include articles that were not within the scope of this study, e.g., that dealt with BPMMs only incidentally or in which the abbreviation BPMM referred to another term. Subsequently, 26 duplicate studies were removed, and the remaining 49 articles were screened based on their full text (see supplementary material C). Papers that did not address the model grounding or potential theoretical approaches were excluded from further analysis. Despite this comprehensive search, the results were again surprisingly sparse. Only five articles referred to any theory: (1) Niehaves et al. [44], (2) Niehaves et al. [45], (3) Pöppelbuß et al. [56], (4) Tapia et al. [75], and (5) van Looy et al. [82]. Among those five papers, four used process life-cycle theory, and two considered convergence theory (see Table 2). Along with the article by Chaghooshi et al. [10] identified during the review of the released BPMMs (Sect. 3.1), it leads to three theoretical approaches that have been referred to. Table 2 summarizes the results.

Overall, the findings of this additional literature search reinforce the impression that the theoretical foundation of BPMMs can be considered weak thus far. None of the publications explicitly addressed the underlying assumptions of the models. Instead, scholars usually resorted to other existing models. This presses the question of how appropriate the few theories mentioned are, which is addressed in the next section.

4 A Closer Look at the Theoretical Approaches Identified

Before evaluating the theoretical approaches, they are briefly introduced in the following. The discussion is limited to the key propositions due to page limitations.

Table 2. Overview of the theories identified

Article by	Convergence theory [38]	Process life-cycle theory [80]	Stage theory [47]
Chaghooshi et al. [10]			•
Niehaves et al. [44]	•		
Niehaves et al. [45]		•	
Pöppelbuß et al. [56]	•	•	
Tapia et al. [75]		•	
van Looy et al. [82]		•	

4.1 Introduction to the Theories

Convergence theory, process life-cycle theory, and stage theory suggest that change is imminent and occurs along a predefined path. This commonality becomes apparent when the core statements of the theories are juxtaposed.

Convergence Theory. The theory posits that the social structures of nations tend to align increasingly as industrialization progresses [85]. Initial differences can be attributed to cultural, political, or economic aspects [39], which later converge due to technological and economic constraints during industrialization. In general, convergence theory assumes that all entities of the same class move toward a general model or an ideal state [56, 69].

Process Life-Cycle Theory. Not to be confused with the homonymous theory from economics that describes people's spending and saving habits over a lifetime, the process life-cycle theory explains how entities develop and evolve [80]. The theory assumes that entities develop linearly and irreversibly along a predefined sequence of phases (or stages) toward an optimal final state [45, 46, 68, 80]. The driving mechanism is "[…] a prefigured program/rule regulated by nature, logic, or institutions" [80, p. 514].

Stage Theory. The theory originates from the model developed by Nolan [47] in the 1970s, which describes the development of IT in organizations in four, later six [48], stages [70]. Stage theory generally proposes a stepwise development of an entity along a predefined and logical path [47]. The stages to be passed through are described by a distinctive set of attributes and the relationships among those attributes. Each stage builds on the previous one [47].

In summary, there are considerable overlaps between the theories. Unlike the other two, convergence theory has its roots in sociology and does not specifically refer to organizations or processes [85]. This is also supported by Niehaves et al. [44, p. 224], who stated that "[…] this perspective is not suitable for the development of BPM and dynamic capabilities in general". Therefore, the convergence theory is excluded from further consideration. The core statements of the process life-cycle theory and stage theory tend to be quite similar, enhanced by several researchers who characterize stage theory as life-cycle theory [46, 68]. However, the question of whether these theories are a suitable theoretical foundation for the development of BPMMs remains open. Therefore, the following section pursues an evaluation.

4.2 Evaluation of the Theories

To date, there is no generally accepted procedure for evaluating the suitability of theories. However, Gieseler et al. [20] suggest six criteria to assess the quality of a theory: (1) consistency, (2) precision, (3) parsimony, (4) generality, (5) falsifiability, and (6) progress. Nonetheless, quality is not synonymous with suitability. Rather, quality is a superordinate concept of which suitability is a partial aspect. The Oxford dictionary [49] defines suitability as "the quality of being right or appropriate for a particular purpose [...]". This refers to the first two quality criteria of consistency and precision. Consistency refers to "correspondence to empirical observations in the laboratory and/or the real world" [20, p. 7]. However, precision requires "clearly defined concepts and operationalizations that allow for little stretching" [20, p. 7]. In case these are not fulfilled, an application of the theory should be questioned (cf. [20]). Therefore, those two criteria were used to assess the suitability of the two remaining theoretical approaches – process life-cycle theory and stage theory.

To identify articles addressing these criteria, a forward search [83] was performed based on the works of the theory's leading proponents, e.g., van de Ven and Poole's work [80] for process life-cycle theory and Nolan's paper [47] for stage theory. The searches were conducted in the databases Web of Science and Google Scholar. We considered only academic literature, i.e., articles published in journals, conference proceedings, or books. Relevant aspects were extracted and assigned to the corresponding criterion, i.e., either consistency or precision. Table 3 summarizes the results.

Interestingly, the authors who draw on these theoretical approaches are at least partially aware of these suitability issues. As an example, Chaghooshi et al. [10, p. 561] stated: "Nolan's stage hypothesis, for instance, stimulated much research that resulted in conflicting findings as regards its empirical validity". Other critical statements can be found in Pöppelbuß et al. [56] and Niehaves et al. [45, pp. 100–101], who concluded that "[...] developmental models for BPM should not adopt a pure life cycle perspective (Van de Ven and Poole, 1995), but should also consider environmental aspects and organizational traits". The various criticisms of the stage theory (see Table 3) lead to doubts about its suitability as a foundation for BPMMs. Process life-cycle theory appears to be a more viable basis for BPMMs. However, some of the theory's core statements and assumptions may prompt some fundamental criticisms of BPMMs elaborated in the following section.

5 Drawing Parallels to Highlight the Necessity to Rethink the Theoretical Foundations of BPMMs

After juxtaposing the theories' assumptions and the fundamental criticisms of BPMMs (Sect. 5.1), the relevance of this paper is addressed and an outlook is provided (Sect. 5.2).

Table 3. Evaluation of the theories' suitability

	Consistency	Precision
Process life-cycle theory	Studies by various researchers have supported the validity of the theory. In many cases, the theory can be found in combination with other process theories (e.g., [2, 18, 28, 46])	The theory is characterized as simple but nevertheless provides precise explanations (e.g., [53, 74]). Scholars refer to the theory and its assumptions as detailed and testable (e.g., [24, 46, 84])
Stage theory	The validity of the theory has been both confirmed and refuted by many empirical studies (e.g., [6, 31]). For example, the S-shaped curve has not been supported (e.g., [35]). Criticisms have been voiced regarding validation studies in terms of their reliability and validity tests of the measurement procedures (e.g., [37]). The stage theory was adapted based on further empirical evidence by Nolan in subsequent years (e.g., [48])	The theory is sometimes described as comprehensive (e.g., [67]). However, the operationalization of the stage model is not publicly available (e.g., [6, 30]). The theory does not adequately define individual terms, such as "technical skills" (e.g., [5]), nor does it explicitly describe the measurement of organizational maturity (e.g., [37]). The theory addresses a fairly complex phenomenon "[…] in a straightforward and clever manner" [30, p. 474]. Researchers have criticized its minimalist approach and described its assumptions as too simplified to be useful (e.g., [21, 30])

5.1 Can the Fundamental Criticisms of BPMMs Be Traced Back to Its Weak Theoretical Foundation?

As already highlighted in the introduction to this paper, some scholars have objected to the models' quality. Accordingly, many BPMMs provide insufficient documentation, which makes their application difficult. Moreover, some strong criticism has questioned the overall usefulness of such models. This refers to their linear, static, absolute nature that reflects a positivist approach to deriving highly accurate prediction models. However, it also oversimplifies reality and gives rise to the problems described in more detail below.

Linear. The central premise of most maturity models is that development proceeds along a predefined, cumulative path, which is well reflected by the numbering of maturity stages [1]. However, empirical evidence for the existence of such a pattern is lacking. Instead, it is reasonable to conclude that the proposed path depends on the subjective, individual perception of the model designers. Such a linear path further contradicts the fact that competitive advantages result from uniqueness and heterogeneity [52]. If all companies follow the same homogenous one-size-fits-all concept while relying on widely acknowledged best practices, it must be considered impossible to outperform competitors. In addition, most maturity models neglect the potential existence of multiple equally advantageous paths [31].

Static. Every model is developed at a certain point in time. Consequently, they represent a specific state of knowledge that is locked into the model, while the conditions, i.e., competitors, customers, and technology, are constantly evolving. If the environment changes, the model's units of analysis have to continually reflect those changes. However, most maturity models do not provide for such permanent, constant change. Applying a static model to a continuously dynamic context will, therefore, most likely not lead to satisfactory results.

Absolute. The basic concept of maturity with a predefined desirable end state to reach is absolute [1]. However, many goals, such as competitiveness, are relative concepts. They depend on the respective context. For instance, businesses do not have to deliver the highest quality possible to be successful. Instead, they only have to be better than competitors. In addition, traditional overarching organizational goals of increased performance and growth do not know an upper limit. Consequently, the normative final stage of the model must be viewed critically since organizational development is continuous and could never be 'complete' as long as a company operates in the market.

While these criticisms are valid for many BPMMs, they do not apply to all of them. Few approaches successfully counteract the aforementioned issues. For instance, some models prescribe a path for a specific capability (e.g., BPM-CF by [64]) and have been updated (e.g., BPM-CF adjusted by [29]). Nevertheless, it is still notable that the three points of criticism are largely similar to the shortcomings of the theoretical foundations derived in the previous section. Both process life-cycle theory and stage theory assume linear relationships, are static, and point to an absolute end state. Such assumptions, however, do not adequately reflect reality. Figuratively speaking, it seems as if the foundation on which the maturity models are based is weak and unstable. As a result of the abductive reasoning process, we, therefore, hypothesize that the highlighted theoretical shortcomings most likely cause some of the BPMMs' fundamental weaknesses. Therefore, it appears not only promising but also necessary to return to the roots, rethink the theoretical basis of maturity models, and look for more suitable alternatives.

5.2   What is the Relevance of This Research and What Are Possible Next Steps Moving Forward?

Compared to other streams of BPMM research, epistemological studies have received comparatively little attention despite their relevance to the longevity of maturity models. The previously published articles by Niehaves et al. [44], Niehaves et al. [45], and Pöppelbuß et al. [56] used case study data to highlight that the theories underlying maturity models do not correspond to the development of organizational capabilities. More than five years have passed since publication. The paper at hand reinforces the results of the previously published article, yet the findings also imply that the few previous contributions have not led to reconsideration in the maturity model domain. BPMMs published in recent years continue to have no or inadequate theoretical foundations. Regardless of such omissions, a strong foundation improves the explanatory power of such artifacts and supports the causal effects of BPM maturity on business performance. Moreover, the lack of theoretical grounding is not limited to BPMMs but also applies to maturity models in general (e.g., [9, 51, 78]). Lasrado et al. [32, p. 5] noted that some model designers do "[...] not conceptually grounding the maturity model characteristics in theory". In addition, they questioned whether those procedure models are adequately supported by theory and provided a new perspective. They suggested set-theoretic methods such as the qualitative comparative analysis (QCA) and the necessary condition analysis (NCA) to conceptualize maturity stages and stage configuration. The novel analytic approach is a counterposition to generic, absolute designs and takes a relative perspective on each case, which reduces the arbitrariness in the model structure. Furthermore, it overcomes the linear structure of BPMMs by allowing for multiple paths toward maturity. Bley [7] showed the approach's applicability, which can be understood as the first step of a paradigm shift. This paper complements their effort for the advancement of maturity models and shows that the increasing complexity of reality requires a refinement of the models' theoretical foundation. Such necessity raises numerous elementary questions about the future of BPMMs, three of which are broached in the following.

First, the set-theoretic approach changes the way maturity levels are derived. Nevertheless, it remains to be seen to what extent maturity levels make any sense at all. Higher maturity levels are not always automatically better, the existence of a single linear path is questionable, and the capabilities associated with a maturity level are constantly changing [15, 45, 56]. A viable alternative might be to consider maturity gaps instead of maturity levels. These maturity gaps arise from the market and customer-specific requirements for processes on the one hand and the status quo on the other.

Second, maturity models lack a time dimension that does justice to dynamic changes. Because of this, developed models are actually already outdated at the time of publication. Furthermore, so-called best practices from today may already be obsolete tomorrow. To account for continuous change, the models would need to have a circular, self-perpetuating component that gives them an evolutionary capability and ensures that the maturity itself can mature.

Third, the current consensus is that models should be descriptive, prescriptive, and comparative. The prescriptive component, i.e., guidance on how to follow the proposed development path, is even emphasized by many authors or criticized if it is missing [54, 76]. However, can there be a satisfactory prescriptive model component when competitive advantage cannot be derived from widely used best practices, and most effective and efficient solutions depend on the individual case? Perhaps it must be admitted that the prescriptive purpose, while desirable, is beyond what a maturity model can accomplish.

To conclude, this section, we provide a brief outlook. First, the results obtained should be validated by considering other domains. Currently, anecdotal evidence indicates that the phenomenon under investigation can also be identified in other domains (e.g., [9, 51, 78]). Accordingly, it may be advantageous not to recommend theories as suitable for one specific domain but to address this issue for maturity models in general. Second, the requirements of the models should be defined to propose suitable theories that can guide their development. There seem to be theories with the potential to meet the requirements better and simplify reality less, such as structural contingency theory [16, 33], diffusion of innovation theory [57], dynamic capability theory [77], and evolutionary theory [17, 43]. The latter seems particularly suitable since it "[...] helps uncover processes through which change happens as well as untangle key relationships among the key factors (e.g., internal, environmental, technological) that affect processes" [79, p. 2]. The answer to the obvious question of which theoretical foundation is best suited goes beyond the purpose of this paper and needs to be answered in a dedicated future effort. It is possible that a combination in the sense of a theoretical multiplicity is most feasible [50].

6 Conclusion

This conceptual work analyzed the literature to show that existing BPMMs in particular and maturity models in general often lack a theoretical foundation. In other words, many researchers blindly adopt the structure of popular existing models without theoretically justifying the fundamental model properties and mechanisms. Moreover, the few theories used need to be critically reflected upon. There are striking analogies between the suitability issues of the theoretical approaches on which previous research has relied and the shortcomings of BPMMs. Abductive reasoning highlighted that a different theoretical foundation appears to be necessary, one that provides a more stable bedrock for developing these models. Economic realities are nonlinear, dynamic, and relative because they depend on the context of the particular object of study. The concept of maturity should therefore reflect this.

A limitation of the paper is that only the 25 BPMMs identified by Tarhan et al. [76] and Felch and Asdecker [19] were analyzed. Despite the limited number of models examined, the results are conclusive considering the lack of theoretical foundation. Furthermore, the literature review for potential theories was limited to BPMM articles. Indications suggest that other domains likewise suffer from a similarly weak theoretical foundation of maturity models. The approach described in the article can be adopted for different domains. Two theories were assessed for their suitability. Although a comprehensive literature review was conducted, Table 3 does not represent an exhaustive list of agreements and criticisms for each theory. Rather, it highlights frequently mentioned, relevant aspects.

Despite the aforementioned limitations, this paper provides several important contributions: First, it raises and reinforces theoretical concerns. Continued critical reflection on the broader theoretical antecedents of BPMMs is essential to avoid "reinventing the wheel" and instead encourage innovative disruption. Second, this research raises elementary questions about the future of BPMMs, which can serve as the fruitful basis of a research agenda for a so far rarely considered literature stream. Third, the article at hand also contributes to the question of how to potentially improve the identified issues. Whether this is a matter of selected separate aspects of maturity models (e.g., a more dynamic model environment with frequent updates) or an actual paradigm shift in the development process remains to be seen and must be the result of future research.

References

1. Andersen, K.N., Lee, J., Mettler, T., Moon, M.J.: Ten misunderstandings about maturity models. In: Eom, S.-J., Lee, J. (eds.) The 21st Annual International Conference on Digital Government Research, Seoul, Korea, pp. 261–266 (2020). https://doi.org/10.1145/3396956.3396980
2. Bayne, L., Purchase, S., Soutar, G.N.: Network change processes for environmental practices. J. Bus. Indus. Market. **36**(10), 1832–1845 (2021). https://doi.org/10.1108/JBIM-02-2020-0094
3. Becker, J., Knackstedt, R., Pöppelbuß, J.: Developing maturity models for IT management. Bus. Inf. Syst. Eng. **1**(3), 213–222 (2009). https://doi.org/10.1007/s12599-009-0044-5
4. Becker, J., Niehaves, B., Pöppelbuß, J., Simons, A.: Maturity models in IS research. In: European Conference on Information Systems, Pretoria, South Africa, pp. 1–12 (2010)
5. Benbasat, I., Dexter, A.S., Mantha, R.W.: Impact of organizational maturity on information system skill needs. MIS Q. **4**(1), 21–34 (1980). https://doi.org/10.2307/248865
6. Benbasat, I., Dexter, A.S., Drury, D.H., Goldstein, R.C.: A critque of the stage hypothesis. Commun. ACM **27**(5), 476–485 (1984). https://doi.org/10.1145/358189.358076
7. Bley, K.: An information systems design theory for maturity models in complex domains. In: Pacific Asia Conference on Information Systems, virtual, pp. 1–14 (2021)
8. Bornmann, L., Nast, I., Daniel, H.-D.: Do editors and referees look for signs of scientific misconduct when reviewing manuscripts? A quantitative content analysis of studies that examined review criteria and reasons for accepting and rejecting manuscripts for publication. Scientometrics **77**(3), 415–432 (2008). https://doi.org/10.1007/s11192-007-1950-2
9. Bvuchete, M., Grobbelaar, S.S., van Eeden, J.: A comparative review on supply chain maturity models. In: International Conference on Industrial Engineering and Operations Management, Pretoria/Johannesburg, South Africa, pp. 1443–1454 (2018)
10. Chaghooshi, A.J., Moradi-Moghadam, M., Etezadi, S.: Ranking business processes maturity by modified Rembrandt technique with considering CMMI dimensions. Iranian J. Manage. Stud. **9**(3), 559–578 (2016). https://doi.org/10.22059/ijms.2016.57543
11. Corley, K.G., Gioia, D.A.: Building theory about theory building: What constitutes a theoretical contribution? Acad. Manage. Rev. **36**(1), 12–32 (2011). https://doi.org/10.5465/amr.2009.0486
12. Dankasa, J.: Developing a theory in academic research: A review of experts' advice. J. Inf/ Sci. Theory Pract. **3**(3), 64–74 (2015). https://doi.org/10.1633/JISTaP.2015.3.3.4
13. de Bruin, T., Rosemann, M.: Using the Delphi technique to identify BPM capability areas. In: Australasian Conference on Information Systems, Toowoomba, Australia, pp. 643–653 (2007)
14. de Bruin, T., Freeze, R.D., Kaulkarni, U., Rosemann, M.: Understanding the main phases of developing a maturity assessment model. In: Australasian Conference on Information Systems, Sydney, Australia, pp. 8–19 (2005)
15. Dijkman, R., Lammers, S.V., de Jong, A.: Properties that influence business process management maturity and its effect on organizational performance. Inf. Syst. Front. **18**(4), 717–734 (2015). https://doi.org/10.1007/s10796-015-9554-5
16. Donaldson, L.: The Contingency Theory of Organizations. SAGE Publications, Thousand Oaks (2001)
17. Dosi, G., Nelson, R.R.: An introduction to evolutionary theories in economics. J. Evol. Econ. **4**, 153–172 (1994). https://doi.org/10.1007/BF01236366
18. Ellwood, P., Williams, C., Egan, J.: Crossing the valley of death: Five underlying innovation processes. Technovation **109**, 1–11 (2022). https://doi.org/10.1016/j.technovation.2020.102162

19. Felch, V., Asdecker, B.: Quo vadis, business process maturity model? Learning from the past to envision the future. In: Fahland, D., Ghidini, C., Becker, J., Dumas, M. (eds.) Business Process Management. Lecture Notes in Computer Science, vol. 12168, pp. 368–383. Springer, Cham (2020). https://doi.org/10.1007/978-3-030-58666-9_21

20. Gieseler, K., Loschelder, D.D., Friese, M.: What makes for a good theory? How to evaluate a theory using the strength model of self-control as an Example. In: Sassenberg, K., Vliek, M.L.W. (eds.) Social Psychology in Action, pp. 3–21. Springer, Cham (2019). https://doi.org/10.1007/978-3-030-13788-5_1

21. Gottschalk, P.: Maturity levels for police oversight Agenciespetter Gottschalk. Police J. Theory, Pract. Principles 82(4), 315–330 (2009). https://doi.org/10.1350/pojo.2009.82.4.472

22. Gottschalk, P., Solli-Sæther, H.: Towards a stage theory for industrial management research. Ind. Manage. Data Syst. 109(9), 1264–1273 (2009). https://doi.org/10.1108/02635570911002315

23. Gregor, S.: The nature of theory in information systems. MIS Q. 30(3), 611–642 (2006). https://doi.org/10.2307/25148742

24. Habib, W.M.: On the change and development in organizations: A critical review of Van de Ven & Poole. SSRN Electron. J. 1995, 1–22 (2008). https://doi.org/10.2139/ssrn.1105250

25. Heinze, P., Geers, D.: Quality management in knowledge intensive business processes – Development of a maturity model to measure the quality of knowledge intensive business processes in small and medium enterprises. In: International Conference on Knowledge Management and Information Sharing, Funchal, Portugal, pp. 276–279 (2009). https://doi.org/10.5220/0002294002760279

26. Jadhav, M., Sapre, G.: The business process maturity model – A tool to assess capability of business process. In: International Conference on Informatics and Semiotics in Organisations, Marrickville, Australia, pp. 458–464 (2009)

27. Jochem, R., Geers, D., Heinze, P.: Maturity measurement of knowledge-intensive business processes. TQM J. 23(4), 377–387 (2011). https://doi.org/10.1108/17542731111139464

28. Kaartemo, V., Coviello, N., Nummela, N.: A kaleidoscope of business network dynamics: rotating process theories to reveal network microfoundations. Ind. Mark. Manage. 91, 657–670 (2020). https://doi.org/10.1016/j.indmarman.2019.01.004

29. Kerpedzhiev, G.D., König, U.M., Röglinger, M., Rosemann, M.: An exploration into future business process management capabilities in view of digitalization. Bus. Inf. Syst. Eng. 63(2), 83–96 (2020). https://doi.org/10.1007/s12599-020-00637-0

30. King, J.L., Krämer, K.L.: Evolution and organizational information systems: an assessment of nolan's stage model. Commun. ACM 27(5), 466–475 (1984). https://doi.org/10.1145/358189.358074

31. King, W.R., Teo, T.S.: Integration between business planning and information systems planning: Validating a stage hypothesis. Decis. Sci. 28(2), 279–308 (1997). https://doi.org/10.1111/j.1540-5915.1997.tb01312.x

32. Lasrado, L.A., Vatrapu, R., Andersen, K.N.: A set theoretical approach to maturity models: Guidelines and demonstration. In: International Conference on Information Systems, Dublin, Ireland, pp. 1–20 (2016)

33. Lawrence, P.R., Lorsch, J.W.: Organization and Environment. Harvard Business School Press, Boston, Massachusetts (1967)

34. Lee, J., Lee, D., Kang, S.: An overview of the business process maturity model (BPMM). In: Chang, K.-C., et al. (eds.) Advances in Web and Network Technologies, and Information Management. LNCS, vol. 4537, pp. 384–395. Springer, Heidelberg (2007). https://doi.org/10.1007/978-3-540-72909-9_42

35. Lucas, H.C., Sutton, J.A.: The stage hypothesis and the S-curve. Commun. ACM 20(4), 254–259 (1977). https://doi.org/10.1145/359461.359472

36. Lynham, S.A.: The general method of theory-building research in applied disciplines. Adv. Dev. Hum. Resour. **4**(3), 221–241 (2002). https://doi.org/10.1177/1523422302043002
37. Mahmood, M.A., Becker, J.D.: Effect of organizational maturity on end-users' satisfaction with information systems. J. Manag. Inf. Syst. **2**(3), 37–64 (1985). https://doi.org/10.1080/07421222.1985.11517736
38. Meyer, J.W., BoliBennett, J., Chase-Dunn, C.: Convergence and divergence in development. Ann. Rev. Sociol. **1**, 223–246 (1975). https://doi.org/10.1146/annurev.so.01.080175.001255
39. Mishra, R.: Welfare and industrial man: A study of welfare in western industrial societies in relation to a hypothesis of convergence. Sociol. Rev. **21**(4), 535–560 (1973). https://doi.org/10.1111/j.1467-954X.1973.tb00496.x
40. Moradi-Moghadam, M., Safari, H., Maleki, M.: A novel model for business process maturity assessment through combining maturity models with EFQM and ISO 9004:2009. Int. J. Bus. Process. Integr. Manage. **6**(2), 167–184 (2013). https://doi.org/10.1504/IJBPIM.2013.054680
41. Müller, B., Urbach, N.: The why, what, and how of theories in IS research. In: International Conference on Information Systems, Milan, Italy, pp. 1–25 (2013)
42. Müller, B., Urbach, N.: Understanding the why, what, and how of theories in IS research. Commun. Assoc. Inf. Syst. **41**, 349–388 (2017). https://doi.org/10.17705/1CAIS.04117
43. Nelson, R.R., Winter, S.G.: An Evolutionary Theory of Economic Change. Belknap Press, Cambridge, London (1982)
44. Niehaves, B., Plattfaut, R., Becker, J.: Business process management capabilities in local governments: A multi-method study. Gov. Inf. Q. **30**(3), 217–225 (2013). https://doi.org/10.1016/j.giq.2013.03.002
45. Niehaves, B., Pöppelbuß, J., Plattfaut, R., Becker, J.: BPM capability development – A matter of contingencies. Bus. Process. Manage. J. **20**(1), 90–106 (2014). https://doi.org/10.1108/BPMJ-07-2012-0068
46. Nielsen, J.F.: Models of change and the adoption of web technologies: Encapsulating participation. J. Appl. Behav. Sci. **44**(2), 263–286 (2008). https://doi.org/10.1177/0021886308314900
47. Nolan, R.L.: Managing the computer resource. Commun. ACM **16**(7), 399–405 (1973). https://doi.org/10.1145/362280.362284
48. Nolan, R.L.: Managing the crises in data processing. Harv. Bus. Rev. **57**(2), 115–126 (1979)
49. Oxford Learner's Dictionary: Definition of Suitability Noun. https://www.oxfordlearnersdictionaries.com/definition/english/suitability. Accessed 3 Mar 2022
50. Park, Y., Fiss, P., El Sawy, O.A.: Theorizing the multiplicity of digital phenomena: The ecology of configurations, causal recipes, and guidelines for applying QCA. MIS Q. **44**(4), 1493–1520 (2020)
51. Patas, J.: Developing individual IT-enabled capabilities for management control systems. In: Mayer, J.H., Quick, R. (eds.) Business Intelligence for New-Generation Managers, pp. 51–66. Springer, Cham (2015). https://doi.org/10.1007/978-3-319-15696-5_5
52. Penrose, E.: The Theory of the Growth of the Firm. JohnWiley, New York (1959)
53. Pentland, B.T.: Building process theory with narrative: From description to explanation. Acad. Manage. Rev. **24**(4), 711–724 (1999). https://doi.org/10.2307/259350
54. Pöppelbuß, J., Röglinger, M.: What makes a useful maturity model? A framework for general design principles for maturity models and its demonstration in business process management. In: European Conference on Information Systems, Helsinki, Finland (2011)
55. Pöppelbuß, J., Niehaves, B., Simons, A., Becker, J.: Maturity models in information systems research: Literature search and analysis. Commun. Assoc. Inf. Syst. **29**, 505–532 (2011). https://doi.org/10.17705/1CAIS.02927
56. Pöppelbuß, J., Plattfaut, R., Niehaves, B.: How do we progress? An exploration of alternate explanations for BPM capability development. Commun. Assoc. Inf. Syst. **36**, 1–22 (2015). https://doi.org/10.17705/1CAIS.03601

57. Rogers, E.M.: Diffusion of Innovations. Free Press of Glencoe, New York (1962)
58. Röglinger, M., Pöppelbuß, J., Becker, J.: Maturity models in business process management. Bus. Process. Manag. J. **18**(2), 328–346 (2012). https://doi.org/10.1108/14637151211225225
59. Rohloff, M.: An approach to assess the implementation of business process management in enterprises. In: European Conference on Information Systems, Verona, Italy (2009)
60. Rohloff, M.: Case study and maturity model for business process management implementation. In: Dayal, U., Eder, J., Koehler, J., Reijers, H.A. (eds.) Business Process Management. LNCS, vol. 5701, pp. 128–142. Springer, Heidelberg (2009). https://doi.org/10.1007/978-3-642-03848-8_10
61. Rohloff, M.: Process management maturity assessment. In: Americas Conference on Information Systems, San Francisco, California, pp. 1–12 (2009)
62. Rohloff, M.: Advances in business process management implementation based on a maturity assessment and best practice exchange. ISeB **9**(3), 383–403 (2010). https://doi.org/10.1007/s10257-010-0137-1
63. Rosemann, M.: The service portfolio of a BPM center of excellence. In: vom Brocke, J., Rosemann, M. (eds.) Handbook on Business Process Management 2. International Handbooks on Information Systems, pp. 381–398. Springer, Heidelberg (2015). https://doi.org/10.1007/978-3-642-45103-4_16
64. Rosemann, M., de Bruin, T.: Towards a business process management maturity model. In: European Conference on Information Systems, Regensburg, Germany (2005)
65. Rosemann, M., de Bruin, T., Hueffner, T.: A model for business process management maturity. In: Americas Conference on Information Systems, New York (2004)
66. Rosemann, M., Recker, J., Vessey, I.: An examination of IS conference reviewing practices. Commun. Assoc. Inf. Syst. **26**, 287–304 (2010). https://doi.org/10.17705/1CAIS.02615
67. Saarinen, T.: Evolution of information systems in organizations. Behav. Inf. Technol. **8**(5), 387–398 (1989). https://doi.org/10.1080/01449298908914568
68. Sabherwal, R., Hirschheim, R., Goles, T.: The dynamics of alignment: Insights from a punctuated equilibrium model. Organ. Sci. **12**(2), 179–197 (2001). https://doi.org/10.1287/orsc.12.2.179.10113
69. Schmitt, C., Starke, P.: Explaining convergence of OECD welfare states: A conditional approach. J. Eur. Soc. Policy **21**(2), 120–135 (2011). https://doi.org/10.1177/0958928710395049
70. Solli-Sæther, H., Gottschalk, P.: The modeling process for stage models. J. Organ. Comput. Electron. Commer. **20**(3), 279–293 (2010). https://doi.org/10.1080/10919392.2010.494535
71. Straub, D.W.: Editor's comments: Why top journals accept your paper. MIS Q. **33**(3), 3–10 (2009). https://doi.org/10.2307/20650302
72. Straub, D.W., Ang, S.: Editor's comments: Readability and the relevance versus rigor debate. MIS Q. **32**(4), 3–8 (2008). https://doi.org/10.2307/25148865
73. Sutton, R.I., Staw, B.M.: What theory is not. Adm. Sci. Q. **40**(3), 371–384 (1995)
74. Swanson, R.A., Holton, E.F., III.: Foundations of Human Resource Development. Berrett-Koehler Publishers, San Francisco (2001)
75. Tapia, R.S., Daneva, M., van Eck, P., Wieringa, R.: Towards a business-IT aligned maturity model for collaborative networked organizations. In: Enterprise Distributed Object Computing Conference Workshops, Munich, Germany, pp. 70–81 (2008). https://doi.org/10.1109/EDOCW.2008.59
76. Tarhan, A., Turetken, O., Reijers, H.A.: Business process maturity models: A systematic literature review. Inf. Softw. Technol. **75**, 122–134 (2016). https://doi.org/10.1016/j.infsof.2016.01.010
77. Teece, D.J., Pisano, G., Shuen, A.: Dynamic capabilities and strategic management. Strateg. Manag. J. **18**(7), 509–533 (1997). https://doi.org/10.1002/(SICI)1097-0266(199708)18:7%3c509:AID-SMJ882%3e3.0.CO;2-Z

78. Thordsen, T., Murawski, M., Bick, M.: How to measure digitalization? A critical evaluation of digital maturity models. In: Hattingh, M., Matthee, M., Smuts, H., Pappas, I., Dwivedi, Y.K., Mäntymäki, M. (eds.) Responsible Design, Implementation and Use of Information and Communication Technology. LNCS, vol. 12066, pp. 358–369. Springer, Cham (2020). https://doi.org/10.1007/978-3-030-44999-5_30

79. Vaast, E., Binz-Scharf, M.C.: Bringing change in government organizations: Evolution towards post-bureaucracy with web-based IT projects. In: International Conference on Information Systems, Paris, France, pp. 1–16 (2008)

80. van de Ven, A.H., Poole, M.S.: Explaining development and change in organizations. Acad. Manage. Rev. **20**(3), 510–540 (1995). https://doi.org/10.2307/258786

81. van Looy, A.: Business Process Maturity. A Comparative Study on a Sample of Business Process Maturity Models. SpringerBriefs in Business Process Management SBPM, pp. 1–86. Springer, Cham (2014). https://doi.org/10.1007/978-3-319-04202-2

82. van Looy, A., Backer, M. de, Poels, G.: Defining business process maturity. A journey towards excellence. Total Qual. Manage. Bus. Excell. **22**(11), 1119–1137 (2011). https://doi.org/10.1080/14783363.2011.624779

83. Webster, J., Watson, R.T.: Analyzing the past to prepare for the future: Writing a literature review. MIS Q. **26**(2), 8–18 (2002)

84. Weick, K.E., Quinn, R.E.: Organizational change and development. Annu. Rev. Psychol. **50**, 361–386 (1999). https://doi.org/10.1146/annurev.psych.50.1.361

85. Williamson, J.B., Fleming, J.J.: Convergence theory and the social welfare sector: A cross-national analysis. Int. J. Comp. Sociol. **18**(3–4), 242–253 (1977). https://doi.org/10.1177/002071527701800303

86. Fettke, P., Zwicker, J., Loos, P.: Business process maturity in public administrations. In: vom Brocke, J., Rosemann, M. (eds.) Handbook on Business Process Management 2. International Handbooks on Information Systems, pp. 485–512. Springer, Heidelberg (2015). https://doi.org/10.1007/978-3-642-45103-4_21

Applying Process Mining in Small and Medium Sized IT Enterprises – Challenges and Guidelines

Mathias Eggert[✉] [ID] and Julian Dyong

University of Applied Sciences Aachen, Eupener Street 70, 52066 Aachen, Germany
eggert@fh-aachen.de

Abstract. Process mining gets more and more attention even outside large enterprises and can be a major benefit for small and medium sized enterprises (SMEs) to gain competitive advantages. Applying process mining is challenging, particularly for SMEs because they have less resources and process maturity. So far, IS researchers analyzed process mining challenges with a focus on larger companies. This paper investigates the application of process mining by means of a case study and sheds light into the particular challenges of an IT SME. The results reveal 13 SME process mining challenges and seven guidelines to address them. In this way, the paper contributes to the understanding of process mining application in SME and shows similarities and differences to larger companies.

Keywords: Process mining · Challenges · Guidelines · SME · Case Study

1 Introduction

Process mining aims to discover, monitor, and improve real processes by extracting event logs from information systems [49]. Process Mining includes process discovery, conformance checking, enhancement, social network mining, case prediction and history-based recommendations [50]. Meanwhile, the application of process mining is widespread and appears in domains, such as healthcare [40], multiple industry sectors (e.g. shipbuilding) [10], retail [22], government agencies [29], even web analytics [36].

So far, the focus of investigating the application of process mining and the derivation of challenges and practical guidelines addressing them lays on large enterprises (e.g. [31, 41, 43]). Small- and medium-sized enterprises (SMEs) differ from large companies regarding the maturity of their processes [15] and their resources, which are more limited [20]. Particularly, IT SMEs have other entrepreneurial opportunities and risks than SMEs in other domains [18]. For example, small IT companies are often managed by highly educated personnel with a high degree of internationalization [18]. Furthermore, small IT companies are confronted with cost pressures, tight delivery schedules and a high personnel attrition [17]. Generally, the IT sector is characterized by significant competitive pressure due to the high degree of internationalization [28]. Therefore, IT SMEs need to adapt even more to a dynamically changing business environment [18, 26]. We hypothesize that these differences might also lead to other challenges and experiences when

© Springer Nature Switzerland AG 2022
C. Di Ciccio et al. (Eds.): BPM 2022, LNCS 13420, pp. 125–142, 2022.
https://doi.org/10.1007/978-3-031-16103-2_11

applying process mining in an SME context. Focusing on the special characteristics of SMEs is quite common in information systems (IS) research [27, 35, 54]. Surprisingly, literature on experiences during the integration and application of process mining in SMEs is limited. So far, solely two recently conducted case studies regarding process mining in SMEs are available [44, 58].

The application of process mining in organizations suffers from unclear success factors [33] and missing implementation guidance in different organizations and domains [33]. So far, solely Stertz et al. [44] discussed challenges for the application and implementation of process mining in SME organizations. However, analyzing challenges of IT SMEs when applying process mining was not a focus of information systems (IS) research, yet. Against this background, the paper at hand answers two research questions:

RQ1: What challenges exist when applying process mining at an IT SME?
RQ2: How can an IT SME be guided to properly address these challenges?

To answer these questions, we conducted a case study at an SME IT vendor of ERP systems in Germany, where we setup three projects to evaluate the applicability of process mining. The results contribute to research in three ways: First, we shed light on SME experiences when applying process mining techniques. Second, based on our observations, we derive major challenges when applying process mining in SMEs and discuss them with the body of knowledge. Third, we provide practical guidelines to master these challenges.

The remainder of this paper is structured as follows. After presenting the foundations and clearly identify the research gap, we describe our research design and the case study setting. Afterwards, we present the project observations and the identified challenges and guidelines. The paper ends with a result discussion and concluding remarks.

2 Foundations

2.1 Process Mining

Process mining is defined as an analysis approach that "aims to discover, monitor, and improve real processes by extracting knowledge from event logs readily available in today's information systems" [49]. To conduct process mining, one needs an event log [50]. Event logs are often gathered from information systems, which manage business processes and the occurrence of events [52]. An event log stores events, which belong to specific instances of a single process [1]. Typically, a process instance is also called case or trace. Each event belongs to a single case and resembles an activity or a task [1].

Three different types of process mining exist: discovery, conformance checking and enhancement. Process discovery focuses on the construction of a process model from an event log by using algorithms that accurately describe the real-life process [53]. Conformance checking describes the case of comparing an existing process model with an event log of the same process [51]. The comparison shows where the real process differs from the modeled process [51]. Enhancement takes an event log and a process model to extend the model using the observed events [51]. It might be used to extend the models with times to show bottlenecks, throughput times, and frequencies [50].

2.2 Small and Medium Sized Enterprises

SMEs differ from large organizations mostly by their size. The Institute for SME research in Bonn, Germany defines SMEs as companies that have less than 500 employees or have a revenue less than 50 million Euro per year [23].

SMEs are described by organizational and leadership characteristics. From an organizational perspective, SMEs are characterized by *limited resources* [2, 47], a *small asset base* [4], a *low formalization level* [20], an *ingrained culture* [6], and *geographical insularity* [20]. From a leadership perspective, SMEs are characterized by low *managerial skills* [20], but deeper *IS/IT knowledge* [3]. Their *attitude and values* describe how managers personally view technology [20]. The *strategic outlook* refers to long-term planning, which is especially constrained in SMEs because its planning is typically short-term oriented [6]. Against these differences, results from research on large enterprises cannot necessarily be generalized to SMEs [38].

Small IT companies differ from small enterprises in other domains regarding entrepreneurial opportunities and risks [18]. For example, small IT companies often have owners with a high level of education, a high degree of internationalization [18] and rely on industry collaboration [17]. Moreover, the IT sector is characterized by significant competitive pressure due to the high degree of internationalization in the IT sector [28]. Therefore, IT companies in general have to adapt to the dynamically changing business environment [18, 26]. Further challenges for small IT companies are cost pressures, tight delivery schedules and a high personnel attrition [17]. Process Mining can help to overcome these challenges. According to Martin et al. [33] process mining can support the digital transformation, enable inter-organizational value creation, facilitate the strategic decision-making and empower organizations to identify business process waste. We argue that small IT companies may benefit from these opportunities when we take into account the before mentioned challenges. Therefore, we hypothesize that a flawless implementation, adoption, and use of process mining is very important for IT SMEs to gain a competitive advantage against other IT companies, which motivates our research.

2.3 Challenges of Process Mining

In order to clarify the research gap, we reviewed relevant process mining literature and searched in particular for case studies that investigate the application of process mining. The literature is analyzed by three characteristics: *case study*, which comprises case study articles, *challenges*, which comprise papers that investigate challenges for applying process mining in organizations, and the characteristic *IT sector*, indicating an IT industry specific focus of the study. In total, we reviewed 19 papers and visualized the results in a Venn diagram (see Fig. 1).

For a rough guidance to structure our literature review we followed vom Brocke et al. [56]. We sourced our literature from Google Scholar, AIS Library and Science Direct and only included journal and conference articles. To search the different databases on challenges, we used the following keywords of which all included the term '*process mining*', concatenated with one of the following keywords: (1) *adoption*, (2) *challenges*,

(3) *implementation*, (4) *use*, (5) *'success factors'*, (6) *experiences*, (7) *success* + *'IT organization'*, (8) *challenges* + *'IT organization'*, (9) *adoption* + *'IT organization'*. To search for exemplary literature on case studies and process mining in the IT sector, we used the following keywords: (I) *industry* (II) *healthcare* (III) *production* (IV) *software development* (V) *incident management*. For the aspect *'IT sector'* and *'Case Study'* we used the representative approach [5]. For literature on challenges of the process mining use in businesses we used the exhaustive approach [5] to gather all possible articles covering this topic. Furthermore, we followed an iterative procedure. The first step comprises the usage of each keyword on one of the three databases. The results yielded by the search were reviewed whether they contained one of the keywords in its title. In special cases, we also incorporated publications where we hypothesized that they might also be relevant for our literature review although they do not contain one of the keywords. We then checked the abstract and the tags of each article to get an understanding of its research goal and determine whether it is relevant for us (e.g. addressing or investigating business related challenges). If the publication did not primarily investigate the use, adoption, or challenges of process mining, we discarded it. If the abstract left us inconclusive, we skimmed the full text of the publication, checked the research goals, and then decided about its inclusion. Articles focusing on research challenges were left out because they do not address business related challenges of process mining. In addition, we solely considered articles from the last ten years as relevant until 2021. Duplicates were omitted.

We identified twelve case studies, in which process mining was applied. Two of them were performed in the IT sector. Lemos et al. [30] used process mining to examine whether the real software development process conforms to the formal specification. They could show that process mining is capable of improving the maturity level of software engineering organizations. Ferreira and Da Silva [16] conducted a case study at an IT vendor to check the processes on conformance with the ITIL guidelines and confirm that process mining may simplify process assessments.

Label	Article
A	[39]
B	[31]
C	[43]
D	[30]
E	[16]
F	[8]
G	[37]
H	[19]
I	[21]
J	[14]
K	[42]
L	[33]
M	[32]
N	[46]
O	[41]
P	[44]
Q	[12]
R	[24]
S	[11]

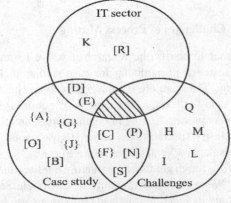

Round brackets (): SME focus
Square brackets []: Focus on large companies
Curved brackets {}: No size information available
No brackets: No focus on specific organization(s)

Fig. 1. Literature search result classification

The only two articles which covered process mining in an IT context without applying a case study research method, were Rubin et al. [42] and Kato et al. [24]. Rubin et al. [42] developed a framework for mining software processes and Kato et al. [24] used process mining techniques to optimize the maintenance of source code. We identified in total ten articles that explicitly identify process mining challenges. Among these ten papers, solely one contains a case study in an SME context [44]. A table containing all our found challenges can be accessed via this Zenodo link: https://doi.org/10.5281/zen odo.6607694. Against this background, we aim at investigating whether these challenges are transferable to an IT SME and whether additional challenges need to be regarded when applying process mining.

3 Research Design

We answer the two research questions by conducting a case study and follow the procedure of Eisenhardt [13]. Figure 2 summarizes the process of preparing and conducting the case study. To increase transparency and validity of the case study results, we followed Dubé and Paré [9] and Keutel et al. [25] as a reference.

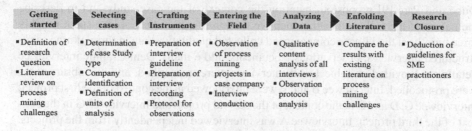

Adapted from Eisenhardt [13].

Fig. 2. Procedure of the research method

Getting Started and Selecting Cases

By conducting a case study, we aim at shedding light into the implementation and application of process mining at an IT SME. We followed the proposal of Dubé and Paré [9] to state the initial research questions (cp. Sect. 1). After defining the research question, we conducted a literature review to capture the status quo of literature on success factors and challenges when implementing and applying process mining (cp. Sect. 2.3). Both, Dubé and Paré [9] as well as Keutel et al. [25] propose a clear description of the selection process of the chosen case. For a suitable case, we setup three core criteria. First, the organization to investigate must belong to the class of SME. Second, the business model must contain the development and integration of software. Third, a suitable case organization needs to have no experiences in applying process mining software so far.

Crafting Instruments and Protocols
We used multiple data sources, because different sources are one of the intrinsic advantages a case study offers [9]. Observations and interviews are the main instrument of the case study at hand. To conduct semi structured interviews, we created an interview guideline based on the results of the literature review in Sect. 2.3. We used the videoconference software Zoom and GoToMeeting which are able to record the interviews. Afterwards, we used Amazon Transcribe to support the transcription process. The language used for the interviews were in seven cases German and in two cases English.

The use of data sources other than interviews is recommended for case studies [25]. In addition to observations and interviews, we also used process documentations to better understand the processes and get an understanding of the process complexity. We exported them as JPEGs and PDFs from the workflow management system and a quality information system. The usage of both observational data and semi-structured interviews is important: Some identified factors cannot be checked properly by observation, e.g., data protection. Others can hardly be captured by interviews, e.g., data quality. Observational data was gathered through protocolling the conduction of three process mining projects performed by the authors.

Entering the Field and Analyzing Data
Dubé and Paré [9] propose to mention the context of the case study and to describe the nature of the collected data, which we thoroughly do in Sect. 4. We spent roughly three months at the IT case company and collected qualitative data in form of nine semi-structured interviews, observations, documents, and e-mails. Demographic interviewee details are provided in Table 1. All interviews were transcribed, and our observations were protocolled. Interviewee B and E were interviewed in the context of the first project, interviewee C, D and E in the context of the second project and interviewee G in the context of the third project. Interviewee A was interviewed independently from the projects. Although two interviews were relatively short, they still provided usable information.

To analyze the transcribed interviews, we applied a qualitative content analysis approach and followed a structured and inductive coding procedure [34]. By 'structured' Mayring [34] means that categories are defined either deductive or inductive. Inductive means that the categories are defined after the screening of the transcribed interviews [34]. Afterwards, so-called 'anchor-examples' are created for every category, which are statements in the analyzed text that exactly fits to one category. Finally, coding rules are defined for each category and form the coding guide. Statements, which fulfill a coding rule are assigned to a certain category.

Enfolding Literature and Reaching Closure
As suggested by Dubé and Paré [9], we frequently use quotes when presenting the results to ensure traceability and objectivity. We deducted challenges and implementation guidelines from both the observations and interviews. In the discussion part, we compare the case study results with the findings from literature, as proposed by Dubé and Paré [9] to increase the confidence of the case findings.

Table 1. Interviewee demographics

Interviewed	Position	Working experience	Gender	Age	Interview duration
A	ERP consultant	10.5 years	Male	25–30	20 min
B	Manager of consulting department	16 years	Male	40–45	8 & 30 min
C	Manager of automation department	25 years	Male	50–55	40 min
D	Project management (automation)	31 years	Male	50–55	13 min
E	Employee at the prj. Mgmt. Office	9 years	Female	35–40	35 min
F	Employee at the concept department	9 years	Female	45–50	30 min
G	Project manager and consultant	11 years	Male	30–35	25 & 35 min

4 Observed Challenges of Applying Process Mining

The IT case company is a medium-sized IT vendor of ERP systems in Germany. The company currently employs round about 480 full time employees with small subsidiaries in other countries. The company runs a distributed ERP system, containing a workflow management system (WFMS), in which several internal processes are implemented. In total, the authors setup three process mining projects, to mine three processes. All projects used the L*-Lifecycle-Model by van der Aalst [48] and used the process mining tool ProM [55] in the version 6.10. Table 2 provides an overview on all challenges we observed during the case study. In the following sub sections, we refer to each challenge and describe its occurrence during the projects.

4.1 Mining the Consulting Request Process

The consulting request process comprises the whole consulting activities from the entry of support or order requests, its processing until its closing. The main goal for the process owner to setup a mining project is to reveal bottlenecks in the process, to optimize it and digitalize these activities. In addition, the process owner wanted to integrate process mining into his digitalization strategy of the process. To address this goal, a shift of manpower is needed (A12): "Then it's also a question of shifting the manpower [...] you haven't gained anything by mining alone, but what you then make out of it. [...] This controlling and thus recording and correcting of these processes, that also has to be done by someone" (Interviewee B). Furthermore, the topic of awareness on process

Table 2. Challenges of applying process mining at an IT SME

ID	Challenge	Description
A1	Data privacy concerns	Concerns regarding the regulatory compliance when storing and deleting event data and process mining results
A2	Outdated workflow/event data	Cases in the event log refer to process versions that are no longer in use
A3	Preparation of event log data	Preprocessing of the data and clearing of errors to prepare the event data for applying process mining
A4	Knowledge gathering (domain)	Gathering of domain and process knowledge in order to interpret the mining results
A5	Computational complexity	The time algorithms need to compute the results and the ability to provide a suitable result
A6	Poor documentation quality	The documentation to create models for conformance checking is unreadable or too high level for conformance checking
A7	Communicating the results	The process mining results are not understandable by the process owner and other stakeholders
A8	Doubting the results of the analysis	Process owners and stakeholders doubt at least a part of the mining results
A9	Selection of the right algorithm	The outputs of many algorithms are not practical and varying the parameters every time is a tedious task
A10	Complexity of the processes	The number of activities, loops, decisions, and arcs in the process to mine increases the confusion with the mining result
A11	Awareness	Creation of the awareness in the organization for process mining, its benefits, and costs
A12	Shifting manpower	The shift of manpower within an organization in order to fulfill process mining tasks
A13	Added value	The missing evidence of the added value by process mining

mining was important (A11): "The challenge is first of all the 'awareness' to really get to grips with the topic; that you really create an understanding for it" (Interviewee B).

When it came to sourcing the event data, the internal administration of the IT case company arose data privacy concerns (A1). "What is always a problem is any kind of deletion periods [...] you do whatever you want to do with it, you analyze it, you generate process models from it, but at some point […] someone must ask: 'Do I still need this data?' and 'Is there a legal basis for me to have this data?'" (Interviewee A).

Roughly one third of all cases had an outdated version of the process as it was implemented previously (A2). This skewed the output of the discovered model because activities and connections are continuously added to the process. For discovering the process models, the algorithms Inductive Miner, ETMd, HeuristicsMiner, Fuzzy Miner, and the DFG miner were applied. The *Inductive Miner* provided a suitable petri net of the process. The other algorithms also provided good results but not on the same scale as the *Inductive Miner*.

4.2 CIM Project Lifecycle

The IT case company uses the CIM project lifecycle process to organize all activities from getting a sales request, create a conceptualization and an individual offer until the ERP system is completely integrated. The process owner was mainly interested in analyzing the process duration in order to optimize the process flow. Again, the event log showed only minor quality issues (A2).

The same five algorithms for analyzing the consulting process were applied. Afterwards, the process is checked for conformance. In this project, the documentation was even more unreadable and the creation of a BPMN model was very time consuming, too (A6). A lot of domain knowledge was necessary to understand the process (A4).

When discussing the results, the activities executed by the sales department were hard to understand for the process owner (A7). "The entire upper [process] branch, which includes recording customer requests, forwarding the product sheet, creating order confirmation […] is absolutely not clear to me" (Interviewee C). The use of the 'Inductive Visual Miner', provided the relative paths, which were interesting for the process owner. Solely about 7% of all cases in the event log entered the implementation phase or in absolute numbers: Only 12 projects were conducted. The rest was aborted after checking whether the customer offer was signed or not. To some extent, the process owner doubts these results of the mined process (A8).

Interviewee C also doubts the general availability of (suitable) event data to perform process mining. The organization is then challenged to define events in a process with the right abstraction and precision of events (A3): "The challenge is, when I have a process that I want to analyze, to define distinctive measuring points in the process, what I want to look at, for example your events. I have to define and create these events in my process if they are not yet there. […] If I have far too many events, I don't see anything afterwards, if I have too few events, I see a lot, but I can't optimize anything, and finding that middle ground when I want to analyze a process, I think that's the challenge" (Interviewee C).

Furthermore, we discovered that the IT case company need evidence regarding the added value of process mining (A13): "So, bringing this tool to the people without any concrete information about the chances of success, is the problem or the challenge" (Interviewee C). Creating awareness (A11) for this added value was also considered as challenging: "It is first of all an investment that a company has to make and to get these investments in a medium-sized company, to get people to do that, will be a challenge. So, to invest time, and money for this tool […]" (Interviewee C).

4.3 Circulation Checklist

The circulation checklist process comprises the migration of a customer ERP system from an older version to a newer version. The goals defined were similar as for the two processes described above: The bottlenecks should be shown in the process model, so the process can be improved later. The analysis reveals that every workflow takes longer than wished by the process owner.

To mine a proper process model was challenging for this process because of its complexity. Thus, the models had to be manually evaluated because every petri net was too large, or the algorithms provided no results after several minutes (A5). The model from the 'Inductive Miner' (default settings) is used, although the model was imprecise and too general. The 'Fuzzy Miner' also produced a result graph, but the ordering of the activities seemed to be wrong in many cases. The DFG-based algorithm produced only a result, when setting the noise threshold to 0.2. Against this background, we conclude that the selection of the optimal mining algorithm is challenging (A9).

During the conformance checking of the process we observed the same difficulties as before. In particular, the retrieval of a reference model was not possible due to its poor as-is documentation (A6). So, we weren't able to check this process on conformance. After enhancing the model with times, no obvious bottlenecks in the process model could be observed.

When presenting the results to the stakeholders, the process owner had many questions regarding the outcomes. He was not able to properly interpret the results, and he did not understand what the mined model was telling and showing him: "Okay based on this analysis, CIM process takes more than the regular so that's the bottleneck.' Maybe there can be a suggestion part [...]" (Interviewee G). The interpretation of the enhanced model was also very difficult and confused the process owner: "Which process is the slowest one? Is it the CIM checking? How do I see which processes take the most time?" (Interviewee G). He criticized the result presentation as too focused on explaining the analysis and not focusing on the results. He was more interested in the outcomes and expected some key factors for optimization and not really on the process of mining itself (A7).

The challenge of creating an awareness for process mining (A11) was also faced in this project. Interviewee G thinks that "the bottleneck is the management; you have to convince them [...]". In addition, we could observe the problem of shifting manpower when performing process mining (A12): "the challenge would be to assign someone and give him this task and let him focusing on to this task" (Interviewee G). Finally, the need for evidence on the added value of process mining was mentioned again (A13): "You know that once you make this process a little bit faster no one will see a difference in revenue. Do you understand my point? No one will see the benefit directly. Of course, we will start processing them faster" (Interviewee G).

5 Guidelines

Besides the identification of process mining challenges for an IT SME, the second research objective of the conducted case study is to deduct guidelines for practitioners. These guidelines are based on the discovered challenges and on statements made by the interviewees. The development of our guidelines followed two steps. First, we checked whether some of the interviewees made a statement, which is useful to formulate guidelines. In some cases, interviewees suggested steps to consider, when implementing, adopting, or using process mining, which were then checked for applicability and meaningfulness. Not every challenge is referenced within our guidelines because we focus on the most relevant ones. We only selected challenges, which were mentioned at least five times across all conducted interviews. Furthermore, if we observed and protocolled a challenge in all three projects, we considered it as important for our guidelines and proposed a possible solution for this.

Second, we decided how to group certain challenges and statements together and how we label each guideline. We tried to convey as much meaning as possible for each guideline. We grouped challenges and/or interview statements according to their main idea if possible. In this way, we aim at covering all main ideas of the challenges from literature and our case study. Based on our observations as well as the interviews, we deduced seven guidelines, which we outline in the following. Thereby, we also refer to the addressed challenges in Table 2.

Begin with Simple Processes: The complexity of the processes regarding the number of activities, loops, parallel executions, and decisions (A10) should be addressed by focusing on simple processes first to get a feeling for process mining. The complexity were a great challenge during our projects.

Focus on Core Functionalities of a Process Mining Software: Interviewee B, C and G elaborated on a potential process mining software and the three classical types of process mining: discovery, conformance checking and enhancement. Topics like prediction or online process mining are not important for practitioners of medium-sized IT organizations. Furthermore, the output of the mining system must be understandable and easy to use (A7). Thus, we recommend choosing a software, which does not use the petri net notation. Rather a software with a simple graph notation should be taken into account.

Create a Comprehensive Knowledge Base: In order to address the available domain knowledge (A4), which was necessary in all observed projects, organizations could use a centralized knowledge platform [43], such as an internal wiki, to bundle relevant information on the reference process and/or existent KPIs. To gather domain knowledge, process participants need to be included within the mining project. Process models should be prepared for an import into a process mining tool. This would be a solution for the challenge of poor-quality process documentations (A7).

Involve Data Protection Stakeholders from the Beginning: Although the awareness on data protection (A1) is rather low in the investigated IT SME, practitioners should not underestimate the value of data protection. The workers council needs to be informed when a process mining software is acquired, the data protection officer to establish rules for process mining, the information security officer to determine where the event data is stored safely, and the process participants need to be asked for approval.

Consider Process Versions When Evaluating Event Data: We observed in all projects that it is challenging to deal with rapidly changing process versions (A2). It is important to consider that challenge because this anomaly in event logs can skew the output of a process mining software. However, practitioners should not do the mistake to simply skip this data. Moreover, transformation rules between process versions need to be developed. The simple deletion of historical event data might hinder valuable results.

Find the Right Balance Between Precision and Abstraction When Creating a Data Set: Not every process in an organization generates event data suitable for process mining. They even might be not digitized yet (A3). If too many events are implemented for a process, one cannot see the relevant parts, if too few events are implemented than one cannot derive measures from the output. To address this challenge, practitioners in IT SMEs can define measuring points, which represent the events later in the log. These measuring points can be implemented in IS, such as ERP systems.

Ensure Top Management Support for Process Mining: The management needs to anchor process mining in the organizational structure and must be willing to shift manpower (A12) or hire experts to conduct process mining. We suggest making that clear before beginning with mining projects. To convince the top management and to show the business value of process mining (A13), practitioners might use success stories of process mining on example processes. The top management is also an integral part in creating awareness for process mining in the organization (A11).

6 Discussion and Conclusion

The paper at hand presents the results of a case study about the application of process mining at an IT SME in Germany. We aim at identifying challenges that arise, when implementing process mining. In total, 13 challenges could be retrieved trough observing three process mining projects. Furthermore, we present guidelines for practitioners to address some of these challenges. The results of the conducted case study share some similarities with already identified challenges, such as data protection [33], inclusion of stakeholders [33, 46], or the availability of event logs [33]. In the following, we discuss the discovered challenges and argue for or against its SME and if possible, IT SME specific causes.

The challenge of data privacy concerns (A1) have been partially discussed in literature [33, 44]. The IT case company seems to have a general distrust regarding event data used by other departments. This can be seen as a so-called "constraining data access barrier" [33], which negatively affects the availability of event logs. Stertz et al. [44] and Martin et al. [33] state that employees respectively workers might have data privacy concerns regarding process mining. Particularly, Interviewee E was concerned about a potential work monitoring but still on a low level. During the interview with a member from the workers council, the aspect regarding a possible rating of employee performance based on process mining results played a big role. Martin et al. [33] call this aspect 'invasive work monitoring'. Although Stertz et al. [44] conducted a focus group study with two SMEs and also discovered the problem of work monitoring, this challenge is probably not SME-specific because many other practitioners mentioned this requirement as Martin et al. [33], Grisold et al. [19] and Eggers and Hein [12] show. No reviewed literature on (IT) SMEs explicitly elaborates on data protection.

Literature on the challenge of outdated workflow event data (A2) is scarce. Solely Homayounfar [21] reported on this challenge based on process mining experiences in hospitals, calling it 'Ad hoc actions and process changes'. We do not perceive this challenge as (IT) SME-specific because processes may change in each organization.

The challenge of the preparation of event log data (A3) is rather less important in both literature and the case study. In the observed case, it was possible to easily download the event data, but we did not see the challenges related to the availability of event logs as mentioned by Smit and Mens [43], Eden et al. [11] as well as Eggers and Hein [12]. We suppose a technical background because all three analyzed processes within our case study are implemented in a workflow management system and therefore the event data is not scattered across several IS. In order to argue for an SME-specific challenge, particularly the contribution by Stertz et al. [44] is of relevance. Stertz et al. [44] discovered that two analyzed SMEs consider the creation of a suitable data basis for process mining as a challenge.

The existence of suitable domain knowledge and its gathering (A4) was discussed by Homayounfar [21], who reported difficulties regarding the understanding of the logged data. While Mans et al. [32] state that the expertise of the process miner is a success factor, Smit and Mens [43] focus more on the challenge of knowledge building and sharing. We perceive the missing of domain knowledge not as (IT) SME specific.

The challenge regarding the computational complexity of some algorithms (A5) was not addressed in any of the reviewed literature. This challenge is closely linked to the selection of suitable algorithms (A9). While no case study reported on this issue, Wang [57] mentioned that one algorithm needed so long that the computation had to be aborted. In our case, evolutionary algorithms needed a very long time to compute. However, we rate this challenge as (IT) SME independent.

The poor quality of the reference models was also considered as challenging (A6). None of the literature from Sect. 2.3 mentions this challenge. The only article elaborating on the possibility of low quality of handmade models is van der Aalst [48]. However, this was just an assumption without concrete consequences. We consider this challenge as SME-specific because the maturity of processes and its documentation in SMEs is typically lower than in large companies [15]. Furthermore, we argue that it could even be specific for IT SMEs. Feldbacher et al. [16] mention that the process maturity of IT SMEs is typically a bit higher than in SMEs of other domains, in which the processes are often not documented at all. This corresponds to our case observations regarding the process documentation, which is already advanced but still too underdeveloped to use it in a process mining context.

Communicating the results of process mining (A7) to people in the organization and the problem of doubting the mining results (A8) were also observed by Martin et al. [33], who call it 'lack of trust in insights'. For the moment, we cannot rate the SME specificity and encourage further investigations of this challenge.

The selection of the right algorithm (A9) is interrelated with the complexity of some algorithms (A5). Wang et al. [57] discuss the efficient selection of discovery algorithms. However, in an SME context, this issue was not discussed yet.

Although the process structure and thus its complexity (A10) was not mentioned very often in related articles, it played a bigger role in our case study. The processes we analyzed are too unstructured to easily interpret them but still maintain a structure that does not allow a usage of declarative process models. We observed that the more complicated a process is, the more difficult it is to mine this process and to interpret the results, which confirms the results of Homayounfar [21]. However, articles that focus on process mining in SMEs did not mention that factor, yet.

The challenge of getting awareness for process mining from the top management (A11) is discussed many times in literature and plays a role in our case study as well. Martin et al. [33] discovered in their Delphi study that a lack of management support is a challenge for organizations when applying process mining. Since SMEs but also IT SMEs usually have limited resources [2, 17, 47], the management must be convinced harder for getting a budget for process mining projects. Thus, we assume a more (IT) SME specific challenge.

Although, the challenge of continuous usage of process mining was also mentioned by Martin et al. [33] and by Grisold et al. [19], the shift of manpower (A12) was not covered yet in the reviewed literature on process mining. We assume an SME specific problem because limited resources are a typical organizational characteristic of SMEs [2, 47]. In addition, Cragg et al. [7] discovered that SMEs rely on external expertise to support their IT function, which additionally indicates an SME-specific process mining challenge. We even consider a shortage in workforce as IT SME specific because IT services are very knowledge-intensive businesses [18] and the acquisition of the newest technical knowledge is also challenging (e.g. by hiring qualified employees) for small IT companies [45]. On top of that, IT SMEs suffer from personnel attrition in general [17]. Companies, which struggle to hire and keep qualified employees, will rather use these employees to gain money instead of using these valuable resources to conduct internal projects, such as a process mining project, which solely yield revenue indirectly. Finally, the challenge of proving added value (A13) was mentioned by Martin [33] as 'elusive business value'. Grisold et al. [19] discovered that process managers need evidence that process mining provides a concrete value, which is in line with our observations. Thus, we perceive that challenge as (IT) SME independent.

We worked as rigorous as possible to reveal valid insights into the application of process mining in SMEs. However, the expressive power of the results is limited. The IT case company we observed belongs to the group of SMEs but with about 480 employees it is at the border to a large company. Furthermore, the interview guideline is grounded on our literature review, which might shrink the room for collecting additional challenges and we could not evaluate the suggested process mining guidelines elsewhere. We encourage researchers to conduct additional case studies in the field of SMEs to extend the knowledge of process mining application in SMEs.

References

1. Bose, R., Jagadeesh Chandra, P., Mans, R.S., van der Aalst, W.M: Wimprove process mining results? In: IEEE Symposium on Computational Intelligence and Data Mining (CIDM). IEEE, pp 127–134 (2013)

2. Boyes, J.A., Irani, Z.: Barriers and problems affecting web infrastructure development: the experiences of a UK small manufacturing business. In: 9th Americas Conference on Information Systems (AMCIS), p. 90 (2003)
3. Caldeira, M.M., Ward, J.M.: Using resource-based theory to interpret the successful adoption and use of information systems and technology in manufacturing small and medium-sized enterprises. Eur. J. Inf. Syst. **12**, 127–141 (2003). https://doi.org/10.1057/palgrave.ejis.300 0454
4. Carbo-Valverde, S., Udell, G.F., Rodríguez-Fernández, F.: Bank market power and SME financing constraints. SSRN J. (2006). https://doi.org/10.2139/ssrn.910226
5. Cooper, H.M.: Organizing knowledge syntheses: a taxonomy of literature reviews. Knowl. Soc. **1**, 104–126 (1988)
6. Cragg, P., Caldeira, M., Ward, J.: Organizational information systems competences in small and medium-sized enterprises. Inform. Manage. **48**, 353–363 (2011)
7. Cragg, P., Mills, A., Suraweera, T.: The influence of IT management sophistication and IT support on IT success in small and medium-sized enterprises. J. Small Bus. Manage. **51**, 617–636 (2013). https://doi.org/10.1111/jsbm.12001
8. Dakic, D., Sladojevic, S., Lolic, T., et al.: Process mining possibilities and challenges: a case study. In: IEEE 17th International Symposium, pp. 161–166 (2019)
9. Dubé, L., Paré, G.: Rigor in information systems positivist case research: current practices, trends, and recommendations. MIS Q. **27**, 597 (2003)
10. Dunzer, S., Zilker, S., Marx, E., Grundler, V., Matzner, M.: The status quo of process mining in the industrial sector. In: Ahlemann, F., Schütte, R., Stieglitz, S. (eds.) WI 2021. LNISO, vol. 48, pp. 629–644. Springer, Cham (2021). https://doi.org/10.1007/978-3-030-86800-0_43
11. Eden, R., Syed, R., Leemans, S.J.J., Buijs, J.A.C.M.: A case study of inconsistency in process mining use: implications for the theory of effective use. In: Polyvyanyy, A., Wynn, M.T., Van Looy, A., Reichert, M. (eds.) BPM 2021. LNCS, vol. 12875, pp. 363–379. Springer, Cham (2021). https://doi.org/10.1007/978-3-030-85469-0_23
12. Eggers, J., Hein, A.: Turning big data into value: a literature review on business value realization from process mining. In: 28th European Conference on Information Systems (ECIS2020) (2020)
13. Eisenhardt, K.M.: Building theories from case study research. AMR **14**, 532 (1989)
14. Er, M., Arsad, N., Astuti, H.M., et al.: Analysis of production planning in a global manufacturing company with process mining. JEIM **31**, 317–337 (2018). https://doi.org/10.1108/JEIM-01-2017-0003
15. Feldbacher, P., Suppan, P., Schweiger, C., Singer, R.: Business process management: a survey among small and medium sized enterprises. In: Schmidt, W. (ed.) S-BPM ONE 2011. CCIS, vol. 213, pp. 296–312. Springer, Heidelberg (2011). https://doi.org/10.1007/978-3-642-23471-2_21
16. Ferreira, D.R., Da Silva, M.M.: Using Process Mining for ITIL Assessment: A Case Study With Incident Management. Bournemouth University (2008)
17. Findikoglu, N.M., Ranganathan, C., Watson-Manheim, M.B.: Partnering for prosperity: small IT vendor partnership formation and the establishment of partner pools. Eur. J. Inf. Syst. **30**, 193–218 (2021)
18. Głodek, P., Łobacz, K.: Transforming IT small business - the perspective of business advice process. Procedia Comput. Sci. **192**, 4367–4375 (2021)
19. Grisold, T., Mendling, J., Otto, M., et al.: Adoption, use and management of process mining in practice. BPMJ **27**, 369–387 (2021)
20. Heidt, M., Gerlach, J.P., Buxmann, P.: Investigating the security divide between SME and large companies: how SME characteristics influence organizational IT security investments. Inf. Syst. Front. **21**(6), 1285–1305 (2019). https://doi.org/10.1007/s10796-019-09959-1

21. Homayounfar, P.: Process mining challenges in hospital information systems. In: Proceedings on Computer Science and Information Systems (FedCSIS) (2012)
22. Hwang, I., Jang, Y.J.: Process mining to discover shoppers' pathways at a fashion retail store using a WiFi-base indoor positioning system. IEEE Trans. Automat. Sci. Eng. **14**, 1786–1792 (2017)
23. Institut für Mittelstandsforschung Bonn: SME definition of the IfM Bonn (2016). https://www.ifm-bonn.org/en/index/definitions/sme-definition-of-the-ifm-bonn. Accessed 08 Mar 2022
24. Kato, K., Kanai, T., Uehara, S.: source code partitioning using process mining. In: Rinderle-Ma, S., Toumani, F., Wolf, K. (eds.) BPM 2011. LNCS, vol. 6896, pp. 38–49. Springer, Heidelberg (2011). https://doi.org/10.1007/978-3-642-23059-2_6
25. Keutel, M., Michalik, B., Richter, J.: Towards mindful case study research in IS: a critical analysis of the past ten years. Eur. J. Inf. Syst. **23**, 256–272 (2014). https://doi.org/10.1057/ejis.2013.26
26. Khatri, N., Baveja, A., Agrawal, N.M., et al.: HR and IT capabilities and complementarities in knowledge-intensive services. Int. J. Hum. Resour. Manage. **21**, 2889–2909 (2010). https://doi.org/10.1080/09585192.2010.528672
27. Kiran, T., Reddy, A.V.: Critical success factors of ERP implementation in SMEs. J. Project Manage. **4**, 267–280 (2019)
28. Kummamuru, S.: HR management challenges of indian IT sector: an application of the viable systems model. ASCI J. Manage. **43** (2014)
29. Leemans, S.J., Poppe, E., Wynn, M.T.: Directly follows-based process mining: exploration & a case study. In: International Conference on Process Mining (ICPM), pp 25–32. IEEE (2019)
30. Lemos, A.M., Sabino, C.C., Lima, R.M.F., et al.: Using process mining in software development process management: a case study. In: IEEE International Conference on Systems, Man, and Cybernetics, pp 1181–1186. IEEE (2011)
31. Lorenz, R., Senoner, J., Sihn, W., et al.: Using process mining to improve productivity in make-to-stock manufacturing. Int. J. Prod. Res. **59**, 4869–4880 (2021). https://doi.org/10.1080/00207543.2021.1906460
32. Mans, R., Reijers, H., Berends, H., et al.: Business process mining success. In: European Conference on Information Systems (ECIS) (2013)
33. Martin, N., et al.: Opportunities and challenges for process mining in organizations: results of a delphi study. Bus. Inf. Syst. Eng. **63**(5), 511–527 (2021). https://doi.org/10.1007/s12599-021-00720-0
34. Mayring, P.: Qualitative Inhaltsanalyse. Grundlagen und Techniken, 12., überarb. Aufl. Beltz, Weinheim (2015)
35. Noudoostbeni, A., Yasin, N.M., Jenatabadi, H.S.: To investigate the success and failure factors of ERP implementation within malaysian small and medium enterprises. In: International Conference on Information Management and Engineering. IEEE (2009)
36. Poggi, N., Muthusamy, V., Carrera, D., Khalaf, R.: Business process mining from E-commerce web logs. In: Daniel, F., Wang, J., Weber, B. (eds.) BPM 2013. LNCS, vol. 8094, pp. 65–80. Springer, Heidelberg (2013). https://doi.org/10.1007/978-3-642-40176-3_7
37. Quintano Neira, R.A., et al.: Analysis and optimization of a sepsis clinical pathway using process mining. In: Di Francescomarino, C., Dijkman, R., Zdun, U. (eds.) BPM 2019. LNBIP, vol. 362, pp. 459–470. Springer, Cham (2019). https://doi.org/10.1007/978-3-030-37453-2_37
38. Raymond, L.: Organizational characteristics and MIS success in the context of small business. MIS Q. **9**, 37 (1985)
39. Rbigui, H., Cho, C.: Purchasing process analysis with process mining of a heavy manufacturing industry. In: 2018 International Conference on Information and Communication Technology Convergence (ICTC). IEEE (2018)

40. Rojas, E., Munoz-Gama, J., Sepúlveda, M., et al.: Process mining in healthcare: a literature review. J. Biomed. Inform. **61**, 224–236 (2016). https://doi.org/10.1016/j.jbi.2016.04.007
41. Rozinat, A., de Jong, I.S., Günther, C.W., et al.: Process mining applied to the test process of wafer scanners in ASML. IEEE Trans. Syst. Man Cybern. C **39**, 474–479 (2009)
42. Rubin, V., Günther, C.W., van der Aalst, W.M.P., Kindler, E., van Dongen, B.F., Schäfer, W.: Process mining framework for software processes. In: Wang, Q., Pfahl, D., Raffo, D.M. (eds.) ICSP 2007. LNCS, vol. 4470, pp. 169–181. Springer, Heidelberg (2007). https://doi.org/10.1007/978-3-540-72426-1_15
43. Smit, K., Mens, J.: Process mining in the rail industry: a qualitative analysis of success factors and remaining challenges. In: BLED 2019 Proceedings (2019)
44. Stertz, F., Mangler, J., Scheibel, B., Rinderle-Ma, S.: Expectations vs. experiences – process mining in small and medium sized manufacturing companies. In: Polyvyanyy, A., Wynn, M.T., Van Looy, A., Reichert, M. (eds.) BPM 2021. LNBIP, vol. 427, pp. 195–211. Springer, Cham (2021). https://doi.org/10.1007/978-3-030-85440-9_12
45. Strohmeier EPaPS: HRM in the digital age – digital changes and challenges of the HR profession. Employee Relations, vol. 36. (2014). https://doi.org/10.1108/ER-03-2014-0032
46. Syed, R., Leemans, S.J.J., Eden, R., Buijs, J.A.C.M.: Process mining adoption. In: Fahland, D., Ghidini, C., Becker, J., Dumas, M. (eds.) BPM 2020. LNBIP, vol. 392, pp. 229–245. Springer, Cham (2020). https://doi.org/10.1007/978-3-030-58638-6_14
47. Thong, J.Y.: Resource constraints and information systems implementation in Singaporean small businesses. Omega **29**, 143–156 (2001)
48. van der Aalst, W.: Process mining: discovering and improving Spaghetti and Lasagna processes. In: IEEE Symposium on Computational Intelligence and Data Mining (CIDM). IEEE (2011)
49. van der Aalst, W.: Process mining. Commun. ACM **55**, 76–83 (2012). https://doi.org/10.1145/2240236.2240257
50. van der Aalst, W., Adriansyah, A., de Medeiros, A.K.A., et al.: Process Mining Mani-festo. Business Process Management Workshops - BPM 2011 International Workshops, 29 Aug 2011, pp.169–194. Revised Selected Papers, Part I 99, Clermont-Ferrand, France (2011). https://doi.org/10.1007/978-3-642-28108-2_19
51. van der Aalst, W., Adriansyah, A., van Dongen, B.: Replaying history on process models for conformance checking and performance analysis. WIREs Data Min. Knowl. Discov. **2**, 182–192 (2012)
52. Aalst, W.M.P.: Process-aware information systems: lessons to be learned from process mining. In: Jensen, K., van der Aalst, W.M.P. (eds.) Transactions on Petri Nets and Other Models of Concurrency II. LNCS, vol. 5460, pp. 1–26. Springer, Heidelberg (2009). https://doi.org/10.1007/978-3-642-00899-3_1
53. Dongen, B.F., Alves de Medeiros, A.K., Wen, L.: Process mining: overview and outlook of petri net discovery algorithms. In: Jensen, K., Aalst, W.M.P. (eds.) Transactions on Petri Nets and Other Models of Concurrency II. LNCS, vol. 5460, pp. 225–242. Springer, Heidelberg (2009). https://doi.org/10.1007/978-3-642-00899-3_13
54. Venkatraman, S., Fahd, K.: Challenges and success factors of ERP systems in australian SMEs. Systems **4**, 20 (2016). https://doi.org/10.3390/systems4020020
55. Verbeek, E.H., Buijs, J.C., van Dongen, B.F., et al.: ProM 6: the process mining toolkit. In: Proceedings of the Business Process Management 2010 Demonstration Track, 14–16 Sept 2010, pp. 34–39. Hoboken NJ, USA (2010)
56. vom Brocke, J., Simons, A., Niehaves, B., et al.: Reconstructing the giant: on the importance of rigour in documenting the literature search process. In: 17th European Conference on Information Systems (ECIS) (2009)

57. Wang, J., Wong, R.K., Ding, J., et al.: Efficient selection of process mining algorithms. IEEE Trans. Serv. Comput. **6**, 484–496 (2013)
58. Zeisler, A., Bernhard, C., Müller, J.M.: Process mining – prerequisites and their applicability for small and medium-sized enterprises. In: Business Intelligence and Analytics in Small and Medium Enterprises. CRC Press (2019)

A Process Mining Success Factors Model

Azumah Mamudu[(✉)] ⓘ, Wasana Bandara ⓘ, Moe T. Wynn ⓘ,
and Sander J. J. Leemans ⓘ

Queensland University of Technology, Brisbane, Australia
azumah.mamudu@hdr.qut.edu.au, {w.bandara,m.wynn,
s.leemans}@qut.edu.au

Abstract. Process mining – a suite of techniques for extracting insights from event logs of Information Systems (IS) – is increasingly being used by a wide range of organisations to improve operational efficiency. However, despite extensive studies of Critical Success Factors (CSF) in related domains, CSF studies of process mining are limited. Moreover, these studies merely identify factors, and do not provide essential details such as a clear conceptual understanding of success factors and their interrelationships. Using a process mining success model published in 2013 as a conceptual foundation, we derive an empirically supported, enhanced process mining critical success factors model. Applying a hybrid approach, we qualitatively analyse 62 process mining case reports covering diverse perspectives. We identify nine process mining critical success factors, explain how these factors relate to the process mining context and analyse their interrelationships with regard to process mining success. Our findings will guide organisations to invest in the right mix of critical success factors for value realisation in process mining practice.

Keywords: Process mining · Success factors · Process mining success · Process mining impact · Case reports

1 Introduction

Process mining (PM) is a research discipline focused on extracting knowledge from event logs readily available in today's business systems to discover, monitor, and improve real processes [1]. Organisations can utilise PM techniques to achieve operational excellence and organisational resilience[1]. In the past decade, the adoption of process mining has expanded considerably [2], evidenced by many use cases reported in industry (e.g. [3]) and academia (e.g. [4]), especially in sectors such as auditing [5], insurance [6], and healthcare [7]. The field has also significantly matured with enhanced capabilities in tools and techniques [8].

According to Gartner[2], the global process analytics market size will grow at a Compound Annual Growth Rate of 50% from US$185 million to US$1.42 billion between

---

[1] https://www.processexcellencenetwork.com/process-mining/articles/why-the-real-value-of-process-mining-lies-in-simulation. Accessed 10th June 2021.

[2] https://www.gartner.com/en/documents/3991229. Accessed 5th June 2021.

© Springer Nature Switzerland AG 2022
C. Di Ciccio et al. (Eds.): BPM 2022, LNCS 13420, pp. 143–160, 2022.
https://doi.org/10.1007/978-3-031-16103-2_12

2018 and 2023. Deloitte's[3] Global Process Mining Survey indicated that 67% of the respondents had started implementing process mining. 87% of non-adopters were considering pilot runs, 83% of "global scale users" intended to expand process mining use, and 84% believed that process mining delivered value to their organisation.

The ongoing growth in PM adoption necessitates further investigation of process mining success, particularly to uncover the complexity and diversity of factors that influence successful project implementation [9]. In this study, a PM initiative is considered a success if it is *effective* (fulfils its objectives) and *efficient* (the relevant activities are completed with the allocated resources such as time, effort and budget). Traditionally, PM research has given more attention to developing tools and techniques [10, 11], with minimal attention to the organisational aspects of PM. This has left areas such as process mining success largely unexplored. Academic discourse on the organisational benefits of process mining is emerging; for example, vom Brocke, Jans, Mendling and Reijers [11] call for research to identify considerations for the adoption, use, and effects of process mining.

One widely used approach in understanding what factors are necessary for success is the study of Critical Success Factors (CSF), originally introduced by Rockart [12]. While many CSF studies exist in related domains, there are very few in the process mining field. These process mining CSF studies identify success factors (e.g., [4]) but provide very little or no contextual interpretation of these factors, their interrelationships, or insights into their level of criticality for organisational success. It has been argued that mere identification of factors, variables and practices without a context-specific understanding of the application of these factors or their interrelationships is ineffective for enabling project success [13]. A better understanding of how CSFs interrelate to directly or indirectly influence success and in what manner they vary in importance over time is argued as essential [14].

This study aims to provide a rich understanding of process mining CSFs in practice. We analysed 62 published case reports to identify and describe CSFs pertinent to the process mining context. To avoid the criticism often received by CSF studies, we sought to derive a PM CSF model that goes beyond a mere list of factors and provides evidence-based interrelationships between these success factors. Such a model provides deeper insights into the combined and integrated influence these CSFs have on attaining PM success.

The subsequent sections of our paper are structured as follows: Sect. 2 discusses the related work on critical success factors in process mining and related domains. Section 3 summarises our study methodology. Section 4 provides the re-specified process mining success factors and contextual explanations. Section 5 presents an enhanced PM success factors model and discusses identified interrelationships, and Sect. 6 concludes the paper. The URL and details of Supplementary Material (an overview of the case reports (A), example quotes from case reports (B) and supporting case evidence for interrelationships (C)) are provided in the Appendix.

---

[3] https://www2.deloitte.com/de/de/pages/finance/articles/global-process-mining-survey-2021.html. Accessed 15th June 2021.

2 Related Work

CSF studies initially gained significant attention after Rockart [12] highlighted their relevance in influencing the information needs of top executives [15]. CSFs are defined as "the limited number of areas in which results, if they are satisfactory, will ensure successful competitive performance for the organisation" [12]. Since then, this concept has been adopted in diverse project-related contexts.

Despite the proliferation of CSF studies, they have been criticised [14] for providing mere lists of factors and lacking a deeper contextual understanding of how these factors may vary in importance over time [13]. Without a contextual understanding, a mere list of factors is ineffective in predicting success or designing interventions that enable success [13]. Fortune and White [14] also argue that CSF studies often do not account for factor interrelationships, although these are "at least as important as the individual factors" [14]. Thus, there is a clear push for CSFs to go beyond lists of factors and provide deeper insights.

CSF studies have been conducted in related domains such as BPM and data mining (e.g., [16] and [17]). Alibabaei, Bandara and Aghdasi [16] propose a holistic BPM success factors framework with nine CSF and related sub-constructs and how they achieve success. The Big Data Analytics (BDA) framework by Grover, Chiang, Liang and Zhang [18] provides a detailed analysis of moderating factors, capabilities, and value realisation potentials for transforming BDA investments into value. Most CSF studies in BPM and data mining hardly explore CSF interrelationships, though this is a commonly criticised aspect of CSF studies [13]. While insights from related domains are valuable, context specificity is essential for a CSF study to be beneficial [13], which points our attention to PM CSF studies.

In the process mining domain, there is recent work highlighting the need to carefully examine the value proposition of process mining (e.g. [11]). However, to the best of our knowledge, very few research studies (e.g., [4] and [19]) explore process mining CSF. The business process mining success model by Mans, Reijers, Berends, Bandara and Prince [19] is the first study on process mining success factors. Published in 2013, it identifies three success measures (model quality, process impact and project efficiency) and six success factors (project management, management support, structured process mining approach, data and event log quality, resource availability and process miner expertise), empirically supported via four case studies. However, the Mans, Reijers, Berends, Bandara and Prince [19] model does not explore CSF interrelationships or their criticality. Syed, Leemans, Eden and Buijs [4] identify four enabling factors for process mining success at the early stages of PM adoption within an organisation: actionable insights, confidence in process mining, perceived benefits, and training and development. However, this study is based on a single case organisation and is specifically focused on the PM adoption stage, thus questioning its generalisability and broader applicability.

The Deloitte Global IT and business executives survey identified 19 PM success factors. The five key factors reported were the need for a cross-departmental alignment between IT and business, good data quality and transformation, clear targets and the value hypothesis, the availability of dedicated resources towards process mining, and the need for leadership commitment. However, as the respondents were all IT and business executives, the results only explain CSF from a high-level organisational perspective with little insight into specific process mining project contexts.

In summary, existing CSF literature in process mining, at best, provides a list of factors. While some try to contextualise, they focus on a single case study organisation at the PM adoption stage; others explain these factors only from a high-level perspective. Potential interrelationships or the level of criticality of the factors are never explored. We aim to address this gap with our re-specified PM CSF model.

3 Study Method

Our study applies a hybrid approach to thematic analysis (i.e. using both inductive and deductive coding) of publicly available process mining case reports, conducted across three phases as outlined below:

Phase 1 focused on deriving a preliminary conceptual base. Given that the Mans, Reijers, Berends, Bandara and Prince [19] model is the most widely known model for process mining, we adopted its CSFs as our a-priori base, as summarised in Table 1.

Table 1. PM success factors from Mans, Reijers, Berends, Bandara and Prince [19]

Construct	Definition
Management Support	The involvement and participation of senior management, and their ongoing commitment and willingness to devote necessary resources and time of senior managers to oversee the process mining efforts
Project Management	The management of activities and resources throughout all phases of the process mining project to obtain the defined project outcomes
Resource Availability	The degree of information available from the project stakeholders during the entire process mining analysis
Process Miner Expertise	The experiences of the person conducting the mining, in terms of event log construction, doing process mining analysis and knowledge of the business processes being mined
Structured Process Mining Approach	The extent to which a process miner uses a structured approach during the entire process mining analysis
Data and Event Log Quality	The characteristics of the raw data and subsequently constructed event logs

Phase 2 tackled re-specifying the model using case reports as the empirical base. We performed a hybrid (inductive and deductive) qualitative analysis of 62 process mining case reports written from the user, tool vendor and practitioner perspectives outlining the success stories, tangible benefits and lessons learnt from over 50 organisations. Since process mining cases focus on applying PM tools within a given context, they are noted for providing detailed insights into PM use and outcomes [20]. Qualitatively

analysing the insights from these cases provides a detailed understanding of PM success factors from a multi-case perspective. Case reports were sourced from "Process Mining in Action" by Reinkemeyer [3], Task Force for Process Mining (TF-PM) online case repository[4] and Business Process Management Cases Vol. 1 and 2 [21, 22]. To the best of our knowledge, these sources constituted the most recent collection as of 5th June 2021. While we do not claim this collection to be the only existing source of PM case reports, we do believe them to be representative. An overview of these 62 case reports is provided in Part A of the Supplementary Material (URL in Appendix).

Coding and analysis occurred in multiple rounds. First (Round 1), using an open-coding approach [23], we inductively extracted all direct and indirect content pertaining to elements that contributed to the project's success by analysing each case report text line-by-line. 453 first-level codes were extracted. These were further analysed in a second coding round (Round 2), moving between deductive and inductive (a hybrid approach) coding [24]. The a-priori model from Phase 1 was used as the initial coding classification scheme where relevant open-codes from Round 1 were re-coded under the a-priori CSFs. Those open-codes that did not fit within the a-priori model were inductively grouped to form new themes. The results from here were exposed to another detailed analysis (Round 3). The resulting (sub-) themes from above were critically analysed and refined to obtain conceptual clarity and parsimony of the identified CSFs. This resulted in our final set of CSFs containing nine themes and 23 related sub-themes, outlined in Table 2 and further explained in Sect. 4.

A coding rulebook was developed to ensure a formalised approach was followed during code extraction [25]. NVivo was used as a qualitative analysis tool to support the coding process. Coder corroborations played a critical role across all rounds of coding. They were essential in forming a unified understanding of identified low-level code groupings, (sub-) themes and descriptions. They also ensured that a credible and high-quality coding process was followed. After the inductive extraction of low-level codes by a primary coder in Round 1, open-codes were discussed and critically reviewed with three secondary coders for alignment to the area of interest (i.e. PM CSFs). The second round of review was conducted after the (sub-) theme extraction phase. Here, coder corroboration aimed to derive consensus on the mapping of lower-level themes to resulting higher-level themes. The third round of corroboration reviewed the forming CSF model as a whole. This focused on ensuring conceptual clarity and parsimony of the nine themes, 23 related sub-themes and their descriptions.

Phase 3 focused on enhancing the re-specified set of PM CSFs. To avoid the critique that CSF studies often provide mere lists of factors without a deeper contextual understanding of how these factors vary in importance over time [13], we identified evidence-based interrelationships between the CSFs and investigated the criticality of these factors, as discussed in Sect. 5.

We identified potential CSF interrelationships in two ways: (a) by noting and separately capturing any identified interrelationships from the case reports during Round 1 coding as 'Relationship nodes'[5] in NVivo and (b) complementing this method with

---

[4] Retrieved from: https://www.tf-pm.org/resources/casestudy. Date: 5th June 2021.

[5] Relationship nodes are special types of nodes that define the connection between two project items.

NVivo's matrix intersection[6] and "near" search queries. Explanations for identified relationships were captured in Memos[7] during the coding process.

Using the case narrative of the identified CSFs, *direct, indirect* and *bilateral* relationships were extracted (see Fig. 1 in Sect. 5.1). The identified interrelationships were further contextualised for PM, applying evidence from the case reports (see Part C of Supplementary Material). A final coder corroboration critically reviewed the evidence supporting each relationship to confirm (a) the existence of each relationship and (b) the nature of the relationship.

4 Re-specified Process Mining Success Factors

While qualitative coding began with the Mans, Reijers, Berends, Bandara and Prince [19] model as a base, our analysis resulted in an extended set of CSF and a re-specified process mining success factors model (see Fig. 1). Key differences between our model and prior work (specifically [19]) are outlined in Sect. 5.

Overall, nine meta-themes (each pertaining to a CSF), with their respective sub-themes, were extracted from our analysis of the 62 reports. These are summarised in Table 2, with a brief description and supporting case-based evidence (i.e., the number of coding references, from how many reports). Example quotes from the case reports are presented in Part B of the Supplementary Material. A detailed explanation of each success factor based on the process mining context, is provided next.

Table 2. Re-specified success factors for process mining

Success factor	Description	Case evidence summary
a. Stakeholder Support and Involvement	Organisational stakeholders' support or involvement in process mining initiatives	61 codes from 29 cases
a.1. Management support	Top-Level Management/Senior Executives support	14 codes from 8 cases
a.2. External stakeholder support	Engagement with external collaborators or industry partners (such as suppliers) who influence an organisation's business process and how they are executed	5 codes from 5 cases
a.3. Subject matter experts (SMEs)	SMEs of a particular business domain who contribute to process mining efforts	26 codes from 17 cases
a.4. User groups	The contribution of ultimate users (such as first-line personnel) to process mining outcomes	6 codes from 5 cases

(*continued*)

---

[6] Matrix intersection is a 2-dimensional table that displays coded content from rows and columns.

[7] Memos allow researchers to capture thoughts and reflections during coding to justify coding choices.

Table 2. (*continued*)

Success factor	Description	Case evidence summary
b. Information Availability	The availability of historical event data and supporting documentation for a process mining initiative	26 codes from 18 cases
b.1. Event data availability	The extent to which historical event data is available for process mining analysis	12 codes from 9 cases
b.2. Availability of contextual information	Access to contextual information such as process models, business rules, policy documents, legal and regulatory requirements that can aid process mining	14 codes from 11 cases
c. Technical Expertise	The various forms of technical skills and experience required to execute process mining projects. Four types of technical expertise were identified:	42 codes from 19 cases
c.1. Process mining expertise	The required know-how needed to execute process mining initiatives and interpret outcomes	6 codes from 5 cases
c.2. Data extraction expertise	The required data analytics expertise for the extraction and integration of event data for process mining	5 codes from 4 cases
c.3. Process analyst expertise	The required expertise for designing, streamlining, and re-engineering business processes	2 codes from 2 cases
c.4. Team configuration	The composition of teams and expert groups involved in process mining projects. Two main configurations namely: **Established units:** An internal team dedicated to executing process mining initiatives. E.g., a Centre of Excellence (CoE) (21 codes from 11 cases) **Ad-hoc units:** A group of experts assembled from different departments within the organisation to execute process mining projects as and when required (8 codes from 5 cases)	29 codes from 14 cases
d. Structured Process Mining Approach	The extent to which an organisation follows a structured approach or technique to execute process mining initiatives	135 codes from 49 cases
d.1. Planning	Identifying questions or project goal(s), selecting business processes to be mined and composing the project team to execute process mining initiatives	32 codes from 21 cases

(*continued*)

Table 2. (*continued*)

Success factor	Description	Case evidence summary
d.2. Extraction	Determining the data extraction scope, extracting event data, and transferring process knowledge between business experts and process analysts	47 codes from 28 cases
d.3. Data processing	Using process mining tools to create views, aggregate events, enrich or filter logs to generate the required insights from event logs	21 codes from 15 cases
d.4. Mining and Analysis	Applying process mining techniques to answer questions and gain insights	23 codes from 18 cases
d.5. Evaluation	Relating analysis results to improvement ideas to achieve project goals	6 codes from 6 cases
d.6. Process improvement and support	Using gained insights to modify the actual process execution	6 codes from 5 cases
e. Data and Event Log Quality	Provisions made for the extraction, preparation, analysis, and data quality considerations of event data for process mining initiatives	84 codes from 45 cases
e.1. Data pre-processing	Provisions for the extraction and preparation of event data from single or multiple sources for process mining based on lessons learnt	61 codes from 40 cases
e.2. Event log quality considerations	The data quality considerations and minimum requirements to be met by event logs for process mining	23 codes from 17 cases
f. Tool Capabilities	The functionalities and features of process mining tools that organisations can use for process mining	67 codes from 35 cases
f.1. Process discovery	Automated process model discovery and process visualisation from event data	27 codes from 20 cases
f.2. Process Benchmarking	Using event data for comparison of process behaviours and process performance	6 codes from 6 cases
f.3. Conformance checking/Compliance	Detection of deviations from process norms using event data	16 codes from 15 cases
f.4. Integration capabilities	Integration of process mining capabilities with other data analytics capabilities	7 codes from 5 cases
f.5. Analytical Scalability	The tool's ability to analyse data for insights into single, multiple and e2e processes	11 codes from 10 cases

(*continued*)

Table 2. (*continued*)

Success factor	Description	Case evidence summary
g. Change Management	The series of activities that ensure that the needed change emanating from process mining results is implemented in the organisation	11 codes from 7 cases
h. Project Management	The management of activities and resources, such as time and cost throughout all phases of the process mining project to obtain the defined project outcomes	9 codes from 8 cases
i. Training	The education and sensitisation of stakeholders on the appropriate execution of process mining initiatives for the intended results	18 codes from 12 cases

a. Stakeholder Support and Involvement

Deep involvement of key stakeholders early on and throughout a process mining project was an important success factor. Such involvement ensured awareness of stakeholder roles and responsibilities in a process mining initiative. It was also instrumental for "coordinative effort and diverse interaction with respective participants" (Case 8) and addressing challenges encountered during process mining. Four stakeholder groups' involvement were identified. First, **management support** – for some organisations, a top-management driven approach to process mining and the right level of management attention proved strategic in systematising processes within the organisation. Also, establishing roles such as Chief Process Officer could focus more on achieving process excellence goals (e.g., Case 9). Close collaboration with **external stakeholders** such as suppliers and other industry partners could facilitate the transfer of process knowledge from one organisation to the other and influence the ability to execute an e2e process mining approach. Contributions from **subject matter experts (SMEs)** such as process owners, process stakeholders, and managers were instrumental for process mining success. Their expertise provides crucial insights (such as business knowledge, deep contextual understanding of the process being mined, and communicating process changes and guidelines) to other stakeholders, which influences the value of process mining outcomes. Furthermore, **user groups** were instrumental as they provided feedback for verifying process mining results and suggesting process improvements. First-line users were also helpful to "uncover additional factors that influence the process, which are often not visible in the data" (Case 13) and identify the exact trouble spots within a process.

b. Information Availability

The availability of information resources such as event data from business systems, detailed workflows, benchmarking and KPI information, and privacy regulations were considered essential. **Event data availability** for process mining was seen to be of utmost importance. While some organisations had teams to ensure that data was "properly prepared and available in the right place at the right time" (Case 25), the availability of such data in other organisations was a hurdle to overcome because there were "constraints in obtaining "accurate" data since this capability was limited to specialised analytics teams" (Case 2). The **availability of contextual information** such as process models documentation, benchmarking information, and other regulatory and compliance requirements enables a clear understanding of the process and ensures that process mining is appropriately done, understanding contextual factors (e.g., Case 11).

c. Technical Expertise

Technical expertise was crucial for executing process mining projects effectively. Whether such expertise was provided in-house (e.g., CoE) or externally (e.g., consultants), these experts were solely responsible for extracting, preparing, analysing, and interpreting analysis results to provide process insights to relevant stakeholders. Four categories of technical expertise were identified. First, **process mining expertise** –the competence of applying process mining tools was an important skill set. Such expertise was essential for applying analytics techniques to extract data insights. When such skills were lacking, some organisations were forced to limit the use of process mining to specific areas (e.g., Case 11). For **data extraction expertise,** initial data engineering expertise facilitated a successful process mining implementation. To maintain daily use of process mining, some organisations sought to "build data engineering and data science expertise from the early beginning during the implementation. This helped us learn during the project phase and to implement new use cases in short time frames without external support later." (Case 12). Also, data scientists knowledgeable in the data source structure and capable of setting up the required project schema for event data extraction enhanced the quality of process mining insights. **Process analyst expertise -** extensive knowledge in traditional process modelling techniques was critical, as it provided the competence to build new process models in-house (e.g., Case 12). Organisations usually relied on a team-based approach when combining expertise for process mining. A sound **team configuration** was crucial and came in two forms: (i) *Established units* – multidisciplinary teams of experts such as business process managers, process analysts and data experts dedicated to undertaking process mining projects. They enable process owners to refine their operations to minimise process variations and implement future process changes. These teams were referred to as; the Centre of Excellence (CoE), Process Excellence Centre (PEC), Process Mining Insights (PMI), Business Process Leadership (BPL) or Process Mining Consulting (PMC) team (e.g., Cases 2, 5 and 12). (ii) *Ad-hoc units* are a temporary group of experts assembled from other departments who possess the needed knowledge and expertise to execute process mining initiatives (e.g., Case 31).

d. Structured Process Mining Approach

As the case reports captured diverse approaches to executing PM initiatives, we adopted the PM^2 framework [26] as a unifying and guiding framework to further analyse meta-themes under (**d**): Structured Process Mining Approach.

Most organisations followed some approach or plan for executing process mining projects. A PM project usually begins with **planning** i.e., specifying a goal or an objective, extracting and analysing event data, interpreting insights, and implementing process improvements. Most process mining projects within organisations are motivated either by a process-related goal, problem or an opportunity that needs attention. Planning also involves considerations about executing process mining projects in tandem with organisational objectives (e.g., Cases 7 and 12). A total of 30 cases reported having engaged in some form of planning. **Extraction** involved taking specific actions with regards to identifying data sources and the mode of extraction. 28 cases reported having engaged in some form of data extraction. Steps taken regarding **data processing** indicated that the nature of the process to be mined influenced the form of logs generated for process mining. 15 cases reported activities related to data processing. **Mining and analysis** usually began with the automated discovery of as-is process models to exploring bottlenecks and process inefficiencies. 56 cases reported having engaged in some form of mining and analysis. **Evaluation** focused on comparing analysis results to improvement ideas to achieve project goals. Six cases reported some form of evaluation. **Process improvement and support** were the actions taken to adjust business processes based on newly gained insights. 26 cases reported some form of process improvement and support such as modifying existing KPIs and changing how processes are optimised (e.g., Case 12).

e. Data and Event Log Quality

Organisations acknowledged the significance of data and event log quality as prerequisites for PM success. Deliberate steps were taken to ensure that event data was of reliable quality. During **data pre-processing,** organisations where data was "structured in a way that closely resembled an event log" minimised "the effort needed to consolidate data" (Case 5). Others with complex data models needed to rely heavily on cross-team collaborations and different technologies to successfully extract event logs for process mining (e.g., Case 5). Organisations also learnt the relationship between data accessibility and valuable insights. Limited data access impaired understanding of the complete flow of activities. For **event log quality considerations**, organisations confirmed high-quality event logs a pre-requisite for obtaining valuable insights into processes. However, there were significant data quality challenges in "pre-processing the data from multiple systems to create high-quality logs" (Case 57). Quality assessment revealed data quality issues such as missing, irrelevant, and misplaced events, granularity and correlation issues, events representing case attributes and diverse activities with the same timestamp (e.g., Cases 3, 15 and 58).

f. Tool Capabilities

Organisations identified key features and capabilities of process mining tools essential for executing process mining projects. Users were keen about the extent to which **integration capabilities** of process mining tools could support existing IT landscapes and other technology such as AI or machine learning techniques (e.g., Case 8). The ability to provide automated **process discovery** and visualisation or process models was a popular feature (in 20 cases), **process benchmarking** (in six cases) and **conformance checking** (in 15 cases) were also highlighted as key capabilities for process mining success. Case 12 also confirmed that "having a realistic view and expectation management of what the tool is capable of" was essential. With **analytical scalability**, organisations were able to analyse a single process or e2e process at various levels of detail and even at high data frequencies (e.g., Case 35).

g. Change Management
Having a well-defined and highly efficient change management approach was critical to accommodate the high rate of continuous change that process mining brings. A change management system was essential for dealing with change across multiple departments, especially in e2e processes (e.g., Case 9). Some organisations (e.g., Case 26) confirmed that the presence of a dedicated individual or team of experts (e.g., a CoE) to lead change management initiatives proved beneficial as they had extensive know-how about digital solutions and organisational processes and could convince end users of the value of process mining.

h. Project Management
Organisations considered the scope, time, and infrastructure resources to support process mining. Organisations that properly managed the implementation of required infrastructural support for PM within reasonable timelines found it essential for its success (e.g., Cases 2, 7 and 10). As the process mining scope widened to an enterprise or global scale, organisations faced further complexities with deployment (e.g., Case 7). It was also discovered that to deploy process mining on a global scale, having a clear governance structure was crucial as it supported the goals, direction, and objectives of the organisation.

i. Training
The case reports indicated that to fully enjoy the benefit of process mining analytical capabilities, end users needed to be trained on how to use the tool. Training occurred either internally (e.g., Case 5) or by external consultants (e.g., Cases 22 and 37). Aside from creating awareness of the usefulness and power of process mining, these educational sessions aimed "to fully engage the true end users and immerse them in the world of Process Mining" (Case 5). They also provided the needed upskilling for technical staff on using PM tools.

5 Discussions

5.1 An Enhanced PM CFS Model

From the re-specified PM success factors in Sect. 4, we present an enhanced PM CSF model. This differentiates our work from existing PM CSF studies (e.g., [19]) in that, not only do we present a more comprehensive set of CSFs from a broad and contemporary case base, but we also identify inherent relationships between the factors to better understand which CSFs to prioritise. Table 3 describes the types of factor relationships identified from the case evidence in Sect. 3.

Table 3. Types of factor relationships identified

Type of relationship	Description
Direct relationship	Capture how one factor can influence another (implying a causal relationship between one CSF and another)
Indirect relationship	Relationships whose outcomes are influenced by either moderating or mediating variables. A moderating variable alters the direction or strength of the relationship between a predictor and an outcome, i.e.; it addresses the "when" or "for whom" a variable most strongly predicts or causes an outcome. A mediating variable is the mechanism through which a predictor influences an outcome, i.e., it establishes the "how" or "why" one variable predicts or causes an outcome variable [27]
Bilateral relationship	A two-way relationship between two CSFs, indicating that they can concurrently influence each other reciprocally

Figure 1 summarises the results, representing a new process mining critical success factors model. Part C of the Supplementary Material provides supporting evidence for each relationship. Overall, 14 relationships were identified, each outlined below.

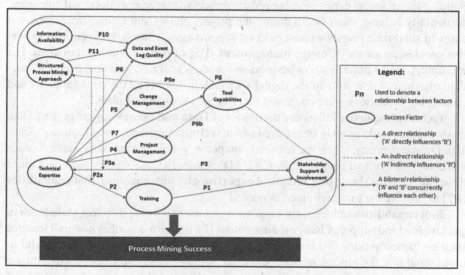

Fig. 1. An enhanced PM CFS model with factor relationships based on case data

Multiple cases indicated how **Training** contributes to **Stakeholder support and involvement [P1]**. Diverse forms of training such as customised classroom training, on-the-job training, online webcasts for specific topics, and open sessions, were conducted across several cases (e.g., Cases 5, 6 and 12). However, all case observations on **P1** related to the training of end users. If and how training may influence other stakeholder groups such as managers, external parties or SMEs was not evidenced from the data.

Furthermore, the case data depicted how **Technical expertise** contributed to **Training [P2]**. Case 2 describes how a small team of data scientists [**Technical Expertise**] ensured that "the end users were trained with the required skills to process mine." These same data scientists also served as "points of contact for other business departments to help them scale out the capability locally" (Case 2). At times (e.g., Case 22), this training was extended to different staff groups (beyond the end users) and embedded with the steps of the **Structured Process Mining Approach [P2a]**. Thus, **Structured Process Mining Approach** could moderate the relationship between **Technical expertise** and **Training**.

A bilateral relationship was observed between **Technical expertise** and **Stakeholder support and involvement**, in particular with the SMEs [**P3**], where the cases vividly explained how the SMEs and the technical teams were "working in parallel" (e.g., Case 26) and how at times the SMEs "set the directions of analysis and conduct" for process mining (e.g., Case 28). P3 was facilitated in a moderating manner by the **Structured process mining approach [P3a]**, where the SMEs were contacted to verify the data and results from process discovery (e.g., Case 50) or while analysing logs together (e.g., Case 41).

Technical expertise enables the overall **Project management [P4]** of process mining initiatives. For example, Case 2 attributes its success to the "huge efficiency gains" obtained from a small, well configured technical team. And Case 12 describes how building the technical expertise enhanced the project management efforts and outcomes, particularly helping them "learn during the project phases and to implement new use cases in short time frames without external support later". Similarly, **Technical expertise** enables the overall **Change management [P5]** of process mining initiatives. For example, Case 14 describes how business and process analysts' technical expertise, particularly their "know-how of the digital solutions in use" (i.e., the configurations and underlying processes), were integrated into change management plans.

Technical expertise influenced the impact of **Data and event log quality [P6]**. Data quality issues could quickly be resolved when **Technical expertise** was high (e.g., Case 10). On the contrary, when the required competency was lacking, data quality issues were very difficult to resolve (e.g., Case 11). Similarly, case studies such as Case 2 illustrate the critical role that **Technical expertise** plays to maximise **Tool capabilities [P7]** to "ensure the tool performed as needed".

Tool capabilities influence (in a mediating manner) how the different **stakeholders get involved** and support **Change management [P8]**. Case 6 describes how tool features such as "benchmarking inside the same process using different context" are "useful to understand how the process owner could make some process improvements to increase the entire process performance", thus increasing the overall interest and acceptance of process mining. Case 10 explains how the tool's seamless integration into the analytics platform helped the users to accept process mining solutions quickly. **Tool capabilities** also played a mediating role between the **Structured process mining approach [P9a]** and the overall **Project management [P9b]**. For example, Case 51 describes how **Tool capabilities** such as automatic process discovery allowed the team (i.e., Project Management) to efficiently execute core steps within the planned process mining approach and "identify improvement opportunities, prioritise them, and achieve benefits."

Information availability can enhance or inhibit **Data and event log quality [P10]**. For example, Case 25 articulates the need for the master data to be "properly prepared and available in the right place at the right time" to enable process mining. Case 43 describes the crippling effect that poor IT controls has on accessing essential data for effective process mining. Case 2 elaborates on similar challenges when access to "accurate data" is limited to only specialised teams.

Structured process mining approach influences **Data and event log quality** considerations **[P11]**. It was seen that the data architecture, data extraction techniques and software tools used by the organisation for process mining influenced future considerations for data and event log quality (e.g., Cases 3 and 4). Once these stages were well-outlined, future efforts were easier. Case 13 states: "given that the data preparation steps are now in place and can be easily repeated on new data, we can now continue to analyse and quantify this process to continuously improve it". Organisations also learnt that 80% of all process mining efforts were focused on data extraction, data preparation and dealing with data quality issues (e.g., Case 15).

Overall, the analysis depicted how some factors are at the core, influencing several other factors. For example, **Technical expertise** influences six other factors, namely **Stakeholder support and involvement, Training, Data and event log quality, Change management, Tool capabilities** and **Project management**. Williams and Ramaprasad [15] describe the value in recognising these direct/indirect relationships, for they assist in determining the order in which the success factors need to be addressed. For example, the findings show that investing in technical expertise will influence many other factors and, hence, be better than investing in other factors such as Project or Change Management efforts.

5.2 Limitations

Our study relies on insights from 62 published case reports to derive our enhanced PM CSF model. Though the 62 cases cover a wide range of PM contexts, we acknowledge that our findings are limited to the information documented in these case reports and are bounded by the scope, bias, and limitations of these reports. Our findings are also exposed to other limitations of secondary data analysis and qualitative research in general, such as possible selection and researcher bias in the case/code selection and overall analysis.

5.3 Contributions

Our theoretical contributions are three-fold. (i) Using state of the art evidence from a wide range of PM success stories, we provide a more comprehensive and in-depth understanding of success factors that extends prior PM CSF research. (ii) We also identify CSF interrelationships and explain which factors have direct, indirect or bilateral influences on attaining process mining success. Finally, (iii) this work provides a sound basis for future research (see Sect. 5.4).

Our PM CSF model is of practical value. The comprehensive details of PM CSF derived from practice enables PM stakeholders to focus on the essential antecedents for PM success and plan accordingly. The summary analysis presented (e.g., the frequency of case data supporting each factor as outlined in Table 2) indicates the different degrees of importance of the factors. The identified interrelationships enable PM project managers and sponsors to determine which CSFs to prioritise when addressing CSFs. Knowledge about which factors have direct, moderating, mediating or bilateral influences on PM project success will also be key when planning PM CSF investments.

5.4 Future Research

Our findings are initial outcomes of an ongoing PhD research. The derived outcomes presented herein are bound by the scope of the analysed case report data. Future work will validate our success factors model using primary data collected from in-depth case studies, specifically to confirm the factor configuration and validate the factor interrelationships. These in-depth studies would also be used to further explore how these factors may vary in importance during PM projects and to identify mechanisms for actualising these CSFs across diverse contexts. Our model could also be extended to integrate success measures and provide deeper insights into a complete nomological net explaining how CSFs create impact in a process mining context. These proposed in-depth case studies can be followed by a quantitative survey (with data from global PM initiatives) to statistically test the success factors and proposed relationships to ascertain their degree of influence on the success of PM initiatives. It is also recommended that, where feasible, an investigation into the extent to which the identified CSFs contributed to failed process mining projects be considered. This would guarantee the presence of these factors as "sufficient conditions" for achieving successful process mining initiatives and provide deeper insights on what constitutes and may influence PM project failure.

6 Conclusion

This study explored critical success factors within the process mining domain. Existing process mining CSF studies (e.g., [19]) do not explore factor interrelationships which is a major criticism in CSF literature [14]. Following a hybrid qualitative analysis approach, our work extends the Mans, Reijers, Berends, Bandara and Prince [19] model by qualitatively analysing evidence from 62 recent case reports from diverse industry settings. Our model presents nine PM Critical Success Factors. In addition to the six CSF from Mans, Reijers, Berends, Bandara and Prince [20], which formed our a-priori model, we identified three new factors: **Change management**, **Tool capabilities** and **Training**. Our analysis confirms that three of the six success factors from Mans, Reijers, Berends, Bandara and Prince [19] still hold true, namely **Structured process mining approach**, **Data and event log quality** and **Project management**. However, we re-specified the scope of the other three: Management support, Resource availability and Process miner expertise, which we now term **Stakeholder support and involvement**, **Information availability** and **Technical expertise**. We presented clear descriptions for each factor, identified sub-factors where necessary and explained how they pertain to the current process mining context. We explore factor interrelationships where we found **nine** direct, **five** indirect (**two** moderating, **three** mediating) and **one** bilateral relationship between the CSFs.

Appendix: Supplementary Material

Supplementary material for this article is available online at https://bit.ly/3qrtrOE. It contains three parts: Part A provides an overview of 62 published case reports, Part B provides example quotes that support success factor explanations, and Part C presents case evidence supporting the identified CSF relationships.

References

1. van der Aalst, W.M.P.: Process Mining: Data Science in Action. Springer, Berlin Heidelberg, Berlin, Heidelberg (2016)
2. Grisold, T., Mendling, J., Otto, M., vom Brocke, J.: Adoption, use and management of process mining in practice. Bus. Process. Manag. J. **27**, 369–387 (2020)
3. Reinkemeyer, L.: Process Mining in Action. Springer, Switzerland (2020)
4. Syed, R., Leemans, S.J.J., Eden, R., Buijs, J.A.C.M.: Process mining adoption. In: Fahland, D., Ghidini, C., Becker, J., Dumas, M. (eds.) BPM 2020. LNBIP, vol. 392, pp. 229–245. Springer, Cham (2020). https://doi.org/10.1007/978-3-030-58638-6_14
5. Jans, M., Alles, M.G., Vasarhelyi, M.A.: A field study on the use of process mining of event logs as an analytical procedure in auditing. Account. Rev. **89**, 1751–1773 (2014)
6. Wynn, M.T., et al.: Grounding process data analytics in domain knowledge: a mixed-method approach to identifying best practice. In: Hildebrandt, T., van Dongen, B.F., Röglinger, M., Mendling, J. (eds.) BPM 2019. LNBIP, vol. 360, pp. 163–179. Springer, Cham (2019). https://doi.org/10.1007/978-3-030-26643-1_10
7. Rojas, E., Munoz-Gama, J., Sepúlveda, M., Capurro, D.: Process mining in healthcare: a literature review. J. Biomed. Inform. **61**, 224–236 (2016)
8. Emamjome, F., Andrews, R., ter Hofstede, A.H.M.: A case study lens on process mining in practice. In: Panetto, H., Debruyne, C., Hepp, Martin, Lewis, D., Ardagna, C.A., Meersman, R. (eds.) OTM 2019. LNCS, vol. 11877, pp. 127–145. Springer, Cham (2019). https://doi.org/10.1007/978-3-030-33246-4_8
9. Zhang, W., Xu, X.: Six Sigma and information systems project management: a revised theoretical model. Proj. Manag. J. **39**, 59–74 (2008)
10. vom Brocke, J., Jans, M., Mendling, J., Reijers, H.A.: Call for papers, issue 5/2021. Bus. Inf. Syst. Eng. **62**(2), 185–187 (2020). https://doi.org/10.1007/s12599-020-00630-7
11. vom Brocke, J., Jans, M., Mendling, J., Reijers, H.A.: A five-level framework for research on process mining. Bus. Inf. Syst. Eng. **63**(5), 483–490 (2021). https://doi.org/10.1007/s12599-021-00718-8
12. Rockart, J.F.: Chief executives define their own data needs. Harv. Bus. Rev. **57**, 81–93 (1979)
13. Bandara, W., Gable, G.G., Tate, M., Rosemann, M.: A validated business process modelling success factors model. Bus. Process. Manag. J. **27**(5), 1522–1544 (2021)
14. Fortune, J., White, D.: Framing of project critical success factors by a systems model. Int. J. Project Manage. **24**, 53–65 (2006)
15. Williams, J., Ramaprasad, A.: A taxonomy of critical success factors. Eur. J. Inf. Syst. **5**, 250–260 (1996)
16. Alibabaei, A., Bandara, W., Aghdasi, M.: Means of achieving business process management success factors. In: 4th Mediterranean Conference on Information Systems. (2009)
17. Sim, J.: Critical success factors in data mining projects. Business Computer Information Systems. University of North Texas, ProQuest Dissertations & Theses Global (2003)
18. Grover, V., Chiang, R.H., Liang, T.-P., Zhang, D.: Creating strategic business value from big data analytics: a research framework. J. Manag. Inf. Syst. **35**, 388–423 (2018)

19. Mans, R., Reijers, H., Berends, H., Bandara, W., Prince, R.: Business process mining success. In: European Conference on Information Systems (2013)
20. Martin, N., et al.: Opportunities and challenges for process mining in organisations: results of a delphi study. Bus. Inf. Syst. Eng. **63**, 511–527 (2021)
21. vom Brocke, J., Mendling, J.: Business process management cases: Digital innovation and business transformation in practice. Springer Nature (2018)
22. vom Brocke, J., Mendling, J. (eds.): Business process management cases. MP, Springer, Cham (2018). https://doi.org/10.1007/978-3-319-58307-5
23. Saldaña, J.: The Coding Manual for Qualitative Researchers. Sage Publications (2013)
24. Swain, J.: A Hybrid Approach to Thematic Analysis in Qualitative Research: Using a Practical Example. SAGE Publications Ltd (2018)
25. DeCuir-Gunby, J.T., Marshall, P.L., McCulloch, A.W.: Developing and using a codebook for the analysis of interview data: an example from a professional development research project. Field Methods **23**, 136–155 (2011)
26. van Eck, M.L., Lu, X., Leemans, S.J., van der Aalst, W.M.: PM^2: A process mining project methodology. In: Zdravkovic, J., Kirikova, M., Johannesson, P. (eds.) Advanced Information Systems Engineering. CAiSE 2015, LNISA, vol. 9097, pp. 297–313. Springer, Cham (2015). https://doi.org/10.1007/978-3-319-19069-3_19
27. Frazier, P.A., Tix, A.P., Barron, K.E.: Testing moderator and mediator effects in counseling psychology research. J. Couns. Psychol. **51**, 115 (2004)

Process Mining

No Time to Dice: Learning Execution Contexts from Event Logs for Resource-Oriented Process Mining

Jing Yang[1]([✉])(iD), Chun Ouyang[1](iD), Arthur H. M. ter Hofstede[1](iD), and Wil M. P. van der Aalst[1,2](iD)

[1] Queensland University of Technology, Brisbane, Australia
roy.j.yang@qut.edu.au
[2] RWTH Aachen University, Aachen, Germany

Abstract. Process mining enables extracting insights into human resources working in business processes and supports employee management and process improvement. Often, resources from the same organizational group exhibit similar characteristics in process execution, e.g., executing the same set of process activities or participating in the same types of cases. This is a natural consequence of division of labor in organizations. These characteristics can be organized along various process dimensions, e.g., case, activity, and time, which ideally are all considered in the application of resource-oriented process mining, especially analytics of resource groups and their behavior. In this paper, we use the concept of execution context to classify cases, activities, and times to enable a precise characterization of resource groups. We propose an approach to automatically learning execution contexts from process execution data recorded in event logs, incorporating domain knowledge and discriminative information embedded in data. Evaluation using real-life event log data demonstrates the usefulness of our approach.

Keywords: Execution context · Resource group · Event log · Process mining · Workforce analytics

1 Introduction

The success of an organization depends on how well its workforce is mobilized and managed [4]. Modern organizations deploy their employees in business processes [5] to deliver on products and services. In order to streamline business processes and to improve employee satisfaction of work [5], process managers need to be able to analyze and understand employee behavior and performance in process execution [8].

Process mining can discover employee-related insights from event log data [14] to support workforce analytics in the context of business processes. Often, employees (human resources) from the same organizational group (such as role, team, department, etc.) share similar characteristics in process execution [12],

© Springer Nature Switzerland AG 2022
C. Di Ciccio et al. (Eds.): BPM 2022, LNCS 13420, pp. 163–180, 2022.
https://doi.org/10.1007/978-3-031-16103-2_13

e.g., resources of the same role are in charge of a subset of process activities. These characteristics may transcend the clustering of activities and manifest in other dimensions including case and time, e.g., resource group taking part in a specific type of cases, working on time shifts, or being in different locations. The connection between resource groupings and groups' characteristics in process execution is a natural consequence of the *specialization* of work, i.e., division of labor in an organization [4]. Some existing work on resource-oriented process mining considers such characteristics [8,9,12] but treats different dimensions separately. Only few [15,17] has considered the characterization of resource groups across various process dimensions holistically.

Execution context [17] is a concept proposed to enable a precise characterization of resource groups, considering various process dimensions. A set of execution contexts is defined by classifying and combining *case types*, *activity types*, and *time types*. Given an event log, applying execution contexts creates a multidimensional view on event data, where subsets of the log can be linked to the signature behavior of resource groups for analyses [16,17]. Figure 1 illustrates the idea. In an insurance company, there are investigators specialized in different types of insurance claims, such as vehicles, business, and health insurance. While these investigators may perform the same field investigation activities in a claim handling process, they are likely to participate in only the type of cases where their expertise fits. If we use merely the activity dimension to characterize resources (Fig. 1a), we will not be able to discover or analyze investigators with specialty that manifests on other process dimensions. But, when we apply execution contexts to view events from multiple dimensions (Fig. 1b), more precise characterization and more dedicated analyses on investigators can be performed.

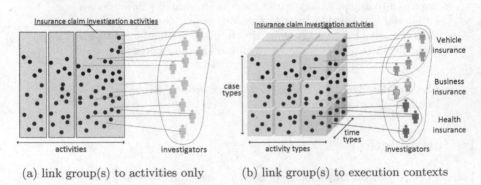

(a) link group(s) to activities only   (b) link group(s) to execution contexts

Fig. 1. Subsets of events (dots) in a log can be linked to groups of resources as a natural consequence of the specialization of work

Our previous work [17] has shown how to manually define execution contexts using prior knowledge about an event log. However, it assumes the availability of relevant domain knowledge. In this paper, we propose an approach that supports *learning* execution contexts from an event log, by exploiting the discriminative information of events embedded in the data rather than relying on domain knowledge. Our approach is built on a customized decision learning algorithm and is capable of deriving categorization rules that can be used to dice event log data to obtain execution contexts. We demonstrate the usefulness of our approach through experiments conducted on real-life event logs.

Our research contributes a solution to the problem of learning execution contexts, thus enhances resource-oriented process mining techniques that focus on analyzing human resources [8] and their groups [12,16,17]. Our work also contributes a method to derive process cube views in multidimensional process mining research [2,13]. Process mining techniques can then be applied to selected sublogs to analyze process variants concerned with certain groups of resources.

2   Related Work

Process mining can be applied on event logs to extract insights about human resources participating in process execution. For many resource-oriented process mining topics, an essential step is to identify the behavioral characteristics specific to resources or groups of resources in process execution. Then, analytics can be conducted on resources for different purposes, e.g., mining resource profiles [8] and mining organizational models [1,12,17]. To achieve that characterization, it is common to consider many process dimensions, e.g., activity [1,8,12], case [8,12], time [8], and location [9].

However, for organizational model mining, much of the literature has not yet employed a holistic view on those various dimensions. Instead, existing work exploits each dimension separately by modeling how resources perform different activities (e.g., [1]), how they hand over between activities (e.g., [3]), or how they participate in the same cases (e.g., [12]). This poses a challenge of mining complex resource groupings where resource characteristics are concerned with multiple dimensions, e.g., employees with the same business role but working different shifts. To address the challenge, we need an approach that jointly exploits multidimensional event log information.

Some recent research on organizational model mining [15,17] has contributed to addressing that literature gap. Van Hulzen et al. [15] propose the notion of "activity instance archetype" to capture contextual factors impacting how activity instances were executed. Activity instance archetypes can be discovered by applying model-based clustering on events enriched with selected attributes. Then, resources are characterized by their execution of activity instance archetypes, and resource groupings concerned with contextual factors can be discovered. In our previous work [17], we propose the notion of "execution context". Execution contexts are built upon categorizing cases, process activities, and time periods. Different from activity instance archetypes [15] (which focus on the homogeneity of activity instances), execution contexts aim at finding a structured, multidimensional way to organize event log data.

Execution contexts can be derived from an event log based on domain knowledge about existing categorization. Resources are characterized by their participation in execution contexts. Then, resource groupings can be discovered and analyzed using event logs, and their conformance can be checked with respect to the logs. However, it remains a problem how such execution contexts can be derived from event log data without relying on available domain knowledge about a process and the event data.

In this paper, we extend our previous work [17] and explore how to automatically learn execution contexts from a given event log. Outside resource-oriented process mining, our work is relevant to the research on multidimensional process mining. A multidimensional data model named event cube was proposed [10] to allow exploiting and integrating different aspects of business process and providing various levels of abstraction on process data to improve business analysis. The notion of process cube [2,13] was later proposed, which provides a more comprehensive view to organize both process models and their data using different dimensions. Process cubes support OLAP (Online Analytical Processing)-like operations dedicated to process mining, therefore enable decomposing large event logs into smaller sublogs to enhance process mining performance and scalability. Our idea of execution contexts resembles that of process cubes regarding the consideration of multidimensionality. As such, our approach to learning execution contexts can be seen as a way to constructing process cube views dedicated to resource-oriented process mining.

3   Preliminaries

A business process consists of tasks conducted in an organization to achieve a business goal [5]. An instance of executing a process is a case [14]. An *event log* (Definition 1) is a set of timestamped events recording how (human) resources performed those tasks (i.e., process activities) within different cases.

Definition 1 (Event Log). *Let $\mathcal{E}$ be the universe of event identifiers, $\mathcal{U}_{Att}$ be the universe of possible attribute names, and $\mathcal{U}_{Val}$ be the universe of possible attribute values. $EL = (E, Att, \pi)$ with $E \subseteq \mathcal{E}$, $E \neq \varnothing$, $Att \subseteq \mathcal{U}_{Att}$, and $\pi : E \to (Att \not\to \mathcal{U}_{Val})$ is an event log. Event $e \in E$ has attributes $dom(\pi(e))$. For an attribute $x \in dom(\pi(e))$, $\pi_x(e) = \pi(e)(x)$ is the attribute value of x for event e.*

Events carry multiple data attributes, i.e., *event attributes* (Definition 2), which can be either categorical or numeric, depending on features of the process and the recording information systems. In this paper, we consider that any event log records at least four standard event attributes: case identifier (*case*), activity label (*act*), timestamp (*time*), and resource identifier (*resource*). Each case identifier corresponds to a unique process execution instance. Specifically, an event attribute is a case attribute if events belonged to the same case share an identical value on that attribute. Case identifier is a case attribute.

Definition 2 (Event Attributes). *Let $\mathcal{C} \subseteq \mathcal{U}_{Val}$, $\mathcal{A} \subseteq \mathcal{U}_{Val}$, $\mathcal{T} \subseteq \mathcal{U}_{Val}$ and $\mathcal{R} \subseteq \mathcal{U}_{Val}$ denote the universe of case identifiers, the universe of activity names, the universe of timestamps, and the universe of resource identifiers, respectively. Any event log $EL = (E, Att, \pi)$ has three special attributes from the set $D = \{case, act, time\}$, referred to as the core event attributes, and a special attribute res, i.e., $D \cup \{res\} \subseteq Att$, such that for any $e \in E$:*

- $D \subseteq dom(\pi(e))$,
- $\pi_{case}(e) \in \mathcal{C}$ is the case to which e belongs,
- $\pi_{act}(e) \in \mathcal{A}$ is the activity e refers to,
- $\pi_{time}(e) \in \mathcal{T}$ is the time at which e occurred, and
- $\pi_{res}(e) \in \mathcal{R}$ is the resource that executed e if $res \in dom(\pi(e))$.

Given a resource $r \in \mathcal{R}$, let $[E]_r = \{\, e \in E \mid res \in dom(\pi(e)) \wedge \pi_{res}(e) = r \,\}$ denote the set of events in the log execution by that resource. $[E]_{\mathcal{R}} = \bigcup_{r \in \mathcal{R}} [E]_r$ is the set of all events in the log that have resource information.

Types describe the categorization of events. We consider *case types*, *activity types*, and *time types* (Definition 3) related to the three core dimensions of process execution. Case types describe the categories of cases, for example, insurance claims can be classified by the type of insurance (e.g., health insurance vs. car insurance). Similarly, activity types categorize activity labels into groups of relevant activities (e.g., claim investigation vs. customer support), and time types categorize timestamps into periods (e.g., weekdays vs. weekends).

Definition 3 (Case Types, Activity Types, and Time Types). *Let $\mathcal{CT}$, $\mathcal{AT}$, and $\mathcal{TT}$ denote the sets of names of case types, activity types, and time types, respectively. The functions $\varphi_{case}\colon \mathcal{CT} \to \mathcal{P}(\mathcal{C})$, $\varphi_{act}\colon \mathcal{AT} \to \mathcal{P}(\mathcal{A})$, and $\varphi_{time}\colon \mathcal{TT} \to \mathcal{P}(\mathcal{T})$ define partitions over $\mathcal{C}$, $\mathcal{A}$, and $\mathcal{T}$, respectively.*

Given an event log $EL = (E, Att, \pi)$, a set of types for one of the process execution dimensions (i.e., case, activity, and time) can be derived based on a set of event attributes, if those attributes are a set of *type-defining attributes* for that dimension (Definition 4). This notion supports deriving types via partitioning the values of some selected event attributes that are related to the categorization of cases, activities, or times. For example, in an insurance claim process event log, we may use case attributes "customer type" and "insurance type" as the type-defining attributes for case types—cases can be categorized into disjoint groups such as ("gold customer", "health insurance") vs. ("silver customer", "car insurance", "boat insurance").

Definition 4 (Type-Defining Attributes). *Let $d \in D$ be a core event attribute. Given an event log $EL = (E, Att, \pi)$, for any $e \in E$, let $X \subseteq dom(\pi(e))$ be some event attributes recorded in the log, $\pi(e) \restriction_X$ the restriction of $\pi(e)$ on X, $V = \{\, \pi(e) \restriction_X \mid e \in E \,\}$ the mappings of the attributes in X recorded in EL.*

X is a set of type-defining attributes for d in EL, if there exists a partition P of V, such that for all $p, q \in P$,

$$p \neq q \Rightarrow \{\, \pi_d(e) \mid e \in E \wedge \pi(e) \restriction_X \in p \,\} \cap \{\, \pi_d(e) \mid e \in E \wedge \pi(e) \restriction_X \in q \,\} = \varnothing,$$

i.e., the partition P corresponds to a partition of the set of distinct values of d recorded in EL.

4 Problem Modeling

In this section, we introduce how we model the problem of learning execution contexts from an event log. We first present the idea of *categorization rules* for defining the classification of case types, activity types, and time types (Sect. 4.1). Then, we discuss how to measure the *quality* of execution contexts with regard to an event log (Sect. 4.2). Finally, we formulate the execution context learning problem based on the notion of categorization rules and the quality measures.

4.1   Categorization Rules

A set of execution contexts specifies a way of partitioning events by defining case types, activity types, and time types. Hence, to learn execution contexts from an event log requires learning those types, i.e., the classification of cases, activities, and times. To this end, we propose to use *categorization rules* to represent types and execution contexts.

A categorization rule is a conjunctive boolean formula (Definition 5) consisting of one or more clauses. Each clause can evaluate an event by its value of some event attribute. For instance, $\sigma = $ customer_type $\in \{$gold$\} \wedge$ amount $\in [10,000, \infty)$ is a categorization rule evaluating a categorical (case) attribute "customer type" and a numeric attribute "amount". Given a set of events, evaluating this rule filters events that record gold customers and amount greater than 10,000.

Definition 5 (Categorization Rule). *Given an event log $EL = (E, Att, \pi)$, let $d \in D$ be a core event attribute, let $X \subseteq Att$ be a set of type-defining attributes for d. $\sigma = \bigwedge_{x \in X} x \in U_x$ is a categorization rule, where $e \in E$ and $U_x \in \mathcal{P}(\mathcal{U}_{Val})$ is a set of attribute values for $x \in X$. For any $e \in E$, σ can be evaluated as follows: $[\![\sigma]\!](e) = true$ if and only if $\pi_x(e) \in U_x$ for all $x \in X$.*

- $[E]_\sigma = \{\, e \in E \mid [\![\sigma]\!](e) \,\}$ *is the set of events in the log satisfying the categorization rule σ.*
- *We introduce a default rule σ_{true} such that $[\![\sigma_{\text{true}}]\!](e) = true$ for all $e \in E$. It follows that $[E]_{\sigma_{\text{true}}} = E$.*
- *Any two categorization rules σ_1 and σ_2 are equivalent, i.e., $\sigma_1 \cong \sigma_2$, if and only if $[E]_{\sigma_1} \doteq [E]_{\sigma_2}$ for any $E \subseteq \mathcal{E}$. Otherwise, we write $\sigma_1 \not\cong \sigma_2$.*

A set of categorization rules can be used to define a set of types on an event log (Definition 6). Consider the example of defining case types. Assume an event log of the insurance claim process has "customer type" as a case attribute. Then the set of rules $\Sigma_1 = \{$customer_type $\in \{$gold$\}$, customer_type $\in \{$silver$\}$, customer_type $\in \{$bronze$\}\}$ can define three case types for this event log, as long as a customer can only be either gold, silver, or bronze. But, for example, another set with two rules, $\Sigma_2 = \{$customer_type $\in \{$gold$\}$, customer_type $\in \{$silver, bronze$\}\}$, would also define case types.

Definition 6 (Define Types by Categorization Rules). *Given an event log $EL = (E, Att, \pi)$, let $d \in D$ be a core event attribute. Σ is a set of categorization rules that define a set of types on d, if and only if:*

1. *for any $\sigma_1, \sigma_2 \in \Sigma$, $\{\, \pi_d(e) \mid e \in [E]_{\sigma_1} \,\} \cap \{\, \pi_d(e) \mid e \in [E]_{\sigma_2} \,\} = \varnothing$; and*
2. *$\bigcup_{\sigma \in \Sigma} \{\, \pi_d(e) \mid e \in [E]_\sigma \,\} = \bigcup_{e \in E} \{\pi_d(e)\}$,*

i.e., the subsets of events satisfying categorization rules in Σ induce a partition of all values of d recorded in EL.

Execution contexts can be defined by three sets of categorization rules that define case types, activity types, and time types, respectively (Definition 7). Given an event log, a set of execution contexts enables (i) projecting the events as data points onto a three-dimensional data space and (ii) partitioning them into sub-logs that can be selected and linked with resources for analyses [16,17].

Definition 7 (Execution Context). *Given an event log $EL = (E, Att, \pi)$, let Σ_{case}, Σ_{act}, and Σ_{time} be three sets of categorization rules that define case types, activity types, and time types, respectively. $CO = \Sigma_{case} \times \Sigma_{act} \times \Sigma_{time}$ is a set of execution contexts defined by the three sets of categorization rules.*

CO specifies a way of partitioning EL. Given an execution context $co = (\sigma_c, \sigma_a, \sigma_t) \in CO$, $[E]_{co} = [E]_{\sigma_c} \cap [E]_{\sigma_a} \cap [E]_{\sigma_t}$ is the set of events in the log having that execution context.

4.2   Quality Measures for Execution Contexts

Given an event log, any categorization rules—as long as they fulfill the requirement (Definition 6)—can be proposed for defining types, resulting in many candidate sets of execution contexts. In this section, we discuss how to measure the quality of execution contexts learned from event logs.

Execution contexts can be applied to characterize resource behavior that concern certain process execution features determined by the specialization of work, a.k.a, division of labor [4]. On the one hand, when specialization is low in a process, resources tend to be interchangeable when performing in process execution, and events they originated are mostly similar. On the other hand, when specialization is high, resources are limited to undertaking specific kinds of tasks, as exhibited by the differences among their originated events. This idea motivates us to consider the following criteria for a set of *good* execution contexts: (i) events originated by the same resource should be partitioned into few execution contexts; and (ii) events in the same execution contexts should be originated by few resources.

Figure 2 illustrates the idea. When resources are considered generalized due to low specialization of work, a small number of execution contexts should be sufficient. When (a group of) resources are highly specialized, it is desired to have dedicated execution contexts for each of them to capture their specific characteristics. We define two quality measures, namely *dispersal* and *impurity*.

Dispersal. Dispersal measures the extent to which events originated by the same resource disperse across different execution contexts (Eq. 1), and yields values in $[0, 1]$. High quality execution contexts have low dispersal, i.e., characterizing

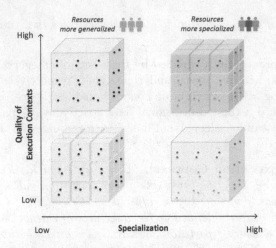

High

Resources more generalized

Resources more specialized

Quality of Execution Contexts

Low

Low   Specialization   High

Fig. 2. Execution context quality: It is desired to use few dedicated execution contexts (cells) to characterize resource behavior recorded in events

the behavior of each individual resource with few execution contexts. Given an event log EL and a set of execution contexts CO,

$$Dis(EL, CO) = \sum_{r \in \mathcal{R}} \left(\frac{|[E]_r|}{|[E]_{\mathcal{R}}|} \times \frac{\sum_{e_1,e_2 \in [E]_r} d_{CO}(e_1, e_2)}{\binom{|[E]_r|}{2}} \right), \tag{1}$$

is the dispersal of CO with regard to EL. In the context of the given execution contexts CO, any event $e \in E$ corresponds to a unique execution context $co^e = (ct^e, at^e, tt^e) \in CO$, for which $e \in [E]_{co^e}$. Then, for any two events $e_1, e_2 \in E$, we define the distance between them using their corresponding execution contexts $co^{e_1} = (ct^{e_1}, at^{e_1}, tt^{e_1})$ and $co^{e_2} = (ct^{e_2}, at^{e_2}, tt^{e_2})$, that is,

$$d_{CO}(e_1, e_2) = \frac{[ct^{e_1} \not\cong ct^{e_2}] + [at^{e_1} \not\cong at^{e_2}] + [tt^{e_1} \not\cong tt^{e_2}]}{ndim}, \tag{2}$$

where $[\varphi]$ is the Iverson bracket that returns 1 if a boolean formula φ holds and 0 otherwise, and $ndim \in \{1, 2, 3\}$ is the number of process dimensions considered in a set of execution contexts. By default, we let $ndim = 3$. However, it is possible that there are not any types defined on a dimension. For example, the case dimension can be omitted if, for any $(ct, at, tt) \in CO$, $ct = \sigma_{\text{true}}$, and thus we have $ndim = 2$. Specifically, if there is only one execution context for all events in a log, then $Dis(EL, CO) = 0$.

Impurity. Impurity measures the extent to which the same execution context contains events originated by different resources. A set of execution contexts is good when most of them contain only events originated by few resources, i.e., characterizing the behavior specific to each individual resource. This is built upon the existing measure of entropy in data mining. Given an event log EL and a set of execution contexts CO,

$$Imp(EL, CO) = \frac{1}{\sum_{r \in \mathcal{R}} p_r \log_2 p_r} \sum_{co \in CO} \left(\frac{|[E]_{\mathcal{R}} \cap [E]_{co}|}{|[E]_{\mathcal{R}}|} \times \sum_{r \in \mathcal{R}} p_{r,co} \log_2 p_{r,co} \right),$$
(3)

is the impurity of CO with regard to EL, where

$$p_r = \frac{|[E]_r|}{|[E]_{\mathcal{R}}|}, \; p_{r,co} = \frac{|[E]_r \cap [E]_{co}|}{|[E]_{\mathcal{R}} \cap [E]_{co}|}$$
(4)

are the relative frequency of events originated by a resource r in terms of the entire log and an execution context co, respectively. Impurity yields a value in $[0, 1]$. If there is only one execution context for all events in a log, then $Imp(EL, CO) = 1$.

Problem Statement. Learning execution contexts is to derive from an event log three sets of categorization rules that define case types, activity types, and time types, respectively, such that the resulting execution contexts have low dispersal and low impurity with respect to the input log.

5   Problem Solution

We propose an approach based on decision trees to solve the problem of deriving categorization rules (see Fig. 3). Below, we elaborate on the approach.

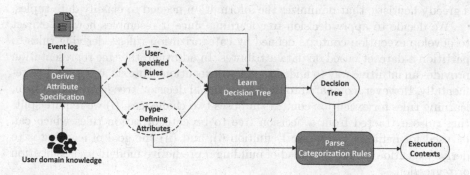

Fig. 3. Approach to deriving categorization rules from an event log to learn execution contexts

5.1   Derive Attribute Specification

Inputs to the approach include an event log and domain knowledge from the user. First, an *attribute specification* (Definition 8) is derived to capture user domain knowledge about the events attributes in the log. An attribute specification comprises (1) X_{case}^{EL}, X_{act}^{EL}, X_{time}^{EL}, which are three sets of type-defining attributes (see Definition 4) regarding case types, activity types, and time types; and optionally, (2) Λ, which is a set of user-specified categorization rules capturing any existing categorization of values for those event attributes. If user-specified categorization rules are unavailable for any event attribute x, we let $\Lambda(x) = \varnothing$.

Definition 8 (Attribute Specification). *Let $EL = (E, Att, \pi)$ be an event log and Σ be the set of all possible categorization rules defined on Att. $S = (X_{case}^{EL}, X_{act}^{EL}, X_{time}^{EL}, \Lambda)$ is an attribute specification on EL for learning categorization rules. $X_{case}^{EL} \subseteq Att$, $X_{act}^{EL} \subseteq Att$, and $X_{time}^{EL} \subseteq Att$ are three disjoint, non-empty sets of type-defining attributes. $\Lambda : Att \nrightarrow \mathcal{P}(\Sigma)$ defines a set of categorization rules for the event attributes in $X_{case}^{EL} \bigcup X_{act}^{EL} \bigcup X_{time}^{EL}$.*

An attribute specification informs how attributes values should be handled in the following step of decision tree learning, and subsequently how the categorization rules extracted from a decision tree specify case types, activity types, and time types.

5.2   Learn Decision Tree

We apply a decision tree induction framework to extract categorization rules from an event log. In decision tree learning, a dataset of multivariate data tuples is iteratively partitioned into smaller subsets by deriving splitting rules on data attributes. The output is represented in a tree structure, where tree nodes hold the subsets of the input data and branches record the disjunctive splitting rules used to obtain the subsets. Decision tree learning is a common solution for classification tasks. For that purpose, splitting rules are often derived following a greedy heuristic that minimizes the information needed to classify data tuples.

We decide to apply decision tree learning since it resembles how we expect to develop execution contexts defined by categorization rules—deriving rules to partition a dataset based on data attributes. In addition, the tree representation provides an intuitive way to understand how execution contexts are derived incrementally. However, compared to the conventional decision tree learning problem, learning rules for execution contexts imposes two challenges: (i) we require splitting rules extracted from a decision tree to be categorization rules which can be used for defining types (see Definition 6); and (ii) the goal of learning is to derive execution contexts instead of building a predictive model for classification or regression.

To address the first issue regarding categorization rules, we choose to construct *Oblivious Decision Trees* (ODT) [7]. An ODT is different from a conventional decision tree in such a way that an ODT's nodes at the same level are constructed by splitting rules based on the *same* data attribute. For any two leaf nodes on an ODT, if we project their data subsets onto a split attribute, then the two projected sets are either disjoint or identical. This feature ensures that a learned ODT can be used to produce categorization rules for defining types.

To address the second issue regarding the learning goal, we use harmonic mean to combine dispersal and impurity and apply a greedy heuristic—whenever there exist several sets of categorization rules as candidates, choose the set that leads to the lowest harmonic mean.

Algorithm 1 describes the customized decision tree learning algorithm. It begins with an empty root node which holds all events in a given log (Line 1). At each iteration, the best split will be found, i.e., finding the best type-defining attribute and its corresponding categorization rules to be applied (Line 3). This selection (FindBestSplit) is based on calculating the harmonic mean of dispersal and impurity. We elaborate on this step later. If that best attribute for splitting can be found, the decision tree is expanded by applying the categorization rules to every leaf node and growing a subtree there. This ensures that the tree is grown as an ODT (Line 5). The decision tree keeps growing either until the next best split cannot be found (Line 6–7) or until the height of the decision tree exceeds a preset maximum value (Line 2). After the iterative tree growth stops, we traverse every level of the current tree and select a subtree to be returned for the subsequent step of parsing categorization rules (Line 10).

Below, we explain the two key operations in the procedure, FindBestSplit (Line 3) and SelectSubTree (Line 10).

FindBestSplit is a sub-procedure for selecting a type-defining attribute and its corresponding categorization rules, i.e., the best split. First, for all type-defining attribute given in the attribute specification ($x \in X_{case}^{EL} \cup X_{act}^{EL} \cup X_{time}^{EL}$), we generate the corresponding categorization rules as candidate splits. When $\Lambda(x) \neq \varnothing$, i.e., there exist user-specified categorization rules for x, we use those rules. Otherwise, we consider x a generic data attribute and apply methods in conventional decision tree learning [6] to infer possible ways to split on x: if x is numeric, apply a histogram-based algorithm; if x is categorical, compute the possible two-subset partitions over its values and sample from all partitions. Note that the attributes are used with replacement, i.e., x can be split more than once in the iterative procedure.

After the candidate splits are generated for every type-defining attribute, we need to evaluate and select one of them to expand the tree. To do this, test how the current tree would be expanded if a candidate split applied. Then, parse the

---

Algorithm 1: The customized decision tree learning algorithm

    input : $EL = (E, Att, \pi)$, an event log; S, an attribute specification;
             H, a constant specifying the maximum height of the tree
    output: a decision tree
1 $root \leftarrow$ CreateTreeNode(E)
2 **for** $h \leftarrow 1$ to H **do**
3      $attr, rules \leftarrow$ FindBestSplit($root, EL, S$)
4   **if** $attr \neq \varnothing$ **then**
5   ExpandTreeOnEveryLeafNode($root, attr, rules$)
6   **else**
7   **break**
8   **end**
9 **end**
10 **return** SelectSubTree($root$)

full set of categorization rules from the "test tree" to determine the execution contexts (see Sect. 5.3). This way, we can obtain CO_x for every candidate split. Finally, the best split can be decided by choosing the candidate whose CO_x would lead to the lowest harmonic mean of dispersal and impurity.

SelectSubTree is a sub-procedure for deciding the subtree to be returned. Here, a subtree refers to a subtree sharing the same root node as the entire decision tree, i.e., any intermediate result obtained during the iterative tree growth process or the complete decision tree. To make the selection, we first apply the elbow method to identify turning points where dispersal and impurity changed significantly. Then, a single subtree can be decided by selecting from the identified turning points.

The values of dispersal and impurity are expected to show opposite trends as a decision tree grows. Initially, all events are placed together (held by the root node), and hence dispersal is 0 while impurity is 1. As the decision tree grows, the number of leaf nodes increases (so is the number of their corresponding execution contexts), which leads to the increase in dispersal and decrease in impurity.

5.3 Parse Categorization Rules

The parsing of categorization rules is to transform the rules recorded on a decision tree into execution contexts that we need (Definition 7). This transformation happens both when we need to evaluate intermediate results (FindBestSplit) and also when we need to obtain the final execution contexts after the decision tree learning stops.

To parse categorization rules, we first follow the conventional way of rule extraction from a decision tree. That is, for each path from the root to a leaf node, a decision rule is formed by conjoining all the rules recorded along the path. Then, for every decision rule obtained, we use the attribute specification as a reference to determine which part of the decision rule is related to case types, activity types, or time types, respectively. Formally, every such decision rule σ can be written as a conjunction $(\sigma_c \wedge \sigma_a \wedge \sigma_t)$, where any of σ_c, σ_a, σ_t can be a default rule (σ_{true}) if no type-defining attributes are included for any of the core event attributes.

As such, we will be able to transform a decision rule related a leaf node of a decision tree into an execution context $co = (\sigma_c, \sigma_a, \sigma_t)$. A set of execution context CO is obtained by parsing the categorization rules for all leaf nodes.

6   Experiments

We implemented the approach and evaluated it through an experiment. The aim is two-fold: (i) to test the feasibility of using our approach to learn execution contexts, and (ii) to demonstrate how the learned execution contexts can be applied for resource-oriented analyses. We share our implementation and the experiment details in an open repository online[1].

---

[1] Implementation and experiment details: https://royjy.me/to/learn-co.

6.1 Experiment Setup

We evaluated our approach on a publicly available, real-life event log dataset, BPIC-15[2], which consists of five event logs that record how five Dutch municipalities performed in a building permit application handling process. We merged them into a single log for the experiment.

Some event attributes were used as type-defining attributes for case types and activity types: *phase* is an event attribute indicating the phase of the process where activities belong to; and *case_parts* is a case attribute indicating the type of project related to the building permit applications. For time types, we derived two type-defining attributes from the original timestamps and appended them to the event logs, i.e., *weekday* and *am/pm* (AM time vs. PM time).

Table 1 describes the experiment dataset and the attribute specification used for learning execution contexts. Specifically, for attribute *case_parts*, we defined the following user-specified categorization rules based on the original description of the data—two rules partition the values of *case_parts* into two subsets, depending on whether a value contains the string 'Bouw' (indicating the case is related to construction) or not.

Table 1. The event log dataset and the attribute specification used for learning execution contexts

Log statistics				Attribute specification		
#events	#cases	#activities	#resources	Att_{case}	Att_{act}	Att_{time}
193453	5599	154	71	{case_parts}	{phase}	{weekday, am/pm}

6.2 Learning Execution Contexts

We applied our approach with the maximum tree height H set to 10 and selected the subtree corresponded to the point where the harmonic mean changed most significantly. Figure 4 illustrates the values of dispersal and impurity per iteration when learning a decision tree and the of changes of their harmonic mean. From Fig. 4a and Fig. 4b, we can observe a clear upward trend of dispersal values and a downward trend of impurity values. This confirms our discussion on the feature of decision tree learning as mentioned before (see Sect. 5.2). There are obvious changes to dispersal and impurity from iteration 5 onwards, which align with the changes of their harmonic mean illustrated in Fig. 4c. We selected iteration 7 as the "elbow" point, where the increase of harmonic mean starts to slow down. We used its corresponding subtree to parse categorization rules and obtained a set of 12 execution contexts.

---

[2] BPIC-15   dataset: https://doi.org/10.4121/uuid:31a308ef-c844-48da-948c-305d167a0ec1.

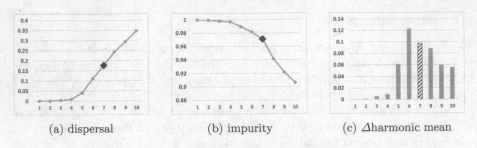

(a) dispersal   (b) impurity   (c) Δharmonic mean

Fig. 4. Dispersal, impurity, and changes of their harmonic mean per iteration

Table 2 shows the learned execution contexts. They are sorted by the number of events they contain. We can see that case type remains the default rule (σ_{true}), which means that attribute *case_parts* was not selected to derive categorization rules. A possible reason is that resources are similar in terms of handling applications concerned with different types of projects. For the activity types, we find the cluster of phases '0', '1', '4', '5' as a single type. These phases are related to the frequently executed activities in the process, where most resources participated in. Finally, the derived time types are only based on partitioning *weekdays*, which show a clear pattern that aligns with common working hours. The other type defining attribute *am/pm* was not used, implying similarities of resource workload in the morning vs. afternoon. These findings about types are consistent with our visual analyses on the behavior of the five municipalities, reported in our previous work [16].

Table 2. The 12 execution contexts learned from log BPIC-15

id	Case type	Activity type	Time type	#events	#res.
1	σ_{true}	($phase \in$ {'0', '1', '4', '5'})	($weekday \in$ {'Mon', 'Tue', 'Wed', 'Thu', 'Fri'})	150666	70
2		($phase \in$ {'3'})	($weekday \in$ {'Mon', 'Tue', 'Wed', 'Thu', 'Fri'})	17836	53
3		($phase \in$ {'2'})	($weekday \in$ {'Mon', 'Tue', 'Wed', 'Thu', 'Fri'})	16451	57
4		($phase \in$ {'8'})	($weekday \in$ {'Mon', 'Tue', 'Wed', 'Thu', 'Fri'})	6871	47
5		($phase \in$ {'0', '1', '4', '5'})	($weekday \in$ {'Sat'})	778	35
6		($phase \in$ {'7'})	($weekday \in$ {'Mon', 'Tue', 'Wed', 'Thu', 'Fri'})	585	32
7		($phase \in$ {'0', '1', '4', '5'})	($weekday \in$ {'Sun'})	149	22
8		($phase \in$ {'3'})	($weekday \in$ {'Sat'})	49	6
9		($phase \in$ {'2'})	($weekday \in$ {'Sat'})	41	6
10		($phase \in$ {'6'})	($weekday \in$ {'Mon', 'Tue', 'Wed', 'Thu', 'Fri'})	15	5
11		($phase \in$ {'8'})	($weekday \in$ {'Sat'})	10	1
12		($phase \in$ {'7'})	($weekday \in$ {'Sat'})	2	2
Total				193453	71

#res.: number of unique resources performed in an execution context

6.3   Applying Execution Contexts

This section reports how the learned execution contexts can be applied to (i) the analysis of resource profiles [8], i.e., describing the behavior of resources in process execution, and (ii) the discovery of organizational models [12], i.e., finding groups of resources having similar characteristics.

Resource Profile Analysis. For an illustration purpose, we selected the 31 resources who performed in the last 6 execution contexts (i.e., id 7–12) and calculated their *activity frequency* [8] with regard to those execution contexts.

Figure 5 shows a heatmap that visualizes the results. Darker colors indicate larger values. We can see that resources exhibit clear differences. Most of them only worked in execution context 7, which is the overtime work ('Sun') on the main phases ('0', '1', '4', '5'). Another distinct pattern is concerned with resources working only in execution contexts 8 and 9. They showed balanced activity frequency regarding the two activity types (phase '3' and '2'). Execution context 11 is specific to a single resource '560519'. This is an interesting observation compared to execution context 4, where the same set of activities performed on weekdays were covered by much more resources (47 of them).

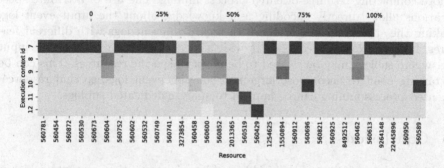

Fig. 5. Visualizing resources' activity frequency given execution contexts 7–12

Organizational Model Discovery. The original AHC method [12] derives an "originator (resource) by activity" matrix, feeds it to agglomerative hierarchical clustering, and gets resource clusters as the output. In our demonstration, we use a "resource by execution context" matrix instead, i.e., count how frequently resource conducted events having specific execution contexts. We keep identical settings on all other steps and apply the silhouette score [11] to evaluate the quality of the outputs. We varied the desired cluster number between 2 and 70.

Figure 6 shows the quality of discovery results, comparing between the use of activity labels vs. execution contexts. The X-axis corresponds to the desired number of resource clusters specified as a parameter, and the Y-axis corresponds to the quality (silhouette score) of the discovery result. We can observe that using execution contexts, compared to the original method, led to better output quality in most situations. This indicates that using the learned execution contexts contributed to uncover resource groups having more distinct characteristics.

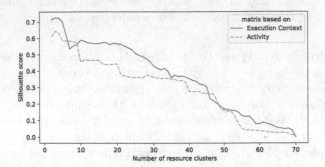

Fig. 6. Quality of discovery results measured by their silhouette scores, per selected number of clusters

Summary. The above applications of our approach show that it can be applied before conducting resource-oriented process mining. The use of attribute specifications allows directly encoding user knowledge about the input event log, making the approach configurable with regard to event logs with different features. Learned execution contexts provide a multidimensional view on the input log, where sublogs may be related to behavior of specific resources. This can be a starting point to decompose large and complex event logs, so that resource-oriented process mining can be applied to analyze dedicated sublogs.

7   Conclusion

In this paper, we discussed the problem of learning execution contexts, which is concerned with characterization of resource groupings in process execution, considering three core process dimensions. We proposed an approach to automatically learn execution contexts using a dedicated decision-tree-based algorithm and tested it on real-life event log data.

Our current work has certain limitations. For one, learned execution contexts are local optimal due to the greedy heuristic and that tree induction is sequential-forward. A possible future direction is to explore searching methods with other heuristics, e.g., simulated annealing, to produce better near-optimal solutions. Second, the experiment used event log data from one process to demonstrate the usefulness of the approach. Future work can look into a more comprehensive evaluation using event logs recording different processes and containing more attributes. Furthermore, the proposed approach can be compared with methods that support the discovery of multidimensional resource characteristics related to resource groupings.

Acknowledgments. The reported research is part of a PhD project supported by an Australian Government Research Training Program (RTP) Scholarship.

References

1. Appice, A.: Towards mining the organizational structure of a dynamic event scenario. J. Intell. Inf. Syst. **50**(1), 165–193 (2017). https://doi.org/10.1007/s10844-017-0451-x
2. Bolt, A., van der Aalst, W.M.P.: Multidimensional process mining using process cubes. In: Gaaloul, K., Schmidt, R., Nurcan, S., Guerreiro, S., Ma, Q. (eds.) CAISE 2015. LNBIP, vol. 214, pp. 102–116. Springer, Cham (2015). https://doi.org/10.1007/978-3-319-19237-6_7
3. Burattin, A., Sperduti, A., Veluscek, M.: Business models enhancement through discovery of roles. In: IEEE Symposium on Computational Intelligence and Data Mining (CIDM), pp. 103–110 (2013)
4. Daft, R.L., Murphy, J., Willmott, H.: Organization Theory and Design. South-Western Cengage Learning, USA (2010)
5. Dumas, M., Rosa, M.L., Mendling, J., Reijers, H.A.: Fundamentals of Business Process Management, 2nd edn. Springer, Heidelberg (2018). https://doi.org/10.1007/978-3-662-56509-4
6. Ke, G., et al.: LightGBM: a highly efficient gradient boosting decision tree. In: Proceedings of the 31st International Conference on Neural Information Processing Systems, NIPS 2017, Red Hook, NY, USA, pp. 3149–3157. Curran Associates Inc., December 2017
7. Kohavi, R., Li, C.H.: Oblivious decision trees graphs and top down pruning. In: Proceedings of the 14th International Joint Conference on Artificial Intelligence, pp. 1071–1077. Morgan Kaufmann Publishers Inc. (1995)
8. Pika, A., Leyer, M., Wynn, M.T., Fidge, C.J., ter Hofstede, A.H.M., van der Aalst, W.M.P.: Mining resource profiles from event logs. ACM Trans. Manag. Inf. Syst. **8**(1), 1:1–1:30 (2017)
9. Reijers, H.A., Song, M., Jeong, B.: Analysis of a collaborative workflow process with distributed actors. Inf. Syst. Front. **11**(3), 307–322 (2009). https://doi.org/10.1007/s10796-008-9092-5
10. Ribeiro, J.T.S., Weijters, A.J.M.M.: Event cube: another perspective on business processes. In: Meersman, R., et al. (eds.) OTM 2011. LNCS, vol. 7044, pp. 274–283. Springer, Heidelberg (2011). https://doi.org/10.1007/978-3-642-25109-2_18
11. Rousseeuw, P.J.: Silhouettes: a graphical aid to the interpretation and validation of cluster analysis. J. Comput. Appl. Math. **20**, 53–65 (1987)
12. Song, M., van der Aalst, W.M.P.: Towards comprehensive support for organizational mining. Decis. Support Syst. **46**(1), 300–317 (2008)
13. Van der Aalst, W.M.P.: Process cubes: slicing, dicing, rolling up and drilling down event data for process mining. In: Song, M., Wynn, M.T., Liu, J. (eds.) AP-BPM 2013. LNBIP, vol. 159, pp. 1–22. Springer, Cham (2013). https://doi.org/10.1007/978-3-319-02922-1_1
14. Van der Aalst, W.M.P.: Process Mining: Data Science in Action, 2nd edn. Springer, Heidelberg (2016). https://doi.org/10.1007/978-3-662-49851-4
15. Van Hulzen, G., Martin, N., Depaire, B.: Looking beyond activity labels: mining context-aware resource profiles using activity instance archetypes. In: Polyvyanyy, A., Wynn, M.T., Van Looy, A., Reichert, M. (eds.) BPM 2021. LNBIP, vol. 427, pp. 230–245. Springer, Cham (2021). https://doi.org/10.1007/978-3-030-85440-9_14

16. Yang, J., Ouyang, C., ter Hofstede, A.H.M., van der Aalst, W.M.P., Leyer, M.: See-
    ing the forest for the trees: group-oriented workforce analytics. In: Polyvyanyy, A.,
    Wynn, M.T., Van Looy, A., Reichert, M. (eds.) BPM 2021. LNCS, vol. 12875, pp.
    345–362. Springer, Cham (2021). https://doi.org/10.1007/978-3-030-85469-0_22
17. Yang, J., Ouyang, C., van der Aalst, W.M.P., ter Hofstede, A.H.M., Yu, Y.: Ordi-
    noR: a framework for discovering, evaluating, and analyzing organizational models
    using event logs. Decis. Support Syst. **158**, 113771 (2022)

A Purpose-Guided Log Generation Framework

Andrea Burattin[1], Barbara Re[2], Lorenzo Rossi[2(✉)], and Francesco Tiezzi[3]

[1] Technical University of Denmark, Kgs. Lyngby, Denmark
andbur@dtu.dk
[2] School of Science and Technology, University of Camerino, Camerino, Italy
{barbara.re,lorenzo.rossi}@unicam.it
[3] Dipartimento di Statistica, Informatica, Applicazioni, University of Florence, Florence, Italy
francesco.tiezzi@unifi.it

Abstract. Process mining is a prominent discipline that collects a variety of techniques fulfilling different mining purposes by gathering information from event logs. This involves the continuous necessity of event logs suitable for testing mining techniques with respect to different purposes. Unfortunately, event logs are hard to find and usually contain noise that can influence the results of a mining technique. In this paper, we propose a framework for generating event logs tailored for different mining purposes, e.g., process discovery and conformance checking. Event logs generation and tuning are made out through business model simulations guided by the mining purpose under consideration. Beyond defining the framework, we implemented it as a tool, which has been also used for the validation of the approach we propose.

Keywords: Process mining · Event log · Log generation · Simulation

1 Introduction

Nowadays, process mining is recognized as an important discipline in extracting non-trivial information from the execution of business processes, thanks to the increasing usage of information systems that record event logs of the deployed processes [2]. The importance of process mining is well recognized also by companies, which appreciate the possibility to gather knowledge from their processes from actual execution data [24].

Process mining is a family of techniques and algorithms that enables to automatically extract information out of event logs recorded during the execution of business processes. The effectiveness and the precision of process mining techniques are strictly related to the reliability of their mining algorithms, whose development requires testing them against different event logs [17], usually coupled with the models that generated them [9]. Mining algorithms extract different

---

Lorenzo Rossi – Main contributor.

© Springer Nature Switzerland AG 2022
C. Di Ciccio et al. (Eds.): BPM 2022, LNCS 13420, pp. 181–198, 2022.
https://doi.org/10.1007/978-3-031-16103-2_14

types of information according to the mining purpose they have to accomplish, e.g., process discovery and conformance checking. Therefore, to test a process mining algorithm it is important to use event logs that suit the purpose for which the algorithm has been devised [22]. For instance, given a family of discovery algorithms that leverages the same set of properties on the logs (e.g., the coverage of the direct following relations for the Alpha miner family [3]), then a fair comparison of the algorithms would require logs where such properties are indeed satisfied. As stated in the literature [7,17,21], each of those purposes heavily relies on the quality, with respect to specific properties, of the event logs given as input to the related mining algorithms.

Obtaining event logs fitting for a purpose is a complex, yet necessary, achievement [6]. Specifically, [7] claims that bad quality logs hamper the use of process mining techniques, thus researchers are encouraged to develop log generators that focus on a specific and explicit mining purpose. Event logs are difficult to find, in particular those directly extracted from deployed IT systems that refer to real-world installations [8]. In this regard, several approaches, e.g., [8,12,15,17,19], propose the automated generation of artificial event logs via the simulation of models in a predetermined language, e.g., BPMN or Petri Net. However, these are *purpose-agnostic*, thus not meant to produce event logs fulfilling properties required for a specific purpose. Instead, they simulate random execution traces, producing every time a different event log. The above-mentioned issue paves the way to the need of answering the following research questions:

RQ1: *Is it possible to define an approach for the automated generation of event logs tailored to different mining purposes?*

RQ2: *Can model simulation be guided to produce event logs that fulfill a mining purpose better than the ones generated with purpose-agnostic simulations?*

To address these research questions, we propose the PURpose-Guided Log gEneration (PURPLE) framework. The main advantages of the PURPLE framework with respect to existing simulators are as follows. PURPLE generates event logs specifically tailored to the purpose of the mining technique under investigation. To shape out an event log, the framework performs a *guided simulation* of the input model that incrementally generates specific execution traces, until the desired purpose is satisfied. The simulation is guided by hints, produced at each step on the basis of the partial log generated up to that moment and the properties required by the mining purpose. Additionally, the framework is meant to *simulate many kinds of business process models* (e.g., BPMN, Petri Net, WF-net). Besides the framework, we provide the PURPLE tool, which implements a BPMN and a Petri-net semantic engine, and addresses mining purposes concerning process discovery and conformance checking. To validate the advancements of our proposal to the state of the art on log generation, we carried out experiments measuring the quality of logs generated by PURPLE for the purposes it supports, and we compared these results with the ones of other log generators.

The rest of the paper is structured as follows. Section 2 provides notions on event logs and Labeled Transition Systems. Section 3 introduces the PURPLE framework. Section 4 presents the PURPLE tool and several instantiations of the framework, while Sect. 5 reports the results of the conducted experiments. Section 6 compares our approach with related works. Finally, Sect. 7 closes the paper discussing assumptions, limitations, and opportunities for future works.

2   Background Notions

This section provides notions we use in the rest of the paper. An **event log** consists of a set of **cases**, each of which refers to some events that can be seen as one possible run of the process. An **event** refers to the execution of a system activity, and it is described by a set of **attributes**. The most common attributes for a recorded event are the *activity name* and the *timestamp*, but also other information can be captured, such as the *resource* involved in the activity execution, or the monetary *cost* associated with it. The sequence of events related to a given case is called **trace**.

Figure 1(a) depicts a system modeled using the BPMN notation [20], Fig. 1(b) shows a table containing an event log fragment with three cases generated by the BPMN model, while Fig. 1(c) reports a *simple event log* [2, Ch. 5], which focuses only on the names of the executed activities. In this situation, an event log can be thought of as a multiset of traces, where a *trace* is a sequence of activity names [2]. The multiplicity of a trace is denoted in the simple event log by a positive integer (omitted when it is equal to 1). A way of generating event logs is through the simulation of a business process model [8]. The main idea is to repeatedly "execute" a model and to record, in a log file, all events observed during the execution. Simulators use the so-called *play-out engines*, like in [4,10], to execute models [2, Ch. 2]. An engine provides the moves a model can perform according to the semantics of the considered modeling language (e.g., the firing rule of Petri Nets [18] or the transition rules of BPMN operational semantics [11]) usually defined by means of Labeled Transition Systems (LTSs).

An **LTS** consists of: *states*, representing the possible system configurations (i.e., the execution states of the model), and *labeled transitions*, corresponding to directed edges connecting states (representing moves in the model execution). Formally, a transition system is a triple $(S, L, \rightarrow)$ where: S, ranged over by s, is a set of states; $L = A \cup \{\tau\}$, ranged over by l, is the union of a set of (visible) activity labels A, ranged over by a, and a special label τ denoting an invisible activity; and $\rightarrow \subseteq S \times L \times S$ is a transition relation. The τ action is used to decorate those transitions of the LTS that do not refer to the performing of an activity included in the model, but refer to the control of the execution flow, e.g., the execution of decisions, that can be neglected in the log generation. In an LTS, we call a state *initial* (resp. *final*) if it does not have incoming (resp. outgoing) transitions. The initial state, labeled s_i, corresponds to the initial configuration of the model, where its execution starts, while a final state, labeled s_f, is an ending configuration, which corresponds to a proper or an improper termination.

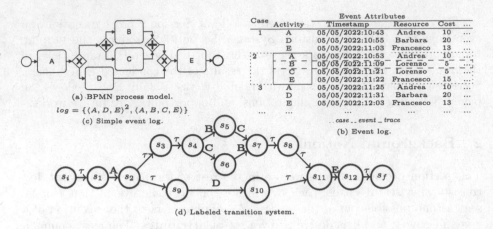

(a) BPMN process model.

$log = \{\langle A, D, E \rangle^2, \langle A, B, C, E \rangle\}$
(c) Simple event log.

(d) Labeled transition system.

Fig. 1. From process model to event log.

Finally, for a given LTS, $(S, L, \rightarrow)$ with $L = A \cup \{\tau\}$, it is possible to characterize: *sub-traces* as sequences of visible labels; *traces* as sequences of visible labels from the initial to a final state; and *logs* as multisets of *traces*. Formally, the sequence of labels $\langle a_1, a_2, \ldots, a_n \rangle$ with $a_1, a_2, \ldots, a_n \in A$ is a **sub-trace** if there exists $\langle l_1, l_2, \ldots, l_m \rangle$ with $l_1, l_2 \ldots, l_m \in L$ such that: *(i)* $\langle a_1, a_2, \ldots, a_n \rangle$ coincides with $\langle l_1, l_2, \ldots, l_m \rangle$ up to occurrences of τ; and *(ii)* $(s_1, l_1, s_2) \in \rightarrow$, $(s_2, l_2, s_3) \in \rightarrow, \ldots, (s_m, l_m, s_{m+1}) \in \rightarrow$ for some $s_1, s_2, \ldots, s_{m+1} \in S$. If s_1 is the initial state and s_{m+1} is a final state, the sub-trace is called **trace**. Figure 1(d) reports the LTS representing the behavior of the model in Fig. 1(a) produced by the BPMN formalization described in [11]. Each configuration of the model, i.e. each marking of tokens, corresponds to a state of the LTS. For example, the initial marking, where there is only one token placed on the start event, corresponds to the state s_i, while the marking obtained by one step of execution from the initial marking, where the token is moved to the sequence edge incoming into the activity A, corresponds to the state s_1. The execution of an activity of the model is rendered in the LTS through a transition labeled by the name of the activity. Traversing the LTS from state s_i to s_f, the sequences of visible labels associated to the transitions represent the execution traces that can be generated from the BPMN model in Fig. 1(a).

3   PURPose-Guided Log gEneration Framework

In this section, we introduce the PURPose-Guided Log gEneration (PURPLE) framework. It is meant to produce, by simulating models, event logs with different properties for targeting different mining purposes. PURPLE supports the simulation of models specified with different languages, by projecting their execution onto a common behavioral model, i.e., an LTS. Figure 2 depicts the components of the PURPLE framework: a **semantic engine**, an **evaluator**, and a guided **simulator**. Except for the simulator that is fixed, the other components can be instantiated with different semantic engines, each one supporting a given modeling language (e.g., BPMN, Petri Net), and with different evaluators, each one tailored to a mining purpose (e.g., process discovery, compliance checking).

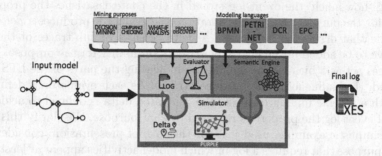

Fig. 2. PURPLE framework components.

Before presenting in detail the PURPLE components, we introduce here the concept of *context* in which the PURPLE components act. The context collects and keeps updated the (even partial) LTS and a log of the model under consideration. It acts as a sort of global variable that the PURPLE components can access/modify during the simulation like in a side-effect function. Notably, at the beginning of a simulation, the context is set to an initial configuration where the log is empty and an LTS contains only the initial state s_i.

We can now define the PURPLE components starting from the **semantic engine**. Being aware of its formal semantics, this component enacts the input model. Given a state of the corresponding LTS (i.e., a model configuration), the semantic engine returns the next reachable states (i.e., the model configurations reachable by one move), and the labels of the transitions leading to them (i.e., the names of the performed activities). For example, considering the LTS in Fig. 1(d) and its state s_4, the semantic engine returns the next reachable states coupled with the labels of the connecting transitions: $\{(B, s_5), (C, s_6)\}$. Relying on different semantic engines, PURPLE can obtain from each business process model with an executable semantics the corresponding LTS [2, Ch. 3], thus gaining in terms of generalizability.

The second component is the **simulator** that is devoted to produce traces from the execution of the input model. By invoking the semantic engine, the simulator component incrementally adds states and transitions to an LTS, then traverses it to produce new traces to be added in the log. Notably, the simulator acts on the LTS and on the log maintained by the context, starting from the initial context. While building the LTS, the simulator generates a partial view of the entire LTS and, at any time it reaches a final state, it stores the visited trace in the log. The peculiarity of this component lies in a guided traversal of the LTS to guarantee the production of traces, and hence a log, that satisfy the desired mining purpose. Indeed, differently from a purely-random simulation, what the framework proposes is a **guided simulation** that takes as input a *guide* suggesting an execution path, or part of it, to follow in the LTS traversal.

Lastly, the **evaluator** component is responsible to evaluate an event log in relation to the peculiarities of the desired mining purpose. More practically, by checking 'how much' the event log stored in the context satisfies the properties needed for the purpose under consideration, the evaluator produces a *delta*, i.e., the guide that drives the simulator. A delta consists of sub-traces of the LTS that have to be added in the log to increase its suitability for the purpose. These sub-traces act as a bias for the simulation indicating the parts of the LTS to be traversed, thus influencing the produced traces. As each evaluator is defined to deal with a specific mining purpose, the generated delta is defined to achieve in the final event log the properties required by that purpose. We clarify this point with a simplistic example, used just for the sake of presentation: considering a mining purpose that requires a log in which model activities appear at least once, the evaluator will select the activities not yet in the log, and will produce a delta containing sub-traces of length one with the labels of the missing activities.

By fixing a modeling language and a mining purpose (hence, a semantic engine and an evaluator, respectively), we get an instantiation of PURPLE ready for producing logs. Providing a model as input, the PURPLE instantiation starts performing the looping four steps routine depicted in Fig. 3. Step (1) loads the input model and sets the initial context. Then, the framework's routine loops between Steps (2) and (3) before producing the final log in Step (4).

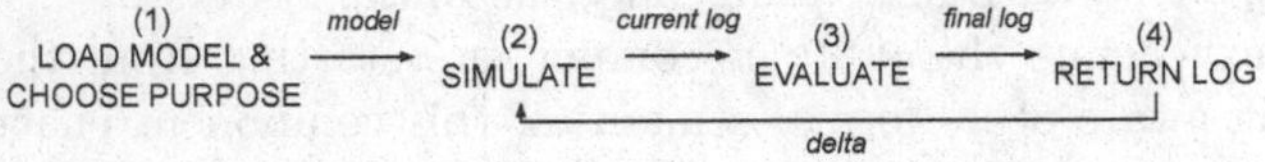

Fig. 3. PURPLE routine.

```
1  SIMULATE(st) :
2    if  st = ⟨⟩
3      return RANDOMSIM()
4    States := FIND(lts, st[1])
5    st := st \ st[1]
6    for  s in States
7      st_p := GETPREFIX(lts, s)
8      t := GUIDEDSIM(st_p, s, st)
9      if  t ≠ ⟨⟩
10       return t
11   return ⟨⟩
```

Listing 1.1. Simulation function for a given context $\langle lts, log \rangle$.

```
1  GUIDEDSIM(st_p, s_curr, st) :
2    if  st = ⟨⟩
3      return FINALISE(st_p)
4    States := FIND(lts, st[1])
5    for  s_next in States
6      if  ∃ (s_curr, st[1], s_next) ∈ →
7        st_p := st_p + st[1]
8        t := GUIDEDSIM(st_p, s_next, st \ st[1])
9        if  t ≠ ⟨⟩
10         return t
11   return ⟨⟩
```

Listing 1.2. Giuded simulation for a given context $\langle lts, log \rangle$.

Step (2) performs the guided simulation of the model taking into consideration the context containing the current LTS and log. As shown in Listing 1.1, the simulation function depends on the input parameter st that is one of the sub-traces in the delta. In case st is empty (i.e., $\langle\rangle$), the function executes the model in a random way (line 3) via the RANDOMSIM() call: starting from the initial state of the LTS contained in the context, it repetitively invokes the semantic engine to know the next states (adding them to the LTS in the context) and chooses one of them randomly until it reaches one of the final states. This happens, for instance, in the case the evaluator has not performed any comparison yet. In case of non-empty delta, instead, the function proceeds by considering st as breadcrumbs to follow for logging a specific trace in the LTS. More in detail, function FIND($lts, st[1]$) (line 4) returns a set of states of the LTS reachable by a transition labeled as the first element in the considered sub-trace, i.e., $st[1]$. In case the found states do not have any successor in the current LTS, the function FIND invokes the semantic engine and adds the results in the LTS. For each found state s (line 6), the simulator calculates a prefix sub-trace st_p that leads to s via function GETPREFIX(lts, s) (line 7). Then, the algorithm calls the recursive function GIUDEDSIM (line 8) to complete the trace with labels corresponding to the remaining part of st, where the first label has been removed (line 5). The guided simulation function, Listing 1.2, takes as input the prefix sub-trace st_p, the current state s_{curr} of the LTS, and the remaining part of the hint of the delta, i.e., the sub-trace st. This function recursively searches for states of the LTS in the context, reachable from s_{curr} through a sequence of transitions labeled by the remaining elements in the hint st (lines 5-10). If a reachable state is found (line 6, where $\rightarrow$ is the transition relation of the LTS), the prefix trace is increased with the label of the connecting transition (line 7, where $+$ denotes the append operator on sub-traces). Then, the function is called recursively on the enriched prefix, the next configuration, and the hint without the first label (line 8). Once st no longer contains labels (line 2), the function enacts the base case (line 3) where function FINALISE(st_p) finalizes the prefix trace logging the labels of the transitions leading to a final state, and returns the entire trace to the calling function.

In Step (3) of the routine, the evaluator uses the context containing the event log produced by the simulator in Step (2). On the basis of the mining purpose, a specific evaluator calculates the delta, and evaluates if the purpose is satisfied. If not, the routine loops back to Step (2) to repeat a new simulation based on the calculated delta. Instead, if the purpose is satisfied, the simulation terminates and the generated event log is given as output in Step (4).

4   PURPLE at work

We present here four instantiations of the PURPLE framework addressing purposes concerning process discovery and conformance checking. These instantiations are described using the PURPLE tool that implements the framework and its routine. Tool, source code, instructions, and examples are available at https://pros.unicam.it/purple/. The PURPLE tool provides two semantic engines that implement a wide subset of the BPMN semantics described in [11], and the Petri-net semantics [18]. Concerning BPMN, PURPLE supports process and collaboration diagrams made up by pools, empty start and end events, message start and end events, terminate end events, intermediate message throw and catch events, tasks, parallel gateways, exclusive gateways, and event-based gateways. The latter engine, instead, supports standard Petri-nets (including particular classes of Petri-nets, such as WF-nets). Moreover, PURPLE implements four evaluators addressing process discovery and conformance checking. Notice that some evaluators may require, besides the log, additional parameters dealing with specific implementation aspects (e.g., a maximum number of traces to generate for ensuring termination). The pseudocode of the four evaluators is available online, at the PURPLE's website, in a companion technical report.

Process discovery in PURPLE. The first instantiation of PURPLE that we consider regards the process discovery. To check the reliability of a discovery algorithm, or to perform a benchmark of different techniques, logs presenting specific characteristics are required. PURPLE implements evaluators addressing two specific discovery purposes: one is devised for algorithms relying on the order relation between activities, such as the Alpha algorithm [3], while the other one is for algorithms relying on frequencies, such as the Heuristics miner [23]. All these purposes can be applied to both BPMN and Petri-net models. In the rest of the section, for the sake of presentation, we consider only BPMN models, but the same reasoning applies to Petri-net ones, with the only difference in the generation of the LTS by means of a different semantic engine.

The aim of the **Process discovery via order relations** purpose is to generate event logs for discovery algorithms that build the output models on the basis of the order relations between activities. These algorithms, e.g., the Alpha family, scans the input event log to find the *footprint* matrix of the original model. Assuming that an activity Y directly follows an activity X ($X > Y$) if and only if there exists a trace in the log where Y appears immediately after X, the footprint matrix can contain three kinds of order relations [3, Def. 3.2]. The sequence relation, denoted by $X \to Y$, holds if and only if $X > Y$ and $Y \not> X$. The parallel relation, denoted by $X \| Y$, means that X directly follows Y and vice versa ($X \| Y \iff X > Y$ and $Y > X$). The last relation, denoted by $X \# Y$, is used when two activities are unrelated, i.e., neither X directly follows Y nor Y directly follows X ($X \# Y \iff X \not> Y$ and $Y \not> X$). Considering the model in Fig. 4(a), the corresponding matrix is provided in Fig. 4(b). To obtain an accurate version of the original model, the input event log has to provide as many order relations as possible to fill the footprint matrix. For instance, logging multiple times the same trace is useless as it always provides the same order relations. This can be achieved with PURPLE through an evaluator that guides the simulation into the discovery of the footprint matrix avoiding to produce duplicates of the same trace. Therefore, PURPLE points at generating the smallest log covering the relations in the footprint matrix.

The simulation step of the routine is triggered at first with an empty *delta*, leading to a random simulation of the model. A possible trace result of the first simulation run maybe $\langle A, B, C, E \rangle$, where the simulator performed tasks A, B, C and E, one after the other modifying the initial context. Specifically, the simulator adds to the initial LTS the states and the transitions discovered by the

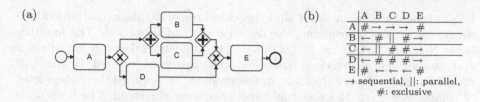

(a)

(b)

	A	B	C	D	E
A	#	→	→	→	#
B	←	#	‖	#	→
C	←	‖	#	#	→
D	←	#	#	#	→
E	#	←	←	←	#

→ sequential, ‖: parallel,
#: exclusive

Fig. 4. An input process model (a), and the related footprint matrix (b).

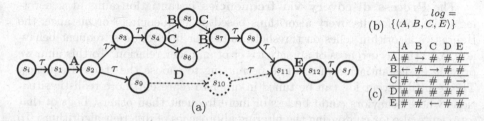

(a)

$log =$
(b) $\{\langle A, B, C, E\rangle\}$

	A	B	C	D	E
A	#	→	#	#	#
B	←	#	→	#	#
C	#	←	#	#	→
D	#	#	#	#	#
E	#	#	←	#	#

(c)

Fig. 5. LTS (a), log (b), and footprint (c) resulting from the first run of simulation.

semantic engine, producing the LTS in Fig. 5(a) to the exclusion of dotted states and transitions which are still to discover. Moreover, it inserts in the empty log the discovered trace, resulting in Fig. 5(b). Notably, to speed up the generation of the entire LTS, the simulation adds to it all the states discovered by the semantic engine, even if they do not take part to the produced trace (see states s_6 and s_9). Then, the evaluator calculates the order relations considering the updated log in the context. The log identifies 3 order relations: $A{\to}B$, $B{\to}C$, and $C{\to}E$; the other activities are still unrelated, thus the resulting footprint matrix is the one in Fig. 5(c). At this point, PURPLE compares the obtained footprint matrix with the one of the original model (Fig. 4(b)) to calculate the missing relations, and produces the delta for the upcoming simulation step. The order relations that are still missing are: $A{\to}D$, $B{\to}E$, $C{\to}B$, and $D{\to}E$. These relations are translated into sub-traces composing the delta as following:$\{\langle A, D\rangle, \langle B, E\rangle, \langle C, B\rangle, \langle D, E\rangle\}$. Since the delta is not empty, this time is crucial to guide the simulator in the search of additional traces containing the missing relations; in doing that, the simulator relies also on the LTS in the context. Considering the first hint of the delta, $\langle A, D\rangle$, the simulator looks for a state with an incoming transition labeled by A, that is state s_2, then it goes forward in the LTS to find a transition labeled by D. Being s_3 already visited, the simulator goes ahead to state s_9 that corresponds to a state in which activity D is enabled. Then, the simulator finalizes the trace until it reaches a final state, logging the trace $\langle A, D, E\rangle$. Instead, considering the hint $\langle C, B\rangle$ of the delta, the simulator has two states with an incoming transition labeled by C, i.e., s_6 and s_7, from which it starts looking for a transition labeled by B. State s_7 leads only to a transition labeled by E, while state s_6 leads to state s_7 with a transition labeled by B. Thus, the simulator follows this latter path in the LTS, logging the trace $\langle A, C, B, E\rangle$. The

LTS produced by the simulator after the second run of simulation corresponds to the one in Fig. 5(a) considering also dotted states and transitions. The resulting log is $\{\langle A, B, C, E \rangle, \langle A, C, B, E \rangle, \langle A, D, E \rangle\}$. The evaluator takes this log as input and assesses that all relations in the footprint matrix are covered, i.e. 100% of completeness is achieved. Notably, in this example, we required the highest level of completeness, but the user could specify a lower threshold. The purpose is satisfied since the log covers all relations and does not contain repeated traces, hence PURPLE produces as output the *.xes* file.

The **Process discovery via frequencies** instantiation aims at generating event logs for discovery algorithms based on frequencies. For instance, the Heuristics algorithm relies on threshold values for filtering less frequent behaviors, e.g., the occurrences of an activity or of an order relation. To this aim, we provide an instantiation of PURPLE permitting to choose the traces frequency. The resulting event log can be tuned in order to represent more realistic situations where behaviors could be less or more frequent than others. Logs of that form suite also for comparing the filtering approaches of different algorithms. To address this purpose, PURPLE extracts the set of traces the model can perform and information regarding the loops. Then, the user specifies the percentage of occurrence for each trace, a threshold value for the maximum number of repetitions of loops, and a minimum number of traces to be produced. Therefore, during the log generation, the evaluator implemented for this purpose compares the occurrences of traces and the thresholds for the loops chosen by the user with the current log, and generates a delta accordingly. In case some of these values are lower than requested, the evaluator passes to the simulator a delta containing the entire traces that are still infrequent in the log. If a trace contains a loop, the evaluator modifies the trace in the delta by repeating the loop (for a random number of times below the given threshold). Then, the delta, which contains only complete execution traces of the input model, guides the simulator from the initial to the final state of the LTS. Once the minimum number of traces in the log is reached, and the requested occurrence percentages are satisfied, PURPLE returns the log *.xes* file.

Conformance Checking In PURPLE. Lastly, we present purposes related to conformance checking, a family of techniques for comparing a model and a log. In particular, we consider techniques based on [2]. They permit to spot differences between the expectation (i.e., the process model) and the reality (i.e., the event log). Alignments explicitly show where deviations are located and which are the involved activities. Computing alignments is an expensive task, especially in presence of models with huge state-space, and there exist different approaches implementing it [5]. To check the reliability of such techniques, or to compare their performances, it is necessary to have logs embedding traces with deviations from the normal behavior, i.e., noisy behaviors. To this end, we propose two instantiations of PURPLE producing event logs from BPMN and Petri-nets with a precise amount of noisy behavior, or with a precise alignment cost.

The **conformance checking via noise frequencies** instantiation generates event logs with the desired percentages of noisy traces. The literature identifies types of noise that can affect a trace in an event log [13]; here we consider the following: *missing head*, a trace without some of the initial events; *missing tail*, a trace without some of the final events; *missing episode*, a trace without some of the intermediate events; *order perturbation*, a trace where some events appear in a wrong order; and *additional event*, a trace in which appears an alien event. This instantiation of the framework takes as inputs a model to simulate, a number of traces to generate, a percentage of occurrence for each type of noise, and a precision in reproducing the noise percentages. Whenever it is invoked, the evaluator sends an empty delta to the simulator to receive back a random trace without noise. Then, it compares the percentage of occurrences for each type of noise in the current log with respect to the requested one. The trace is hence modified introducing the type of noise farthest from the requested occurrence. In case of *missing head*, *missing tail*, or *missing episode*, PURPLE removes a random number of events from the head, from the middle, or from the tail of the trace, respectively. In case of *order perturbation* it swaps two or more events in the trace, while in case of *additional event* it inserts an event named differently from every activity name in the model. Once the evaluator finds the desired noise percentages and number of traces, it returns the final log.

The **conformance checking via fixed align cost** purpose aims at generating event logs with a precise amount of noise that involves a specific cost for the alignment. Roughly speaking, the alignment cost indicates the number of deviations between the model and the log. An alignment cost equal to zero indicates a perfect match between the log and the model, while higher costs indicate the presence of non-compliant behaviors. Synchronous moves between trace and model cost zero, while moves that can be performed only in the model or only in the trace usually cost 1. The same trace can be aligned to the model following different execution paths and leading to different costs; the one to consider for calculating the mean value is the lowest i.e., the optimal alignment. The overall alignment cost is the average of the optimal alignments for each trace in the log. Considering the model in Fig. 4 (a), a noisy trace could be $\langle B, C, E \rangle$, where the event labeled with A lacks. By aligning this trace through the path $\langle A, D, E \rangle$, only the last event matches, thus we have to perform two moves in the trace and two moves in the model that cost in total 4. While following the path $\langle A, C, B, E \rangle$ and $\langle A, B, C, E \rangle$, the alignment costs are respectively 3 and 1. Therefore, the optimal alignment cost to consider is the lowest one, i.e. 1.

Here, PURPLE takes as input a model, a desired alignment cost, a log size, and a precision in reproducing the exact alignment cost. Before evaluating the current log, the framework extracts from the model the set of traces that can be produced, and uses them later for calculating the alignment costs. Then, similarly to the previous purpose, the evaluator receives from the simulator traces without noise, perturbs them with a type of noise, and updates the reached alignment cost. Every time a noisy trace is added to the current log, the evaluator calculates the optimal alignment cost computing the minimum among the Levenshtein distances [16] between the noisy trace and traces previously extracted from the model.

5 Validation

In this section, we present a list of experiments on the presented instantiations of the framework, using the corresponding implementations in the PURPLE tool. The experiments are carried out by means of synthetic and real(istic) BPMN and Petri-net models, respectively generated by PLG2 (https://plg.processmining. it/) or obtained from the literature. The models contain start/end events, activities, and XOR/AND gateways; their dimension ranges from a minimum of 8 to a maximum of 53 elements. Concerning their topology, they are both structured and unstructured, and some of them contain loops. Any further information about the models and the artifacts generated during the experiments is available at https://bitbucket.org/proslabteam/validation/. Notably, the aim of this validation is to show the suitability of the framework in addressing mining purposes of different kinds. In each experiment, we use as a measure a quality criterion for the event logs, set on the basis of the purpose to address. When possible, we compare the results of these measurements with the ones achieved by reference tools, such as PLG2, BIMP (https://bimp.cs.ut.ee/), and the ProM (https:// www.promtools.org/) plugin of the GED methodology [14]. We selected these tools among the ones found in the literature (we refer to Sect. 6 for a comprehensive review of tools for log generation) using as inclusion criteria: the availability of an operating software to be used for the experiments, and the possibility of tailoring the produced logs to the mining purpose under analysis.

Table 1. Process discovery via order relations validation results.

Model	El.	Traces	PURPLE	Coverage with 1000 traces			Coverage with min traces		
				BIMP	GED	PLG2	BIMP	GED	PLG2
p0	10	3	100%	63%	100%	100%	63%	100%	100%
p1	11	3	100%	63%	100%	100%	63%	63%	75%
p2	12	5	100%	75%	100%	100%	75%	100%	100%
p3	17	5	100%	83%	100%	100%	83%	92%	100%
p4	21	10	100%	61%	100%	100%	56%	89%	100%
p5	27	10	100%	74%	91%	91%	70%	91%	83%
p6	34	14	100%	39%	69%	100%	39%	69%	94%
p7	40	76	100%	24%	68%	97%	24%	68%	93%
p8	49	226	100%	6%	49%	99%	6%	49%	97%
p9	53	41	100%	25%	54%	99%	25%	50%	89%

Regarding the **process discovery via order relations,** the comparison measure we use to assess event logs quality is *coverage*, i.e., the percentage of activity relations provided in the log with respect to the entire set of relations present in the model. In this regard, we ran the logs generation setting to 1000 the number of traces to produce by the tools, except for PURPLE since it stops autonomously the simulation once the purpose is satisfied. In a second experiment, for each input model we decreased the number of traces to be produced to the amount of traces that PURPLE needs to cover the entire footprint matrix. Notably, both kinds of experiments have been repeated 10 times for each model, but, for the sake of presentation, the results reported in the following consider the worst results achieved by PURPLE and the average results achieved by the other tools. For each of the considered process model, we obtained eight event logs, two from each tool, and we compared them with respect to the coverage of the footprint matrix. Table 1 summarizes the results of this comparison. The first two columns, *Model* and *El.*, contain the name of the process model and the number of its elements, respectively. The third column, *Traces*, reports the number of traces autonomously generated by PURPLE that permit to cover the entire footprint matrix as reported in column 4. Columns from 5 to 7 show the percentages of activity relations covered by BIMP, GED, and PLG2, respectively, using a threshold of 1000 traces to be generated. The last three columns provide results for analogous experiments where the values of column 3 are used as threshold for the traces to be generated. Being guided by the evaluator, PURPLE covered entirely the relations matrix for each of the considered process models. Instead, the other tools show worse results, especially in the case of bigger models containing many parallel or exclusive branches, as such models involve higher numbers of order relations. Indeed, a model with n activities to be executed in parallel implies having $n(n-1)$ relations to discover, while a model with n activities in sequence (one after the other) shows just $n-1$ relations. For instance, model $p8$ has six parallel split gateways and one exclusive split gateway with 3 levels of nesting, and the resulting footprint matrix contains 699 relations to be discovered. The results achieved using the number of traces generated by PURPLE as threshold show that, on average, BIMP covers the 6% of the footprint matrix, PLG2 the 97%, and GED the 49%. When we increase the number of traces to produce, the results get slightly better for PLG2 which reaches the 99% of coverage, while they remain unchanged for BIMP and GED.

For what concerns the **process discovery via frequencies**, we use as quality measures the *error* in reproducing the desired percentages of occurrence for the trace variants, and the number of repetitions of each loop in the model. For this instantiation, a comparison between PURPLE and other tools would be unfair, since none of the other tools permits to customize the trace frequencies. Therefore, we run the simulations only on PURPLE. To this aim we used a set of models that contain loops, using a random value for the trace frequencies, the loop repetition thresholds fixed to 5, and the number of traces set to 10000. We analyzed the resulting logs using ProM to extract the occurrences of each trace variant and the number of loop repetitions. The results are presented in Table 2. We report the dimension of the input model, the number of generated traces, and the error. For each model, PURPLE reproduces the correct number of trace variants, keeping the loop repetitions under the selected threshold. These results were expected since the evaluator always provides deltas that

Table 2. Process discovery via frequencies results.

Model	El.	Traces	Loops repetition avg.	Error
p10	8	10000	3.2	0%
p11	10	10000	3.2	0%
p12	19	10000	2.9	0%
p13	25	10000	2.7	0%
p14	38	10000	2.9	0%

force the simulator to follow a precise execution trace in the LTS. Thus, the simulator produces exactly the log required by the user, avoiding errors.

For the **conformance checking via noise frequencies**, we compare the event logs generated by PURPLE and PLG2, as the latter permits to choose percentages of noise. We compare event logs with 5000 traces and the 10% of noised traces for each type of noise, i.e., 500 for missing head, 500 for missing tail, 500 for missing episode, 500 for order perturbation, and 500 for additional event. Finally, we analyze the logs to calculate the *error* in reproducing the desired occurrence rate for each type of noise.

Table 3 reports the results of the comparison. It shows that PURPLE produces always the exact number of noised traces, while PLG2 produces fewer noised traces than requested. On average, the error in the logs of PLG2 is equal to 20,8%, meaning that around 500 noised traces over 2500 are missing. The bigger lack results in reproducing traces with order perturbation, probably

Table 3. Conformance checking via noise frequencies results.

Model	El.	Traces	Error	
			PLG2	PURPLE
p25	10	5000	20,6%	0%
p26	11	5000	22,5%	0%
p27	12	5000	21,0%	0%
p28	17	5000	21,3%	0%
p29	21	5000	18,8%	0%

because PLG2 swaps also activities that are in parallel, so that the resulting trace is still compliant with the model. This problem is avoided in PURPLE, because it checks if the noised trace is compliant or not with the model before adding it to the log.

With respect to the **conformance checking via fixed align cost**, we evaluate only the logs of PURPLE, as no other tool supports this purpose. Here we set the desired alignment cost to 3 for each simulated model and a log size of 2000 traces, then we use the resulting logs and the input models to calculate via ProM the real costs for the alignments.

Table 4 puts in comparison, for each considered model, the required and the obtained alignment costs. The results show that the generated logs have alignment costs very close to the expectations. Overall, the error percentage made by the tool is on average equal to 2.7%. This discrepancy depends on the fact that the tool generates noised traces in order to make the log converge to the

Table 4. Conformance checking via fixed align cost results.

Model	El.	Traces	Alignment cost		
			Required	Obtained	Error
p30	6	2000	3	3.03	1%
p31	18	2000	3	2.91	3%
p32	27	2000	3	2.93	3%
p33	35	2000	3	2.91	2.3%
p34	43	2000	3	2.89	4.3%

required alignment cost, but before reaching it the simulation is stopped because the requested number of traces to produce is reached.

6   Related Works

This section discusses the most relevant works on the generation of artificial event logs. In describing them, we put the focus on the main features of PURPLE, such as the generation of event logs tailored to a desired mining purpose from models specified in different modeling languages. Thus, we mainly consider which kinds of event logs these approaches can generate, and which models they support.

Esgin and Karagoz present in [12] a solution to the problem of unlabeled event logs [1] proposing a synthetic event log generation approach. The generation of event logs can be tuned according to four parameters: the activity priority, an unexpected process termination probability, a noise threshold, and a branching probability for the choice gateways. Apart from these options, the simulation performs random executions of the input Petri-net. Differently from us, the approach supports only Petri-nets, cannot handle different mining purposes, and is not implemented in a tool.

Kataeva and Kalenkova propose in [15] grammar rules generating well-structured WF-Nets from which to produce logs. With respect to PURPLE, this work strongly limits the kind of logs that can be produced. Indeed, it handles just well-structured WF-Nets; moreover, logs cannot be tuned for specific mining purposes. In the same fashion, Burattin presents in [8] a tool, called Process Log Generator (PLG2), that creates well-structured BPMN models, and produces event logs from their simulation. To produce artificial models, PLG2 combines different control-flow patterns, via context-free grammar according to options like the number of gateways, or the presence of noise. With respect to PURPLE, this approach relies on random executions of the input model and works only with BPMN. Similarly, Alves and Günter propose in [17] a tool for the generation of event logs through the simulation of colored Petri-nets. They point out the issues related to the use of real-life event logs to fine tune mining algorithms, and how the incompleteness of an event log or the presence of noise can compromise the evaluation of the mining algorithms. Also this approach cannot tune the logs to produce since it relies on a random simulation of the input model.

Mitsyuk et al. face in [19] the problem of defining and generating logs from collaborative processes. They use an executable BPMN semantics supporting a subset of elements from the standard notation, such as tasks (also send/receive), sub-processes, parallel and exclusive gateways and cancellation events. Moreover, they consider the data perspective, as data objects can store single data values used for driving exclusive choices. The result is a log generator integrated in the ProM framework that produces random event logs in *.xes* files. With respect to our work, they can simulate communication between processes; however their approach only deals with a single modeling language and cannot be tuned for specific purposes. Stocker and Accorsi introduce in [21] an approach for generating event logs for a specific purpose, i.e., testing security properties. They present a tool, called *SecSy*, that generates logs from the simulation of a Petri-net in a specific scenario. The simulation performs random execution of the model, then it applies transformations to the generated event log. These transformations remove or insert activities and modify traces in order to violate security properties. Compared to ours, this approach takes care of just one specific purpose for which the produced event logs are tuned, and of just one modeling language (i.e., Petri-net).

Finally, Jouck and Depaire present in [14] a log generation approach specific for the comparison of discovery algorithms. They produce, and then simulate, a population of well-structured models from selected workflow patterns, to ensure the presence of specific activities order relations chosen by the user. Compared to ours, this work uses process trees to produce logs and, apart from process discovery, further purposes are not taken in consideration.

Summing up, differently form the PURPLE framework, the works mentioned above mainly focus on generating random event logs without focusing on specific mining purposes. Moreover, they limit the simulation to single modeling languages, and also to structured models. Lastly, some of them produce logs in non-standard formats, jeopardizing the compatibility with process mining tools.

7 Concluding Remarks

The presented work proposes a novel framework, PURPLE, to generate event logs via guided simulation of business models. PURPLE is meant to deal with several modeling languages and different mining purposes, as well as to ensure that the produced event log brings properties related to the selected mining purpose. Along with the definition of PURPLE, we present framework instantiations addressing the generation of event logs tailored to four purposes. These instantiations and two semantic engines for BPMN and Petri-Net are implemented in the PURPLE tool we provide. The analysis of the related works and the comparison we conducted between the existing log generators show that PURPLE is able, better than the others, to tune the simulation to the mining purpose.

In conclusion, both research questions presented in the Introduction can be positively answered. Concerning **RQ1**, we provided a general framework for the automated generation of event logs tailored to different mining purposes, as well as several instantiations of it, thus proving the feasibility of the approach. Regarding **RQ2**, we experimented our solution considering different purposes, and it proved to be more effective compared to purpose-agnostic simulators.

Assumptions and Limitations. We formalize the PURPLE framework under the assumption of simple event logs, which contain only activity names. Consequently, the PURPLE framework focuses mainly on control-flow aspects. In particular, the delta inherits this assumption as the sub-traces included in the delta are lists of activity_names. Notably, this still allows defining mining purposes and evaluators that guide the simulation according to aspects of some other model perspectives. For instance, one can define an evaluator producing log on the basis of the cost of the activities tailored to what-if analysis techniques. Nevertheless, handling traces with just the activity names results in a limitation to the variety of mining purposes and evaluators that can be defined on top of PURPLE. For example, simple logs do not deal with the resource perspective needed for social network analysis purposes, or the data perspective for decision mining purposes. Moreover, even if PURPLE produces event logs containing timestamps, they correspond to the moments the tool records the events. The user cannot influence timestamps, e.g., setting activity durations and delays between activities.

Regarding the delta definition in terms of sub-traces, another concern is that it cannot guide the simulator toward more abstract or generic behaviors. For instance, the delta cannot suggest the simulator to look for traces where a loop is repeated a casual number of times or where an event follows another not directly as some events may appear in the middle. Instead, by defining the delta using a language for expressing a set of traces (e.g., regular expressions), we could make more complex queries on the LTS, and thus address more purposes.

Future Works. As future works, we intend to pursue the development of the PURPLE framework, both from the theoretical and the practical point of view. We aim to formalize the PURPLE framework and its components in order to investigate its formal properties. Moreover, we intend to define and implement other evaluators, in order to handle other mining purposes and take into account other model perspectives, like data and multi-party communication. This can, for instance, give the chance to the user to generate event logs with different data quality issues in order to test approaches and algorithms dealing with them. Regarding the tool, we aim at parallelizing the computations of the simulation by handling more than one hint of the delta at the same time, and we plan to implement a debugging console for spotting useful information on simulations, including possible tool anomalies.

Acknowledgments. Work partially supported by the Italian MIUR PRIN projects *Seduce* n. 2017TWRCNB and *Fluidware* n. 2017KRC7KT, and the INdAM GNCS 2020 project *Sistemi reversibili concorrenti: dai modelli ai linguaggi*.

References

1. van der Aalst, W.: Matching observed behavior and modeled behavior: An approach based on Petri nets and integer programming. Decis. Support Syst. **42**(3), 1843–1859 (2006)
2. van der Aalst, W.: Process Mining: Data Science in Action. Springer, Berlin (2016). https://doi.org/10.1007/978-3-662-49851-4
3. van der Aalst, W., Weijters, T., Maruster, L.: Workflow mining: Discovering process models from event logs. Knowl. Data Eng. **16**(9), 1128–1142 (2004)
4. Abdul, B., Corradini, F., Re, B., Rossi, L., Tiezzi, F.: UBBA: unity based BPMN animator. In: CAiSE Forum. LNCS, vol. 350, pp. 1–9. Springer, Cham (2019). https://doi.org/10.1007/978-3-030-21297-1_1
5. Adriansyah, A.: Aligning observed and modeled behavior. Ph.D. thesis, Mathematics and Computer Science (2014). https://doi.org/10.6100/IR770080
6. Andrews, R., van Dun, C., Wynn, M., Kratsch, W., Röglinger, M., ter Hofstede, A.: Quality-informed semi-automated event log generation for process mining. Decis. Support Syst. **132** ,(2020)
7. Bose, R., Mans, R., van der Aalst, W.: Wanna improve process mining results? In: 2013 IEEE Symposium on Computational Intelligence and Data Mining, pp. 127–134. IEEE (2013)
8. Burattin, A.: PLG2: multiperspective process randomization with online and offline simulations. In: BPM Demo Track. vol. 1789, pp. 1–6. CEUR-WS.org (2016)
9. Cios, K., Pedrycz, W., Swiniarski, R., Kurgan, L.A.: Data Mining: A Knowledge Discovery Approach. Springer, New York (2007). https://doi.org/10.1007/978-0-387-36795-8

10. Corradini, F., Muzi, C., Re, B., Rossi, L., Tiezzi, F.: MIDA: multiple instances and data animator. In: BPM Demo. vol. 2196. CEUR-WS.org (2018)
11. Corradini, F., Muzi, C., Re, B., Rossi, L., Tiezzi, F.: Formalising and animating multiple instances in BPMN collaborations. Inf. Syst. **103** (2022)
12. Esgin, E., Karagoz, P.: Process profiling based synthetic event log generation. In: IC3K. vol. 1, pp. 516–524. SCITEPRESS (2019)
13. Günther, C.: Process mining in flexible environments. Ph.D. thesis, Technische Universiteit Eindhoven - Industrial Engineering and Innovation Sciences (2009)
14. Jouck, T., Depaire, B.: Generating artificial data for empirical analysis of control-flow discovery algorithms: a process tree and log generator. Bus. Inf. Syst. Eng. **61**(6), 695–712(2019)
15. Kataeva, V., Kalenkova, A.: Applying graph grammars for the generation of process models and their logs. In: Young Researchers' Colloquium on Software Engineering, vol. 8, pp. 83–87. HSE University (2014)
16. Levenshtein, V.: Binary codes capable of correcting deletions, insertions, and reversals. Soviet Phy. doklady. **10**, pp. 707–710 (1966)
17. de Medeiros, A., Günther, C.: Process mining: using CPN tools to create test logs for mining algorithms. In: Practical Use of Coloured Petri Nets and CPN Tools, vol. 576, pp. 177–190. University of Aarhus (2005)
18. Meseguer, J., Montanari, U., Sassone, V.: On the semantics of petri nets. In: Cleaveland, W.R. (ed.) CONCUR 1992. LNCS, vol. 630, pp. 286–301. Springer, Heidelberg (1992). https://doi.org/10.1007/BFb0084798
19. Mitsyuk, A., Shugurov, I.S., Kalenkova, A., van der Aalst, W.: Generating event logs for high-level process models. Simul. Model. Pract. Theory **74**, 1–16 (2017)
20. OMG: Business Process Model and Notation (BPMN V 2.0) (2011)
21. Stocker, T., Accorsi, R.: SecSy: security-aware synthesis of process event logs. Enterp. Model. Inf. Syst. Archit. 71–84 (2013)
22. van Dongen, B.F., Alves de Medeiros, A.K., Wen, L.: Process mining: overview and outlook of petri net discovery algorithms. In: Jensen, K., van der Aalst, W.M.P. (eds.) Transactions on Petri Nets and Other Models of Concurrency II. LNCS, vol. 5460, pp. 225–242. Springer, Heidelberg (2009). https://doi.org/10.1007/978-3-642-00899-3_13
23. Weijters, A., van Der Aalst, W., De Medeiros, A.: Process mining with the heuristics miner-algorithm. In: TU/e, Tech. Rep. vol. 166, pp. 1–34 (2006)
24. Yang, H., Park, M., Cho, M., Song, M., Kim, S.: A system architecture for manufacturing process analysis based on big data and process mining techniques. In: IEEE International Conference on Big Data, pp. 1024–1029 (2014)

Conformance Checking with Uncertainty via SMT

Paolo Felli, Alessandro Gianola, Marco Montali, Andrey Rivkin[✉],
and Sarah Winkler

Free University of Bozen-Bolzano, Bolzano, Italy
{pfelli,gianola,montali,rivkin,winkler}@inf.unibz.it

Abstract. Logs of real-life processes often feature uncertainty pertaining the recorded timestamps, data values, and/or events. We consider the problem of checking conformance of uncertain logs against data-aware reference processes. Specifically, we show how to solve it via SMT encodings, lifting previous work on data-aware SMT-based conformance checking to this more sophisticated setting. Our approach is modular, in that it homogeneously accommodates for different types of uncertainty. Moreover, using appropriate cost functions, different conformance checking tasks can be addressed. We show the correctness of our approach and witness feasibility through a proof-of-concept implementation.

1 Introduction

Process mining is a well-established field of research at the intersection between BPM and data science. The vast majority of process mining tasks assumes that their input event data provide an accurate and complete digital footprint of reality [20]. In many settings, this is an unrealistic assumption: events may be missing or totally/partially wrongly recorded, due to various factors such as human errors, faulty loggers, errors in the acquisition of events (e.g., through sensors), etc. To mitigate this issue, two lines of research emerged lately. The first deals with methodologies and techniques to improve the quality of event data, thus handling uncertainty in the data preparation phase [21]. The second aims instead at incorporating the management of uncertainty within the process mining tasks themselves, leading to a new generation of process mining techniques where process models [1,4,13,18] and/or event logs [8,17] explicitly address different kinds of uncertainty.

Surprisingly enough, the latter has received much less attention from the community. In this work, we aim at contributing to the advancement of process mining on uncertain data, considering in particular the problem of conformance checking [7]. Specifically, our contribution is twofold:

1. We introduce a framework for data-aware conformance checking over uncertain logs, through a suitably extended notion of alignment. The framework employs Data Petri nets [14] for reference process models, and addresses event logs incorporating sophisticated forms of uncertainty, pertaining the recorded timestamps, data values, and/or events. Notably, the framework comes with a generic cost function whose components can be flexibly instantiated to homogeneously account for a variety of measures required for computing optimal alignments.

© Springer Nature Switzerland AG 2022
C. Di Ciccio et al. (Eds.): BPM 2022, LNCS 13420, pp. 199–216, 2022.
https://doi.org/10.1007/978-3-031-16103-2_15

2. We devise a corresponding operational counterpart to effectively attack the problem of computing alignments and their costs. Instead of relying on ad-hoc algorithmic techniques, our approach builds on and extends [11] to encode the problem into the well-established automated reasoning framework of SMT. This allows us to employ state-of-the-art SMT solvers.

To handle uncertainty in the log, we follow the approach in [17], where the log is explicitly enriched with annotations reflecting the degree and nature of uncertainty. Such annotations may be derived from operational characteristics of the information system recording the event data (considering its logging precision and reliability), and/or by directly attaching them to the generated events. For instance, the log may be enriched with explicit details on the coarseness or precision of an automatic logging device (such as a sensor); alternatively, uncertainty-related annotations may be derived from domain knowledge on the precision and frequency of a specific human activity. In particular, our framework accounts for four main types of uncertain event data.

- *Uncertain events:* these are recorded in a log trace but come with a known *confidence value*, capturing the degree of (un)certainty about the fact that a recorded event actually happened at all during the process execution.
- *Uncertain timestamps:* due to coarseness of the logging activity, events are in general not totally ordered, but come with a fixed range of possible timestamp values. This calls for considering multiple possible orderings and treating a log trace as *a set* of events rather than a sequence.
- *Uncertain activities:* this pertains events whose reference activity is not certainly known. Hence, the event comes with a candidate set of possible activities (each with its own confidence value).
- *Uncertain data values:* in the execution of data-aware processes, for instance due to sensor precision, event data attributes may come with both coarseness and ambiguity. Specifically, the log may only record a set of possible values or an interval for a given attribute, requiring all possible values to be considered.

We stress that the notion of confidence used here should not be confused with that of probability: it measures the degree of trust in the recorded behaviour, which has nothing to do with the likelihood/frequency of such a behaviour.

To account for these different types of uncertain event data, we borrow from [17] and adapt to our data-aware setting the notion of *realization*. A realization of a log trace with uncertainty is an *ordered sequence* of events in which the uncertainty of all types of event data as above is resolved. Our task then concretely becomes as follows: given a Data Petri net and a log trace with uncertainty, find some realization of that trace that admits an *optimal alignment*, i.e., an alignment of minimal cost among all possible realizations for that log trace. Differently from [17], the confidence values of the original trace are used as an essential component for measuring the cost incurred in selecting realizations.

Crucially, since we are in a data-aware setting, a log trace may correspond to infinitely many possible realizations. This is handled symbolically thanks to our SMT-based approach.

The rest of the paper is organized as follows. First, in Sect. 2 we recall the required preliminaries. Then, in Sect. 3 we fix the shape of traces in event logs with uncertainty and the notion of alignments. In Sect. 4 we detail the cost components that must be considered in the setting with uncertain even data and that we use to define the conformance checking task. We discuss separately one main cost component: the notion of data-aware alignment cost function (in Sect. 4.1). In Sect. 5 we illustrate our SMT-based encoding and we report on the implementation. We conclude in Sect. 6.

2   Preliminaries

In this section we recall data Petri nets (DPNs) and their execution semantics, and the main notions of the machinery behind our approach, namely SMT.

2.1   Data Petri Nets

We use Data Petri nets (DPNs) for modelling multi-perspective processes, adopting the same formalization as in [11,14]. For lack of space, in what follows we only recall the definitions and notation required for our technical development, referring the reader to [11,14] for further details.

Let V be a set of *process variables*, each with a type and an associated domain: booleans (type $\texttt{bool}$), integers ($\texttt{int}$), rationals ($\texttt{rat}$) or strings ($\texttt{string}$). We consider two disjoint sets of annotated variables $V^r = \{v^r \mid v \in V\}$ and $V^w = \{v^w \mid v \in V\}$ to be read and written by process activities, as explained below. Based on these, we define constraints according to the grammar for c:

$$c ::= v_b \mid b \mid n \geq n \mid r \geq r \mid r > r \mid s = s \mid c \wedge c \mid \neg c \quad s ::= v_s \mid t$$
$$n ::= v_z \mid z \mid n + n \mid -n \qquad\qquad\qquad r ::= v_r \mid q \mid r + r \mid -r$$

where $v_b \in V_{\texttt{bool}}$, $b \in \mathbb{B}$, $v_s \in V_{\texttt{string}}$, $t \in \mathbb{S}$, $v_z \in V_{\texttt{int}}$, $z \in \mathbb{Z}$, $v_r \in V_{\texttt{rat}}$, and $q \in \mathbb{Q}$. Standard equivalences apply, hence disjunction (i.e., $\vee$) and comparisons $>$, $\neq, <, \leq$ can be used as well ($\texttt{bool}$ and $\texttt{string}$ only support (in)equality). The set of constraints over variables V is denoted $\mathcal{C}(V)$. These form the basis for expressing conditions on the values of variables that are read and written during the execution of process activities. For instance, a constraint $(v_1^r > v_2^r)$ dictates that the current value of variable v_1 is greater than the current value of v_2. Similarly, $(v_1^w > v_2^r + 1) \wedge (v_1^w < v_3^r)$ requires that the new value given to v_1 (i.e., assigned as a result of the execution of the activity to which this constraint is attached) is greater than the current value of v_2 plus 1, and smaller than v_3.

Definition 1 (DPN). *A tuple* $\mathcal{N} = (P, T, F, \ell, A, V, guard)$ *is a Petri net with data (DPN), where:*

- (P, T, F, ℓ) *is a Petri net with two non-empty disjoint sets of places P and transitions T, a flow relation $F : (P \times T) \cup (T \times P) \rightarrow \mathbb{N}$ and a labeling function $\ell : T \rightarrow A \cup \{\tau\}$, where A is a finite set of activity labels and τ is a special symbol denoting silent transitions;*

- *V is a set of typed process variables; and*
- *guard: $T \to \mathcal{C}(V)$ is a guard assignment (for $t \in T$ with $\ell(t) = \tau$ we assume that guard(t) does not use variables in V^w).*

As customary, given $x \in P \cup T$, we use ${}^\bullet x := \{y \mid F(y, x) > 0\}$ to denote the *preset* of x and $x^\bullet := \{y \mid F(x, y) > 0\}$ to denote the *postset* of x.

To assign values to variables, we consider a *state variable assignment*, i.e., a total function α that assigns a value (of the right type) to each variable in V. A *state* in a DPN $\mathcal{N}$ is a pair (M, α) constituted by a marking $M: P \to \mathbb{N}$ for the underlying Petri net (P, T, F, ℓ), plus a state variable assignment α. Therefore, a state simultaneously accounts for the control flow progress and for the current values of all variables in V, as specified by α.

Given $\mathcal{N}$, we fix one state (M_I, α_0) as *initial*, where M_I is the initial marking of the underlying Petri net (P, T, F, ℓ) and α_0 specifies the initial value of all variables in V. Similarly, we denote the final marking as M_F, and call *final* any state of $\mathcal{N}$ of the form (M_F, α_F) for some α_F.

We now define when a Petri net transition may fire from a given state (M, α). Informally, a *transition firing* is a couple (t, β) where $t \in T$ and β is a function used to determine the new values of variables after the transition has fired. The step yields a new state (M', α'), and is denoted $(M, \alpha) \xrightarrow{(t_n, \beta_n)} (M', \alpha')$. A transition firing is *valid* in a state (M, α) when t is enabled in M and α satisfies the constraint associated to t. The formal definition can be found, e.g., in [11, 14].

Based on this single-step transition firing, we say that a state (M', α') is *reachable* in a DPN with initial state (M_I, α_0) iff there exists a sequence of valid transition firings of the form $\mathbf{f} = \langle (t_1, \beta_1), \ldots, (t_n, \beta_n) \rangle$ such that $(M_I, \alpha_0) \xrightarrow{(t_1, \beta_1)}$ $\ldots \xrightarrow{(t_n, \beta_n)} (M', \alpha')$. Moreover, such a sequence $\mathbf{f}$ is called a *process run* of $\mathcal{N}$ if $(M_I, \alpha_0) \xrightarrow{\mathbf{f}} (M_F, \alpha_F)$ for some α_F, i.e., if the run leads to a final state. As in [11, 15], we restrict to DPNs where at least one final state is reachable.

We denote the set of transition firings of a DPN $\mathcal{N}$ by $\mathcal{F}(\mathcal{N})$, and the set of process runs by $Runs(\mathcal{N})$.

Example 1. Let $\mathcal{N}$ be as shown (with initial marking $[p_0]$ and final marking $[p_3]$). $Runs(\mathcal{N})$ contains, e.g., $\langle (a, \{x^w \mapsto 2\}), (b, \{y^w \mapsto 1\}), (c, \{x^r \mapsto 2, y^r \mapsto 1\}) \rangle$ and $\langle (a, \{x^w \mapsto 1\}), (b, \{y^w \mapsto 1\}), (d, \{y^r \mapsto 1, x^r \mapsto 1\}) \rangle$, for $\alpha_0 = \{x, y \mapsto 0\}$.

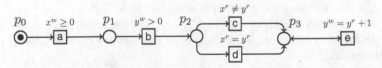

2.2 Satisfiability Modulo Theories (SMT)

The classic propositional satisfiability (SAT) problem amounts to, given a propositional formula φ, either find an assignment ν under which φ evaluates to true, or detect that φ is unsatisfiable. E.g., given the formula $(p \lor q) \land (\neg p \lor r) \land (\neg r \lor \neg q)$, a satisfying assignment is $\nu(p) = \nu(r) = \top, \nu(q) = \bot$. The SMT problem [3] is an extension of

SAT that consists of establishing satisfiability of a formula φ whose language enriches propositional formulas with constants and operators from one or more theories $\mathcal{T}$ (e.g., arithmetics, bit-vectors, arrays, uninterpreted functions). In this paper, we only consider the theories of linear integer and rational arithmetic ($\mathcal{LIA}$ and $\mathcal{LQA}$). For instance, the SMT formula $a > 1 \wedge (a + b = 10 \vee a - b = 20) \wedge p$, where a, b are integer and p is a propositional variable, is satisfiable by the assignment ν such that $\nu(a) = \nu(b) = 5$ and $\nu(p) = \top$. Another important problem studied in the area of SMT and relevant to this paper is the one of Optimization Modulo Theories (OMT) [19]. The OMT problem asks, given a formula φ, to find a satisfying assignment of φ that minimizes or maximizes a given objective expression. SMT-LIB [2] is an initiative aiming at providing an extensive on-line library of benchmarks and promoting the adoption of common languages and interfaces for SMT solvers. In this paper, we employ the SMT solvers Yices 2 [10] and Z3 [9].

3 Event Logs with Uncertainty and Alignments

Let ID be a finite set of event identifiers, A be a finite set of activity labels, and TS be a totally ordered set of possible timestamps (for simplicity, we use $\mathbb{N}$).

Definition 2. *An event with uncertainty is a tuple* $ue = \langle \text{ID}, conf, \text{LA}, \text{TS}, \alpha \rangle$ *s.t.*

- *ID* $\in ID$ *is an event identifier;*
- $0 < conf \leq 1$ *expresses the confidence that the event actually happened. We say that the event is an uncertain event whenever* $conf < 1$;
- *LA* $= \{b_1 : p_1, \ldots, b_n : p_n\}$ *is a finite, non-empty subset of activity labels* $b_i \in A$, *each associated to a confidence value* $0 < p_i \leq 1$ *so that* $\sum_{i=1}^{n} p_i = 1$;
- *TS is either a finite set of timestamps in* TS *or an interval over* TS;
- *with some abuse of notation,* α *is a (possibly partial) function returning for variables in* V *a finite set of values in the domain of* v *or an interval over such domain (if* v *is of type* int *or* rat*).*

Given an event $ue = \langle \text{ID}, conf, \text{LA}, \text{TS}, \alpha \rangle$, we denote its components by $\text{ID}(ue)$, $conf(ue)$, $\text{LA}(ue)$, $\text{TS}(ue)$ and $\alpha(ue)$, respectively.

Note that we do not associate confidence values to timestamps, along the lines of [17]. We also do not consider timestamp values following any kind of distribution, e.g., a normal distribution, as this would make the encoding in Sect. 5 computationally too challenging.

Definition 3. *A* log *trace with uncertainty* **ue** *is a finite set of events with uncertainty, such that all event identifiers are unique.*

Thus, there is no fixed order among the events in a trace with uncertainty. An *event log L* is a multiset of log traces with uncertainty.

Example 2. Consider $\mathcal{N}$ from Example 1. For simplicity, we use natural numbers for timestamps. The following are three possible traces with uncertainty:

$$\textbf{ue}_1 = \{\langle \#_1, .25, \{a:1\}, [0\text{-}5], \{x \mapsto \{2,3\}\}\rangle, \ \langle \#_2, .9, \{b:.8, c:.2\}, \{2\}, \{y \mapsto \{1\}\}\rangle$$
$$\textbf{ue}_2 = \{\langle \#_3, 1, \{a:1\}, \{0\}, \{x \mapsto [1, 6.5]\}\rangle, \ \langle \#_4, 1, \{b:1\}, \{2\}, \{y \mapsto \{1\}\}\rangle,$$
$$\langle \#_5, 1, \{c:1\}, \{3\}, \emptyset\rangle\}$$
$$\textbf{ue}_3 = \{\langle \#_6, 1, \{a:1\}, \{2\}, \{x \mapsto \{6\}\}\rangle, \ \langle \#_7, 1, \{b:1\}, \{2\}, \{y \mapsto \{1\}\rangle\}\}$$

For instance, $\textbf{ue}_1$ has two events with uncertainty: $\#_1$ and $\#_2$. The former is uncertain (confidence 0.25), has event label a (with confidence 1), timestamp interval $[0,5]$ and a variable assignment such that x is assigned to either 2 or 3. Also $\#_2$ is uncertain, has label b or c (with associated confidence values 0.8 and 0.2, respectively), timestamp 2 and variable assignment $y = 1$. Another example of an uncertain event is $\#_3$ in $\textbf{ue}_2$, where x takes a value from the interval $[1, 6.5]$.

An activity label $b \in A$ is *admissible* for an event with uncertainty *ue* iff it is consistent with LA(*ue*), i.e., if there is some p such that $(b, p) \in$ LA(*ue*). Admissibility of timestamp and variable values is defined similarly.

Intuitively, given a log trace with uncertainty $\textbf{ue}$, a *realization* of $\textbf{ue}$ is a sequence $\textbf{e} = \langle e_1, \ldots, e_n \rangle$ of events corresponding to a possible sequentialization of *a subset of* the events with uncertainty in $\textbf{ue}$ that is consistent with their uncertain timestamps, and in which only one possible value is chosen for event labels and variable assignments. The remaining events with uncertainty in $\textbf{ue}$ but not in $\textbf{e}$ are simply discarded.

An event without uncertainty, or simply *event*, is a tuple (ID, b, $\hat{\alpha}$), where ID is again an event identifier, $b \in A$ is an activity label, and $\hat{\alpha}$ is a special variable assignment that assigns to each variable $v \in V$ a *single* value of the correct type. Given an event $e = $ (ID, b, $\hat{\alpha}$), we denote its components by ID(e), *lab*(e) and $\hat{\alpha}(e)$, respectively. These events are akin to the standard notion of events in conformance checking literature, extended with variable assignments as in [11], with the addition of identifiers (which are needed to relate them to the corresponding event with uncertainty in the log, as explained later). The set of all possible such events is denoted by $\mathcal{E}$.

Definition 4 (Realization). *A sequence* $\textbf{e} = \langle e_1, \ldots, e_n \rangle$ *of events as above is a* realization *of a log trace with uncertainty* $\textbf{ue}$ *if there is a subset* $\{ue_1, \ldots, ue_n\} \subseteq \textbf{ue}$ *and a sequence of timestamps* $t_1 \leq t_2 \leq \cdots \leq t_n$ *such that for each* $i \in [1, n]$:

(i) t_i *is admissible for* ue_i, *hence defining an ordering on* $\textbf{e}$;
(ii) ID(e_i) = ID(ue_i);
(iii) *lab*(e_i) = b *with* b *admissible for* ue_i;
(iv) $\hat{\alpha}(e_i)(v) \in \alpha(ue_i)(v)$ *for all* v *such that* $\alpha(ue_i)(v)$ *is defined.*

Moreover, we impose that for every $ue \in \textbf{ue}$ *with conf*(*ue*) = 1 *there is an event* $e \in \textbf{e}$ *with* ID(e_i) = ID(ue_i), *namely a realization cannot discard events in the log that are not uncertain.*

A realization of a trace with uncertainty $\textbf{ue}$ is thus a possible sequentialization of (a subset of) the events with uncertainty in $\textbf{ue}$ in which a single, admissible timestamp value, activity label and value for variables are selected from the corresponding event with uncertainty $ue \in$ **ue** with $\text{ID}(e) = \text{ID}(ue)$. We denote that e is a realization of **ue** by writing $e \in \mathcal{R}(\textbf{ue})$. Events in a realization e are no longer associated with confidence values (which remain in **ue**).

Note that $\mathcal{R}(\textbf{ue})$ cannot be empty, as it is always possible to select $\{t_1, \ldots, t_n\}$ as in Definition 4: even if two events cannot be ordered because they admit the same single timestamp, both orderings are accounted for by different realizations. $\mathcal{R}(\textbf{ue})$ can be infinite if data variables are assigned by **ue** to intervals over dense domains.

Example 3. Consider the trace with uncertainty $\textbf{ue}_1$ in Example 2. It has 13 realizations, since the first event has two possible variable assignments, the second event has two possible labels; moreover, the two events can be ordered in both ways and in addition each event can also be removed (as they are uncertain).

Two possible realizations of $\textbf{ue}_1$ are $e' = \langle\langle \#_1, a, \{x \mapsto 2\}\rangle, \langle \#_2, b, \{y \mapsto 1\}\rangle\rangle$ and $e'' = \langle\langle \#_2, c, \{y \mapsto 1\}\rangle, \langle \#_1, a, \{x \mapsto 3\}\rangle\rangle$. Note that these realizations differ in the order of the two events, label selection and variable assignments.

We focus on a conformance checking procedure to construct an *alignment* of a log trace e (that is a realization of a log trace with uncertainty **ue**) w.r.t. the process model (i.e., the DPN $\mathcal{N}$), by matching event labels in the log trace against transition firings in the process runs of $\mathcal{N}$. However, when constructing an alignment, not every event in the log trace can always be put in correspondence with a transition firing, and vice versa. Therefore, as customary, we consider a special "skip" symbol $\gg$ and the extended set of events $\mathcal{E}^\gg = \mathcal{E} \cup \{\gg\}$ and, given $\mathcal{N}$, the extended set of transition firings $\mathcal{F}^\gg = \mathcal{F}(\mathcal{N}) \cup \{\gg\}$.

Given a DPN $\mathcal{N}$ and a set $\mathcal{E}$ of events (without uncertainty) as above, a pair $(e, f) \in \mathcal{E}^\gg \times \mathcal{F}^\gg \setminus \{(\gg, \gg)\}$ is called *move*. A move (e, f) is called: (i) *log move* if $e \in \mathcal{E}$ and $f = \gg$; (ii) *model move* if $e = \gg$ and $f \in \mathcal{F}(\mathcal{N})$; (iii) *synchronous move* if $(e, f) \in \mathcal{E} \times \mathcal{F}(\mathcal{N})$. Let $Moves_\mathcal{N}$ be the set of all such moves. We now show how moves can be used to define alignments of realizations.

For a sequence of moves $\gamma = \langle (e_1, f_1), \ldots, (e_n, f_n)\rangle$, the *log projection* $\gamma|_L$ of γ is the subsequence $\langle e_1', \ldots, e_i'\rangle$ of $\langle e_1, \ldots, e_n\rangle$ that is in $\mathcal{E}^*$ and is obtained by projecting away from γ all $\gg$ symbols. Similarly, the *model projection* $\gamma|_M$ of γ is the subsequence $\langle f_1', \ldots, f_j'\rangle$ of $\langle f_1, \ldots, f_n\rangle$ such that $\langle f_1', \ldots, f_j'\rangle \in \mathcal{F}(\mathcal{N})^*$.

Definition 5 (Alignment). *Given $\mathcal{N}$, a sequence of moves γ is a* complete alignment *of a realization* e *if $\gamma|_L = $ e and $\gamma|_M \in Runs(\mathcal{N})$.*

Example 4. Consider the realization $e' = \langle\langle \#_1, a, \{x \mapsto 2\}\rangle, \langle \#_2, b, \{y \mapsto 1\}\rangle\rangle$ from Example 3. The following are examples of possible complete alignments of e' with respect to the DPN from Example 1:

$\gamma^1_{e'}$	$\#_1$		$\#_2$		$\gg$
	a	$x^w \mapsto 2$	b	$y^w \mapsto 1$	c

$\gamma^2_{e'}$	$\#_1$		$\#_2$		$\gg$
	a	$x^w \mapsto 5$	b	$y^w \mapsto 1$	c

$\gamma^3_{e'}$	$\#_1$		$\gg$		$\#_2$
	a	$x^w \mapsto 2$	b	$y^w \mapsto 2$	d

We denote by $Align(\mathcal{N}, e')$ the set of all complete alignments for e' w.r.t. $\mathcal{N}$.

As shown in Example 4, some alignments are more fitting than others: for instance, they can have mismatching variable assignments (e.g., in the first move of $\gamma^2_{e'}$) and label matching (e.g., in the third move of $\gamma^3_{e'}$). This will be captured by the cost function, described next.

4　Costs and Optimal Alignments

In this paper we do not wish to restrict to specific cost functions, and therefore fix only a *cost schema* which leaves several elements arbitrary. We however illustrate the cost components and describe one possible instantiation of said schema, which we use in the encoding in Sect. 5. The overall cost schema for alignments is shown in Fig. 1.

$$\mathfrak{K}(\gamma_{\mathbf{e}}, \mathbf{ue}) = \overbrace{\sum_{i\in[1,n]} \underbrace{\kappa(e_i, f_i)}_{\substack{\text{data-aware}\\ \text{alignment cost (Sec. 4.1)}}} \otimes \underbrace{\theta(e_i, \mathbf{ue})}_{\text{confidence cost}}}^{\text{alignment cost } \kappa_A(\gamma_{\mathbf{e}}, \mathbf{ue})} + \overbrace{\sum_{e\in\mathbf{ue}, e\notin\mathbf{e}} \kappa_{\mathbf{ue}}(e)}^{\text{event removal cost } \kappa_R(\mathbf{e}, \mathbf{ue})}$$

Fig. 1. Structure of the cost of an alignment $\gamma_{\mathbf{e}} = \langle(e_1, f_1), \ldots, (e_n, f_n)\rangle$ of a realization $\mathbf{e}$ of a trace with uncertainty $\mathbf{ue}$. The cost associated to the selection of $\mathbf{e}$ is given by $\kappa_R(\mathbf{e}, \mathbf{ue})$ plus, at each step, the additional penalty given by $\theta(e_i, \mathbf{ue})$ according to $\otimes$.

We first give the intuition. The general idea is that, as we are not merely interested in finding a cost-minimal alignment for an arbitrary realization as in [17], i.e., without considering the confidence associated to the selection of realizations, we impose a confidence cost on realizations *in addition* to the cost of aligning them, as illustrated in Fig. 1. As a result, the cost $\mathfrak{K}(\gamma_{\mathbf{e}}, \mathbf{ue})$ of an alignment $\gamma_{\mathbf{e}}$ with respect to an uncertain trace $\mathbf{ue}$ is the sum of two costs:

1) The alignment cost $\kappa_A(\gamma_{\mathbf{e}}, \mathbf{ue})$ measures the quality of the alignment $\gamma_{\mathbf{e}}$ for the realization $\mathbf{e}$. As customary in the conformance checking literature, it is based on a mapping $\kappa \colon Moves_{\mathcal{N}} \to \mathbb{R}^+$ that assigns a cost to every move $(e_i, f_i) \in \gamma_{\mathbf{e}}$. In Sect. 4.1 we will discuss in more detail how this function κ can be defined.

In addition, for synchronous moves and log moves, this cost is combined with a confidence penalty that depends on $conf(e_i)$ and on the confidence value p associated to the activity label $b = lab(e_i)$ according to the event with uncertainty ue so that $\mathrm{ID}(e_i) = \mathrm{ID}(ue)$, i.e., $(b, p) \in \mathrm{LA}(ue)$. Intuitively, this imposes a penalty for selecting b as the activity chosen for e_i in the realization $\mathbf{e}$ of $\mathbf{ue}$.

We do not fix a specific calculation of this penalty, but keep it parametric and denote it as $\theta(e_i, \mathbf{ue})$. The cost of an alignment $\gamma_{\mathbf{e}}$ can then be defined as:

$$\kappa_A(\gamma_{\mathbf{e}}, \mathbf{ue}) = \sum_{i=1}^n \kappa(e_i, f_i) \otimes \theta(e_i, \mathbf{ue})$$

where $\otimes$ denotes an arbitrary operator to combine the two costs.

For instance, in Sect. 5 we assume, for a realization $\mathbf{e}$ of $\mathbf{ue}$ and alignment $\gamma_{\mathbf{e}} = \langle(e_1, f_1), \ldots, (e_n, f_n)\rangle$:

$$\kappa(e_i, f_i) \otimes \theta(e_i, \mathbf{ue}) = \begin{cases} \kappa(e_i, f_i) & \text{if } e_i = \gg, \text{ otherwise:} \\ \theta(e_i, \mathbf{ue}) & \text{if } \kappa(e_i, f_i) = 0 \\ \kappa(e_i, f_i) \cdot (1 + \theta(e_i, \mathbf{ue})) & \text{if } \kappa(e_i, f_i) > 0 \end{cases}$$

in which we fix $\theta(e_i, \mathbf{ue}) = (1 - conf(e_i)) + (1 - p)$, where b is the label of e_i, i.e., $b = lab(e_i)$, and p is the confidence value associated to b, i.e., $(b, p) \in \text{LA}(\mathbf{ue})$.

Intuitively, in this definition of $\kappa_A(\gamma_\mathbf{e}, \mathbf{ue})$, the cost of model moves is simply (a data-aware extension of) the usual alignment cost, which we define in Sect. 4.1. Otherwise, the cost includes a penalty for having selected $lab(e_i)$ in the realization $\mathbf{e}$ of $\mathbf{ue}$. Such penalty decreases the more we are confident about the selected activity among the possible activities associated to the event with uncertainty. Other definitions of θ and $\otimes$ are however possible.

2) The event removal cost $\kappa_R(\mathbf{e}, \mathbf{ue})$ measures the cost of selecting the subsets of the events in $\mathbf{ue}$ that appear in $\mathbf{e}$, discarding the remaining (uncertain) events. Although we do not wish to restrict to a specific function κ_R, a reasonable option is to assume it to be based on a mapping $\kappa_\mathbf{ue} \colon \mathcal{E} \to \mathbb{R}_{\geq 0}$ that assigns a removal cost to each event, proportionally to the confidence value $conf(ue)$ for $ue \in \mathbf{ue}$ so that $\text{ID}(e) = \text{ID}(ue)$. Hence, the total event removal cost can be computed as:

$$\kappa_R(\mathbf{e}, \mathbf{ue}) = \sum_{e \in \mathbf{ue}, e \notin \mathbf{e}} \kappa_\mathbf{ue}(e)$$

For instance, in Sect. 5 we will take $\kappa_\mathbf{ue}(e)$ to be precisely $conf(ue)$, for ue as above, when such a confidence value is less than 1, and equal to infinity otherwise (to prevent events that are not indeterminate to be discarded from realizations). Other definitions of κ_R are however possible. Again, according to these expressions, the cost of selecting $\mathbf{e}$ as a realization of $\mathbf{ue}$ results from $\kappa_R(\mathbf{e}, \mathbf{ue})$ for removed events plus, at each step, a penalty $\theta(e_i, \mathbf{ue})$ for not having discarded e_i but having selected one admissible label among those associated to the uncertain event in $\mathbf{ue}$ with the same ID.

Example 5. Consider again the trace with uncertainty $\mathbf{ue}_1$ from Example 3:

$$\mathbf{ue}_1 = \{\langle \#_1, .25, \{\mathsf{a} : 1\}, [\text{0-5}], \{x \mapsto \{2, 3\}\}\rangle, \; \langle \#_2, .9, \{\mathsf{b} : .8, \mathsf{c} : .2\}, \{2\}, \{y \mapsto \{1\}\}\rangle\}$$

and three of its possible realizations $\mathbf{e}_1 = \langle\langle \#_1, \mathsf{a}, \{x \mapsto 3\}\rangle\rangle$, $\mathbf{e}_2 = \langle\langle \#_2, \mathsf{b}, \{y \mapsto 1\}\rangle\rangle$ and $\mathbf{e}_3 = \langle\langle \#_2, \mathsf{c}, \{y \mapsto 1\}\rangle\rangle$, where in all cases one of the two events was removed. If we adopt the specific implementation of cost functions exemplified above (and used in our encoding in Sect. 5), we have that $\kappa_R(\mathbf{e}_2, \mathbf{ue}_1) > \kappa_R(\mathbf{e}_1, \mathbf{ue}_1)$ since $conf(\#_2) > conf(\#_1)$. Similarly, the difference between $\mathbf{e}_2$ and $\mathbf{e}_3$ is only in the activity chosen for $\#_2$, therefore the cost of selecting $\mathbf{e}_2$ is smaller than that for $\mathbf{e}_3$, because the confidence associated to activity b is greater than the one associated to c; hence $\theta(\langle \#_2, \mathsf{b}, \{y \mapsto 1\}\rangle, \mathbf{ue}_1) < \theta(\langle \#_2, \mathsf{c}, \{y \mapsto 1\}\rangle, \mathbf{ue}_1)$.

Definition 6 (Cost of alignments). *Fixed the two arbitrary cost functions κ_A and κ_R introduced above, given $\mathcal{N}$, a trace with uncertainty $\mathbf{ue}$ that has realization $\mathbf{e} = \langle e_1, \ldots, e_n \rangle$ and an alignment $\gamma_\mathbf{e} = \langle (e_1, f_1), \ldots, (e_n, f_n) \rangle \in Align(\mathcal{N}, \mathbf{e})$, the cost of $\gamma_\mathbf{e}$ w.r.t. $\mathbf{ue}$, denoted $\mathfrak{K}(\gamma_\mathbf{e}, \mathbf{ue})$, is obtained as shown in Fig. 1:*

$$\mathfrak{K}(\gamma_\mathbf{e}, \mathbf{ue}) = \kappa_A(\gamma_\mathbf{e}, \mathbf{ue}) + \kappa_R(\mathbf{e}, \mathbf{ue}).$$

An alignment γ_e is *optimal for* e if $\kappa_A(\gamma_e, \textbf{ue})$ is minimal among all complete alignments for e, i.e., there is no $\gamma_e' \in Align(\mathcal{N}, e)$ with $\kappa_A(\gamma_e', \textbf{ue}) < \kappa_A(\gamma_e, \textbf{ue})$. Similarly, given $\mathcal{N}$ and a trace with uncertainty **ue**, we say that γ_e is *optimal for* **ue** if $\mathfrak{K}(\gamma_e, \textbf{ue})$ is minimal among all possible realizations of **ue**, i.e., there is no other realization $e' \in \mathcal{R}(\textbf{ue})$ and alignment $\gamma_{e'} \in Align(\mathcal{N}, e')$ so that $\mathfrak{K}(\gamma_{e'}, \textbf{ue}) < \mathfrak{K}(\gamma_e, \textbf{ue})$.

Definition 7 (Conformance checking). *Given $\mathcal{N}$, the* conformance checking task *for a trace with uncertainty* **ue** *is to find a realization* e *of* **ue** *and an alignment* γ_e *that is optimal for* **ue**.

Multiple realizations e and optimal alignments γ_e may exist for **ue**, though the minimal cost is unique for a given cost function. The *conformance checking task for an unordered log* consists of the conformance checking task for all its traces.

Note that we can easily formulate the task of finding the lower-bound on the cost of possible alignments among all realizations (as in [17]), given **ue**, by simply imposing $\kappa_R(e, \textbf{ue}) = 0$, $\theta(e, \textbf{ue}) = 1$ and by taking $\otimes$ as product: this corresponds to impose no cost for selecting an arbitrary realization, thus simply returning one that has minimal alignment cost κ.

In the remainder, we discuss separately the definition of alignment cost κ.

4.1 Data-Aware Alignment Cost Function

We use a generalized form of a cost function to measure the conformance between a realization and a process run in $Runs(\mathcal{N})$, i.e., to define $\kappa: Moves_\mathcal{N} \to \mathbb{R}_{\geq 0}$ used in Definition 6. As in [11], we parameterize this by three penalty functions:

$$P_L : \mathcal{E} \to \mathbb{N} \qquad P_M : \mathcal{F}(\mathcal{N}) \to \mathbb{N} \qquad P_= : \mathcal{E} \times \mathcal{F}(\mathcal{N}) \to \mathbb{N}$$

called *log move penalty*, *model move penalty* and *synchronous move penalty*, respectively. Intuitively, $P_L(e)$ gives the cost that has to be paid for a log move e; $P_M(f)$ penalizes a model move f; and $P_=(e, f)$ expresses the cost to be paid for a synchronous move of e and f. By suitably instantiating $P_=$, P_L, and P_M, one can obtain conventional cost functions [11]: the Levenshtein distance [5,6], standard cost function for multi-perspective conformance checking [14,15].

Then, the data-aware cost function $\kappa: Moves_\mathcal{N} \to \mathbb{R}_{\geq 0}$ we adopt in Definition 6 is simply defined as $\kappa(e, f) = P_L(e)$ if $f = \gg$, $\kappa(e, f) = P_M(f)$ if $e = \gg$, and $\kappa(e, f) = P_=(e, f)$ otherwise.

Data-Aware Cost Component of $P_=$. Crucially, for DPNs we typically consider a data-aware extension of the usual distance-based cost function for synchronous moves. Indeed, given an event $e = (\text{ID}, b, \hat{\alpha})$ of a realization and a transition firing $f = (t, \beta)$, we want $P_=(e, f)$ to compare also the values assigned to variables by $\hat{\alpha}$ and β. For instance, in Example 4, the alignment $\gamma_{e_1}^2$ is so that its first (synchronous) move has a mismatch between the value assigned to variable x by the event #1 (i.e., $\hat{\alpha}(\#_1)(x) = 2$) and transition firing $(a, \{x^w \mapsto 5\})$. Various data-aware realizations of $P_=$ have been already addressed in the literature [11,15].

Example 6. Consider again the trace with uncertainty $\mathbf{ue}_1$ from Example 5, i.e., $\mathbf{ue}_1 = \{\langle \#_1, .25, \{\mathtt{a}:1\}, [0\text{-}5], \{x \mapsto \{2,3\}\}\rangle, \quad \langle \#_2, .9, \{\mathtt{b}:.8, \mathtt{c}:.2\}, \{2\}, \{y \mapsto \{1\}\}\rangle\}$. Assume to fix P_M, P_L to be as usual in the standard cost function, as illustrated in [11], namely $P_L(b, \alpha) = 1$; $P_M(t, \beta) = 0$ if t is silent (i.e., $\ell(t) = \tau$) and $P_M(t, \beta)$ equal to 1 plus the number of variables written by $guard(t)$ otherwise. For $P_=$, assume a data-aware extension (of the $P_=$ used to match the standard cost function [11]) defined as: $P_=(\langle \mathrm{ID}, b, \hat{\alpha}\rangle, (t, \beta)) = |\{v \mid \hat{\alpha}(v) \neq \beta(v^w)\}| / |V|$ if b is the label of t, i.e. $b = \ell(t)$, and $P_=(\langle \mathrm{ID}, b, \hat{\alpha}\rangle, (t, \beta)) = \infty$ otherwise. Then, if we instantiate cost functions as in Example 5 (also used in our encoding in Sect. 5), the optimal alignment of $\mathbf{ue}_1$ w.r.t. the DPN $\mathcal{N}$ depicted in Example 1 is $\gamma_{e'}^1$, as shown in Example 4 (of cost 2.05).

Further, if we consider the task of finding the lower-bound on the cost of optimal alignments for any realization of $\mathbf{ue}_1$ (as discussed below Definition 7), then this is 1 and it is given as well by the realization e' and $\gamma_{e'}^1$.

5 Encoding

In this section we describe our SMT encoding, obtained as the result of 4 steps:

(1) represent the process run, the trace realization, and the alignment symbolically by a set of SMT variables;
(2) set up constraints Φ that express optimality of the alignment;
(3) solve Φ to obtain a satisfying assignment ν;
(4) decode the process run, trace realization, and optimal alignment γ from ν.

The same procedure was followed in [11], with important differences. In step (1), we now need to represent both the process run and also the trace realization, which is complicated by the fact that the order of the events is not fixed. Moreover, the cost functions are defined differently, as described in Sect. 4. These changes also affect the decoding in step (4).

Similarly to earlier SAT-based approaches [6,11], we aim to construct a symbolic representation of both a process run and an alignment, that are subsequently concretized using an SMT solver. Since the symbolic representation depends on a finite set of initial variable declarations (and thus must be finite), we need to fix upfront an upper bound on the size of the process run. This upper bound, and even its existence, depends on the cost function of choice. The Lemma below shows how a (coarse) upper bound can be established for the cost model from Sect. 4, where the cost function is the standard one as in Example 6.

Lemma 1. *Let $\mathcal{N}$ be a DPN and $\mathbf{ue}$ a trace with uncertainty that has m_1 certain and m_2 uncertain events. Let $\langle f_1, \dots, f_n \rangle$ be a run of $\mathcal{N}$ such that $c = \sum_{j=1}^{n} P_M(f_j)$ is minimal, and k the length of the longest acyclic sequence of silent transitions in $\mathcal{N}$. Then there is an optimal alignment γ for $\mathbf{ue}$ such that the length of $\gamma|_M$ is at most $(4m_1 + 2m_2 + c) \cdot k$.*

The proof of this lemma can be found in [12]. Note that, in case the model admits loops that entirely consist of silent transitions, then there can be infinitely many optimal alignments that are not bounded in length (as such loops can be repeated arbitrarily many times without incurring in any additional penalty on the alignment cost). Thus, the above lemma shows only *existence* of an optimal alignment within that bound, but in general the bound does not apply to *all* optimal alignments.

5.1   Encoding the Process Run

Assuming that the process run in the optimal alignment has length at most n, we use the following SMT variables to represent this run:

(a) transition step variables S_i for $1 \leq i \leq n$ of type integer; if $T = \{t_1, \ldots, t_{|T|}\}$ then it is ensured that $1 \leq S_i \leq |T|$, so that S_i is assigned j iff the i-th transition in the process run is t_j;

(b) marking variables $M_{i,p}$ of type integer for all i, p with $0 \leq i \leq n$ and $p \in P$, where $M_{i,p}$ is assigned k iff there are k tokens in place p at instant i;

(c) data variables $X_{i,v}$ for all $v \in V$ and i, $0 \leq i \leq n$; the type of these variables depends on v, with the semantics that $X_{i,v}$ is assigned r iff the value of v at instant i is r; we also write X_i for $(X_{i,v_1}, \ldots, X_{i,v_k})$.

Note that variables (a)–(c) encode all information required to capture a process run of a DPN with n steps. They will be used to represent the model projection of the alignment γ. To encode the process run, we use the constraints

$$\varphi_{run} = \varphi_{init,fin} \wedge \varphi_{trans} \wedge \varphi_{enabled} \wedge \varphi_{mark} \wedge \varphi_{data}$$

where the subformulas above reflect requirements to the solution as follows:

– The initial and final markings M_I and M_F, and the initial assignment α_0 are respected:

$$\bigwedge_{p \in P} M_{0,p} = M_I(p) \wedge \bigwedge_{v \in V} X_{0,v} = \alpha_0(v) \wedge \bigwedge_{p \in P} M_{n,p} = M_F(p) \qquad (\varphi_{init,fin})$$

– Transitions correspond to transition firings in the DPN:

$$\bigwedge_{1 \leq i \leq n} 1 \leq S_i \leq |T| \qquad (\varphi_{trans})$$

– Transitions are enabled when they fire:

$$\bigwedge_{1 \leq i \leq n} \bigwedge_{1 \leq j \leq |T|} (S_i = j) \rightarrow \bigwedge_{p \in {}^\bullet t_j} M_{i-1,p} \geq |{}^\bullet t_j|_p \qquad (\varphi_{enabled})$$

where $|{}^\bullet t_j|_p$ denotes the multiplicity of p in the multiset ${}^\bullet t_j$.

– We encode the token game:

$$\bigwedge_{1 \leq i \leq n} \bigwedge_{1 \leq j \leq |T|} (S_i = j) \rightarrow \bigwedge_{p \in P} M_{i,p} - M_{i-1,p} = |t_j{}^\bullet|_p - |{}^\bullet t_j|_p \qquad (\varphi_{mark})$$

where $|t_j{}^\bullet|_p$ is the multiplicity of p in the multiset $t_j{}^\bullet$.

– The transitions satisfy the constraints on data:

$$\bigwedge_{1 \leq i < n} \bigwedge_{1 \leq j \leq |T|} (S_i = j) \rightarrow guard(t_j)\chi \wedge \bigwedge_{v \notin write(t_j)} X_{i-1,v} = X_{i,v} \qquad (\varphi_{data})$$

where the substitution χ uniformly replaces V^r by X_{i-1} and V^w by X_i. Above, $write(t)$ denotes the set of variables that are written by $guard(t)$.

5.2 Trace Realization Constraints

Next, we describe how an admissible realization for a given trace with uncertainty **ue** is encoded. To this end, additional variables are needed. Let $\mathbf{ue} = \{ue_1, \ldots, ue_m\}$ such that $ue_i = \langle \text{ID}, conf, \text{LA}, \text{TS}, \alpha \rangle$ for each $1 \leq i \leq m$, with $\text{LA} = \{b_1 : p_1, \ldots, b_{N_i} : p_{N_i}\}$. We use the following sets of variables for all i:

(d) a boolean *drop variable* drop_{ue_i} expressing whether the event is absent in the real-
    ization; it must satisfy $\text{drop}_{ue_i} \implies (ue_i.conf < 1)$, i.e., it can only be assigned
    true for uncertain events with confidence below 1,
(e) an integer *activity variable* A_{ue_i} that expresses which of the labels $b_1, \ldots, b_{N_i}$ is
    taken, so it must satisfy $1 \leq \text{A}_{ue_i} \leq N_i$, and
(f) *trace data variables* D_{v,ue_i} of suitable type for all $v \in V$ that satisfy either that
    $\bigvee_{c \in ue.\alpha} \text{D}_{v,ue_i} = c$ if $\alpha(ue)$ is a set, or $l \leq \text{D}_{e_i} \leq u$ if $\alpha(ue) = [l, u]$ is an interval.

If each uncertain event in **ue** has a single, distinct timestamp, we call **ue** *sequential*, and assume it is ordered by time as $\langle ue_1, \ldots, ue_n \rangle$. If **ue** is not sequential, we need the following additional variables: For all i, $1 \leq i \leq m$:

(g) a *time stamp variable* T_{ue_i} to express when event ue_i happened, with the constraint
    $\bigvee_{t \in \text{TS}} \text{T}_{ue_i} = t$ if $\text{TS}(ue_i)$ is a set, or $l \leq \text{T}_{e_i} \leq u$ if $\text{TS}(ue_i) = [l, u]$ is an interval,
(h) an integer *position variable* P_{ue_i} to fix the position of ue_i in the realization,
(i) an integer *item variable* L_j that indicates the j-th element in the realization, i.e.,
    L_j has value $\text{ID}(ue_i)$ if and only if the j-th event in the trace with uncertainty is
    ue_i; we thus issue the constraint $\bigvee_{i=1}^{m} \text{L}_j = \text{ID}(ue_i)$ to fix the range of L_j, for all
    $1 \leq j \leq m$.

The formula φ_{trace} consists of the range constraints in (d)-(i), in addition to

$$\bigwedge_{i=1}^{m} \bigwedge_{j=1}^{m} (\text{P}_{ue_i} < \text{P}_{ue_j} \implies \text{T}_{ue_i} \leq \text{T}_{ue_j}) \wedge (\text{T}_{ue_i} < \text{T}_{ue_j} \implies \text{P}_{ue_i} < \text{P}_{ue_j})$$
$$\bigwedge_{i=1}^{m} \bigwedge_{j=1}^{m} \text{L}_i = \text{ID}(ue_j) \iff \text{P}_{ue_j} = i$$

so as to require that, first, the positions assigned to uncertain events by P_{ue_j} is compatible with the time stamps assigned by T_{ue_j} and, second, that the P_{ue_j} variables work as
an "inverse function" of the L_i.

5.3 Encoding the Cost Function

To encode the alignment and its cost we use, additionally:

(j) distance variables $\text{d}_{i,j}$ of type integer for $0 \leq i \leq m$ and $0 \leq j \leq n$, where $\text{d}_{i,j}$ is
    the alignment cost of the prefix $\mathbf{e}|_i$ of the log trace realization $\mathbf{e}$ and prefix $\mathbf{f}|_j$ of
    the process run $\mathbf{f}$, both of which are yet to be determined.

The search for an optimal alignment is based on a notion of *edit distance*, similar as
in [6,11]. More precisely, we assume that the data-aware alignment cost $\kappa(e_i, f_i)$ in
Fig. 1 can be encoded using a *distance-based* cost function with *penalty functions* P_L,
P_M, and $P_=$ as discussed in Sect. 4.1. Recall that $P_=$ is assumed to be data-aware,

i.e., to take into account the mismatching variable assignments between the events in realizations and transition firings in process runs. Intuitively, such functions assess the degree of "closeness" between a process run and a log trace. We assume that there are SMT encodings of these penalty functions that use variables (a)–(i), denoted as $[P_=]_{i,j}$, $[P_M]_j$, and $[P_L]_i$.

Moreover, we assume that there are encodings of the event removal cost function $[\kappa_{\mathbf{ue}}]_i$ and the confidence cost function $[\theta_{\mathbf{ue}}]_i$, defined for the i-th element of the log trace realization. We then consider the following constraints for $i, j > 0$:[1]

$$d_{0,0} = 0 \qquad d_{i,0} = \min([P_L]_i \cdot [\theta_{\mathbf{ue}}]_i, [\kappa_{\mathbf{ue}}]_i) + d_{i-1,0} \qquad d_{0,j} = [P_M]_j + d_{0,j-1}$$

$$d_{i,j} = \min \begin{cases} ite([P_=]_{i,j} = 0, [\theta_{\mathbf{ue}}]_i, [P_=]_{i,j} + [P_=]_{i,j} \cdot [\theta_{\mathbf{ue}}]_i) + d_{i-1,j-1} \\ [P_L]_i \cdot [\theta_{\mathbf{ue}}]_i + d_{i-1,j} \\ [\kappa_{\mathbf{ue}}]_i + d_{i-1,j} \\ [P_M]_j + d_{i,j-1} \end{cases} \qquad (\varphi_\delta)$$

This encoding constitutes an *operational way* for computing the cost function represented in Fig. 1, where the components $\kappa_{\mathbf{ue}}$ and θ are distributed to single moves, which at the same time allows us to use the encoding schema based on the edit distance. The inductive case $d_{i,j}$ is computed so as to locally choose the move with minimal cost. In particular, the first and the second line of the case distinction correspond exactly to the specific instantiation of the expression $\kappa(e_i, f_i) \otimes \theta(e_i, \mathbf{ue})$ exemplified in Sect. 4. For instance, the cost penalty $\kappa(e_i, f_i) \cdot (1 + \theta(e_i, \mathbf{ue}))$ in case $\kappa(e_i, f_i) > 0$ (see Sect. 4) corresponds here, in the *ite* construct, to the cost penalty $[P_=]_{i,j} + [P_=]_{i,j} \cdot [\theta_{\mathbf{ue}}]_i$ in the *else* statement. The expression $d_{m,n}$ encodes then the cost of the complete alignment, which will thus be used as the minimization objective.

The encodings of the penalties, as well as $[\kappa_{\mathbf{ue}}]_i$ and $[\theta_{\mathbf{ue}}]_i$, also depend on the choice of the respective functions. For those exemplified in Sect. 4, one can define $[\kappa_{\mathbf{ue}}]_i$ as a (nested) case distinction on the element from $\mathbf{ue}$ that is chosen for the i-th position (represented with variable L_i – see Sect. 5.2):

$$[\kappa_{\mathbf{ue}}]_i = ite(L_i = \text{ID}(ue_1) \wedge \text{drop}_{ue_1}, conf(ue_1), \ldots$$
$$ite(L_i = \text{ID}(ue_m) \wedge \text{drop}_{ue_m}, conf(ue_m), \infty) \ldots)$$

A similar case distinction can be done for $[\theta_{\mathbf{ue}}]_i$, also exemplified in Sect. 4.

5.4 Solving and Decoding

We use an SMT solver to obtain a satisfying assignment ν for the following constrained optimization problem:

$$\varphi_{run} \wedge \varphi_{trace} \wedge \varphi_\delta \quad \text{minimizing} \quad d_{m,n} \qquad (\Phi)$$

For a satisfying assignment ν for (Φ), we construct the process run $\mathbf{f}_\nu = \langle f_1, \ldots, f_n \rangle$ where $f_i = (t_{\nu(\mathbf{s}_i)}, \beta_i)$, assuming that the set of transitions T consists of $t_1, \ldots, t_{|T|}$

---

[1] We assume that P_L is always positive, otherwise, a case distinction using *ite* is also required in the second line.

in the ordering already used for the encoding. The transition variable assignment β_i is obtained as follows: Let the state variable assignments α_j, $0 \le j \le n$, be given by $\alpha_j(v) = \nu(X_{j,v})$ for all $v \in V$. Then, $\beta_i(v^r) = \alpha_{i-1}(v)$ and $\beta_i(v^w) = \alpha_i(v)$ for all $v \in V$. Moreover, we construct a realization $e_\nu = \langle e_1, \ldots, e_k \rangle$ by ordering the events in **ue** according to T_{ue_i}, dropping those where drop_{ue_i} is true, and fixing the label and data values to A_{ue_i} and D_{ue_i}, respectively. Finally, let the (partial) alignments $\gamma_{i,j}$ be defined as follows, for $i, j > 0$:

$$\gamma_{0,0} = \epsilon \qquad \gamma_{0,j+1} = \gamma_{0,j} \cdot (\gg, f_{j+1})$$

$$\gamma_{i+1,0} = \begin{cases} \gamma_{i,0} \cdot (e_{i+1}, \gg) & \text{if } \nu(\delta_{i+1,0}) = \nu([P_L]_{i+1} \cdot [\theta_{ue}]_{i+1} + \delta_{i,0}) \\ \gamma_{i,0} & \text{if } \nu(\delta_{i+1,0}) = \nu([\kappa_{ue}]_{i+1} + \delta_{i,0}) \end{cases}$$

$$\gamma_{i+1,j+1} = \begin{cases} \gamma_{i,j+1} \cdot (e_{i+1}, \gg) & \text{if } \nu(\delta_{i+1,j+1}) = \nu([P_L]_{i+1} \cdot [\theta_{ue}]_{i+1} + \delta_{i,j+1}) \\ \gamma_{i,j+1} & \text{if } \nu(\delta_{i+1,j+1}) = \nu([\kappa_{ue}]_{i+1} + \delta_{i,j+1}) \\ \gamma_{i+1,j} \cdot (\gg, f_{j+1}) & \text{if otherwise } \nu(\delta_{i+1,j+1}) = \nu([P_M]_{j+1} + \delta_{i+1,j}) \\ \gamma_{i,j} \cdot (e_{i+1}, f_{j+1}) & \text{otherwise} \end{cases}$$

5.5 Correctness

The next results state that the constructed alignment satisfies our conformance checking task as in Definition 7. The formal proofs are omitted for reasons of space, but can be found in the extended version [12]. It is however easy to see that our encoding matches the same definitions as in Sects. 2 and 3, and the cost functions in Sect. 4.

Lemma 2. *For any satisfying assignment ν to (Φ), (a) f_ν is a process run, and (b) e_ν is a realization of **ue**.*

This lemma shows that the decoding provides both a valid process run and a trace realization. Next we demonstrate that the decoded alignment is optimal.

Theorem 1. *Let $\mathcal{N}$ be a DPN, **ue** a log trace with uncertainty and ν a solution to (Φ) as in Sect. 5.4. Then $\gamma_{m,n}$ is an optimal alignment for **ue**.*

Moreover, as explained in Sect. 4 (after Definition 7), we can easily capture the additional task of computing the lower-bound on the optimal cost of alignments of realizations for a given trace with uncertainty, as considered in [17]. By taking advantage of the modularity of our framework, this simply amounts to set $\kappa_{ue} = 0$ and $\kappa(e_i, f_i) \otimes \theta(e_i, \text{ue}) = \kappa(e_i, f_i)$, thus ignoring all confidence values specified in **ue**. This allows us to freely select, without any penalty, the realization of **ue** that has the minimal alignment cost.

Lemma 3. *For $\mathcal{N}$, **ue** as above and $\gamma_{m,n}$ the alignment decoded from a satisfying assignment ν for (Φ) as in Sect. 5.4, there is no realization e of **ue** and alignment γ for e such that $\kappa(\gamma) < \kappa(\gamma_{m,n})$.*

Note that in contrast to the approach in [17], our approach entirely avoids any explicit construction of realizations, which is a huge benefit for the overall performance.

5.6   Implementation

As a proof of concept, the uncertainty conformance checking approach described in this paper was implemented in cocomot – a Python command line tool that was originally designed for data-aware conformance checking without uncertainties [11]. It uses pm4py (https://pm4py.fit.fraunhofer.de/) to perform parsing tasks, and the SMT solvers Yices 2 [10] and Z3 [9].

The tool takes as input two files: a DPN in .pnml format and a log in .xes, specified using the XES extension for uncertain data described in [16]. The command line option -u triggers the use of the uncertainty module, and the tool outputs the optimal alignment as well as its cost. Based on the encoding in Sect. 5, the tool employs the two cost functions mentioned in Example 6 to achieve two different tasks: Using the first cost function that takes confidence values into account, the cost of the optimal alignment can be interpreted as an expectation value of the best alignment cost for all realizations (parameter -u fit). Using the second cost function, a lower bound on the cost of the optimal alignment among all realizations is computed (parameter -u min). More information on the tool usage, the format for specifying uncertain logs, execution options and further details, together with the source code, can be found on the tool website: https://github.com/bytekid/cocomot.

Although the presented encoding shows that the overall theoretical complexity of our approach does not change with respect to the one reported in [11] (that is, the problem of finding the optimal alignment for logs with uncertainty is NP-complete), experimental evaluations are required so as to assess the feasibility of the encoding in practical scenarios. More specifically, we plan to enrich publicly available logs for multi-perspective conformance checking [15] with uncertainty information, as done in [17].

6   Conclusions

In this work we have proposed an extension of the foundational framework for alignment-based conformance checking of data-aware processes studied in [11], to support logs with different types of uncertainties in events, timestamps, activities and other attributes. To account for all possible combinations of uncertainties in a trace, we rely on a notion of realization to fix one of its possible *certain* variants. However, given that there are potentially infinitely many realizations, performing the conformance checking task on each of them is not feasible.

To attack this problem, we considered a version of the conformance checking task aimed at searching for the best alignment among all possible realizations. This has been achieved by introducing an involved cost model that incorporates traditional alignment-related penalties together with extra costs accounting for the selection of specific realizations. Although these cost components are presented as arbitrary and can in fact be tailored to specific settings and assumptions, we have provided a concrete instantiation and its corresponding encoding.

Thanks to the modularity of our conformance cost definition, we have also shown how we can accommodate different conformance checking tasks for logs with uncertainty, including those studied in the literature [17].

The theoretical underpinning of our approach is SMT solving. Our work is the first one to employ techniques based on *satisfiability* of formulae modulo suitable logical theories for solving data-aware conformance checking tasks with uncertainty, and to leverage well-established solvers to handle them. The approach was implemented in the `cocomot` tool.

In future work, we plan to investigate further more involved notions of uncertain logs, and conduct an experimental evaluation of our approach and implementation. To this end, instead of considering artificially generated logs, one first step is to compile a benchmark for data-aware conformance checking of uncertain logs, which is currently not available.

Acknowledgments. This research has been partially supported by the UNIBZ projects VERBA, MENS, WineID, SMART-APP and by the PRIN 2020 project PINPOINT.

References

1. Alman, A., Maggi, F.M., Montali, M., Peñaloza, R.: Probabilistic declarative process mining. Inf. Syst. **109** (2022)
2. Barrett, C., Fontaine, P., Tinelli, C.: The SMT-LIB Standard: Version 2.6. Technical report (2018), http://smtlib.cs.uiowa.edu/language.shtml
3. Barrett, C., Tinelli, C.: Satisfiability Modulo theories. In: Handbook of Model Checking, pp. 305–343. Springer, Cham (2018). https://doi.org/10.1007/978-3-319-10575-8_11
4. Bergami, C., Maggi, F.M., Montali, M., Peñaloza, R.: Probabilistic trace alignment. In: Proceedings of ICPM 2021, pp. 9–16. IEEE (2021)
5. Boltenhagen, M., Chatain, T., Carmona, J.: Encoding conformance checking artefacts in sat. In: Di Francescomarino, C., Dijkman, R., Zdun, U. (eds.) BPM 2019. LNBIP, vol. 362, pp. 160–171. Springer, Cham (2019). https://doi.org/10.1007/978-3-030-37453-2_14
6. Boltenhagen, M., Chatain, T., Carmona, J.: Optimized SAT encoding of conformance checking artefacts. Computing **103**(1), 29–50 (2020). https://doi.org/10.1007/s00607-020-00831-8
7. Carmona, J., van Dongen, B.F., Solti, A., Weidlich, M.: Conformance Checking - Relating Processes and Models. Springer, Cham (2018). https://doi.org/10.1007/978-3-319-99414-7
8. Chesani, F., et al.: Compliance in business processes with incomplete information and time constraints: a general framework based on abductive reasoning. Fundam. Inform. **161**(1–2), 75–111 (2018)
9. de Moura, L., Bjørner, N.: Z3: an efficient SMT solver. In: Ramakrishnan, C.R., Rehof, J. (eds.) TACAS 2008. LNCS, vol. 4963, pp. 337–340. Springer, Heidelberg (2008). https://doi.org/10.1007/978-3-540-78800-3_24
10. Dutertre, B.: Yices 2.2. In: Biere, A., Bloem, R. (eds.) CAV 2014. LNCS, vol. 8559, pp. 737–744. Springer, Cham (2014). https://doi.org/10.1007/978-3-319-08867-9_49
11. Felli, P., Gianola, A., Montali, M., Rivkin, A., Winkler, S.: CoCoMoT: conformance checking of multi-perspective processes via SMT. In: Polyvyanyy, A., Wynn, M.T., Van Looy, A., Reichert, M. (eds.) BPM 2021. LNCS, vol. 12875, pp. 217–234. Springer, Cham (2021). https://doi.org/10.1007/978-3-030-85469-0_15
12. Felli, p., Gianola, A., Montali, M., Rivkin, A., Winkler, S.: Conformance checking with uncertainty via SMT (extended version). Technical report (2022). https://arxiv.org/abs/2206.07461
13. Leemans, S.J.J., van der Aalst, W.M.P., Brockhoff, T., Polyvyanyy, A.: Stochastic process mining: earth movers' stochastic conformance. Inf. Syst. **102** (2021)

14. Mannhardt, F.: Multi-perspective process mining. Ph.D. thesis, Technical University of Eindhoven (2018)
15. Mannhardt, F., de Leoni, M., Reijers, H.A., van der Aalst, W.M.P.: Balanced multi-perspective checking of process conformance. Computing **98**(4), 407–437 (2015). https://doi.org/10.1007/s00607-015-0441-1
16. Pegoraro, M.: Process mining on uncertain event data (extended abstract). In: Proceedings of ICPM-D 2021, pp. 1–2. CEUR (2021)
17. Pegoraro, M., Uysal, M.S., van der Aalst, W.M.P.: Conformance checking over uncertain event data. Inf. Syst. **102** (2021)
18. Polyvyanyy, A., Kalenkova, A.A.: Conformance checking of partially matching processes: an entropy-based approach. Inf. Syst. **106** (2022)
19. Sebastiani, R., Tomasi, S.: Optimization modulo theories with linear rational costs. ACM Trans. Comput. Log. **16**(2), 12:1–12:43 (2015)
20. van der Aalst, W., et al.: Process mining manifesto. In: Daniel, F., Barkaoui, K., Dustdar, S. (eds.) BPM 2011. LNBIP, vol. 99, pp. 169–194. Springer, Heidelberg (2012). https://doi.org/10.1007/978-3-642-28108-2_19
21. Wynn, M.T., Sadiq, S.: Responsible process mining - a data quality perspective. In: Hildebrandt, T., van Dongen, B.F., Röglinger, M., Mendling, J. (eds.) BPM 2019. LNCS, vol. 11675, pp. 10–15. Springer, Cham (2019). https://doi.org/10.1007/978-3-030-26619-6_2

Process Mining Practice

The Dark Side of Process Mining. How Identifiable Are Users Despite Technologically Anonymized Data? A Case Study from the Health Sector

Friederike Maria Bade[✉], Carolin Vollenberg, Jannis Koch, Julian Koch, and Andre Coners

South Westphalia University of Applied Sciences, Haldener Str. 182, 58095 Hagen, Germany
bade.friederikemaria@fh-swf.de

Abstract. Over the past decade, process mining has emerged as a new area of research focused on analyzing end-to-end processes through the use of event data and novel techniques for process discovery and conformance testing. While the benefits of process mining are widely recognized scientifically, research has increasingly addressed privacy concerns regarding the use of personal data and sensitive information that requires protection and compliance with data protection regulations. However, the privacy debate is currently answered exclusively by technical safeguards that lead to the anonymization of process data. This research analyzes the real-world utility of these process data anonymization techniques and evaluates their suitability for privacy protection. To this end, we use process mining in a case study to investigate how responsible users and specific user groups can be identified despite the technical anonymization of process mining data.

Keywords: Process mining · Privacy measures · Healthcare sector · Hospital information system

1 Introduction

Healthcare providers, especially hospitals, are under increasing pressure from policy-makers and patient advocates to manage rising healthcare costs while improving the quality of care. The lack of efficiency due to poorly coordinated processes is considered a fundamental problem in achieving cost and quality goals [1, 2]. Since most of the information flow is mapped through the Hospital Information System or occurs "through informal communication, unsystematic processes, and uncontrolled access to information" [3], information deficits can often occur at interfaces. Therefore, a patient's data must always be recorded and updated in concrete terms so that efforts and risks can be reduced on the one hand and quality can be increased on the other hand. In addition, the involvement of various disciplines and departments in the care process of a patient hospital journey aggravates a continuously guaranteed information flow of patient data. The different departments involved in a patient's journey often have only little insight

© Springer Nature Switzerland AG 2022
C. Di Ciccio et al. (Eds.): BPM 2022, LNCS 13420, pp. 219–233, 2022.
https://doi.org/10.1007/978-3-031-16103-2_16

into what is happening in other disciplines or departments [1, 4]. The coordination of the involved stakeholders is often "hampered by informal communication, unsystematic processes, and uncontrolled access to information" [5]. This often leads to errors, mistakes, and confusion in information flows.

Process mining is an increasingly used technique in the field of information systems and is used to analyze and improve processes by using event logs. Different organizations from various industries have gained interest and use process mining in other cases and applications [3, 6]. Process mining has already also been successfully applied in healthcare and has helped to provide various insights to improve healthcare processes [1, 7]. However, the purely administrative processes in healthcare areas that require exclusively IT-based processing, such as care documentation, appointment scheduling, or billing processes, have so far only been addressed peripherally in the case of process mining [8, 9]. Therefore, process mining is also an emerging research area within the healthcare sector. Process mining is a growing research topic, which has not only increased in research interest in the last decade but also more recently [10, 11]. However, the main focus of research is on the technical implementation of process mining. At this point, research focuses on developing new algorithms and improving existing process mining techniques [12]. Research on practical aspects of process mining, as well as the adoption of the technology and the influences of process mining on the organization and employees, is less available [12]. While the benefits of process mining are widely recognized, the scientific community also expresses concerns about the irresponsible use of personal data within process mining and the aspects of anonymization of the data used by process mining. Thus, the ethical and legal issues are also of great interest and importance to researchers acutely. Therefore, it is becoming increasingly crucial for scientific efforts in this area to address privacy and confidentiality issues in process mining.

According to Pika et al. [13], Grishold et al. [12], Mannhardt et al. [14], or vom Brocke et al. [15], research should also specifically address these issues in the future and should put more effort into the examination of the privacy issues and risks, especially in sensitive areas like healthcare.

However, the debate on how far anonymized data of process mining can be identifiable has so far been answered and addressed by the use of technical data transformation techniques to anonymize process analysis data [4, 9, 13, 16, 17]. Therefore, considerable research and practical efforts have been made in recent years to develop, implement, and integrate appropriate privacy and confidentiality protection techniques in process mining. Though, this consideration of privacy aspects has been treated here still too inadequately and one-sidedly, because on the one hand it only considers the personal data of, e.g., patients, but not those of the process executors, and on the other hand, it only refers to the measures of data protection and privacy of obvious personal data via usernames [18].

Our research starts here and shows that privacy debates beyond the usual personal data are important and need to be part of ethical and technical discourse by showing that process mining can provide a way to provide role-based and personal information, e.g., when data has been changed and despite technically anonymized data. Here, we want to clarify to what extent process mining can attribute the execution of process steps (deviation from the target process) or the change of data to a specific originator. Especially we want to examine if the user's identification in the existing system (here: Hospital Information System) is explicitly technically prevented by the system. Therefore our research question is the following:

RQ: Can process mining provide identification of users in explicitly technically anonymized systems and assign errors directly to them?

We have been able to explore an approach that allows, using process mining data, to identify deviations from the standard process and user routines to associate users or groups of users to data changes in the system. The case of the hospital in West Germany with its hospital information system presented in this study gave us the unique opportunity to investigate how technically anonymized process mining data can be used, despite anonymization, to identify sources of error among assigned users to investigate whether the technical security measures are sufficient to protect the privacy of the users. The data basis used for this purpose was worked out with the responsible persons and users and the hospital's ethics committee in an experimental setup so that no ethical or legal concerns could arise for our research itself.

2 Background

Process mining is an emerging and essential technology in business process management (BPM). Particularly in information systems, the interest lies in how technologies can change and optimize processes. In the last decade, there has been a significantly increased interest in process mining, both in research and in practice [6]. Due to the increasing amount of available event data in organizations' easily accessible information systems, a variety of opportunities arose to analyze and optimize processes using information from this event data [2]. Process mining aims to gain a traceable overview of a process, to provide insights into a process and the actual process flow, and to support improvements [2, 3]. However, the awareness of privacy issues, and thus the ethical issue of using process mining and with this the use of personal data, has increased significantly [10, 19].

In the context of the application, process mining enables the discovery, verification of conformities, and improvement of processes [2]. In the context of discovery, the corresponding technique of process mining helps in the discovery of the real process by creating a process model using event logs of the process. The conformance checking type of process mining compares the evaluated real process model with the predefined process model and reveals whether the reality matches the predefined model [2, 6]. The enhancement type of process mining "aims to modify or extend the a priori model. For example, by using timestamps in the event logs, the model can be extended to show bottlenecks, service levels, lead times, and frequencies" (van der Aalst, 2012) [2, 20].

Process mining encompasses techniques used to analyze and optimize processes. These techniques provide data-driven methods of process analysis that focus on the evaluation and extraction of information from event logs - information stored in IT systems about individual and actual process steps. Event logs store information such as the entity, e.g. a person or device, that performs or triggers an activity. In addition, event logs store timestamps of an event or data elements recorded with an event. These event logs may contain direct and indirect identifiers of personal data and may disclose the user's personal data or groups of users. The discussion on revealing this data increases due to the new General Data Protection Regulation. In this regard, it can be mentioned that data protection in hospitals is exceptionally high, as sensitive data is involved. There is a lively discussion and debate in the scientific discourse on anonymization techniques of datasets collected through data mining, such as perturbation, anonymization and cryptography. There are various privacy preserving techniques such as Generalisation, Suppression, Distortion, Swapping or Masking. Each technique pursues the goal of transforming personal data. Thus, it is expected to reduce the original information in the dataset by a certain amount. In particular, there are leading research efforts; as per Murhy [21], the consensus is that the anonymization technique of Suppression is the most efficient and resource-saving solution of all [21].

For the underlying process mining technologies, several privacy-preserving techniques on the research side attempt to anonymize the data collected and processed in the event log [16].

Pika et al. [13] already analyzed and evaluated existing privacy approaches to anonymize process data for process mining. They tested the suitability of three different approaches, confidentiality framework, PRESTA, and differential privacy model for event logs. The analysis showed a trade-off between privacy and utility. The methods that maintain higher data utility for process mining purposes (e.g., encryption) do not provide strong privacy protection.

Process mining has also gained a lot of interest in recent years, especially in the healthcare sector [4, 17, 22–24]. Hospitals, in particular, face the challenge of streamlining their processes and the documentation of patient data. The processes in hospitals are characterized by the fact that several departments may be involved in the process of patient care. These different organizational departments often have their own specific IT applications, add different information to the hospital information system, or need different information about a patient. As a result, problems often arise in obtaining data related to healthcare and hospital processes in particular.

Hospital processes are characterized by a high degree of complexity, and the extremely flexible implementation of patient care, which always depends on the needs and condition of the individual patient and their individual treatment. In addition, as already mentioned, hospitals involve various actors, staff from different disciplines, and diverse departments. This setting, therefore, offers great potential for the use of process mining due to the involvement of different actors, the various existing systems, including the hospital information system, and the huge amount of available data and event logs [1, 4]. However, data security and anonymization of these data, especially in healthcare, is a high priority and therefore regulated by high standards, laws, and guidelines. In particular, personal data must be anonymized to ensure data security and to protect personal rights.

Process mining can be used to identify and quantify activity patterns that reflect how users act in processes [13]. Despite the use of privacy transformation techniques to anonymize data, the use of process mining offers the possibility to provide both anonymized and non-anonymized data about user information and thus personal data. However, this also depends on the given conditions of the existing system. It is only possible if the user data is available in the existing information system, only then can process mining provide transparency about which process steps are performed by whom and at what time [19]. Patient and staff identities could be revealed with the help of background knowledge and the event log.

If process mining is used within an information system that does not allow for a clear user assignment because only anonymized data is available or no user assignment is made, process mining with the existing event logs cannot contribute to the identification of users, e.g. to the identification of vulnerabilities. Thus, despite technical security measures to anonymize the data, the question should be asked whether the use of process mining is ethically correct, as conclusions can still be drawn about patients and staff through the background information on the process. Especially when the process flows in practice often do not correspond to the predefined process models, it is possible to get the link to the particular employee. As process flows in practice often do not correspond to the predefined process models, but differ in their actual execution, which leads to a "significant gap between what is prescribed or supposed to happen and what happens in reality" (Mans et al. 2008) [24] the assignment of identification of personal resources could be possible with particular background information. In this way, the weak points can also be identified, poor quality, loss of time, and higher process costs can be avoided. Only an accurate assessment of reality can help in reviewing process models to optimize business processes [25]. Still, this accurate assessment of reality process steps by process mining can also be critical in terms of anonymization of personal data.

3   Methodology

To achieve sustainable results in the rather explorative nature of our study, we use a case study approach. Since we want to gain an understanding of the anonymization and identifiability of users or groups of users in anonymized event logs, as per our formulated RQ, a case study approach, according to Yin, is most appropriate [26]. As it allows us to study a phenomenon in a firmly grounded context by using triangulation of different data sources – Process Mining data, field observations, interviews, and documentation – to gain insights, this approach is well suited for our RQ [26].

The case study method is suitable for gaining a thorough and detailed understanding of factors (such as anonymization and identifiability in our case) [27], and it involves the use of case organizations to prove existing theories of process mining anonymization systems based on empirical evidence. In addition to the primary process mining data collected, we use data triangulation from event logs, database schemas, and development and execution logs generated by the respective PM application systems to provide scientific rigor [20].

Figure 1 shows an overview of our case study approach, which consists of four main phases: case selection, data collection, data analysis, and case conclusion.

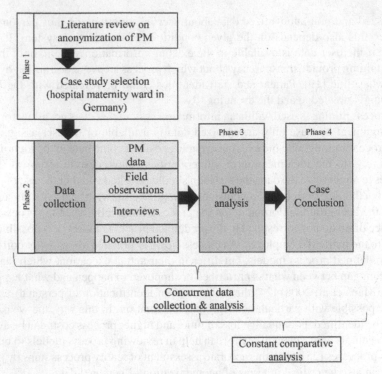

Fig. 1. Research method.

3.1 Case Selection

To arrive at a representative selection of objects of study, we deliberately chose a hospital maternity ward in Germany to cover a broad spectrum of particularly sensitive and especially anonymous process data and event logs. In this context, it was key that the users and data originators were informed in advance about the data use and analysis and gave their consent to our research. The organization investigated in this study is a large hospital group in Germany with more than 20,000 employees. This case study organization comprises more than 100 facilities, including more than 20 hospital sites with over 6,000 beds, above 30 residential and nursing care facilities with more than 3,000 nursing places, more than 30 medical care centers, and about 10 outpatient care services. Over one million patients are cared for annually in our case study organization.

In our case study presented here, we focus on the data collection process at birth, where the first data of a new life is collected and entered into the Hospital Information System as core data for each newborn. During a hospital stay, however, there are always changes in the master data recognized, which lead to unintentional errors and consequential (negative) effects. These irregularly executed processes of master data manipulation ultimately have an extremely negative impact on the performance, conformance, and compliance of process fulfillment within the whole treatment of the newborn. Further, this affects the work of the medical staff in the different departments a newborn is transferred to during hospital stay. But it also affects the newborn and can somehow disturb or affect the newborn's mother. For example, if not all information about the newborn is available after the release from the hospital, examinations must be repeated in the hospital; this means additional work for medical staff, higher financial expenses for the hospital, and an unnecessary burden and stressful process for the young mother and the newborn as well.

The treatment of a newborn in the studied maternity ward and the regarded master data process takes into account, among other things, the course of a premature baby from birth in the delivery room to the time of transfer to the regular pediatric ward and discharge from the hospital as demonstrated in Fig. 2. After the birth of a premature child, the newborn is initially physically located in the delivery room. The child's first master data is entered into the Hospital Information System by the doctor or the participating obstetrician. Following that, the child gets to the premature care department of the neonatology unit. There, the premature child gets special care from a specialized team of doctors and nurses as it may need extra help, e.g., breathing or eating. All data (e.g., weight or length of the newborn) and information about the treatment of the newborn are entered into the Hospital Information System by the different specialized team members involved. Later it is transferred to the regular perinatal department of the center before the child leaves the center and gets to the pediatric department of the hospital, is being discharged home, or transferred to another hospital.

Because the child and the patient master data are entered, used, and passed through several different departments, the medical staff in the various departments of the hospital, considered in the case study, frequently recognized changes, missing master data, or wrong adjustments of the master data and had problems assuming the relevant data, e.g., assigning the premature child, or often had to do extra administrative work by asking other staff to obtain information. In addition, often, the staff had to search for patient data because the system does not list the data or incorrect data are recorded, e.g., the name of the premature child or the accurate date of birth. In concrete terms, data loss and changes between the individual departments occurred over time again and again. Accordingly, these documentation processes of premature child's master data (e.g., the name of the newborn) and the additional according to data (e.g., the treatments, the parameters of treatment, and continuously raised parameters of the newborn) were selected for our research, by using process mining to proof the anonymization of the user or user groups that are responsible for changes in master data.

Along the described process of treating a premature child (cf. Fig. 2), the mentioned documentation processes were selected as the focus of our research on using process mining to examine and identify the user or user groups responsible for changes in master data. Our application of the process mining methodology is summarized in three steps according to van Dongen et al. (2005).

In step 1, we defined the scope of the extraction by screening out the granularity of the data and the most significant possible observation period as well as associated attributes from the Hospital Information System. This dataset went back a total of 20 months. In step 2, the event logs were analyzed by applying process discovery and conformance checking methods with the process mining solution *ProM* [28]. In step 3, the discovered process model was evaluated using the fitness, precision, and generalizability measures proposed by van Dongen et al. (2005). Finally, in step 4, these data were anonymized using a common and established technical protection measure within the *ProM* program package, specifically including all resource names (proper names, usernames, user abbreviations, logins), case IDs, and roughly detailed timestamps (months and years).

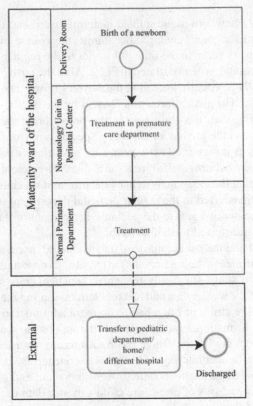

Fig. 2. The delivery process of a premature child.

3.2 Data Collection and Data Analysis

In our case study, we followed the established approach of Yin and enrich our research approach with more flexible data collection (field observations, 18 interviews, 5 programmer documentations, and 2 data documentations of the Hospital Information System) in addition to the primary process mining data [26]. To gain a sound and low-threshold insight into the existing process landscape, we used pre-built connectors, in particular XESame [29, 30], which contained a readymade data mapping template and allowed us to generate the standard XES event log formats from the HIS underlying SQL database. To gain insights into the process of the master data documentation during the birth and treatment of a premature child, we have collected data implementing process mining in the hospital studied. We have used process mining to review this process and get depth insights into the as-is process. Further, the aim was to avoid errors and extra work and deliver master data correctly. As semi-structured interviews allow us to gain a depth understanding of the whole documentation process and the role of process mining in this context, we chose this method as our primary data collection. For this purpose, we developed an interview protocol that was largely based on the collection of process mining data and included, for example, the single-cell process steps and possible deviations or inconsistencies. The collected and analyzed process mining data thus semantically specified the interview structure used. The interviews allowed us to observe the participants and gain insights on their non-verbal behavior as well. One author observed the typical input process of master data and characteristic interaction processes within the HIS and observed the system-based recording of planned and delivered care as well as the software-based documentation of the individual steps of care planning. In addition, interaction with the documentation file in HIS was observed to understand the context of the documentation process, including the documentation folder and templates.We collected a total of 340 min of interview material, which was transcribed verbatim, and computer coded by two authors according to the methodology of Flick's depth analysis [31]. The used interview pool consisted of 8 respondents: 3 Nurses, 2 Assistant Doctors, the Head of Patient Management, a staff member of the clinical IT, and an external Application Manager of the Hospital Information System as demonstrated in Table 1.

Table 1. Overview of data collection.

Participant	Data scope	Participant's role
Respondent 1	(2 × semi-structured interview in total 70 min)	Nurse
Respondent 2	(4 × semi-structured interview in total 60 min)	Nurse
Respondent 3	(1 × semi-structured interview in total 20 min)	Nurse
Respondent 4	(1 × semi-structured interview in total 20 min)	Assistant doctor
Respondent 5	(2 × semi-structured interview in total 30 min)	Assistant doctor
Respondent 6	(6 × semi-structured interview in total 110 min)	Staff member clinical IT systems/interface manager
Respondent 7	(1 × semi-structured interview in total 15 min)	Application manager hospital information system (extern)
Respondent 8	(1 × semi-structured interview in total 15 min)	Head of patient management

4   Results

The processes of the Hospital Information System extracted by process mining contained 3,913 historized complete executions of the examined master data. After reviewing the correctness of the studied process, we found that in about 17% (n = 661), the master data did not comply with the compliant-process flow and were permanently changed later than the initial entry. Concerning the research question already introduced, the following observation was made: In 71% (n = 469) of these error cases, the cause of the error could be assigned to a specific identifiable user or user role by the other users running the process, even though the system and process mining had anonymized the data through technical protection measures. Since the other users involved in the process knew each other's work methods and routines, the users involved were able to identify different user roles and assign errors to specific users. For example, only a few qualified users have access to particular program parameters when entering legal regulations according to the Narcotics Act as well as release processes, a certain sequence of operations that are required for the specific work assignment of the persons, or the assignment of defined shift sequences that are linked to individual qualification profiles of persons.

Based on the recorded traces, we have so far been able to determine that the occurrence of these irregular process executions or irregular changes to the master data has certain activity patterns. In doing so, we could establish a tangible link between certain execution routines and the erroneous master data manipulations and changes. This enabled it to identify a user group or, in many cases (n = 188), even the specific process executer via the corresponding process knowledge of the process owners. This became possible based on the detailed sequence of process steps with the help of process mining. Within that, the errors in the process were made possible to be assigned by certain individual process sequences or execution sequences to the individually unique user assignments, although only anonymized data were available.

As an example, we present here the process variants identified so far. Each of which led to the most frequent manipulations and changes of master data in the case study data. This exemplary identified process flow is shown in Fig. 3 and consists of the defined, collected process steps above.

As can be seen in Fig. 3, certain process steps are executed with different frequencies. This allowed us to directly assign certain sequence combinations and variations to specific users or user roles via certain combinations in the process variance or execution, taking into account process knowledge and contextualizing the execution variations. For example, internal staff performs the steps in a different sequence than external temporary staff. Intensive and Aesthesia nurses work through the processes in a different sequence depending on the duty roster and assigned nurse manager. We were able to assign deviations from the standard process found with process mining as these execution routines to a certain user group (in this case the ward doctors) based on their type and sequence of process steps. By looking at the detailed sequences of the process steps, process mining was thus able to clearly show which user role was responsible for the errors in the process, although only anonymized data was available.

This shows how security measures can be circumvented without much effort. Despite anonymizing advanced identifiers such as decoupling usernames and timestamps or generally making usernames unidentifiable to ensure user anonymity, we were able to demonstrate that individual users could be identified. Because process mining is becoming increasingly important and thus widespread, especially in healthcare, our research results are of great importance for the upcoming process mining use in practice

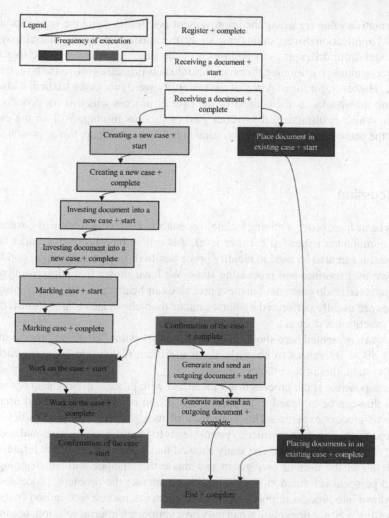

Fig. 3. Master data process in process mining.

and research. They believe in the sufficiency of technical safety measures previously assumed in the literature in research.

We suggest, therefore, a solution approach to overcome this lack of anonymization by only examining process sections instead of the overall processes. If the counter-measures considered safe in the literature were taken and supplemented by making the complete processes unrecognizable, the desired anonymization could be achieved. As part of the authors' proposed solution, an attempt was made to achieve some degree of decontextualization. This decontextualization of the process flow was achieved, among other things, by breaking the process flow down into smaller substeps and isolating it instead of viewing it as a whole. On the one hand, this led to the desired anonymization of the underlying process executors, but on the other hand, it also significantly limited the informative value regarding the improvement possibilities and the recognition gain of the PM application. Since, in this case study, the master data in the dataset may contain changes from different users and much of the deanonymization is due to activity pattern recognition, partitioning the processes into sub-processes proved to be extremely effective. However, for the majority of the maternity ward processes studied, it also had significant drawbacks, as the context of the overall process was lost for process optimization, which is ultimately the overall goal of process mining. Still, in the case of keeping the personal data and anonymization in process mining up, this approach seems suitable.

5 Discussion

From its basic idea, process mining technology enables the reconciliation of governance, risk, or compliance issues. At the user level, this article has shown that under certain conditions, it can also be used to identify processes in d. In particular, these conditions are known user routines and processing steps. We have shown that administrative-like tasks characterize downstream business processes in healthcare. These administrative processes are usually performed by nurses rather than physicians (e.g., obstetricians or general practitioner doctors).

The work presented here shows that, contrary to popular belief that process mining does not allow inferences to be made about user data if they are anonymized in the existing system, this is nevertheless possible.

This is possible if the process flows identified with process mining also reveal user routines that can be assigned to a specific user group or even an individual user with the help of process expertise. This evaluation of user routines and process deviations using process mining in anonymized systems leads to identifying unintentional and user-induced changes in core data. The study showed that these routines allow inferences to be made about the user or user group that made the changes without mapping user data and personal information in systems. To counteract the problem, processes were broken down into process snippets to prevent the activity pattern recognition enabled by process mining. Since this solution can only be a temporary interim solution, because the contextuality that is lost in the process can have disadvantages, as already mentioned, this naturally raises new questions for the theory as an implication. For example, to design security measures that demonstrably do not have the shortcomings we have pointed out, it would have to be possible to exclude activity pattern recognition. On the other hand, it must be ensured that there is no decontextualization as a consequence not to impair the effectiveness of process mining.

We will address these questions in the future, and they need to be addressed by other research teams and researchers as well.

Our research also has considerable added value for practical applications. We have demonstrated that the technical protection measures, which are mostly found to be sufficient in the literature, have serious weaknesses. As already mentioned, we were able to take most processes out of context and make deanonymization impossible. However, there were still processes that offered such a high identification potential that even these measures were insufficient. Department heads can only handle release processes in particular. If errors were to occur here, technical anonymization would not be possible by nature. This raises the question of the extent to which process mining may be used in companies that deal with very sensitive data. Although the measures basically protect the data of uninvolved parties (usually customers and/or clients), the activities of the employees are accessible to everyone via process mining, which means that the measures are not sufficient to protect the privacy of the users.

Our findings are, of course, principally limited, as they are initially restricted in the work presented here to a single case study and the data collected and available there. However, we estimate the transferability of the results outside the case study context used here to be very high since medical procedures, as well as information systems, are largely used internationally and must comply with international standards throughout. We also have to assume that other process mining technologies and used procedures may provide different results in anonymization quality. Nevertheless, the identifiability of process participants via mere variations and combinations of process flows and steps is, of course, technically not solvable without including further steps. We are also aware that our interviews varied in scope depending on the interviewee. We did include this in the coding, but care should still be taken to make the interviews more consistent in subsequent research.

References

1. Mans, R.S., Schonenberg, M.H., Song, M., van der Aalst, W.M.P., Bakker, P.J.M.: Application of process mining in healthcare – a case study in a dutch hospital. In: Fred, A., Filipe, J., Gamboa, H. (eds.) BIOSTEC 2008. CCIS, vol. 25, pp. 425–438. Springer, Heidelberg (2008). https://doi.org/10.1007/978-3-540-92219-3_32
2. van der Aalst, W., et al.: Process mining manifesto. In: Daniel, F., Barkaoui, K., Dustdar, S. (eds.) BPM 2011. LNBIP, vol. 99, pp. 169–194. Springer, Heidelberg (2012). https://doi.org/10.1007/978-3-642-28108-2_19
3. van der Aalst, W.M.P.: Process Mining. Data Science in Action. Springer, Heidelberg (2016). https://doi.org/10.1007/978-3-662-49851-4
4. Rovani, M., Maggi, F.M., de Leoni, M., van der Aalst, W.M.: Declarative process mining in healthcare. Expert Syst. Appl. **42**, 9236–9251 (2015). https://doi.org/10.1016/j.eswa.2015.07.040
5. Fredriksen, E., Martinez, S., Moe, C.E., Thygesen, E.: Communication and information exchange between primary healthcare employees and volunteers - challenges, needs and possibilities for technology support. Health Soc. Care Community **28**, 1252–1260 (2020). https://doi.org/10.1111/hsc.12958
6. Ghasemi, M., Amyot, D.: Process mining in healthcare: a systematised literature review. Int. J. Electron. Healthc. **72** (2016). https://doi.org/10.1504/IJEH.2016.078745
7. Erdogan, T.G., Tarhan, A.: A goal-driven evaluation method based on process mining for healthcare processes. Appl. Sci. **8**, 894 (2018). https://doi.org/10.3390/app8060894
8. Mans, R., Reijers, H., Wismeijer, D., van Genuchten, M.: A process-oriented methodology for evaluating the impact of IT: a proposal and an application in healthcare. Inf. Syst. **38**, 1097–1115 (2013). https://doi.org/10.1016/j.is.2013.06.005
9. Martin, N., et al.: Recommendations for enhancing the usability and understandability of process mining in healthcare. Artif. Intell. Med. **109**, 101962 (2020)
10. Munoz-Gama, J., et al.: Process mining for healthcare: characteristics and challenges. J. Biomed. Inform. **127**, 103994 (2022). https://doi.org/10.1016/j.jbi.2022.103994

11. Eggers, J., Hein, A., Böhm, M., Krcmar, H.: No longer out of sight, no longer out of mind? How organizations engage with process mining-induced transparency to achieve increased process awareness. Bus. Inf. Syst. Eng. **63**(5), 491–510 (2021). https://doi.org/10.1007/s12 599-021-00715-x

12. Grisold, T., Mendling, J., Otto, M., vom Brocke, J.: Adoption, use and management of process mining in practice. BPMJ **27**, 369–387 (2020)

13. Pika, A., Wynn, M.T., Budiono, S., ter Hofstede, A.H.M., van der Aalst, W.M.P., Reijers, H.A.: Privacy-preserving process mining in healthcare. Int. J. Environ. Res. Public Health **17**, 1612 (2020). https://doi.org/10.3390/ijerph17051612

14. Mannhardt, F., Koschmider, A., Baracaldo, N., Weidlich, M., Michael, J.: Privacy-preserving process mining. Bus. Inf. Syst. Eng. **61**, 595–614 (2019). https://doi.org/10.1007/s12599-019-00613-3

15. vom Brocke, J., Jans, M., Mendling, J., Reijers, H.A.: A five-level framework for research on process mining. Bus. Inf. Syst. Eng. **63**(5), 483–490 (2021). https://doi.org/10.1007/s12599-021-00718-8

16. Rafiei, M., van der Aalst, W.M.P.: Privacy-preserving data publishing in process mining. In: Fahland, D., Ghidini, C., Becker, J., Dumas, M. (eds.) BPM 2020. LNBIP, vol. 392, pp. 122–138. Springer, Cham (2020). https://doi.org/10.1007/978-3-030-58638-6_8

17. Loxton, M.: Process Mining in Healthcare. (2016, Unpublished)

18. Weise, M., Kovacevic, F., Popper, N., Rauber, A.: OSSDIP: open source secure data infrastructure and processes supporting data visiting. Data Sci. J. **21** (2022). https://doi.org/10.5334/dsj-2022-004

19. Nuñez von Voigt, S., et al.: Quantifying the re-identification risk of event logs for process mining. In: Dustdar, S., Yu, E., Salinesi, C., Rieu, D., Pant, V. (eds.) CAiSE 2020. LNCS, vol. 12127, pp. 252–267. Springer, Cham (2020). https://doi.org/10.1007/978-3-030-49435-3_16

20. Houghton, C., Casey, D., Shaw, D., Murphy, K.: Rigour in qualitative case-study research. Nurse Res. (2013)

21. Murthy, S., et al.: A comparative study of data anonymization techniques. In: 5th International Conference on Big Data Security on Cloud (BigDataSecurity), IEEE International Conference on High Performance (2019)

22. Rojas, E., Munoz-Gama, J., Sepúlveda, M., Capurro, D.: Process mining in healthcare: a literature review. J. Biomed. Inform. **61**, 224–236 (2016). https://doi.org/10.1016/j.jbi.2016.04.007

23. Mans, R.S., van der Aalst, W.M.P., Vanwersch, R.J.B.: Process Mining in Healthcare: Evaluating and Exploiting Operational Healthcare Processes. Springer, Cham (2015). https://doi.org/10.1007/978-3-319-16071-9

24. Mans, R.S., Schonenberg, M.H., Leonardi, G., Panzarasa, S., Quaglini, S., van der Aalst, W.M.P.: Process mining techniques: an application to stroke care. Stud. Helath Technol. Inform. **136**, 573–578 (2008)

25. van der Aalst, W.M.P., Günther, C.W.: Finding structure in unstructured processes: the case for process mining. In: Basten, T., Juhás, G., Shukla, S.K. (eds.) Seventh International Conference on Application of Concurrency to System Design, ACSD 2007, Proceedings, Bratislava, Slovak Republic, 10–13 July 2007. IEEE Computer Society, Los Alamitos (2007). https://doi.org/10.1109/acsd.2007.50

26. Yin, R.K.: Case Study Research and Applications: Design and Methods. SAGE Publications, Thousand Oaks (2017)

27. Yin, R.K.: Qualitative Research from Start to Finish. Guilford Publications, New York (2015)

28. Dongen, B.F., Medeiros, A.K.A., Verbeek, H.M.W., Weijters, A.J.M.M., Aalst, W.M.P.: The ProM framework: a new era in process mining tool support. In: Ciardo, G., Darondeau, P. (eds.) ICATPN 2005. LNCS, vol. 3536, pp. 444–454. Springer, Heidelberg (2005). https://doi.org/10.1007/11494744_25

29. Verbeek, H.M.W., Buijs, J.C.A.M., van Dongen, B.F., van der Aalst, W.M.P.: Xes, xesame, and prom 6. In: Soffer, P., Proper, E. (eds.) CAiSE Forum 2010. LNBIP, vol. 72, pp. 60–75. Springer, Heidelberg (2011). https://doi.org/10.1007/978-3-642-17722-4_5

30. Günther, C.W., van der Aalst, W.M.P.: A generic import framework for process event logs. In: Eder, J., Dustdar, S. (eds.) Business Process Management Workshops, pp. 81–92. Springer Berlin Heidelberg, Berlin, Heidelberg (2006). https://doi.org/10.1007/11837862_10

31. Flick, U.: Triangulation. In: Mey, G., Mruck, K. (eds.) Handbuch Qualitative Forschung in der Psychologie, pp. 185–199. Springer, Wiesbaden (2020). https://doi.org/10.1007/978-3-658-26887-9_23

Analyzing How Process Mining Reports Answer Time Performance Questions

Carlos Capitán-Agudo[1]([✉]) [iD], María Salas-Urbano[1] [iD], Cristina Cabanillas[1,2] [iD], and Manuel Resinas[1,2] [iD]

[1] SCORE Lab, Universidad de Sevilla, Seville, Spain
{ccagudo,msurbano,cristinacabanillas,resinas}@us.es
[2] I3US Institute, Universidad de Sevilla, Seville, Spain

Abstract. The advances in process mining have provided process analysts with a plethora of different algorithms and techniques that can be used for different purposes. Previous research has studied the relationship between these techniques and business questions, but how process analysts use them to answer specific questions is not fully understood yet. We are interested in discovering how process analysts respond to specific business questions related to time performance. We have coded 110 answers to time performance questions in more than 60 process mining reports. As a result, we have identified 55 different operations with 137 variants used in them. We have analyzed the types of answers and their similarities, and examined how contextual information as well as existing process mining support may affect them. The results of the study provide an overview of the current state-of-practice to answer time performance questions and unveil opportunities to improve process mining tools and the way these questions are answered.

Keywords: Process mining · Time performance · Qualitative analysis · Quantitative analysis · BPI Challenge · Grounded Theory

1 Introduction

Many process mining techniques and tools have been developed in the last years to assist the discovery, monitoring and improvement of business processes based on the event logs provided by the information systems that support them [1]. Each technique usually targets specific aspects of the processes, such as the existence and order of the process activities [15], the assignment and distribution of process participants [4], or the time performance of the process execution [18].

The importance acquired by process mining has also led to the development of methodologies, guidelines and case studies on how to perform process mining. These have mainly concentrated on understanding or guiding the use of process mining from a global perspective but they have not explored extensively how process mining analysts use these techniques to respond to specific business questions [14]. A better understanding of this matter can help to identify limitations in the approaches followed by the analysts to answer such questions. A good example of this are the limitations derived from the widespread use of the directly-follows graph as a way to analyze process execution [2]. It may also help to find common patterns that facilitate the building of reference guidelines to support them in their task. Finally, it may ease the identification of gaps between the features of process mining tools and what is actually done by analysts.

In this paper, we conduct a systematic analysis of process mining reports aiming to discover what process mining operations (e.g. filtering, data manipulation, graphical representation) are used by process analysts to address a specific type of business questions, namely, time performance questions; and how these operations are related to each other as well as to the questions. Time performance questions refer to aspects like cycle time, waiting time, or bottlenecks. They constitute one of the most recurrent problems in process mining projects [10].

The data source of our analysis are the process mining reports submitted to the BPI Challenge (BPIC for short), an annual competition since 2011. Every year the challenge organizers publish a real-life event log provided by an organization together with specific business questions posed by the organization, so that the solutions provide them added value. Participants answer these questions or perform other complementary analyses and submit them in a report. Considering time performance questions, this includes 62 reports belonging to 4 different BPIC with a total of 110 answers. There are several reasons for choosing BPIC as the source of our study. First, they provide different perspectives on how to analyze the same data for the same question, which makes the answers more comparable. Second, they cover several analyst profiles: academics, students, and professionals. Third, the analyses in the reports are not undirected, but driven by specific questions posed by the organization, which is aligned with the way process mining is used in practice [10]. Fourth, as the reports are page limited, they collect the most important and conclusive information, avoiding distractions with irrelevant information. Finally, all reports analyzed are publicly available, which helps with the traceability and replicability of the results.

We have analyzed these reports applying a four-step methodology following a mixed-methods research approach similar to the one in [14]. The results obtained include a catalogue of 55 operations and 137 variants that provides an overview of the current state-of-practice to answer time performance questions. The study also gives insights about the type of answers that can be found in the reports, how the context of the analysis (pursued goal, log and authors' profile) affects the characteristics of the answers, and what is the observed impact of current state of the art on the answers. These results can be useful to further improve process mining tools and the way in which questions are addressed by analysts.

The paper is structured as follows. Section 2 outlines the literature related to this work. Section 3 describes the methodology followed to conduct the analysis. Section 4 provides details of the analysis and the findings. Section 5 summarizes the conclusions drawn and directions for future work.

2   Related Work

Methodologies and guidelines to do process mining have been developed over the last 10 years. Several methodologies define high-level stages, inputs, outputs and activities that should be performed in a process mining project. Examples are the Process Diagnostic Method [3], the L^* life-cycle model [1] and PM^2 [10], which identifies 6 stages in a process mining project: planning, extraction, data processing, mining & analysis, evaluation, and process improvement & support. Similarly, [12] provides guidelines to support organizations in systematically using process mining techniques aligned with Six Sigma. Due to their broad scope, these methodologies are intentionally open in terms of which techniques can be used to address specific questions. Instead, we are interested in understanding the details of how time performance-related questions are addressed in practice. The methodologies are useful though, to frame the context of the research presented in this paper: the mining & analysis and, partially, the data processing stage.

Another workstream has focused on analyzing published process mining case studies to provide different perspectives on how process mining is used in practice. For instance, [11] assesses the maturity of the field from a practical viewpoint by considering the diffusion of tools and the thoroughness of the application of process mining methodologies over the years. However, it does not cover the specific details of how the questions are answered in the case studies except for an enumeration of 7 process mining techniques used. Other studies focus on a specific field. For instance, [21] discusses healthcare case studies according to 11 main aspects, but the level of abstraction is similar to [11]. A similar analysis applied to the BPIC reports is performed in [16]. The authors focus on the methods, tools, and techniques used in the reports submitted by the participants. None of these papers links the analysis techniques used to the business questions addressed nor discusses the context in which these techniques are applied.

The closest work to ours is [14] and [27]. Klinkmüller et al. [14] qualitatively analyze BPIC reports to understand how process analysts perform their work. The focus is put on visual representations and their information needs for all types of questions. We complement this research with a narrower but deeper analysis. We focus on identifying all specific low-level operations that are used to answer time performance questions. Because of that, the operations identified in our paper are more fine-grained, which brings a more precise understanding about how questions related to time performance are addressed. Zerbato et al. [27] conduct an empirical study to understand how analysts perform a process mining task. Their study focuses on the initial exploratory phase of process mining where analysts examine and understand an event log. It reveals that the 12 analysts who participated in the research follow different behavior patterns when exploring event logs with Disco, and identifies some typical operations to carry out process mining. We complement this research by focusing on specific business questions and looking beyond the exploratory phase. Specifically, we identify operations related to time performance that have been performed by analysts with different profiles (i.e. various organizations and countries) making use of several process analysis tools (e.g. Disco, ProM and Celonis).

Moreover, specific techniques and visualizations have been developed to analyze the time perspective of a business process, its cycle time, and its bottlenecks (e.g. [13,17,20,22,25]). With respect to them, this paper helps to understand if they are used in practice and the context in which they could replace some of the more general techniques used in the BPIC reports.

3 Research Methodology

We apply a methodology similar to the one proposed in [19], which follows a mixed-method approach that combines qualitative and quantitative research methods. We first perform a qualitative coding similar to the one proposed in [14]. This coding allows us to quantitatively analyze the operations in the BPIC reports. With this study we want to answer 4 research questions:

RQ1: What operations are used to answer the questions on time performance? We aim to identify the analysis operations frequently used in time performance analysis.

RQ2: What types of answers to time performance questions can be identified? We aim to discover categories in the answers provided by the authors of the reports depending on the operations used. This can inspire future process analysts.

RQ3: How does the context affect the similarity of the answers to time performance questions? The context involves the specific goal pursued in the question, the event log analyzed and the authors' profile. We aim to find commonalities and differences regarding these aspects to understand how they affect the answers.

RQ4: What is the observable impact of the current state of the art on the answers to time performance questions? We aim to understand how the existing tool support and literature help and limit the answering of the questions.

The steps of the methodology are described next. Details and materials are available at our repository [5].

3.1 Step 1: Data Collection

As we focus on questions related to time performance in business processes, the first step was to review which BPIC had business questions concerning this perspective. We found a total of 7 questions related to time in 4 editions: 2015 [6], 2017 [7], 2019 [8] and 2020 [9]. In BPIC 2020, we noticed that many operations and data of the first question were reused to answer the second one, so we have considered them as the same question. We classified these questions attending to their goal in: *differences*, whose goal is to find differences between the throughput

Table 1. Questions related to time performance in the BPI challenges since 2011. The column of BPIC (ID) represents the selected questions and the identifiers of the questions used to refer them (e.g. 2015-Q5 whose identifier is C15, represents question 5 of challenge 2015).

BPIC (ID)	Type	Question	Answers
2015-Q5 (C15)	Differences	Where are differences in throughput times between the municipalities and how can these be explained?	9
2017-Q1 (C17)	Differences, fragments	What are the throughput times per part of the process, in particular the difference between the time spent in the company's systems waiting for processing by a user and the time spent waiting on input from the applicant as this is currently unclear?	21
2019-Q2 (C19)	Fragments	What is the throughput of the invoicing process, i.e. the time between goods receipt, invoice receipt and payment (clear invoice)? To answer this, a technique is sought to match these events within a line item, i.e. if there are multiple goods receipt messages and multiple invoices within a line item, how are they related and which belong together?	12
2020-Q1, 2020-Q2 (C20A)	Fragments, differences	What is the throughput of a travel declaration from submission (or closing) to paying?, Is there are difference in throughput between national and international trips?	20
2020-Q4 (C20B)	Fragments	What is the throughput in each of the process steps, i.e. the submission, judgement by various responsible roles and payment?	17
2020-Q5 (C20C)	Bottlenecks	Where are the bottlenecks in the process of a travel declaration?	18
2020-Q6 (C20D)	Bottlenecks	Where are the bottlenecks in the process of a travel permit (note that there can be multiple requests for payment and declarations per permit)?	13
Total	-	-	110

of different processes; *fragments*, whose goal is to calculate the throughput of parts of the process; and *bottlenecks*, whose goal is to find bottlenecks. A question can be related to more than one goal (cf. Table 1).

We considered only the reports that answer the selected questions, specifically, those that have a specific section dedicated to respond to a question. As a result, 62 reports and 110 different answers were included in the analysis: 9 of 9 reports in 2015, 21 of 24 in 2017, 12 of 15 in 2019, and 20 of 37 in 2020. The number of answers to questions in 2020 varies because not every report provided an answer to every question. Additionally, the reports were grouped according to the authors' profile: students, professionals, and academics. The distribution of reports and answers in these profiles per year can be found in our repository [5].

3.2 Step 2: Coding

We followed an inductive category development based on several coding iterations. The way in which these iterations were performed was inspired by the Grounded Theory methodology [23]. First, we applied open coding to the answers to the questions provided in the BPIC reports. This involved reading the answers and marking them with annotations to derive codes. During this initial phase, we noted that each report answered the time performance questions using their own specific terms, but these terms referred to the same concepts. Thus, we had

Table 2. Examples of coding

Text in the report	Annotation	Operation	Variant
For the former case, filtering was performed by designating A_Pending as forbidden and O_Cancelled as end activity	Filter traces depending on the lack of A_Pending and O_Cancelled as end point	Filter traces	Filter traces by activities
By filtering all cases that did not have a project number in Disco	Filtering of traces without project number in disco	Filter traces	Filter traces by organizational units

to unify the vocabulary to compare the annotations of different answers more easily, since our purpose was to discover commonalities among the answers. To do so, we created a *key concept code* where we related different terms to a unique concept. For instance, in some cases the authors referred to the total execution time of the process as *throughput* but the implementations provided calculate the time required to complete (a part of) a process (*cycle time*), so we decided to rename it to *cycle time*. The name *throughput* was kept in the cases where the number of activities or process instances per time unit is calculated.

Afterwards, we grouped the annotations of the answers by some time performance questions that we sampled to better handle the annotations depending on their similarity. Once the annotations were grouped, we could identify their corresponding operations and detect the same operations from different reports. For instance, in two different reports of the BPIC 2020 we found the two similar annotations shown in Table 2. In both annotations the authors are filtering traces, despite using different criteria. Therefore, we grouped them into an operation called *Filter traces*. This way, we created an *operation code* to avoid defining similar operations with different names. This also made us notice that the implementations of some operations had the same purpose but were performed over different variables. We call them *variants*. For example, the aforementioned annotations are two variants of how to filter traces, since one is filtering by activities and the other by organizational units. Thus, we labelled them as variants of *Filter traces* as depicted in Table 2.

The coding process was iterative and finished when no more operations and variants were obtained from the reports. In total, we found 55 operations with 137 variants. To mitigate the bias of one researcher having to identify the codes, during the whole process two authors annotated and coded a subset of the reports independently and then shared the results. In case of disagreement, the four authors discussed the differences to reach a consensus. Moreover, we categorized the operation codes into 6 types based on their goals as explained in Sect. 4.1.

3.3 Steps 3 and 4: Dataset Creation and Quantitative Analysis

Next, we handcrafted a dataset that relates the operations and variants identified to the answers in which they appear. We also included metadata related to the question, year, category, and type. The resulting dataset has 955 actions and 110 answers, where an action is an execution of an operation variant.

Finally, we performed a quantitative analysis of the dataset to answer the research questions. Specifically, we analyzed the dataset using frequency distributions and descriptive statistics to answer RQ1. In order to respond to RQ2 we analyzed the answers depending on the number of performed operations and we applied KMeans clustering of the answers in the reports according to the operations used in them. To answer RQ3, we used the Sørensen-Dice coefficient [24] between pairs of answers to find similarities. This index $DSC(A, B) = (2|A \cap B|)/(|A| + |B|)$ measures the similarity between two sets A and B, where 0 indicates two totally different sets and 1 two equal sets. Furthermore, we used 45 of the 72 measures described in [26] to retrieve properties about the event logs that could help us to cluster them to understand how they can affect the answers. We excluded those measures that had problems during their computation (e.g., too long execution times). Finally, to answer RQ4, we checked the existing tool support in process mining tools as well as related literature. More details of this step are provided in Sect. 4. The codes in bold in the first column of Table 1 will be used therein for the sake of brevity.

3.4 Threats and Limitations

First, as it often happens in qualitative research, there could be personal bias because the annotations and coding rely on a subjective interpretation of the description that appears in the report. We mitigate this threat as discussed above, but a residual risk remains.

Second, the conclusions of this study are based on reports that address 8 time performance questions. These questions deal with typical temporal problems, such as bottlenecks or differences in throughput time between processes. Although the sample is representative, it does not cover all possible questions.

Third, the reports analyzed could be done by the same organization and hence, be more similar to each other than otherwise. We checked the organizations of the reports and found that there was a predominant organization with 10 reports of 62. However, the similarity using the DSC index between the reports of this organization is smaller than the similarity with the reports belonging to participants from other organizations (0.12 and 0.16, respectively). Thus, we concluded that including them would not bias the results.

Finally, the analysis is based only on the reported answers. This has two implications. First, the order in which the operations appear in the report may be different from the order in which the analysts performed them. To partially mitigate this risk, we ignore the operations order for our analysis. Second, not all operations used by the analysts may appear in the report. Some of them might not give relevant results and be omitted in the report, and others might have been removed because of space restrictions. This risk cannot be fully avoided with our study design. However, we can safely assume that the operations that appear in the report are those that the authors found more relevant. Furthermore, as long as one of the operations appears in one answer, it is considered in our analysis.

Table 3. Classification of operations sorted in descending order of frequency. In bold are those with a frequency higher than the average (17.05).

Operation (absolute frequency - number of variants - number of questions)

OPERATIONS TO ANALYZE TIME:

Calculate cycle time (152-12 - 7), **Find bottlenecks (63 - 5 - 6)**, **Compare cycle time (30 - 1 - 7)**, **Calculate waiting time (27 - 1 - 2)**, **Calculate throughput (18 - 1 - 3)**, Calculate processing time (10 - 1 - 3), Compare throughput (2 - 1 - 1), Compare waiting time with processing time (2 - 1 - 1), Analyze cycle time depending on the events (1 - 1 - 1), Calculate intervals of time of the traces (1 - 1 - 1)

OPERATIONS TO MANIPULATE THE DATA:

Filter traces (86 - 7 - 7), **Group traces (58 - 12 - 6)**, Preprocess the traces of the logs (11 - 1 - 5), Filter events (11 - 4 - 5), Group activities (8 - 4 - 3), Filter activities (9 - 4 - 5), Filter sub-processes (6 - 2 - 1), Filter variants depending on frequency (2 - 1 - 1), Preprocess the events of the logs (2 - 1 - 1), Group events by attributes (1 - 1 - 1), Group events by time (1 - 1 - 1), Group organizational units (1 - 1 - 1), Group sub-processes (1 - 1 - 1)

OPERATIONS TO CALCULATE STATISTICS

Calculate number of elements (76 - 7 - 7), **Calculate percentages (55 - 4 - 6)**, **Calculate statistics (36 - 4 - 6)**, **Calculate frequency (25 - 7 - 7)**, Calculate average of activities per trace (3 - 1 - 3)

OPERATIONS TO REPRESENT THE PROCESS GRAPHICALLY:

Represent process map (47 - 2 - 7), **Represent bar charts (36 - 6 - 6)**, **Represent histograms (32 - 3 - 7)**, **Represent temporal series (25 - 4 - 5)**, Represent heat maps of cycle time and an attribute (6 - 1 - 3), Represent linear tendency of cycle time with respect an attribute (5 - 1 - 2), Represent scatter plot of cycle time and an attribute (5 - 1 - 4), Represent circular charts of attributes of the traces (3 - 1 - 1), Represent box plots of cycle time (3 - 1 - 1), Represent density diagram of cycle time (2 - 1 - 2), Represent lineal distribution of an attribute by traces (2 - 1 - 1), Represent correlation graph of variables (1 - 1 - 1)

OPERATIONS TO IDENTIFY ELEMENTS IN THE DATA:

Identify attributes (34 - 3 - 6), Identify resources (10 - 3 - 3), Identify transitions by cycle time (10 - 1 - 4), Identify organizational units (9 - 3 - 1), Identify activities (8 - 4 - 4), Identify roles (7 - 2 - 3), Identify traces by cycle time (2 - 1 - 2), Identify specific sub-processes (1 - 1 - 1), Identify impact of bottlenecks by organizational unit (1 - 1 - 1)

OTHERS:

Calculate dates of the development of activities of resources (2 - 1 - 1), Assign resource to each activity (1 - 1 - 1), Apply techniques of machine learning (1 - 1 - 1), Apply decision trees (1 - 1 - 1), Discover happy path of the process (1 - 1 - 1), Discover process maps (1 - 1 - 1)

Despite these limitations, we believe that the use of the BPIC reports also brings relevant advantages as discussed earlier. Furthermore, we think that the analysis conducted provides relevant insights that can be used as a starting point to improve our understanding of how questions are answered in practice.

4 Results

Next, we describe how we have addressed the research questions defined in Sect. 3 and the results obtained.

4.1 RQ1: Operations Used to Answer Time Performance Questions

We identified 55 different operations and 137 variants and classified them in 6 groups according to their purpose (cf. Table 3). The operations that do not fit in any of these groups are classified as *others*. Table 3 also shows the absolute frequency of each operation, the number of variants identified for each operation and the number of questions for which at least one answer uses each operation. Most of the operations (35) have only one variant. The others have between 2 and 7 variants, except *Group traces* and *Calculate cycle time (CT)* with 12 variants each. In the following, we outline how operation variants are defined.

The *operations to analyze time* focus on the temporal analysis of the process, such as calculating and comparing cycle time and waiting time, or finding bottlenecks. Two operations have more than one variant. *Calculate CT* can be implemented for different process elements (e.g. the whole process or pairs of events) and considering either all traces or subsets of them. *Find bottlenecks* varies depending on where to look for the source of the bottleneck (e.g. activities or process fragments) and the criteria used to consider that a bottleneck is happening (e.g. activities that exceed the average cycle time of all activities).

The *operations to manipulate data* reorganize the traces or the events from the log, including their filtering, grouping, and preprocessing. Concerning their variants, filters and groupings are applied on some process element (e.g., traces or activities) and implemented depending on a condition related to a temporal performance measure, an attribute of the event log, or another process perspective. For instance, traces can be filtered depending on the existence of activities and activities can be grouped according to certain thresholds of cycle time.

The *operations to calculate statistics* give numerical insights applying descriptive statistics, such as counts, proportions and frequencies. Regarding their variants, *Calculate number of elements*, *Calculate percentages* and *Calculate frequency* are implemented depending on the process element to which they apply, like calculating the number, percentage or frequency of each activity, or to calculate the number of values that an attribute takes or the frequency with which one of them occurs. As for *Calculate statistics*, this operation is applied to cycle time, throughput and activities.

The *operations to represent the process graphically* show visual insights of the process by creating process maps, bar charts, or histograms among others. Their variants are based on what is being represented (e.g. cycle time), or the values of some attribute (e.g. *Represent process map with CT*). Additionally, *Represent temporal series* and *Represent bar charts* vary depending on the process element. Temporal series are also used to represent throughput.

The *operations to identify elements in the data* find a specific aspect of the process and its context, such as process fragments, activities, attributes (e.g. roles and resources) based on some condition. In this case, the variants represent the conditions used to identify the elements (e.g. *Identify attributes by CT*).

4.2   RQ2: Types of Answers

We have analyzed the answers from two different perspectives. First, we have compared the average number of total and distinct operations per questions, which are collected in Table 4. We believe that there can be two factors related to the difference between questions: (i) the question itself, e.g. C19 is broader and hence, requires more operations to give a proper answer; and (ii) the fact

Table 4. Average total and different operations per question and per authors' profile: academics (ACA), students (STU), professionals (PRO)

Average operations	C15	C17	C19	C20A	C20B	C20C	C20D	ACA	STU	PRO	Total
Total	9.55	10.90	15.42	9.40	5.41	8.38	3.15	6.60	9.75	9.12	8.68
Different	6.55	6.33	8.25	5.60	3.64	4.44	2.84	4.51	6.03	5.29	5.29

that BPIC 2020 has several questions related to time, while in other challenges all aspects related to time are focused on one question.

Second, we have performed a clustering analysis using the KMeans clustering algorithm to discover categories of similar answers. The input was a boolean matrix where the rows are answers, the columns are operations, and the cells represent whether an operation is used in an answer or not. Since it was not clear how many clusters can be expected, we evaluated the results with different numbers of clusters (from 2 to 9 clusters) and the best results were obtained with 4 clusters. The Average Silhouette Width is not high (0.12), indicating that the clusters are unstructured. This is expected because of the high variation that has been found between the answers as we detail later. Nevertheless, the clustering provides a useful classification of the answers in 4 broad answer categories, whose distribution among the questions is depicted as pie charts in Fig. 1.

The Exhaustive Answer. It includes 17 answers that perform an exhaustive analysis of temporal performance aspects. It includes the longest answers with an average number of 9 different operations and 19 steps. Almost all answers use *Calculate CT* and a significant number of answers *Find bottlenecks*, too. They also frequently apply several manipulation operations like filters and groupings, and compute statistics, percentages and frequencies. Finally, the answers of this category also represent graphical information in a higher proportion than the other answer categories, especially using bar charts, histograms or process maps.

The Difference Finder Answer. It includes 29 answers whose main focus is to find differences in the performance of different process variants. It includes average-sized answers with 5 different operations and 8 steps on average. Almost all answers use *Calculate CT* and *Filter traces*. However, the main difference with the other groups is the use of *Compare CT*, which appears in 65% of the answers (compared to less than 11% in the other categories). They usually do not have a graphical representation, being histograms the most frequently used (25%).

The Manipulatory Answer. It includes 8 answers that use manipulation operations like *Filter traces* and *Group traces* in a significantly higher proportion than in the other categories. They are also characterized by the lower use of *Calculate CT* and the higher use of operations to calculate statistics. Also, unlike the other categories, temporal series are used in 60% of the answers to represent the data. In terms of size, this group also includes large answers with an average number of 8 different operations and 16 steps.

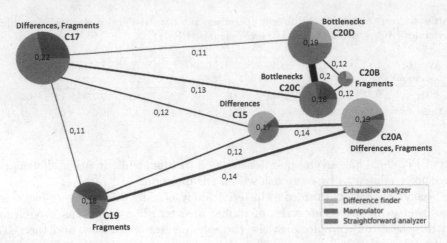

Fig. 1. Graph of most similar relationships between questions

The Straightforward Answer. It includes 56 short answers that are characterized by the low use of manipulation operations, especially filters, which is much higher in the other categories. Their use of *Calculate CT* is significant (70%) but not as widespread as in the first two categories. *Find Bottlenecks* also appears with the same frequency, which is similar to the frequency it appears in the first category. As for the representation, this category uses process maps like the exhaustive analyzer but in a lower proportion. Concerning the size, this category includes the shortest answers with 3.8 different operations and 4.4 steps on average.

4.3   RQ3: Effect of the Context on the Answers

In this section, we study how 3 contextual elements (question objectives, log characteristics and authors' profile) can influence the answers. We have not studied the difference between winner and non-winner reports because our sample only constitutes around 30% of the total number of questions. Therefore, we cannot assume that winning a BPIC is directly related to the answers we have analyzed.

Effects of Question Objectives and Logs. To analyze the effects of questions and logs on the answers, we used 3 elements to characterize them. Specifically, we used 5 logs of BPIC 2015, 5 logs of BPIC 2020, 1 log of BPIC 2017 and 1 log of BPIC 2019. First, we assigned one or more goals (difference, fragments or bottlenecks) to every question based on their description as discussed in Sect. 3.1. Second, we computed the similarity between the logs used in each question. To this end, we used most of the measures described in [26] to retrieve properties about them. Then, we used these measures to group the event logs using a KMeans clustering algorithm. We evaluated the results using different number of clusters with The Average Silhouette Width, and the best ones were obtained with 3 groups:

one for the logs used in 2015 and 2020, one for the log in 2017 and a third group for the log in 2019. Finally, to find the similarities between answers, we computed the Sørensen-Dice coefficient DSC_v for every pair of answers in our dataset considering two variants of the same operation as different. We also compared the similarity between each pair of questions Q_x and Q_y by computing the average of the DSC_v of all pairs of answers that respond to Q_x and Q_y, respectively.

The results obtained are summarized in Fig. 1, which depicts a graph whose nodes are the questions and the edges are those pairs whose average DSC_v exceeds the average DSC_v of the whole dataset, which is 0.11. The label of each edge shows the average DSC_v and its width is proportional to this value. Similarly, the label of each node shows the average DSC_v between the answers of that question, and its size is proportional to that value. In addition, we have added other labels to the nodes to include the question objectives, and the nodes are positioned in the figure based on the similarity between the logs used in each question, i.e. the closer are two nodes, the more similar their logs are.

The results show that the greatest similarity on average occurs between pairs of answers that belong to the same question except for C20B, which is not amongst them; and the pair (C20C, C20D), which ranks above 6 pairs of answers of the same question. For the pairs of answers of the same question, these results make sense because if the question is the same and the challenge is the same, the answers are expected to be more similar between them. The fact that the pair (C20C, C20D) is ranked high also makes sense because both questions refer to bottlenecks in the same challenge. Therefore, although they refer to different event logs, a number of authors performed almost the same analysis for each of them, which significantly increases the similarity between these questions. Regarding C20B, we believe that the diversity in the answers could be caused by the logs used to answer this question. The reason is that, unlike the other questions in which all authors analyze the same logs, in this case the question involved several logs and not all authors decided to use the same set of logs.

Another interesting insight is that the answers of C20A are more similar to the answers of questions that share similar objectives than to the answers of the other questions of BPIC 2020. As a matter of fact, if we consider all questions belonging to BPIC 2020 together (as if it were a single question), it is more similar to other challenges than to itself. This suggests that the objective is more important than the event log in terms of similarity.

Question objectives are not the only factor that affects the similarity between answers, though. For instance, both C20B and C17 are linked to C20C and C20D, but they do not share any objective. This is also evident from the fact that the predominant answer type in the four of them is *straightforward analyzer*. For C17 the reason of this similarity stems from the fact that many of its answers try to find bottlenecks in the process even though it was not a clear objective in the question. Instead, for C20B the similarity is because of the length of its answers and the fact that it is in the same BPIC as C20C and C20D.

Another relevant aspect related to C17 is that its log is the only one that allows to properly calculate the waiting time of the process since it records the beginning and the end of the activities by means of attribute *lifecycle:transition*.

Table 5. Comparison between the common frequent operation variants per questions with the same objective: differences (DIFF), fragments (FRAG), bottlenecks (BTL). The color intensity represents the percentage of questions in which the variants exceed the average case frequency: 0%, (0, 25]%, (25, 50]%, (50, 75]% and (75, 100]%, where white represents 0% and pure black represents 100%.

	Variants	DIFF	FRAG	BTL
1	Calculate CT of the whole process for all traces			
2	Calculate CT of the whole process for each subset of traces			
3	Calculate CT of the whole process for a subset of traces			
4	Calculate CT of a fragment of the process for a subset of traces			
5	Calculate CT for all pairs of events for all traces			
6	Find activities as bottlenecks applying temporal performance criteria			
7	Find sub-processes as bottlenecks applying temporal performance criteria			
8	Find sub-processes with incorrect orders with respect to the happy path as bottlenecks			
9	Compare CT			
10	Filter traces by activities			
11	Filter traces by attributes			
12	Group traces depending on attributes			
13	Calculate number of traces			
14	Calculate number of events			
15	Calculate number of activities			
16	Calculate percentage of traces			
17	Calculate Statistics of CT			
18	Represent process map with CT			
19	Represent histograms of CT			
20	Identify values of attributes			
21	Identify transitions by CT			

Let us look now into the details of the operation variants that are common to different questions. Table 5 shows the common frequent operation variants per questions with the same objective. The columns group the questions by objective as detailed in Table 1, but we have included C17 in the 3 categories since the operation *Find bottlenecks* is used in 85.7% of its answers.

Regarding the variants of *Calculate CT* (rows 1–5), if the goal is to find *differences* in the process, it is frequent to calculate the cycle time of the whole process for all or some subset of traces, indistinctly. It also involves the comparison of the resulting values as support to this task (row 9). Instead, the analysis of time performance in process *fragments* frequently involves the calculation of cycle time for subsets of traces. As for the *bottlenecks* objective, it is common to calculate cycle time computed for all pairs of events and to carry out several variants of *Find bottlenecks* (rows 6–8). Manipulation operation variants (rows 10–12) are most used within the *fragments* objective, although *Filter traces by activities* is the only variant that is used with great frequency in all the questions analyzed in this research. Regarding the variants of the *Calculate statistics* operations (rows 13–17), these are frequently used to find *differences* and to analyze the cycle time of *fragments* of the process. The most common is to carry out *Calculate number of traces* and *Calculate percentage of traces*, respectively. Finally, concerning the *graphical representations* (rows 18–19), we observe that if the goal is to find *bottlenecks*, it often involves *Representing a process map of CT.* Instead, if the goal is to find *differences* or to analyze the cycle time of *fragments* of the process, *Histograms of CT* is the most frequent representation.

Effects of the Authors' Profile. An analysis of the answers grouped by the category of the analysts (academics, professionals, students) shows differences in both the number of operations and the number of different operations in each answer (cf. Table 4). Specifically, the average number of operations in academics is significantly lower than in students. For professionals, the values are in the middle. These results suggest that academics tend to be more precise with their reports and include only the most relevant information, whereas professionals and particularly students, tend to include more information.

Another difference lies in the types of operations used. The most used operations by the 3 profiles are *operations to analyze time.* However, the second most used type of operations for academics and students are *operations to manipulate data*, while professionals use *operations to calculate statistics.*

Finally, we also look into the operations that appear more frequently in the answers of each category. Specifically, we consider the operations whose absolute frequency exceeds the average frequency of each category and focus on operations that are common only to pairs of profiles. On the one hand, both students and professionals use the same graphical representations (*Represent process map, Represent bar charts* and *Represent histograms*) and one statistical measure (*Calculate percentages*). On the other hand, academics and professionals share two operations: *Calculate statistics* and *Identify attributes.* Regarding the unique operations per category, academics use more specific representations to analyze cycle time, such as *Represent density of CT* or *Represent box plots of CT.* In contrast, professionals have a greater interest in operations at the level of activity and event, such as *Analyze CT depending on the events* and *Calculate dates of the development of activities of resources.*

4.4   RQ4: Impact of the Current State of the Art on the Answers

We have observed that existing tools might be influencing two aspects of the answers, specifically, the finding of bottlenecks and the visualization of cycle time. First, if we look at the operations that analyze time, we observe that the operation *Find bottlenecks* is usually implemented in a naive way with the variant *Find activities as bottlenecks applying temporal performance criteria*, which highlights those activities whose cycle time is higher than the average or that lead to process executions whose cycle time is higher than the average. This approach disregards resource contention problems, which are the usual cause of bottlenecks. This naive implementation may be due to the lack of advanced mechanisms to detect bottlenecks in typical process mining tools. To mitigate this problem, there are proposals in the literature like [22] that provide more advanced tools to detect them. However, there could be more factors influencing this aspect such as the lack of awareness or data. Something similar occurs with the operations that calculate cycle time for all pairs of events, which do not take parallelism into account and can lead to wrong conclusions as discussed in [2].

Second, the huge majority of visual representations used to depict time performance information are either general purpose visualizations (histograms or bar charts) or process maps with performance information. In fact, only some of the representations used by academics go beyond the visualizations commonly provided by process mining tools. This contrasts with the current state of the art, which includes approaches like [17,20,25] that highlight aspects that are relevant to answer the questions depicted in Table 1. The reason may be again that these approaches are not well-known beyond academia and are not integrated in the software tools used for the analysis.

5 Conclusions

The results of this work provide an overview of the current state-of-practice to answer time performance questions and can be useful as a comparison framework to evaluate the fitness of process mining tools for addressing them. For this purpose, it is important to note that the catalogue covers only the operations used specifically to answer time performance questions, but before addressing these questions it might be necessary to perform discovery and familiarization activities that may require additional tasks as discussed in [14]. This study can also be useful to identify opportunities to improve the way in which questions are answered. Aligned with the findings in [14], the study shows that the comparison of the cycle time of different subsets of traces is extremely common either explicitly with the operation *Compare CT*, or implicitly by applying *Calculate CT* for different subsets of data. In fact, *Calculate CT* is the only operation that is used more than once on average in each answer. For process mining tools this means extending their current ability to quickly apply filters, to visualize and compare the results of applying different filters at the same time.

As a next step, we plan to extend the analysis to investigate specific ordering of certain operations by searching for dependencies in the process mining reports, and more exhaustively compare the catalogue of operations with the support provided by current process mining tools.

Acknowledgements. This work has been funded by grants RTI2018-100763-J-I00 and RTI2018-101204-B-C22 funded by MCIN/AEI/10.13039/501100011033/ and ERDF A way of making Europe; grant P18-FR-2895 funded by Junta de Andalucía/FEDER, UE; and grant US-1381595 (US/JUNTA/FEDER,UE).

References

1. van der Aalst, W.M.P.: Process Mining - Data Science in Action, 2nd edn. Springer, Berlin (2016). https://doi.org/10.1007/978-3-662-49851-4
2. van der Aalst, W.M.P.: A practitioner's guide to process mining: Limitations of the directly-follows graph. Procedia Comput. Sci. **164**, 321–328 (2019)
3. Bozkaya, M., Gabriels, J., van der Werf, J.M.: Process diagnostics: a method based on process mining. In: eKNOW, pp. 22–27 (2009)

4. Cabanillas, C., Ackermann, L., Schönig, S., Sturm, C., Mendling, J.: The RALph miner for automated discovery and verification of resource-aware process models. Softw. Syst. Model. **19**(6), 1415–1441 (2020). https://doi.org/10.1007/s10270-020-00820-7

5. Capitán-Agudo, C., Salas-Urbano, M., Cabanillas, C., Resinas, M.: BPI challenge analysis: how are time performance questions answered, March 2022. https://github.com/isa-group/bpi-challenge-performance-analysis

6. van Dongen, B.: BPI Challenge 2015. 4TU.ResearchData, May 2015. https://doi.org/10.4121/uuid:31a308ef-c844-48da-948c-305d167a0ec1

7. van Dongen, B.: BPI Challenge 2017. 4TU.ResearchData, February 2017. https://doi.org/10.4121/uuid:5f3067df-f10b-45da-b98b-86ae4c7a310b

8. van Dongen, B.: BPI Challenge 2019. 4TU.ResearchData, January 2019. https://doi.org/10.4121/uuid:d06aff4b-79f0-45e6-8ec8-e19730c248f1

9. van Dongen, B.: BPI Challenge 2020. 4TU.ResearchData, March 2020. https://doi.org/10.4121/uuid:52fb97d4-4588-43c9-9d04-3604d4613b51

10. van Eck, M.L., Lu, X., Leemans, S.J.J., van der Aalst, W.M.: PM^2: a process mining project methodology. In: CAiSE, pp. 297–313 (2015)

11. Emamjome, F., Andrews, R., ter Hofstede, A.H.: A Case Study Lens on Process Mining in Practice. In: OTM Conferences. pp. 127–145 (2019)

12. Graafmans, T., Turetken, O., Poppelaars, H., Fahland, D.: Process mining for six sigma. Bus. Inf. Syst. Eng. **63**(3), 277–300 (2021)

13. Hompes, B.F.A., Maaradji, A., Rosa, M.L., Dumas, M., Buijs, J.C.A.M., Aalst, W.M.P.v.d.: Discovering causal factors explaining business process performance variation. In: CAiSE, pp. 177–192 (2017)

14. Klinkmüller, C., Müller, R., Weber, I.: Mining process mining practices: an exploratory characterization of information needs in process analytics. In: BPM, pp. 322–337 (2019)

15. de Leoni, M., van der Aalst, W.M.P., Dees, M.: A general process mining framework for correlating, predicting and clustering dynamic behavior based on event logs. Inf. Syst. **56**, 235–257 (2016)

16. Lopes, I.F., Ferreira, D.R.: A survey of process mining competitions: the BPI challenges 2011–2018. In: BPM Workshops, pp. 263–274 (2019)

17. Low, W.Z., van der Aalst, W.M.P., ter Hofstede, A.H.M., Wynn, M.T., De Weerdt, J.: Change visualisation: analysing the resource and timing differences between two event logs. Inf. Syst. **65**(Supplement C), 106–123 (2017)

18. Maggi, F.M.: Discovering metric temporal business constraints from event logs. In: Johansson, B., Andersson, B., Holmberg, N. (eds.) BIR 2014. LNBIP, vol. 194, pp. 261–275. Springer, Cham (2014). https://doi.org/10.1007/978-3-319-11370-8_19

19. Revoredo, K., Djurica, D., Mendling, J.: A study into the practice of reporting software engineering experiments. Emp. Softw. Eng. **26**(6), 1–50 (2021). https://doi.org/10.1007/s10664-021-10007-3

20. Richter, F., Seidl, T.: TESSERACT: time-drifts in event streams using series of evolving rolling averages of completion times. In: BPM, pp. 289–305 (2017)

21. Rojas, E., Munoz-Gama, J., Sepúlveda, M., Capurro, D.: Process mining in healthcare: A literature review. J. Biomed. Inform. **61**, 224–236 (2016)

22. Senderovich, A., et al.: Conformance checking and performance improvement in scheduled processes: a queueing-network perspective. Inf. Syst. **62**, 185–206 (2016)

23. Stol, K., Ralph, P., Fitzgerald, B.: Grounded theory in software engineering research: a critical review and guidelines. In: ICSE, pp. 120–131 (2016)

24. Sørensen, T.: A method of establishing groups of equal amplitude in plant sociology based on similarity of species content and its application to analyses of the vegetation on Danish commons. Am. J. Plant Sci. **5**, 1–34 (1948)
25. Wynn, M.T., et al.: ProcessProfiler3D: a visualisation framework for log-based process performance comparison. Decis. Support Syst. **100**(Supplement C), 93–108 (2017)
26. Zandkarimi, F., Decker, P., Rehse, J.R.: Fig4PM: a library for calculating event log measures. In: ICPM Doctoral Consortium and Demo Track, pp. 27–28 (2021)
27. Zerbato, F., Soffer, P., Weber, B.: Initial insights into exploratory process mining practices. In: BPM Forum, pp. 145–161 (2021)

Process Mining of Knowledge-Intensive Processes: An Action Design Research Study in Manufacturing

Bernd Löhr[✉][iD], Katharina Brennig[iD], Christian Bartelheimer[iD],
Daniel Beverungen[iD], and Oliver Müller

Faculty of Business Administration and Economics, Paderborn University,
Warburger Street 100, 33098 Paderborn, Germany
{bernd.loehr,katharina.brennig,christian.bartelheimer,daniel.beverungen,
oliver.mueller}@uni-paderborn.de

Abstract. Existing process mining methods are primarily designed for processes that have reached a high degree of digitalization and standardization. In contrast, the literature has only begun to discuss how process mining can be applied to knowledge-intensive processes—such as product innovation processes—that involve creative activities, require organizational flexibility, depend on single actors' decision autonomy, and target process-external goals such as customer satisfaction. Due to these differences, existing Process Mining methods cannot be applied out-of-the-box to analyze knowledge-intensive processes. In this paper, we employ Action Design Research (ADR) to design and evaluate a process mining approach for knowledge-intensive processes. More specifically, we draw on the two processes of product innovation and engineer-to-order in manufacturing contexts. We collected data from 27 interviews and conducted 49 workshops to evaluate our IT artifact at different stages in the ADR process. From a theoretical perspective, we contribute five design principles and a conceptual artifact that prescribe how process mining ought to be designed for knowledge-intensive processes in manufacturing. From a managerial perspective, we demonstrate how enacting these principles enables their application in practice.

Keywords: Process mining · Knowledge-intensive processes · Action design research · Design principles

1 Introduction

Process mining enables the data-driven analysis and continuous improvement of business processes in organizations. 61% of decision makers view process mining as the most relevant technology to improve business processes and use or plan to use it in the next twelve months [6]. Process mining methods and tools are inspired by and adapted from data science, and thus rely on the availability of large amounts of structured data in the form of event logs [1, p. 15–20].

© Springer Nature Switzerland AG 2022
C. Di Ciccio et al. (Eds.): BPM 2022, LNCS 13420, pp. 251–267, 2022.
https://doi.org/10.1007/978-3-031-16103-2_18

This, however, implies that process mining methods and tools need to be adapted to less structured processes that are human-centric and subject to substantial degrees of knowledge and flexibility. These processes have been termed knowledge-intensive processes (KIPs) [8,27]. Due to the high degree of creative, flexible, and knowledge-intensive activities, overly automating or standardizing KIPs is neither feasible nor desirable [8,23,24,27]. Further, while explicit process knowledge can be codified and made available for process mining, implicit knowledge—or tacit knowledge—often evades codification and is not available for process mining [8,30]. Because KIPs often lack extensive digital event logs, many of the common process mining techniques cannot be applied without adaption. Unsurprisingly, practitioners and scientists recently identified *"limited data access," "high cost of data preparation," "issues with data quality," "lack of intuitiveness and guidance,"* and *"difficulties to understand the process mining output"* as challenges for process mining, all relating to KIPs [28]. Contrasting these challenges, analyzing and improving KIPs is paramount for organizations, because they often constitute an organization's competitive advantage [24].

We present two cases of manufacturing companies that strive to use process mining for analyzing and improving their complex product innovation and engineer-to-order (ETO) processes. Our findings build on and extend previous research on applying process mining in unstructured processes. Other KIPs—such as diagnostic processes in the healthcare domain—have been addressed by the process mining community [22] and share some characteristics with KIPs in manufacturing. Nonetheless, we also identified crucial differences, including the use of guidelines and protocols as fairly precise prescriptions for instantiating healthcare processes. Product innovation and ETO processes, on the other hand, display higher degrees of creativity and do not follow such detailed prescriptions on a product engineering level [22], even though they are often comprised of a stage-gate structure that distinguishes creative work from quality gates and guides the control flow [7]. Further, while healthcare KIPs clearly focus on the recovery of patients [22], manufacturing KIPs do not always follow such well-defined goals but often target an anonymous market and are subject to various goals that depend on a specific product and market [8].

This paper presents findings from an Action Design Research (ADR) project on developing process mining methods and tools for analyzing KIPs in manufacturing. Our main contribution is a set of theoretically ingrained and empirically evaluated design principles that can serve as blueprints for researchers and practitioners to apply process mining in similar contexts (i.e., other KIPs). Our proposed principles question taken-for-granted knowledge on the design and applicability of process mining methods and extend it for the context of KIPs. Amongst others, we propose that focusing on process-external goals, using small data instead of big data, and restricting process analysis to process instances that are conceptually similar are unique characteristics of applying process mining to KIPs. Building on the design principles, we instantiate a process analytics tool as an exemplary IT artifact to analyze KIPs in manufacturing.

The paper is structured as follows. In Sect. 2, we review process mining fundamentals and previous applications on KIPs. In Sect. 3, we describe our

ADR method in detail. In Sect. 4, we present field evidence from KIPs in two manufacturing cases and identify specific requirements for the design of a process mining artifact. We then propose five design principles for process mining of KIPs and instantiate them with a process analytics tool. In Sect. 5, we discuss implications for the theory and practice before concluding the paper in Sect. 6.

2   Related Research

2.1   Knowledge-Intensive Processes

Business processes are assumed to be intentionally designed and to represent standard operating procedures in organizations [2]. A process is a *"collection of inter-related events, activities and decision points that involve a number of actors and objects, and that collectively lead to an outcome that is of value to at least one customer"* [10, p. 6–7]. Generally, the Business Process Management (BPM) literature assumes that a desirable and often mandatory way of conducting work in organizations exists and can be inscribed into process models, implying that deviations from to-be processes are viewed as deviant behavior that must be prevented [18]. However, a process in which a considerable number of activities depend on the application of tacit human-centered knowledge [30], is a KIP [8,27]. KIPs are competitive processes for organizations [24] **(F4)** and strongly rely on human creativity [3] **(F1)**. They are characterized by high degrees of complexity and flexibility [13,23,38] **(F1)**, and of uncertainty regarding their in- and output [8] **(F5)**.

KIPs can neither be fully automated nor standardized [23], since human-centered and knowledge-intensive activities represent an integral part of the process [8,27] **(F3)**. As value-adding and competitive activities within a process, they require flexibility to support innovation [24]. Thus, the sequence of tasks in KIPs is often not prescribed in process models, but can evolve during their instantiation. Each instantiation is thus highly dependent on the knowledge, decisions, and actions of single or multiple actors executing process activities [8,13,17,23] **(F1 & F6)**.

In knowledge-intensive activities, actors apply *"knowledge about, in and derived from a process"* [13, p. 223] *"in order to achieve organizational goals and create value"* [27, p. 4]. The knowledge applied by actors can be distinguished in two categories. Explicit knowledge can be easily codified, stored, and communicated, whereas tacit knowledge *"derives from experience, mental models [sic] and perspectives which cannot be easily formalized or shared through an externalization process"* [8, p. 32]. Thus, tacit knowledge often remains hidden and is subject to the actors' willingness to document and share it [8] **(F3)**. Further, it remains unclear whether a process instance, based on the decisions taken, was successful, due to the inability to codify tacit knowledge appropriately [8] **(F7)**.

In manufacturing application scenarios, KIPs, such as product development and product innovation [13], are performed by highly-trained knowledge workers such as engineers, while they also require cross-discipline collaborations for realizing products [12]. Especially the processes which are part of the product lifecycle management rely on the derivation of knowledge of integrated information to support the required engineering and manufacturing tasks [12]. They represent core processes, which are the base for the value creating production and sales processes and are thus essential for manufacturing companies.

2.2 Process Mining of Knowledge-Intensive Processes

Process mining bridges the gap between BPM and data science by leveraging log data from numerous information systems, including ERP- or BPM-systems. These data are aggregated to an event log to be analyzed with process mining methods and tools, enabling analysts to discover and improve processes by identifying and eliminating deviations [1]. Event logs consist of process-related data and need to satisfy certain properties to enable the application of process mining [21]. There exist different maturity levels for event logs defined by the amount of data available and their quality. A certain maturity level is required to obtain meaningful results from process mining [1,21].

Concerning KIPs, event logs that display a sufficient maturity level are often unavailable [17,23], since deviations from predefined process sequences are commonplace [27]. Due to this lack of data, KIPs cannot be easily understood, analyzed, nor captured [24], because understanding the knowledge dimension of the process and considering the role of human-centered knowledge is paramount [8] (F2). Beyond data availability constraints, process behavior of KIPs is often dependent on human decisions and elements external to the process, and is influenced by other elements and factors inside the organizations [13].

Nevertheless, few approaches have been proposed to apply process mining in KIPs. Rhemus et al. [34] provide a framework to guide the modeling of KIPs considering their characteristics. Drawing on the results, Gronau and Weber [17] developed a tool to model and analyze KIPs, focusing on the transfer and implementation of person-bound knowledge. Because KIPs' properties differ from those of standard processes, organizations employ various IT artifacts to support their execution. In case the artifacts are not process-aware, Pérez-Castillo et al. [33] designed a dynamic analysis that retrieve a log during runtime, including instance-specific domain knowledge. Khanbabaei et al. [24] applied data mining techniques to extract patterns hidden in KIPs. They present a model that recommends improvement opportunities by identifying the behavior of competitive and knowledge-intensive processes. Also, data mining techniques like clustering or classification have been applied to extract information from KIPs to analyze their behavior [24]. Benner-Wickner et al. [4] developed an approach of agenda-driven case management in the context of KIPs, focusing on mining and recommending templates for new cases, while Terziev et al. [37] proposed a recommender system that uses case-related information from similar cases.

In manufacturing, process mining has been combined with case-based reasoning to provide guidance for product innovation processes, using synthetic data [5]. Further, process mining was employed to identify and analyze maintenance activities in cellular manufacturing [11]. Most process mining papers focus on process discovery, while only a few studies applied conformance checking, often focusing on specific context such as shipbuilding or special machine engineering [11]. Another knowledge-intensive domain often addressed is healthcare, with process mining being applied in patient treatment, diagnostics, and organizing tasks [22]. However, processes in healthcare refer to a case-based execution of predefined process strands, while manufacturing processes are less standardized. Moreover, the often regarded clinical pathway always implies the health of the patient as the most important objective, while KIPs in manufacturing follow different aspects such as the quality of the product, the compliance to standards or the satisfaction of the customer [8].

3 Research Method

Design Science Research (DSR) is one of the fundamental research paradigms in Information Systems (IS) [26]. DSR is on a dual mission, to solve important organizational problems through the development of innovative IT artifacts [26] to abstract generalized knowledge and contribute theories for *"design and action"* to the IS knowledge base [14, 15]. A research method that is particularly suited to serve this dual mission is ADR. Rooted in both action and design research, ADR is *"a research method for generating prescriptive design knowledge through building and evaluating ensemble IT artifacts in an organizational setting"* [36, p.40]. ADR focuses on the role of the organizational context for shaping the research process as well as the IT artifact. Sein et al. [36] describe ADR comprising of four phases: (1) the problem formulation, (2) the central building, intervention, and evaluation, (3) reflection and learning, and (4) formalization of learning. All phases follow their individual principles, which in general emphasize the close cooperation of scientists and practitioners, while each has an equal impact on the shape and function of the artifact.

In this paper, we apply ADR to design and evaluate a process mining tool for KIPs in manufacturing as an ensemble artifact [26]. The IT artifact is an instantiation that demonstrates the feasibility and effectiveness of the models and methods contained [26]. As envisioned by ADR, we conducted several interventions through workshops and semi-structured interviews to integrate practitioners into the research process to have a significant impact on the artifact's design. The core ADR team consisted of five researchers and eight highly experienced practitioners from the BPM and IT departments of two anonymized companies who are process owners and managers of a product innovation process and an ETO process. The core team was involved in all phases of the ADR process.

In the first iteration of our ADR project, we started by identifying relevant manufacturing KIPs for the application of process mining by conducting workshops and modeling process maps (c.f. Fig. 1). In the following iteration, we strove to better understand the contexts by conducting semi-structured interviews, focusing on day-to-day work, interactions with co-workers, other business units, and the information systems used. Moreover, we questioned the interviewees about their perception of data-driven BPM methods and elucidated how

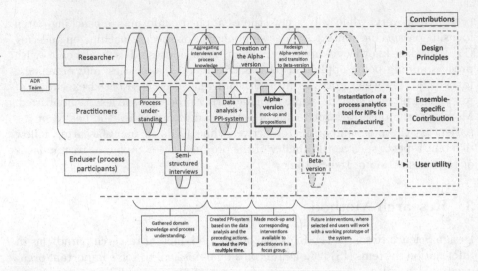

Fig. 1. Action design research applied [36]

process mining could support them. Targeted interviewees ranged from operational workers up to higher management roles to examine the process from various perspectives. Each interviewee received a short introduction into process mining as most participants did not have prior experience with process mining due to the so far limited applicability on their processes. We conducted 27 interviews, accumulating to 40 h of raw audio data. We learned that each company had their own characteristics, while they also shared similarities even referring to different processes, allowing us to narrow down the use case.

We further analyzed the information systems and their underlying data structure in the second iteration to identify the amount and quality of data available and evaluate their maturity level for process mining. Also, we documented process performance indicators (PPIs) for each use case, which ultimately helped us to agree on specific goals to be addressed by the IT artifact's design. In total, we conducted 49 workshops comprising more than 125 h of direct interactions with the companies.

Based on aggregating the information from the workshops and available literature, we developed four propositions (c.f. Table 2) that informed the design of an initial mock-up of a process mining tool for manufacturing KIPs (c.f. Fig. 2 for the final version). We subsequently shared the mock-up with the practitioners in an intervention. To evaluate the propositions and the initial mock-up of our tool, we conducted confirmatory focus groups with the core teams from both companies consisting of two consecutive steps. In step 1, we presented the propositions and requested the practitioners' feedback on the propositions' feasibility and usefulness. They were asked to note their remarks on a whiteboard, while mapping them to the propositions and signaling, whether the feedback was positive, negative or neutral. After discussing the general propositions, we continued with step 2, in which the practitioners evaluated the mock-ups of a concrete instantiation of a process mining tool for KIPs in manufacturing. We applied the same procedure as in the first round. Each step of the evaluation took 90 minutes and resulted in valuable feedback that informed us how to design the artifact and how to abstract the gained knowledge in five design principles. This state of the artifact represents the Alpha-Version that is typically built and evaluated in close cooperation with the practitioners but is not yet presented to the end user at this early conceptual stage [36]. Hence, in line with other ADR papers (e.g., [19]), the roll-out, application, and evaluation of the artifact to operational business (so called Beta-Version) needs to be pursued in future interventions of the ADR project.

4 Results

4.1 Description of the Application Contexts

Product innovation at Industrial Connectors Inc. *Industrial Connectors Inc.* specializes in the production of connection appliances such as industrial connectors. Their products range from fairly simple terminals, which are produced in huge volumes up to smart devices that require many and diverse components, and specialized software **(I1)**. We identified the product innovation process at *Industrial Connectors Inc.* as a KIP [13,23,38]. It is characterized by a stage-gate structure that is typical for product innovation processes [7] **(I2)**. The process participants are highly-trained knowledge workers (i.e., engineers), who need to cooperate in all phases of the process [13]. Reaching a gate at the end of each phase, the project manager checks and approves the process's progress, permitting a transition into the next stage. Often, this evaluation must conform with the four-eye-principle, involving additional participants. Process execution is supported by a PLM-system that also implements features of a BPM-system, such as providing an execution engine for orchestrating the tasks [10, p. 347–353]. During the interviews it became apparent, that workarounds have been developed using the PLM-system additionally as a knowledge repository **(I6)**. The process is designed to allow for flexibility in each stage by performing the activities in any order. Thus, engineers are responsible to execute all activities based on their own experience and judgement [8,13,17,23] **(I1)**. As these processes require considerable man power, knowledge, and time, and have a significant impact on the future product portfolio [24], few instances are being conducted simultaneously, with roughly ten innovation processes completed per year **(I8)**.

Engineer-to-Order at Fluid Processing Ltd. *Fluid Processing Ltd.* specializes in manufacturing industrial machines for fluid processing. We studied the order fulfillment process in its ETO version, that serves for customer-specific manufacturing of machines. Engineering complex machines is a KIP, involving products worth millions of euros that are tailored to individual application contexts of different customers **(I1)**. The ETO process is more rigid than the product innovation process at *Industrial Connectors Inc.* and contains semi-fixed milestones comparable to gates in a stage-gate process **(I2)**. The milestones differ in each process instance, depending on individual customer requests. One of the milestones always included is *production release*, after which no further changes can be made to the final product. Changes, however, do still occur regularly and often require individual implementations. These late changes can result from internal causes—such as construction mistakes—or external causes—such as unexpected additional customer requests. Any way, handling post production-release change requests is a significant challenge for *Fluid Processing Ltd.* as the different business units still work independently and communicate rarely **(I4)** although handling changes requires knowledge work performed by highly-skilled professionals, flexibility, and strong communication skills [8,13,17,23].

Table 1. Case comparison

	Industrial Connectors Inc.	**Fluid Processing Ltd.**
Process	Product Innovation Process	Engineer-to-Order Process
Structure	Fixed Stage-Gate Structure	Variable Gate Structure
Process Model	Process adapted on a case by case basis by the project manager.	Process depends on machine type and customer requests.
Process Tasks	Knowledge driven and variable process tasks. Only few non-knowledge-driven tasks.	Variable and knowledge intensive treatment of post production changes. Many non-knowledge-driven tasks for handling the instances.
System Support	PLM-System as BPM-System.	No distinct BPM-System. ERP-System orchestrates parts of the process.
Number of instances	10 instances per year	700 instances per year, where 350 instances are with changes
Typical duration	12 months typical duration	6-7 months typical duration

The cases exhibit similarities and differences (cf. Table 1). Both refer to a manufacturing context, involving the design, configuration, and update of high-tech equipment for industrial applications. Consequently, both processes are product-focused and represent core processes for the organizations, making them subject to strong quality assurance procedures, specialized knowledge, high customer orientation and individualization, low frequencies, heterogeneous data **(I8)**, group decisions, and complex software applications. Still, the contexts and processes also differ to some extent. Compared to *Industrial Connectors Inc.*, process execution at *Fluid Processing Ltd.* is not supported by a workflow management tool. Instead, an ERP-system is used to handle order processing and for documenting any changes made to products. Nevertheless, both companies still lack the storage and distribution of knowledge as some process executions are handled analogously **(I5)**. In line with that, both inherit data integrity issues, where either information is heavily aggregated or is not individually captured (**I3**). Further, *Industrial Connectors Inc.* targets an anonymous market with their products, while *Fluid Processing Ltd.* produces and designs according to specific customer requests. Nevertheless, they are both missing external evaluation possibilities since they do not store information to measure the process quality (**I7**). Also, the product innovation process features just about ten instances per year, while the ETO process comprises 350 instances with actual changes (**I8**).

4.2   Initial Design Propositions and Prototype Artifact

Based on the initial propositions we derived from our analysis of the related literature and the insights from our use cases (cf. Table 2), we designed a prototype IT artifact to evaluate with the practitioners and research partners. This prototype was a set of mock-ups of a process analytics tool that is focused on the needs of KIPs. Particularly, the prototype featured two views, distinguishing a process level from an instance level. Each screen contained items that represented aspects of the initial design propositions. The prototype visualized the propositions, making them accessible for an evaluation of our project team. Similar approaches have long been advocated in Design Thinking, a design approach that proposes creating conceptual artifacts early to evaluate and improve the design in a cyclic approach [9].

4.3   Evaluation

Based on the initial propositions and the mock-up, we conducted an evaluation with practitioners of the core ADR-team, which enabled us to develop the resulting design principles (cf. Table 2) and IT artifact. Participating practitioners had theoretical knowledge about process mining, but have not yet applied it to processes they have worked with so far. Further, the participating scientists ranged from seasoned users to beginners. The mock-ups and the initial propositions were received positively, especially proposition 1. Thus, no major alterations were made to derive **DP1**. Proposition 2 caused some minor concern (**E2**), as the challenges of handling unstructured data and perils of bad data quality were highlighted, if process participants were forced to enter decisive meta-data in every instance. Practitioners at *Industrial Connectors Inc.* (**E3**), however, were optimistic about the automation potential, provided by proposition 2. Regarding proposition 3, both focus groups noted (**E4**) that different persons and business units will likely attempt to implement different goals. Balancing these interests will be crucial to ensure the long-term support of these actors. The feedback **E2**, **E3** and **E4** led us to perform minor changes to the mock-up and the resulting **DP2** and **DP3**.

Proposition 4 was subjected to the most intensive discussion. First (**E5**), the practitioners recognised that it is important to identify and store individual solutions of the process. Practitioners of *Fluid Processing Ltd.* added that (**E6**) consequences of past non-codified decisions and solutions cannot clearly be identified. So, automatically extracting past solutions might lead to wrong assumptions. Additional feedback regarded the term *similarity* (**E7**). The informants stated that similarity is dependent on the particular situation; for instance, in some cases the machine type can be decisive for a process, whereas in other cases the goals articulated by the customer have an overwriting relevance. Based on this feedback, we introduced **DP4** and **DP5**, which cover the identification of and comparison with similar instances, offering recommended actions that were successful in the past, and a data field to provide feedback for stimulating organizational learning.

4.4 Resulting Design Principles

Subsequently, we present the resulting design principles. Table 2 exhibits how the principles are rooted in the initial propositions, the related literature, and the empirical data retrieved from our use cases and evaluation. We structure the design principles in line with Gregor et al. [16], presenting the *aim, implementer & user*, the *mechanism*, and the *rationale* of each design principle. Because all design principles are built on the premise to make process mining applicable in the context of KIPs in manufacturing (cf. Subsect. 4.1), we did not specifically include the *context* in each design principle. We further defined three different roles that are addressed in the design principles as users or implementers. These roles are composed of the organization, a process analyst (e.g., process manager) and process participants (e.g., process engineer, process executor) [10, p. 24–26].

Table 2. Derivation of initial Propositions and final Design Principles

Insights from Cases	Findings from Literature	Original Propositions	Impacting Feedback	Resulting DP
(I1) Diverse control flows and contents (I2) Stage-gate approach (I3) Data integrity issues	(F1) Diverse control flows and contents (F2) Data integrity issues	**Proposition 1** - Enable neglecting the control flow and distinguish decision- from knowledge-intensive activities.	—	DP1
(I4) Separate work in business units (I5) Lack of knowledge storage (I6) Knowledge distribution via workarounds	(F3) Failure to use (unstructured) domain knowledge and data	**Proposition 2** - Incorporate internal and external domain-specific knowledge.	E2, E3	DP2
(I7) Missing external evaluation	(F4) KIPs are competitive processes	**Proposition 3** - Identify and define labels to control the process quality.	E4	DP3
(I1) Diverse control flows and contents (I8) Limited data availability	(F1) Diverse control flows and contents (F5) Uncertain input and output (F6) Difficult decision-making (F7) Lack of insights on decisions' success	**Proposition 4** - Focus on small (right) data that impacts the process behaviour.	E7	DP4
			E5, E6	DP5

The first three design principles build the foundation for **DP4** and **DP5** and are based on the fundamental properties of KIPs (cf. Sect. 2 & Table 2). Applying process mining in KIPs is often subject to limitations induced by heterogeneous and incomplete process logs that exhibit variable data features [29]. Therefore, **DP1** (cf. Table 3) focuses on decomposing the process into several stages and gates to enable analyzing the gates by means of process mining while maintaining flexibility within the stages. Thereby, process quality is ensured at the gates and for the knowledge-intensive activities the necessary flexibility is preserved [29,35]. The decomposition can be performed manually or by using algorithmic approaches. Besides theoretical justification, empirical support for this principle comes from our use cases, where the processes follow a stage-gate structure (cf. Table 1) and only limited information on individual events is available.

DP2 and **DP3** focus on enhancing the event log with both domain-specific knowledge as well as process-external goals. In order to complete a process successfully, knowledge flows and transfers between system and process participants are necessary [17]. KIPs typically involve unstructured domain knowledge, which is often not codifiable [32]. Our interviews revealed that the different business units involved in a process often work independently and communicate rarely. To cope with this potential lack of knowledge sharing, process participants at *Industrial Connectors Inc.* developed workarounds using the PLM-system as a process knowledge repository, re-purposing project functions of the system that were originally intended for other tasks. Therefore, to enable improved contextualization, we propose that (unstructured) domain data and knowledge should be codified [31] and added directly to the event log (cf. Table 3) (see **DP2**). Further, goals of process instances can vary and decisions on process goals are often based on incomplete knowledge, because information systems only represent limited information, see **DP2**. However, KIPs represent competitive key processes for organizations [24] that have to satisfy process-external goals, like profitability or customer satisfaction. Therefore, process-external goals should be defined as dependent variables and labels for process quality, and added as attributes to the event log (cf. Table 3). To make these goals measurable, the required knowledge and data need to be linked to the event log. Still, deriving these labels from the event log in our use cases was difficult because external evaluations were often not available (see **DP3**).

Further, it is difficult to predict the behavior of KIPs [23,27], because of their diverse control flows and contents [8,13,17,23], limited data availability, and a high variance in their structure, causing uncertainty of their inputs and outputs [8]. From these insights, we could derive **DP4** (cf. Table 3) that aims to generate information and knowledge on similar instances available. The evaluation **(E5)** and **(E6)** (described in Subsect. 4.3) constitute the foundation for **DP5** (cf. Table 3). Related research has identified that complex decisions require human judgement, experiences, and creativity [27], but decisions are difficult to make since individuals are subject to bounded rationality. Thus, making decisions based on the comparison of similar instances can offer valuable assistance and allows effective learning and inference from suitable past data. Executing those prescriptive actions and evaluating their results enables providing feedback and learning from past decisions [25,31].

Table 3. Design pinciples

Design Principle 1: Decision- vs. Knowledge-intensive Activities	
Implementer	Organizations using process mining for KIPs
User	Process participants & process analyst
Aim	To balance between quality insurance and flexibility in performing KIPs.
Mechanism	Distinguish decision- from knowledge-intensive activities.
Rationale	We draw on Seidel's et al. [35] framework of *pockets of creativity* that aims at conceptualizing creativity within business processes.

Design Principle 2: Capture Domain-Specific Knowledge	
Implementer	Organizations using process mining for KIPs
User	Process participants
Aim	To capture process-relevant domain knowledge.
Mechanism	Enhance the event log with relevant unstructured data (e.g., codifiable knowledge, including business documents, drawings, notes).
Rationale	In *adaptive case management*, actionable knowledge is collected from process participants and required information for processing the case is stored [31]. The idea has already been transferred to KIPs by Herrmann and Kurz [20].

Design Principle 3: Define Process-External Goals	
Implementer	Organizations using process mining for KIPs
User	Algorithms
Aim	To learn relationships between process execution and higher-order business goals.
Mechanism	Annotate event logs with process-external goals.
Rationale	Goals of process instances can vary and decisions on process goals are often based on incomplete knowledge. Considering *adaptive case management* [31], we extend this idea to also take the integration of external factors into account.

Design Principle 4: Retrieve Process Knowledge	
Implementer	Organizations using process mining for KIPs.
User	Process participants
Aim	To make experiences from past process instances accessible.
Mechanism	Retrieve and analyze similar past instances.
Rationale	The principle is grounded in the first phases of the *case-based reasoning* lifecycle [25,31]. The idea has been used in process mining by Berriche et al. [5].

Design Principle 5: Derive Actionable Interventions	
Implementer	Organizations using process mining for KIPs
User	Process participants & process analyst
Aim	To use experiences from past process instances to perform new instances.
Mechanism	Consider and implement prescriptive actions derived from similar instances, and provide reasoning on decisions to evaluate the effects that past decisions had on process goals.
Rationale	In *case-based reasoning*, reusing information from similar past cases, revising proposed solutions, and retaining experiences can support solving new problems [25,31]. We extend this idea by taking the integration of recommended actions into consideration.

4.5   Manifestation of the Design Principles in the IT Artifact

After formulating the design principles, we used them to re-design the initial mock-up of the process analytics tool for manufacturing KIPs. The tool still features a process and an instance level. The process level covers information about a process in general, including a process model, PPIs, and an overview of process instances currently running. In contrast, the instance level focuses on one particular process instance, covering instance-specific PPIs, and specific stage-gate information. Figure 2 shows one perspective of the instance level, for the product innovation process at *Industrial Connectors Inc.*. The process is represented by stages and gates that are set manually, because the process is already well-designed in this regard. Further, stage-specific information cover an overview of completed and non-completed activities, aiming to foster flexibility and creativity within a stage **(DP1)**.

To incorporate domain-specific knowledge, the artifact provides the possibility to upload important stage-specific documents, e.g. modification masters in form of text documents at *Fluid Processing Ltd.*, and additional information, e.g. part numbers at *Industrial Connectors Inc.*, at the instance level. Further, gate-related metadata that cover the dimensions of time, complexity, quality and costs of the respective process instance can be documented **(DP2)**. Since the process can have unique structures, the PPIs shown will adjust at the instance level and indicate the internal **(DP2)** resp. process-external goals **(DP3)** for the specific stage resp. gate. Selecting a specific gate, the artifact allows analysts to set resp. goals and their weighting. This applies to the entire process instance, but it can always be adjusted once a new gate is reached, to allow for settings new goals or weights **(DP3)**. By considering the selected process properties and the defined goals, comparable process instances to the resp. case are identified, displayed and clustered **(DP4)**. Additionally, different actions are recommended based on available data and can be selected by the analyst. Once a recommended action took place, the artifact provides the possibility to evaluate that action, to train the system further **(DP5)**.

5   Discussion

Process mining relies on the availability of large amounts of structured process-related data, provided as event logs [21]. However, our findings expose that existing process mining tools and methods cannot only be applied to standardized and highly digitized processes [21], but also to KIPs that exhibit higher degrees of complexity and flexibility, while comprising a small number of instances [8,13]. For this purpose, traditional assumptions on process mining need to be reconsidered and updated. Information such as unstructured domain data are often ignored in event logs, although they can provide relevant insights and enable improved contextualization [32]. Our findings highlight the need to enhance event logs of KIPs with internal and external (unstructured) domain data and knowledge to define and measure process-external goals to control quality. To integrate this information, process mining can be enhanced by adapting the concept of

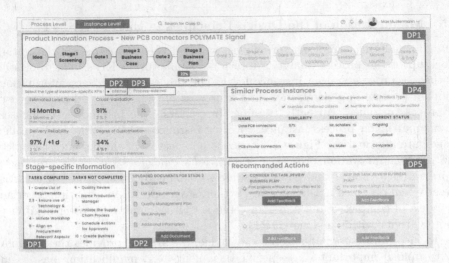

Fig. 2. Exemplary perspective of the mock-up (Instance level)

adaptive case management systems from healthcare [4,31] that aims to collect knowledge from process participants and to store information that might be important for processing a similar case in the future. Thus, knowledge can be shared effectively among process participants [31].

Also, the sequences of activities in KIPs are not prescribed on purpose [8,13], yet their results have to conform to a certain level of quality. Focusing on the external results of a process is, therefore, more important than considering the control flow of a process in too much detail. Through distinguishing decision- from knowledge-intensive activities—inspired by the concept of pockets of creativity by Seidel et al. [35]—data mining techniques such as clustering or classification can be applied [24]. As the sequences of activities in KIPs can evolve during the process, many different process variants are generated [8,13,17,23]. To deal with the high variety of KIPs, similar process variants must be identified based on other process properties than the control flow to improve decision making. In this case, case-based reasoning proves useful as a problem-solving paradigm [31,37]. Case-based reasoning can utilize specific knowledge from the past, to identify a similar case and mobilize related information for solving a new problem. It enables incremental, sustained learning through retaining new experiences each time a problem has been solved, making them immediately available in addressing new problems that might arise in the future [31].

Finally, the identification of deviations in process execution through conformance checking is also different for KIPs. Deviations are currently often viewed as not-intended, negative behavior of process participants and eliminated if detected [18]. However, deviations in KIPs can occur regularly due to their less structured properties. Thus, when performing conformance checking in KIPs, deviations should not be eliminated *per se* but rather investigated in more detail to leverage bottom-up process innovation through workarounds [2].

6   Conclusion and Future Work

We conducted an Action Design Research study with two manufacturing companies, focusing on product innovation and engineer-to-order as two knowledge-intensive processes. We grounded our work on 27 semi-structured interviews and 49 workshops, culminating in an evaluation with two focus groups. While related literature has provided first insights on using process mining for KIPs in other domains, we posit that product- and innovation-related processes in the manufacturing industries exhibit conceptual differences that require adjusting process mining methods and tools to this domain.

We developed five design principles and a prototypical process analytics tool, serving as blueprints for applying process mining of KIPs in manufacturing. Specifically, we proposed to distinguish decision- from knowledge-intensive activities, enhance event logs with domain knowledge, use process-external goals to measure process success, compare instances with similar instances only, and use small data for predictive and prescriptive process analytics.

Other researchers can use our design principles as prescriptive knowledge to design and evaluate their own process analytics tools for KIPs in a manufacturing context. Our next steps for research will be to analyze field data from the two companies to identify the analytic and predictive power of our approach. To make the data fit for the analysis, we will label past process instances, mobilizing them to analyse current process instances. Also, we strive to validate and extent our findings with a third company, to improve their external validity further.

Acknowledgements. This research and development project is funded by the Ministry of Economic Affairs, Innovation, Digitalization, and Energy of the State of North Rhine-Westphalia (MWIDE) as part of the Leading-Edge Cluster, Intelligente Technische Systeme OstWestfalenLippe (it's OWL) and supervised by the project administration in Jülich (PtJ). The responsibility for the content of this publication lies with the authors.

References

1. van der Aalst, W.: Process Mining - Data Science in Action. Springer, Heidelberg (2016). https://doi.org/10.1007/978-3-662-49851-4
2. Alter, S.: A workaround design system for anticipating, designing, and/or preventing workarounds. In: Gaaloul, K., Schmidt, R., Nurcan, S., Guerreiro, S., Ma, Q. (eds.) CAISE 2015. LNBIP, vol. 214, pp. 489–498. Springer, Cham (2015). https://doi.org/10.1007/978-3-319-19237-6_31
3. Bahrs, J., Müller, C.: Modelling and analysis of knowledge intensive business processes. In: Althoff, K.-D., Dengel, A., Bergmann, R., Nick, M., Roth-Berghofer, T. (eds.) WM 2005. LNCS (LNAI), vol. 3782, pp. 243–247. Springer, Heidelberg (2005). https://doi.org/10.1007/11590019_28
4. Benner-Wickner, M., Brückmann, T., Gruhn, V., Book, M.: Process mining for knowledge-intensive business processes. In: Proceedings of the 15th International Conference on Knowledge Technologies and Data-driven Business, pp. 1–8. ACM, Graz Austria (2015)

5. Berriche, F.Z., Zeddini, B., Kadima, H., Riviere, A.: Combining case-based reasoning and process mining to improve collaborative decision-making in products design. In: 2015 IEEE/ACS 12th International Conference of Computer Systems and Applications (AICCSA), pp. 1–7. IEEE, Marrakech, Morocco (2015)

6. Blum, J.: Research Shows Majority of Business Decision-Makers to Use, or Evaluating, Process Mining this Year (2022)

7. Cooper, R.G.: Stage-gate systems: a new tool for managing new products. Bus. Horiz. **33**(3), 44–54 (1990)

8. Di Ciccio, C., Marrella, A., Russo, A.: Knowledge-intensive processes: characteristics, requirements and analysis of contemporary approaches. J. Data Seman. 4(1), 29–57 (2014). https://doi.org/10.1007/s13740-014-0038-4

9. Dolata, M., Schwabe, G.: Design thinking in is research projects. In: Brenner, W., Uebernickel, F. (eds.) Design Thinking for Innovation, pp. 67–83. Springer, Cham (2016). https://doi.org/10.1007/978-3-319-26100-3_5

10. Dumas, M., La Rosa, M., Mendling, J., Reijers, H.A.: Fundamentals of Business Process Management. Springer, Heidelberg (2018). https://doi.org/10.1007/978-3-662-56509-4

11. Dunzer, S., Zilker, S., Marx, E., Grundler, V., Matzner, M.: The Status Quo of Process Mining in the Industrial Sector. In: Ahlemann, F., Schütte, R., Stieglitz, S. (eds.) WI 2021. LNISO, vol. 48, pp. 629–644. Springer, Cham (2021). https://doi.org/10.1007/978-3-030-86800-0_43

12. El Kadiri, S., Kiritsis, D.: Ontologies in the context of product lifecycle management: state of the art literature review. Int. J. Prod. Res. **53**(18), 5657–5668 (2015)

13. Eppler, M.J., Seifried, P.M., Röpnack, A.: Improving knowledge intensive processes through an enterprise knowledge medium. In: Proceedings of the 1999 ACM SIGCPR conference on Computer personnel research - SIGCPR 1999, pp. 222–230. ACM Press, New Orleans (1999)

14. Gregor, S.: The nature of theory in information systems. MIS Q. **30**(3), 611–642 (2006)

15. Gregor, S., Hevner, A.R.: Positioning and presenting design science research for maximum impact. MIS Q. **37**(2), 337–355 (2013)

16. Gregor, S., Kruse, L., Seidel, S.: Research perspectives: the anatomy of a design principle. J. Assoc. Inf. Syst. **21**, 1622–1652 (2020)

17. Gronau, N., Weber, E.: Management of knowledge intensive business processes. In: Desel, J., Pernici, B., Weske, M. (eds.) BPM 2004. LNCS, vol. 3080, pp. 163–178. Springer, Heidelberg (2004). https://doi.org/10.1007/978-3-540-25970-1_11

18. Hadasch, F., Maedche, A., Gregor, S.: The influence of directive explanations on users' Iusiness process compliance performance. Bus. Process Manag. J. **22**(3), 458–483 (2016)

19. Haj-Bolouri, A., Rossi, M.: Proposing design principles for sustainable fire safety training in immersive virtual reality. In: Proceedings of the 55th Hawaii International Conference on System Sciences (2022)

20. Herrmann, C., Kurz, M.: Adaptive case management: supporting knowledge intensive processes with IT dystems. In: Schmidt, W. (ed.) S-BPM ONE 2011. CCIS, vol. 213, pp. 80–97. Springer, Heidelberg (2011). https://doi.org/10.1007/978-3-642-23471-2_6

21. van der Aalst, W., et al.: Process mining manifesto. In: Daniel, F., Barkaoui, K., Dustdar, S. (eds.) BPM 2011. LNBIP, vol. 99, pp. 169–194. Springer, Heidelberg (2012). https://doi.org/10.1007/978-3-642-28108-2_19

22. IEEE Task Force on Process Mining: Process-oriented data science for healthcare alliance: process mining for healthcare: characteristics and challenges. J. Biomed. Inf. **127**, 1–15 (2022)
23. Isik, O., Van den Bergh, J., Mertens, W.: Knowledge intensive business processes: an exploratory study. In: 2012 45th Hawaii International Conference on System Sciences, pp. 3817–3826. IEEE, Maui, HI, USA (2012)
24. Khanbabaei, M., Alborzi, M., Sobhani, F.M., Radfar, R.: Applying clustering and classification data mining techniques for competitive and knowledge-intensive processes improvement. Knowl. Process Manag. **26**(2), 123–139 (2019)
25. Kolodner, J.L.: An introduction to case-based reasoning. Artif. Intell. Rev. **6**(1), 3–34 (1992)
26. March, S.T., Smith, G.F.: Design and natural science research on information technology. Decis. Support Syst. **15**(4), 251–266 (1995)
27. Marjanovic, O., Freeze, R.: Knowledge Intensive business processes: theoretical foundations and research challenges. In: 2011 44th Hawaii International Conference on System Sciences, pp. 1–10. IEEE, Kauai, HI (2011)
28. Martin, N., Fischer, D.A., Kerpedzhiev, G.D., Goel, K., Leemans, S.J.J., Röglinger, M., van der Aalst, W.M.P., Dumas, M., La Rosa, M., Wynn, M.T.: Opportunities and Challenges for Process Mining in Organizations: Results of a Delphi Study. Bus. Inf. Syst. Eng. **63**(5), 511–527 (2021). https://doi.org/10.1007/s12599-021-00720-0
29. Nguyen, H.: Stage-based Business Process Mining. In: CAiSE-Forum-DC. pp. 161–169 (2017)
30. Nonaka, I., Takeuchi, H.: The Knowledge-Creating Company: How Japanese Companies Create the Dynamics of Innovation, vol. 105. Oxford University Press, New York (1995)
31. Osuszek, Ł, Stanek, S.: Case based reasoning as an element of case processing in adaptive case management systems. Ann. Comput. Sci. Inf. Syst. **6**, 217–223 (2015)
32. Pentland, B.T., Recker, J., Wolf, J., Wyner, G., Wolf, J.: Bringing context inside process research with digital trace data. J. Assoc. Inf. Syst. **21**(5), 1214–1236 (2020)
33. Pérez-Castillo, R., Weber, B., de Guzmán, I.R., Piattini, M.: Process mining through dynamic analysis for modernising legacy systems. IET Softw. **5**(3), 304–319 (2011)
34. Remus, U., Lehner, F.: The role of process-oriented enterprise modeling in designing process-oriented knowledge management systems. In: Designing Process-Oriented Knowledge Management Systems. 2000 AAAI Spring Symposium, Bringing Knowledge to Business Processes, pp. 1–7 (2000)
35. Seidel, S., Müller-Wienbergen, F., Rosemann, M.: Pockets of creativity in business processes. Commun. Assoc. Inf. Syst. **27** (2010)
36. Sein, M.K., Purao, S., Rossi, M., Henfridsson, S.: Action design research. MIS Q. **35**(1), 37–56 (2011)
37. Terziev, Y., Benner-Wickner, M., Brückmann, T., Gruhn, V.: Ontology-based recommender system for information support in knowledge-intensive processes. In: Proceedings of the 15th International Conference on Knowledge Technologies and Data-driven Business, pp. 1–8. ACM, Graz Austria (2015)
38. Vaculin, R., Hull, R., Heath, T., Cochran, C., Nigam, A., Sukaviriya, P.: Declarative business artifact centric modeling of decision and knowledge intensive business processes. In: 2011 IEEE 15th International Enterprise Distributed Object Computing Conference, pp. 151–160. IEEE, Helsinki, Finland (2011)

Process Mining Practices: Evidence from Interviews

Francesca Zerbato[1(✉)], Pnina Soffer[2], and Barbara Weber[1]

[1] Institute of Computer Science, University of St. Gallen, St. Gallen, Switzerland
{francesca.zerbato,barbara.weber}@unisg.ch
[2] University of Haifa, Mount Carmel, 3498838 Haifa, Israel
spnina@is.haifa.ac.il

Abstract. Process mining provides organizations with methods and techniques to extract knowledge from event logs. In their work, process analysts can draw from the wealth of techniques developed over the years by researchers and professionals. Still, there is limited understanding of how process mining is used in practice and, in particular, how individual analysts approach the analysis stage. Towards filling this gap, we conducted semi-structured interviews with 37 practitioners and academics working with process mining. Based on the results of the interviews, we characterize common analysis strategies and examine related challenges and factors affecting their use in practice. Our findings contribute to an improved understanding of process mining practices and provide a solid empirical basis for future research developing guidance and support addressing the practical needs of process analysts.

Keywords: Process mining · Interview study · Mining and analysis stage · Analysis strategies

1 Introduction

Process mining provides organizations with data-driven methods and techniques to extract knowledge from process execution data in the form of event logs [1].

Over the last two decades, process mining as a field of research has grown in maturity, leading to the development and consolidation of techniques and tools for analyzing event logs [10]. However, research has mainly emphasized technical work giving less attention to empirical studies focusing on understanding how process mining is used in practice.

With the increasing adoption of process mining techniques among practitioners, a growing number of empirical studies have explored how process mining is used in organizational settings [3]. For example, Grisold et al. [10] investigated process mining adoption through the eyes of managers, while Martin et al. [14] elicited a broad set of opportunities and challenges of using process mining in organizations. Still, the work practices of analysts, including the strategies they follow to conduct their work tasks, remain largely unexplored [11].

© Springer Nature Switzerland AG 2022
C. Di Ciccio et al. (Eds.): BPM 2022, LNCS 13420, pp. 268–285, 2022.
https://doi.org/10.1007/978-3-031-16103-2_19

Process mining projects include different stages, starting from project plan-ning and data extraction and progressing until the results are utilized for process improvement and support [7]. Typically, these stages unfold in many iterations, leading to different activities, goals, and, potentially, challenges, which, in turn, may require different kinds of support. In this work, we focus on one specific stage - the *mining & analysis* stage (henceforth *analysis*), in which process ana-lysts apply process mining techniques to event logs to address analysis questions and gain insights from the data. Thus, we leave the other stages out of our scope, focusing on gaining deep insights into the practices of process mining analysis.

We aim to complement existing research by taking an *individual perspective* [3] and understanding from analysts how they work in the analysis stage. Indeed, we believe that learning how individual analysts conduct process mining analyses can help develop support to address the practical needs of process analysts and stake-holders involved in the analysis stage. In particular, we ask ourselves the following research question (RQ): **What are common strategies used in the analy-sis stage?**, where by strategy we refer to "an approach, a manner or a means to achieve a certain intention" [15]. To address this question, we conducted an inter-view study with 37 academics and practitioners working with process mining in different organizations. The interviews were designed in the context of a broader observational study, where we used a process mining task as an anchor to let par-ticipants reflect upon a concrete analysis and share their work experiences.

In this paper, we present the results of these interviews, which allowed us to look at the analysis stage from the retrospective thoughts and reflections of our participants and learn what strategies they apply in their daily work. Our findings include (i) a characterization of analysis strategies, organized into four main phases representing intermediate analysis goals; (ii) examples of recurring challenges associated with each phase; (iii) a set of factors influencing the use of the strategies in practice. By raising awareness about the work practices of process analysts, our findings can help both process analysts and business stake-holders to reflect upon their (joint) work and learn about possible difficulties. In addition, they provide a solid empirical basis for designing methods and tools to support process analysts, guiding several directions for future research.

The paper is structured as follows. Section 2 provides an overview of related work. Section 3 presents the research method followed to design, conduct and ana-lyze the interviews. Section 4 reports our findings. Section 5 reviews our results and discusses the limitations of our work, outlining directions for future research.

2 Related Work

Our work falls into the stream of research investigating process mining practice from a general perspective, i.e., not tied to a specific organizational setting.

Following the increasing uptake of process mining in the industry, more and more researchers have investigated the use of process mining in practice using a variety of research methods and looking at the effects of process mining adoption from different angles [3].

One group of related papers has looked into the practical use of process mining by using published empirical studies, case studies or process mining reports as their main source of knowledge. For example, Thiede et al. [21] explored the use and maturity of process mining technology in organizations from a service perspective. Emamjome et al. [9] investigated the diffusion and maturity of tools and methodologies from case studies discovering, for example, that the thoroughness of the analysis stage, among others, has not improved over the years. Also using published case studies as a source, Koorn et al. [12] focused on understanding the goals and methods of evaluations in which domain experts are involved. Then, based on their findings, they proposed six strategies for the qualitative evaluation of findings in process mining projects. Klinkmüller et al. [11] focused on examining visual representations from BPI Challenge reports to understand the relationships between domain problems captured by analysis questions and the information needs of process analysts. While these works provide valuable insights on a number of aspects relevant to process mining practices, they rely on case study and scientific reports, which often contain little or no information about the dynamics of the *process of process mining.*

Another group of works has focused on investigating the use of process mining in practice through explicit reporting from experts working in the field, who were directly involved in empirical studies. Syed et al. [20] conducted an interview study with 9 stakeholders in the context of a Dutch pension fund, identifying challenges and enablers of process mining adoption. Their findings touch upon different levels, going from tensions between teams to user challenges with learning tools. Grisold et al. [10] conducted a focus group study with 22 process managers to investigate the benefits and challenges of process mining adoption in organizations. Eggers et al. [8] conducted a multiple case study to investigate how organizations engage with the process transparency created by process mining to achieve increased process awareness. The study, which included 24 semi-structured expert interviews among the different data sources, revealed seven mechanisms that employ process mining to achieve increased process awareness on different levels. Recently, Martin et al. [14] conducted a Delphi study with 40 experts to explore the opportunities and challenges of process mining adoption in organizations. Some of the key challenges they discovered, which also emerged in [10,20], include elusive business value and unclear organizational anchoring. Our work also falls within this second group of studies. However, while the papers presented above reveal how process mining is used within organizations, we take an *individual* perspective [3].

The only paper so far that looks at process mining from an individual perspective is our previous work [22], which presents the results of a pilot study combining behavioral data and interviews to understand patterns and strategies of the initial exploration phase. In contrast, this paper, which is based on a newly collected data set, considers the whole analysis stage, focusing on strategies that analysts apply in their general work practices and considering factors and challenges affecting their use.

3   Research Method

This section describes the design of our study, the data collection and analysis.

Study Design. To improve our understanding of strategies used in the analysis stage (cf. RQ in Sect. 1), we designed an interview study following the empirical standards for qualitative surveys [17]. The interview was part of a broad observational study where participants engaged in a realistic process mining task. In the context of this paper, the process mining task served as an anchor to let participants work and reflect upon a concrete analysis they could use as a reference to share their work experiences. To ensure familiarity with event log analysis, we defined two requirements for participating in the observational study: (i) having analyzed at least two real-life event logs in the two years prior to the study and (ii) being knowledgeable of at least one of the process mining tools available for the task. In addition, for the interview study, we considered participants having experience in process mining projects aimed at analyzing process data for a customer to ensure that they had gained sufficient practical experience with event log analysis. Such requirements allowed us to exclude beginners from our population of interest but still include participants with different backgrounds (e.g., academics vs. practitioners) and varying levels of experience and expertise, which are needed to gain a broad understanding of analysis strategies.

Materials. The process mining task was designed to observe participants as they analyze an event log guided by a high-level question. Specifically, we used the road traffic fine management event log [6] and asked participants to investigate circumstances (scenarios) for not paying a fine and, if possible, identify potential reasons for doing so. The event log was ready to be analyzed so that participants could fully focus on the analysis stage. For the analysis, participants had at their disposal bupaR[1], Celonis[2], Disco[3], Pm4Py[4], ProM[5], and SQL, which we selected considering the top-six tools used in BPI Challenge reports published until 2020[6]. We also prepared an online form to collect the participants' answers to the task question. Finally, we developed the interview guide following a semi-structured approach [16], organizing questions into four main themes related to the analysis stage: (i) activities and artifacts; (ii) goals; (iii) strategies; and (iv) challenges. Each question was formulated twice, first concerning the process mining task and then generalizing to the work practices of the participants. All the materials were pilot tested with two researchers external to the author team and adjusted based on their feedback.

Data Collection Procedure. We invited participants in our professional networks, ensuring diversity in their affiliation, job role and position, and tool knowledge.

---

[1] bupaR: https://bupar.net.
[2] Celonis: https://www.celonis.com.
[3] Fluxicon Disco: https://fluxicon.com/disco/.
[4] PM4Py: https://pm4py.fit.fraunhofer.de.
[5] ProM: https://www.promtools.org/doku.php.
[6] BPI Challenges: https://www.tf-pm.org/competitions-awards/bpi-challenge.

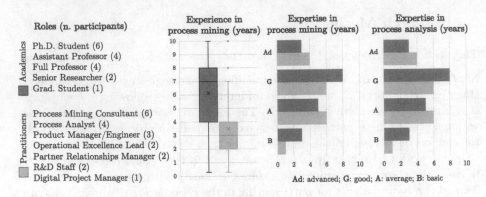

Fig. 1. Job role, process mining experience and expertise of the 37 participants.

Eleven participants were additionally recruited with the help of five participants to include in our sample users of all the five process mining tools available for the task. The participation requirements were explicitly stated in the invitation email, along with a description of the study purpose and procedure. The first author collected the data between May and July 2021 via virtual sessions with the participants. We discontinued the data collection when we achieved data saturation [19], i.e., we noticed that responses repeated across interviews.

Before each session, we collected demographic characteristics with a questionnaire and screened participants for participation requirements, process mining experience and expertise, and project experience. On the appointed day, we instructed participants about the task and gave them access to a remote desktop environment with the study materials and tools. The task was silently supervised by one author, who was available to support with questions and technical issues. Then, we asked participants to report their answers in an online form. Finally, we conducted the semi-structured interviews following our guide and prompting interviewees to share anecdotes and work experiences within their current organizations [16]. All the sessions were conducted in English and recorded.

Data Validation and Analysis. Overall, 41 people participated in our study. We averaged 83 min of recordings per participant, including the process mining task and the interviews, which we fully transcribed verbatim. For this paper, we excluded four participants from the analysis. Two participants did not conduct the task prior to the interview for personal and technical reasons. Two others did not have experience with process mining projects aimed at analyzing process data for a customer, which we considered relevant to our research question. Thus, in this paper, we focus on the interviews of the 37 remaining participants, i.e., 20 practitioners and 17 academics working in 29 different organizations. Their interviews lasted 1136 min in total and 31 min on average.

Figure 1 gives an overview of the 37 participants. All the participants meet the participation requirements and have experience with at least one process mining project requiring analyzing process data for a customer. The participants have a variety of job roles, such as process mining consultant, process analyst, Ph.D.

student, and full professor, among others. Overall, the participants reported an average of 4.5 years of process mining experience. Most of them additionally indicated experience in related fields, with some indicating more than 6 years of experience in process analytics (14/37), business intelligence (17/37), and data science (15/37). More than half of the academics (11/17) reported also having worked in the process mining industry for two years on average.

Given the exploratory nature of our research, we followed an inductive approach for the analysis. More specifically, we relied on the qualitative coding guidelines in [18] and coded the interview transcripts in several rounds. First, we engaged in an initial coding round, analyzing each participant individually and fragmenting the text using in-vivo and process coding [18] to identify core concepts and steps related to strategies. Then, we used focused coding to refine and aggregate them into categories, considering codes supported by at least 10% of the participants. For example, we included strategy "choose tool" as part of "determine analysis approach", as we recognized that selecting the "right" tool is part of planning the analysis approach. Finally, we relied on axial coding to focus on the most frequent categories and find relationships among them until we achieved saturation. One author coded all the data in several iterations. The other authors checked the codes independently to ensure consistency. Throughout the analysis, all the authors revised and refined the codes collaboratively in several meetings. Our analysis led to 16 strategies related to the analysis stage. Factors affecting the use of the strategies in practice also emerged from our coding. Then, we organized the strategies into four main phases and iterated once more through the data with this structure in mind to code examples of common challenges related to each phase. The interested reader may find the documents with details about data collection and analysis based on the empirical standard on qualitative surveys [17] on: https://doi.org/10.5281/zenodo.6644982.

4 Findings

In this section, we report on our findings of strategies related to the analysis stage in Sect. 4.1 and factors affecting their use in practice in Sect. 4.2.

4.1 Phases and Strategies of the Analysis Stage

From our analysis, we derived 16 strategies that our participants apply in practice, some of which they also used in the process mining task (cf. Sect. 3). Based on our coding, we organized the strategies into four main phases representing intermediate analysis goals: **understand, plan, analyze**, and **evaluate**.

Understanding covers strategies to understand the problem and make sense of the needs of business stakeholders, the business domain, and the data. Planning includes strategies to devise an analysis plan. Analyzing refers to strategies to execute the analysis by applying process mining techniques within tools. Evaluating covers strategies to verify and validate analysis artifacts and findings. The identified phases resemble known phases from the problem-solving literature [4].

Since the analysis unfolds as a highly iterative and flexible process [7], phases are not strictly performed in order, but analysts rather cycle through them in several iterations, potentially skipping some phases.

In the following, we break down the *analysis* stage within the context of each phase and describe the strategies and their rationale with example statements from our participants, whose numeric IDs (p#) are reported in parentheses. For each strategy, we also indicate the number of participants explicitly mentioning it in parentheses. Then, for each phase, we summarize example challenges emerging from our interviews. A summary of the strategies is provided in Table 1.

Understand. Our participants referred to understanding as a crucial initial step of any analysis, covering (S1) problem understanding, (S2) domain understanding, and (S3) data understanding.

(S1) Understand the problem (6/37) refers to understanding the problem under analysis, including business stakeholders' needs and how process mining can be enacted within a specific organizational context. Problem understanding helps analysts orient themselves in the problem space and assess if stakeholder needs can be addressed with process mining because *"sometimes the problem is not directly a process problem but is a more statistical nature problem"* (p34). Also, it allows them to *"define the objective of the analysis"* (p1), *"translate that into an analysis that enables them* [the stakeholders] *to achieve that objective"* (p14), and figure out which type of process mining to apply.

(S2) Understand the domain (16/37) refers to gaining knowledge about the domain in which the processes under analysis are enacted, including *"which processes are in place"* (p39), what is the business meaning of the activities, who are the actors involved in the process, and what business rules are in place. Domain understanding *"helps with putting everything into context"* (p20) and *"knowing where to search"* (p14), and it seems necessary to extract and validate the data and define analysis questions that are in line with stakeholder needs.

Usually, problem and domain understanding occur during workshops with stakeholders, such as process owners and domain or IT experts, where analysts can *"ask about the problem"* and get *"an introduction to the business"* (p9). However, some participants (6/37) said that they *"don't always have the domain knowledge"* (p18) and, at times, *"lack interaction with business people"* (p34). Thus, they are used to deriving domain understanding from available documentation, such as *"secondary data and reports issued by the organization"* (p30).

(S3) Understand the data (19/37) concerns learning the data structure, the attribute values, and the related data models. Analysts aim to know what the data contains, how it is *"prepared and formatted"* (p3), and *"how it interacts [...] how it changes when I just filter"* (p9). They also inspect the data to know *"what is possible to analyze"* (p18). Some participants (6/37) mentioned focusing on the data structure to learn the main components of the log and how they can create new features to use later in the analysis. Data understanding often builds upon raw event data, accessed with process mining tools, spreadsheets, databases, and visual analytic software. Two expert participants (2/37) reported always starting from disaggregated data *"because sometimes you miss some other*

Table 1. Strategies S1-S16. **Grp**: highlights if the strategy was reported *mostly* by a specific group of participants, e.g., [P] practitioners or [E] experts; otherwise we write "all". **#P**: number of participants explicitly mentioning the use of a strategy.

	Id	Strategy Description	Grp	#P	Example Statements
UNDERSTAND	S1	**Understand the problem:** understand stakeholders' needs and how process mining can support	all	6	'So, this is my way. So, opening my ears for problems. When I... I'm aware of problems, I try to determine if it is a process problem." (p34)
	S2	**Understand the domain:** learn about the domain in which the processes to analyze are enacted.	all	16	"...I try to get a grip on what the process is all about, what the pitfalls are, understand exactly the business rules and the domain..." (p11)
	S3	**Understand the data:** learn about the data structure, the attributes, and related data models.	all	19	'I would directly learn from the event log, the SQL [...] I would try to learn much more from the raw data before going to the map." (p37)
PLAN	S4	**Formulate analysis questions:** define questions based on stakeholders' needs such that they can be answered with process mining.	all	11	'We discuss what we get with the partner. [...] And with that and an overview of the field, for example, I would try to make some sense or some questions on my own." (p32)
	S5	**Prioritize analysis directions:** prioritize analyses based on the foreseen value of the findings.	[P]	11	"...like is this really an issue worth pursuing? [...] like if my bolts are rather inexpensive [...] then that might not be the biggest problem..." (p14)
	S6	**Determine analysis approach:** evaluate which approach suits the analysis best.	[E]	8	"And that's the first step. I can choose the analysis because that's part of the strategy to adapt the analysis based on the question." (p9)
	S7	**Map the question to the data:** identify the main "players" for the analysis and link them to the data	all	11	'I also will spend a lot of time on understanding the questions and map the questions to the data. How do I define a situation..." (p2)
	S8	**Make hypotheses:** hypothesize about possible answers for the question or explanations for the stakeholders' problems.	all	15	"...this is basically this CRISP thing, right? It's the data understanding and then... iterating upon and creating, like, looking at picking out hypothesis right from the question..." (p12)
ANALYZE	S9	**Understand the process:** learn about the process control flow using statistics or visualizations.	all	32	"...understand the different activities and the control flow[...] So, some understanding of the process will always be the first step" (p18)
	S10	**Discover patterns:** identify patterns in the data that can suggest new hypotheses or analysis directions.	all	20	'What I'm looking for is patterns. [...] Something that tells me 'ok, here is, uh, this is something that's worth investigating further.'" (p31)
	S11	**Classify and compare cases:** create groups of cases based on attributes or KPIs and compare them.	[P]	9	"...by defining these kind of KPIs, so I need a kind of central number, which helps me to understand if the situation is good or bad. " (p33)
	S12	**Look for correlations:** search for correlations among different process characteristics, e.g., attributes	all	7	'So, I would drill down in those attributes and see if I can find a correlation between a reason, between an attribute and the scenario..." (p17)
	S13	**Focus on narrow scopes of interest:** dig deep into selected parts of the event log deemed interesting.	all	24	'If I see some interesting point in the process model or in the event log, I go deeper and deeper and try to understand it." (p29)
	S14	**Test hypotheses:** check hypotheses using analysis algorithms and tools.	all	15	'Then I'll go to the conformance page and usually that validates that can help me validate or disprove my hypothesis." (p13)
EVALUATE	S15	**Verify artifacts/findings:** check findings' or artifacts' correctness	all	5	'My main steps were, first, to verify that process map generation was seemed to be correct." (p11)
	S16	**Validate artifacts/findings with stakeholders:** evaluate artifacts or findings with stakeholders.	all	18	'We always present our results to our customers, to our partners, to discuss the results with them to check if it makes sense..." (p22)

parts in looking at the aggregated levels" (p24), whereas a less experienced one said that going into *"the raw data is like going down a rabbit hole"* (p13). Some participants (4/37) mentioned looking also at data quality. Although this is typically part of pre-processing, analysts check data quality when starting the analysis *"to know more about the* [provided] *log and the fields"* (p38).

Recurring challenges of the understanding phase are (i) the lack of domain knowledge and (ii) the limited availability of business stakeholders, as summarized by one participant: *"The second part of the challenge is the same, this is... the lack of understanding of the business, of the business process and, maybe, a lack of interaction with business people."*(p34). Indeed, *"the event log per se could be meaningless unless there is some semantics attached"* (p8) and, thus, analysts *"need somebody who has this business understanding"* (p25). Still, even if stakeholders are available, *"the biggest challenge is also getting those people to help"* and understand from a methodological viewpoint how *"to involve and interact with the domain experts, to feed my process with their information"* (p9).

Plan. Planning concerns coming up with a plan for conducting the analysis, including finding a *direction* for the analysis by (S4) defining analysis questions and (S5) prioritizing analysis directions, giving a *structure* to the analysis by (S6) determining an analysis approach, and finding concrete *entry points* for the analysis by (S7) mapping the question to the data and (S8) making hypotheses.

(S4) Formulate analysis questions (11/37) entails deriving questions from higher-level stakeholder needs and formulating them in suitable way for process mining. Analysts define questions to have a *"clear direction to look into the data"* (p33). Indeed, *"if you understand the question deeply, then you can go to that special part of the analysis that you need to answer that question"* (p6). Questions also help to *"avoid getting lost in the complexity"* (p33) as *"this could be dangerous because you have to keep focus"* (p26). Analysts formulate questions independently or during workshops with stakeholders, at times using standard business hypotheses and pre-defined analyses as templates to *"break the ice [...] so that you don't start with an empty piece of paper"* (p33).

(S5) Prioritize analysis directions (11/37) refers to prioritizing analysis directions based on the foreseen value or impact of the findings. Prioritizing helps analysts find analyses that are valuable for the stakeholder, rather than focusing on things *"already known to the customer"* (p11) or *"the small ones that could be accidental"* (p7) because *"you can spend hours and hours doing something that doesn't have an impact"* (p24). To prioritize, analysts combine different criteria, such as the monetary value of the objects involved in the process or indications about process execution frequency and time performance. This strategy was reported by 11 participants, of which nine were practitioners (cf. Table 1).

(S6) Determine analysis approach (8/37) consists of evaluating which method suits the analysis best, given an analysis question or lack thereof, the tools available, and the desired outcomes. Our participants, especially experts, reported following a *"structured approach"* (p33) that they *"adapt based on the question"* (p9) or *"the* [end] *user group"*, because *"if it's done too complex, then people won't use it"* (p17). One part of determining the approach is also selecting the tools to use for the analysis. Analysts do so based on the tools' strengths, the artifacts that the tools allow them to create, and the need for customization, e.g., *"depending on the questions, I will select a different tool, and with a different tool. I will go for a different thing"* (p7). Six participants reporting this strategy indicated "advanced" expertise, while two have "good" expertise (cf. Table 1).

(S7) Map the question to the data (11/37) allows analysts to break the question down into relevant "players" for the analysis that can be linked to key "data objects" in the event log, serving as concrete entry points for the analysis, e.g., *"I'm immediately looking at those event attributes because they were or they seem to be the most central data object for the guiding questions"* (p18).

(S8) Make hypotheses (15/37) refers to conjecturing about (i) explanations for the problem at hand, e.g., *"I try to create a hypothesis which explains the situations or the main problem for the initiative"* (p27), (ii) possible answers for the analysis questions, e.g., *"trying to build hypotheses related to the general question"* or (iii) observations in the data. Analysts often make hypotheses based on their experience, belief, or (limited) evidence in the data. Participants report *"picking out hypothesis right from the question"* (p12) to find entry points for the analysis, e.g., *"I look for the research questions and I start a preliminary hypothesis of... how it could look like or what patterns I could find"* (p7).

From our interviews, it emerged that analysts often plan their analysis very close to the question to avoid *"getting lost in complexity"* (p33) and to *"have a structure at hand"* (p9). Still, formulating questions is perceived as difficult. Indeed, *"it is very often hard to identify the correct question"* (p36), and *"often there would be a lack of concrete questions"* (p12) and *"you do not really know what you actually look out for"* (p37). While using pre-defined analyses seems to help analysts formulate initial questions, the lack of a concrete "structure" for planning the analysis and of methodologies that can help elicit questions in collaboration with stakeholders remains a challenge.

Analyze. In the *analyze* phase, process mining and visual analytic techniques are applied. Common strategies are (S9) understanding the process, (S10) discovering patterns, (S11) classifying and comparing cases, (S12) looking for correlations, (S13) focusing on narrow scopes of interest, and (S14) testing hypotheses.

(S9) Understand the process (32/37) concerns making sense of the control flow, including activities and their relations, variants, the happy path, the process structuredness, and the desired flow. It helps analysts to *"get familiar with the data and the process"* (p28) and *"characterize process behavior"* (p37), assess if additional data or artifacts are needed for the analysis and check their understanding of the process from stakeholders or documents, e.g., *"trying to confirm also my initial feeling on how the process looks like"* (p17). Process understanding is mostly supported by visualizations of directly-follows graphs (DFGs), variants, and dotted charts. DFGs seem to be by far the most used as they are *"very straightforward"* (p10) and allow one *"to imagine what is the problem you have inside the process"* (p26). Still, some expert participants (5/37) advised using DFGs carefully because *"they are easily misinterpreted"* (p39), and if *"you start with a filtered view, it can be really misleading"* (p12).

(S10) Discover patterns (20/37) aims to identify patterns and relationships among observed phenomena, scenarios, or outliers that help analysts make (new) hypotheses or find new analysis directions, e.g., *"you first try to understand what's going on, try yourself to look at meaningful patterns, high-frequent patterns"* (p39). Some participants (6/37) also reported trying to *"explore beyond a question"* (p30) and *"keep an eye open for things that stand out"* (p19) to *"attract those who are going to listen to my analysis afterward"* (p5).

(S11) Classify and compare cases (9/37) concerns partitioning the event log into *"subsets that are meaningful"* (p31) or *"dimensions that could be relevant"* (P24) to describe and classify clusters of cases and compare them to find differences between groups, e.g., *"I want to see the category versus the non-category"* (p31). Analysts classify and compare cases based on data attribute values, enumerated or numeric, or a custom KPI, i.e., a *"kind of central number, which helps me to understand if the situation is good or bad"* (p33). Comparing groups of cases is a way to search for correlations and potential root causes. This strategy was reported by nine participants, of which eight were practitioners (cf. Table 1). In particular, the term "KPI" was explicitly used only by practitioners involved in improvement initiatives where KPIs help monitor *"later on when I implement improvement measures, if the situation has really improved"* (p33).

(S12) Look for correlations (7/37) concerns looking for relations among different process characteristics, for example, by combining control-flow-related characteristics with data attributes or performance metrics to find *"influencing factors"* (p33). Correlations usually serve as hints for generating new hypotheses or for finding potential root causes of a problem, e.g., *"Usually, the root cause is somewhere hidden in the attributes. So, we have some pattern [...] and then you want to know why, but it's really seldom that it's because of the actual process flow. Usually, you have to correlate it with attributes"* (p37).

(S13) Focus on narrow scopes of interest (24/37) concerns focusing on specific parts of the log that capture the analysts' attention or are suggested by stakeholders. Analysts describe "drilling down" or "deep-diving" into the data with the help of filters to focus on interesting things, such as patterns, scenarios, issues, specific process behavior, or, simply, something *"that sticks out"* (p17). Usually, they narrow their scopes of interest to explain observations or spot inconsistencies, e.g., *"you know from a benchmark where the critical process steps are and you can deep-dive into those and see if it's really an issue"* (p23).

(S14) Test hypotheses (15/37) concerns gathering evidence in the data to confirm or disprove a hypothesis. Analysts test ideas when attempting to answer analysis questions, usually checking if hypotheses can be *"backed up by real data"* (p11). Hypotheses are tested with tools and algorithms such as conformance checking. Our participants described testing hypotheses by *"applying filters, looking at the results from different points of view"* (p34) or searching for *"violations to the hypotheses"* (p8). They also mention keeping track of *"tests to present later"* to show to stakeholders *"what we found and rejected"* (p32).

As reported by our participants, the main challenge related to the analysis phase is the lack of techniques for identifying causality, which makes it difficult to "*get real value from the analysis*" (p12). Indeed, "*understanding the real why is a big question*" (p1), and with process mining, "*you don't get the clear reason out of it [...] but it gives you indications for further deep dives*" (p33). Moreover, not knowing the root causes of a problem makes it difficult to recommend solutions, e.g., "*you found... Like the root cause or where it may be, but then what is the next step? So, process mining is like not helping you to solve the issue*" (p17).

Evaluate. Evaluation entails verifying (S15) and validating (S16) analysis artifacts and findings, helping analysts to determine the end of an analysis iteration.

(S15) Verify artifacts/findings (5/37) is about checking the correctness and accuracy of analysis artifacts, such as filters, dashboards, and visualizations, or (intermediate) results, usually by comparison with the original data. Participants reported checking the logic of filters to ensure that they "*build something tested, validated on like a raw table*" (p14), combining different tools "*to understand just for sanity check if we have the same insights in both tools*" (p27), or writing scripts to check "programmatically" if things do not "add up."

(S16) Validate artifacts/findings with stakeholders (18/37) concerns evaluating analysis artifacts or findings together with business stakeholders. Validation helps analysts to confirm if analysis artifacts and findings "*make sense in terms of domain knowledge*" (p22) for the stakeholders, who can assess if "*their questions were answered*" (p16) or if "*they see any additional things which might be interesting*" (p33). While this strategy is often used at the end of an analysis iteration, it can also occur spontaneously, for example, if analysts want to validate their hypotheses or discover something interesting during the analysis.

Besides the lack of domain knowledge, which also affects evaluation, a challenge related to this phase is the lack of tool support for validating findings, e.g., "*there is very little support for validation. So, you see something that stands out, is it actually true?*" (p19). Analysts can combine different tools for "sanity-check", but they mainly depend on stakeholders for validating what they observe "*because the data will not tell you that the answer is invalid*" (p12). Still, "*how do you evaluate your results with domain experts?*" (p16) remains an open question.

4.2 Factors Affecting the Use of Strategies in Practice

From our analysis, we derived four *factors* affecting the use of strategies in practice: Ⓠ the analysis questions, Ⓡ the analyst's role, Ⓢ the availability of business stakeholders, and Ⓣ the tools used. Here, we reflect on such factors, reporting the strategies they affect and representative examples from our interviews.

Ⓠ **Analysis questions (S4, S6, S7, S8, S9, S10, S11).** Often, strategies "*depend on the question*" (p15), especially when it comes to planning and analyzing. In some cases, stakeholders "*have limited knowledge of what is possible*" (p39) and may not pose questions, making analysts formulate questions starting from pre-defined analyses or start with "*finding out patterns in the event log*" (p36). If present, questions are reported to be specific or broad, affecting how analysts plan their analysis, e.g., "*If that's a clear question, that is for us very nice, because we can just dive deep straight away. And then a pipeline is* *always to start exploring [...] If the question is not so clear [...] then we have a more fixed scheme"* (p39). In particular, specific questions seem easier to be mapped to the data and allow analysts to spend little or no time on hypotheses making, pattern discovery, and, even, process understanding. Indeed, in this setting, *"you know what to watch out for"* (p37) and *"you might not even go to a process model or process mining tool"* (p19). Ultimately, questions can be of many kinds, such as "statistical" or about "deviations", affecting what analyses can be done, e.g., *"If there's a simple statistic question [...] statistical analysis will be enough"* (p2).

Ⓡ **Analyst's role (S2, S3, S12):** Among our interviewees, we could discern between "generalists", i.e., analysts who oversee all process mining stages, and "specialists", i.e., analysts specialized in specific steps such as creating event logs or dashboards based on stakeholders' needs, e.g., *"I'm very technical... most time I spend in event collections or setting up the data model, setting up the activities, instead of running the analytics"* (p24). Generalists often analyze the processes enacted within the company for which they work, thus gaining domain and data knowledge as part of their work experience, e.g., *"speaking about roles and capabilities, we own everything in our team [...] we know all the data, we create the event logs [...] we do all this ourselves, including the analysis"*. By contrast, specialists tend to work in teams. Thus, they may gather domain and data understanding from team members in closer contact with stakeholders or even work on the data without much domain knowledge, focusing on its structure. We also observed that analysts with a technical background in data engineering are more comfortable than others with strategies requiring deep data understanding and manipulation because with *"a relational database background, it's probably not as tricky. But if you come from, like, just a BI background, you probably aren't thinking in this way. And so, I think the aggregation can be the challenge"* (p14).

Ⓢ **Availability of business stakeholders (S1, S2, S4, S8, S16).** Stakeholders' availability mainly affects problem and domain understanding and validation strategies. Indeed, analysts organize *"interactive sessions with the data owner to understand things"* (p34) and, if they are in close touch, also validate intermediate results, e.g., *"For me, it's not hard to talk during analysis [...] I am accustomed to getting some feedback. So my goal usually would be to do the analysis, get feedback, adjust it."* (p15). Stakeholders' availability can also influence planning. Indeed, if stakeholders are available, they can contribute to the generation of questions and hypotheses, e.g., *"I can do things with the guy in real-time and, usually, it triggers questions. And some questions I can try to understand live with the process mining tool, some questions will need to be worked on after the interactive session."* (p34). When access to stakeholders is limited, analysts tend to focus on time performance and conformance questions, as they are the *"most common perspectives that one can look at while doing process mining without having additional information about the context"* (p36).

Ⓣ **Tools (S3, S9, S10, S11, S12).** Tools can influence data and process understanding, as well as what analyses can be planned and executed. Here, we distinguish between tools allowing (raw) data access and manipulation, tools supporting visual analyses and interactive exploration, and tools supporting custom analyses. Tools supporting raw data access and manipulation allow analysts to *"open the log file"* (p11) and *"understand, like the sort of data structure first"* (p14). Such tools seem to better support strategies, such as S11 and S12, where data manipulation and querying is needed, e.g., *"I wanted to delve in and do the subset analysis, but it was taking me a long time to go from generating a subset to producing the XES, to loading it and looking at it and say, 'oh, that's not quite right'. Then I went to [...] because it does all this processing out-of-the-box"* (p31). However, these tools are not necessarily process mining tools. Analysts report preferring tools that allow creating visualizations and animations when they engage in exploration or *"want to do a quick filter or get a quick idea"* (p19), e.g., *"I feel that the animation button [...] is key for me to understand, have better clarity on the identification of the process flow"* (p20). During planning, analysts choose which tools they will use. The possibility of creating custom analyses seems to be one of the criteria considered when making this choice, e.g., *"I wanted a way of taking very complicated data and just pushing a button to get my PowerPoint"* (p31).

5 Discussion and Conclusion

From interviews with 37 practitioners and academics working with process mining, we have derived 16 strategies related to the analysis stage and have examined recurring challenges and factors affecting their use in practice. These findings contribute to our understanding of the work practices of individual process analysts, with a specific focus on the analysis stage and its needs.

Our participants reported some challenges related to analysis strategies. One is the lack of domain knowledge, which affects both domain and data understanding (S2, S3) as well as the validation of results (S17), and is often connected to the limited ⓢ availability of business stakeholders (e.g., domain experts) and, also, the Ⓡ analyst's role. While this challenge is renowned in the field [7,14], our participants stress the need for more concrete advice to work interactively and involve domain experts *"because only they know exactly what is there"* (p33). In this direction, our characterization of analysts into generalists and specialists could inspire research on how to organize work at the group level [3]), considering roles and skill sets of process mining teams. Indeed, the lack of clearly defined roles and responsibilities seem to be a cause for collaborative tensions [20].

Our results also show that the generation and refinement of Ⓠ analysis questions is a persistent challenge of the planning phase and also an influential factor for the analysis phase. Indeed, many participants reported that *"the analysis should always follow the question"* (p36) but that identifying the "right" question remains difficult. In their work, Emamjome et al. [9] have discovered that the thoroughness of question formulation as the first phase of process mining projects has improved over the years, indicating ⓢ increased interactions with business stakeholders. Still, our interviewees remark that *"specific methodologies that can help to define more effectively a research question"* (p36) are missing.

Analysts rely on their experience to deal with different kinds of questions or lack thereof, sometimes using the pre-defined analyses available in some tools as a starting point. We think that pre-defined analyses could inspire the development of guidance for the less experienced. For example, they could be used as a starting point to develop questions catalogs or checklists adapted to specific use cases or event log features. This could be realized by collecting and refining frequently posed questions such as those identified by Mans et al. for healthcare processes [13] or the domain problems in [11]).

Regarding the analysis phase, our participants mentioned the lack of support of process mining tools for identifying causality being a challenge, limiting them in developing recommendations for solving stakeholders' problems (which relates to C.17 in [14]). The research community has picked up on this challenge, and recent work on causal machine learning provides promising results towards addressing it [2]. Still, more research is needed to support the identification of root causes and the consequent development of practical recommendations.

Last but not least, from our analysis, we have discovered some factors affecting the practical use of strategies (cf. Sect. 4.2). We believe that further investigation of these factors and, potentially, other individual and organizational factors could enrich our understanding of process mining practices and help us explain which strategies are suitable in a given context and why. Such an understanding could then be used to inform the development of concrete process mining guidance covering several aspects, e.g., analysis questions.

Limitations. Since they are based on retrospective interviews, our findings are subject to validity threats typical of interview studies [16]. First, the participants' behavior could have been influenced by the interviewer's presence and behavior (reactivity). To mitigate this risk, we asked questions using the wording defined in the interview guide, guaranteed anonymity of the answers, and recorded only the audio of the interviews, i.e., the videos of interviewees and interviewer were turned off. Second, the participants' answers could have been biased, e.g., by the study setting (respondent bias). To mitigate this risk, we developed our interview guide, considering questions about individual work practices, i.e., focusing on "general" actions, which did not require participants to name specific companies or share sensitive information. Also, participants had no prior knowledge of the study and the interview questions. Third, the data collection and analysis could have been biased by the researcher's personal opinions and interpretations (researcher bias). To mitigate this risk, we conducted all the interviews following the prepared guide, which had been pilot tested with researchers external to the author team as advised in [5] and we periodically met to review and discuss the coding process. In addition, the interviewer did not have any personal or working relationships with the participants.

Regarding the generalizability of our results, we would like to note that the set of strategies presented in this paper was derived from explicit statements of our participants, who may not have reflected upon all strategies they apply in practice. Thus, this set may not be complete. We cannot exclude that other strategies can emerge in different settings, for example, when anchored to different tasks, interviewing specific user groups, or focusing on certain application domains. Still, our sample included participants with varied backgrounds and experience levels who use process mining in different sectors, including healthcare, food processing, and insurance. Moreover, we considered only strategies that were reported by at least 4 participants, suggesting that most of them are relevant to different contexts.

We foresee several directions for future work. As a first direction, we plan to extend our findings by triangulating them with behavioral data. In particular, we aim to analyze the behavioral data collected during the process mining task to allow for a finer-grained characterization of strategies, including concrete steps and specific process mining techniques used to implement these steps. Such an analysis will also let us look deeper into factors and, potentially, explain how they affect the use of the strategies. In addition, we envision investigating strategies applied in other stages of a process mining project, such as the *extraction* and *data processing*. Last but not least, we hope that the strategies we identified in this paper will inspire research on developing actionable support for process mining practitioners.

Acknowledgment. We would like to thank our participants for taking the time to participate in the study and sharing their valuable experience. We also thank N. Soltani for her help with the transcriptions.

Funding Information. This work is part of the ProMiSE project, funded by the Swiss National Science Foundation under Grant No.: 200021_197032. P. Soffer was partly funded by the Israel Science Foundation as part of the Eco4PPM project, grant agreement 2005/21.

References

1. Van der Aalst, W.M.: Process Mining: Data Science in Action, pp. 3–23. Springer, Heidelberg (2016). https://doi.org/10.1007/978-3-662-49851-4 (2016)
2. Bozorgi, Z.D., Teinemaa, I., Dumas, M., La Rosa, M., Polyvyanyy, A.: Process mining meets causal machine learning: discovering causal rules from event logs. In: 2nd International Conference on Process Mining (ICPM), pp. 129–136 (2020). https://doi.org/10.1109/ICPM49681.2020.00028
3. vom Brocke, J., Jans, M., Mendling, J., Reijers, H.A.: A five-level framework for research on process mining. Bus. Inf. Syst. Eng. **63**(5), 483–490 (2021). https://doi.org/10.1007/s12599-021-00718-8
4. Carlson, M.P., Bloom, I.: The cyclic nature of problem solving: an emergent multidimensional problem-solving framework. Educ. Stud. Math. **58**(1), 45–75 (2005). https://doi.org/10.1007/s10649-005-0808-x
5. Chenail, R.: Interviewing the investigator: strategies for addressing instrumentation and researcher bias in qualitative research. Qual. Rep. **16**(1), 255–262 (2011). https://doi.org/10.46743/2160-3715/2011.1051
6. De Leoni, M., Mannhardt, F.: Road Traffic Fine Management Process. Eindhoven University of Technology, Dataset (2015)

7. van Eck, M.L., Lu, X., Leemans, S.J.J., van der Aalst, W.M.P.: PM^2: a process mining project methodology. In: Zdravkovic, J., Kirikova, M., Johannesson, P. (eds.) CAiSE 2015. LNCS, vol. 9097, pp. 297–313. Springer, Cham (2015). https://doi.org/10.1007/978-3-319-19069-3_19

8. Eggers, J., Hein, A., Böhm, M., Krcmar, H.: No longer out of sight, no longer out of mind? How organizations engage with process mining-induced transparency to achieve increased process awareness. Bus. Inf. Syst. Eng. 63(5), 491–510 (2021). https://doi.org/10.1007/s12599-021-00715-x

9. Emamjome, F., Andrews, R., ter Hofstede, A.H.M.: A case study lens on process mining in practice. In: Panetto, H., Debruyne, C., Hepp, M., Lewis, D., Ardagna, C.A., Meersman, R. (eds.) OTM 2019. LNCS, vol. 11877, pp. 127–145. Springer, Cham (2019). https://doi.org/10.1007/978-3-030-33246-4_8

10. Grisold, T., Mendling, J., Otto, M., vom Brocke, J.: Adoption, use and management of process mining in practice. Bus. Process Manage. J. 27 (2020). https://doi.org/10.1108/BPMJ-03-2020-0112

11. Klinkmüller, C., Müller, R., Weber, I.: Mining process mining practices: an exploratory characterization of information needs in process analytics. In: Hildebrandt, T., van Dongen, B.F., Röglinger, M., Mendling, J. (eds.) BPM 2019. LNCS, vol. 11675, pp. 322–337. Springer, Cham (2019). https://doi.org/10.1007/978-3-030-26619-6_21

12. Koorn, J.J., et al.: Bringing rigor to the qualitative evaluation of process mining findings: an analysis and a proposal. In: 3rd International Conference on Process Mining (ICPM), pp. 120–127. IEEE (2021). https://doi.org/10.1109/ICPM53251.2021.9576877

13. Mans, R.S., van der Aalst, W.M.P., Vanwersch, R.J.B., Moleman, A.J.: Process mining in healthcare: data challenges when answering frequently posed questions. In: Lenz, R., Miksch, S., Peleg, M., Reichert, M., Riaño, D., ten Teije, A. (eds.) KR4HC/ProHealth -2012. LNCS (LNAI), vol. 7738, pp. 140–153. Springer, Heidelberg (2013). https://doi.org/10.1007/978-3-642-36438-9_10

14. Martin, N., et al.: Opportunities and challenges for process mining in organizations: results of a Delphi study. Bus. Inf. Syst. Eng. 63(5), 511–527 (2021). https://doi.org/10.1007/s12599-021-00720-0

15. Nurcan, S., Etien, A., Kaabi, R., Zoukar, I., Rolland, C.: A strategy driven business process modelling approach. Bus. Process Manage. J. (2005). https://doi.org/10.1108/14637150510630828

16. Padgett, D.K.: Qualitative Methods in Social Work Research, vol. 36. Sage, Thousand Oaks (2016)

17. Ralph, P., et al.: Empirical standards for software engineering research. arXiv preprint arXiv:2010.03525 (2021)

18. Saldaña, J.: The Coding Manual for Qualitative Researchers. Sage, Thousand Oaks (2015)

19. Saunders, B., et al.: Saturation in qualitative research: exploring its conceptualization and operationalization. Qual. Quant. 52(4), 1893–1907 (2018). https://doi.org/10.1007/s11135-017-0574-8

20. Syed, R., Leemans, S.J.J., Eden, R., Buijs, J.A.C.M.: Process mining adoption. In: Fahland, D., Ghidini, C., Becker, J., Dumas, M. (eds.) BPM 2020. LNBIP, vol. 392, pp. 229–245. Springer, Cham (2020). https://doi.org/10.1007/978-3-030-58638-6_14

21. Thiede, M., Fuerstenau, D., Barquet, A.P.B.: How is process mining technology used by organizations? A systematic literature review of empirical studies. Bus. Process Manage. J. **24**(4), 900–922 (2018). https://doi.org/10.1108/BPMJ-06-2017-0148
22. Zerbato, F., Soffer, P., Weber, B.: Initial insights into exploratory process mining practices. In: Polyvyanyy, A., Wynn, M.T., Van Looy, A., Reichert, M. (eds.) BPM 2021. LNBIP, vol. 427, pp. 145–161. Springer, Cham (2021). https://doi.org/10.1007/978-3-030-85440-9_9

Analytics

Measuring Inconsistency in Declarative Process Specifications

Carl Corea[1]([☒]), John Grant[2], and Matthias Thimm[3]

[1] Institute for IS Research, University of Koblenz-Landau, Koblenz, Germany
ccorea@uni-koblenz.de
[2] University of Maryland, College Park, USA
grant@cs.umd.edu
[3] Artificial Intelligence Group, University of Hagen, Hagen, Germany
matthias.thimm@fernuni-hagen.de

Abstract. We address the problem of measuring inconsistency in declarative process specifications, with an emphasis on linear temporal logic on fixed traces (LTL$_{ff}$). As we will show, existing inconsistency measures for classical logic cannot provide a meaningful assessment of inconsistency in LTL in general, as they cannot adequately handle the temporal operators. We therefore propose a novel paraconsistent semantics as a framework for inconsistency measurement. We then present two new inconsistency measures based on these semantics and show that they satisfy important desirable properties. We show how these measures can be applied to declarative process models and investigate the computational complexity of the introduced approach.

Keywords: Inconsistency measurement · LTL · Declare

1 Introduction

Linear temporal logic (LTL) is an important logic for specifying the (temporal) behavior of business processes in the form of *declarative process specifications* [1,18]. The underlying idea is that time is represented as a linear sequence of states $T = (t_0, ..., t_m)$, where t_0 is the designated starting point. At every state, some statements may be true. Temporal operators specify properties that must hold over the sequence of states. For example, the operator **X** (*next*) means that a certain formula holds at the next state. Likewise, the operator **G** (*globally*) means that a certain formula will hold for all following states. Note that we assume the sequence to be finite, i.e., we consider a linear temporal logic over finite traces (LTL$_f$) [5,18].

Traditionally, model checking has been used to verify that a particular model—that is, the assignment of truth values for statements over the time sequence—satisfies the requirements. However, a problem in this use case arises

This work has been partially supported by the Deutsche Forschungsgemeinschaft (Grant DE 1983/9-1).

C. Di Ciccio et al. (Eds.): BPM 2022, LNCS 13420, pp. 289–306, 2022.
https://doi.org/10.1007/978-3-031-16103-2_20

if the set of formulas is *inconsistent*, i.e., contains contradictory specifications. In such a case, the set of specifications cannot be applied for its intended purpose of process verification. For example, consider the two sets of formulas $\mathcal{K}_1$ and $\mathcal{K}_2$ (we will formalize syntax and semantics later):

$$\mathcal{K}_1 = \{\mathbf{X}a, \mathbf{X}\neg a\} \qquad\qquad \mathcal{K}_2 = \{\mathbf{G}a, \mathbf{G}\neg a\}$$

Both $\mathcal{K}_1$ and $\mathcal{K}_2$ are inconsistent, as they demand that both a and $\neg a$ hold in (some) following state, which is unsatisfiable. This calls for the analysis of such inconsistencies, to provide insights for inconsistency resolution.

In classical logic, all inconsistent sets are equally bad [12]. However, considering again the two sets, intuitively, $\mathcal{K}_2$ is "more" inconsistent than $\mathcal{K}_1$: The inconsistency in $\mathcal{K}_1$ only affects the next state, while the inconsistency in $\mathcal{K}_2$ affects all following states. This is an important insight that could prove useful for debugging or re-modelling LTL$_f$ specifications or LTL$_f$-based constraint sets in general such as Declare. While there have been some recent works that can *identify* inconsistent sets in declarative process specifications [2,3,20], those works cannot look "into" those sets or compare them. In this work, we therefore show how to distinguish the *severity* of inconsistencies in LTL$_f$, specifically, LTL$_{ff}$.

A scientific field geared towards the quantitative assessment of inconsistency in knowledge representation formalisms is *inconsistency measurement* [8,22], and therefore represents a good candidate for this endeavour. Inconsistency measurement studies measures that aim to assess a *degree* of inconsistency with a numerical value. The intuition here is that a higher value represents a higher degree of inconsistency. Such measures can provide valuable insights for debugging inconsistent specifications, e.g., to determine whether certain sets of formulas are more inconsistent than others. As we will show, existing measures are currently not geared towards LTL$_f$ and temporal operators, and therefore cannot provide a meaningful analysis. Therefore, the main goal of this paper is to develop a new approach for measuring inconsistency in linear temporal logic. To frame this problem, we introduce a variant of LTL$_f$, which we coin linear temporal logic on *fixed* traces LTL$_{ff}$ (cf. Sect. 2.2).

Our contributions are as follows. We formalise the problem of measuring inconsistency in LTL$_{ff}$ and propose a rationality postulate that should be met by quantitative measures applied to this setting (Sect. 2). We show that existing inconsistency measures do not satisfy this property, and propose an approach for measuring inconsistency based on a novel paraconsistent semantics for LTL$_{ff}$ (Sect. 3). We then show how our approach can be applied for measuring inconsistency in declarative process models (Sect. 4). For evaluation, we investigate the computational complexity of central aspects regarding inconsistency measurement in LTL$_{ff}$ (Sect. 5). A conclusion is provided in Sect. 6. Proofs for technical results can be found in a supplementary document provided online.

2   Preliminaries

The traditional setting for inconsistency measurement is that of propositional logic. For that, let At be some fixed propositional signature, i.e., a (possibly infinite) set of propositions, and let $\mathcal{L}(\mathsf{At})$ be the corresponding propositional language constructed using the usual connectives $\wedge$ (*conjunction*), $\vee$ (*disjunction*), and $\neg$ (*negation*). A literal is a proposition p or negated proposition $\neg p$.

Definition 1. *A knowledge base* $\mathcal{K}$ *is a finite set of formulas* $\mathcal{K} \subseteq \mathcal{L}(\mathsf{At})$. *Let* $\mathbb{K}$ *be the set of all knowledge bases.*

For a set of formulas X we denote the set of propositions in X by $\mathsf{At}(X)$.

Semantics for a propositional language is given by *interpretations* where an interpretation ω on At is a function $\omega : \mathsf{At} \to \{0,1\}$ (where 0 stands for false and 1 stands for true). Let $\Omega(\mathsf{At})$ denote the set of all interpretations for At. An interpretation ω *satisfies* (or is a *model* of) an atom $a \in \mathsf{At}$, denoted by $\omega \models a$, if and only if $\omega(a) = 1$. The satisfaction relation $\models$ is extended to formulas in the usual way. For $\Phi \subseteq \mathcal{L}(\mathsf{At})$ we also define $\omega \models \Phi$ if and only if $\omega \models \phi$ for every $\phi \in \Phi$. Furthermore, for every set of formulas X, the set of models is $\mathsf{Mod}(X) = \{\omega \in \Omega(\mathsf{At}) \mid \omega \models X\}$. Define $X \models Y$ for (sets of) formulas X and Y if $\omega \models X$ implies $\omega \models Y$ for all ω.

Let $\top$ denote any tautology and $\bot$ any contradiction. If $\mathsf{Mod}(X) = \emptyset$ we write $X \models \bot$ and say that X is *inconsistent*.

2.1 Inconsistency Measurement

Inconsistency as defined above is a binary concept. To provide more fine-grained insights on inconsistency beyond such a binary classification, the field of inconsistency measurement [22] has evolved. The main objects of study in this field are *inconsistency measures*, which are quantitative measures that assess the degree of inconsistency for a knowledge base $\mathcal{K}$ with a non-negative numerical value. Intuitively, a higher value reflects a higher degree, or severity, of inconsistency. This can be useful for determining if one set of formulas is "more" inconsistent than another. Let $\mathbb{R}_{\geq 0}^{\infty}$ be the set of non-negative real values including ∞. Then, an inconsistency measure is defined as follows.

Definition 2. *An inconsistency measure* $\mathcal{I}$ *is any function* $\mathcal{I} : \mathbb{K} \to \mathbb{R}_{\geq 0}^{\infty}$.

To constrain the desired behavior of concrete inconsistency measures, several properties, called *rationality postulates*, have been proposed. A well-agreed upon property is that of *consistency*, which states that an inconsistency measure should return a value of 0 iff there is no inconsistency.

Consistency (**CO**) $\mathcal{I}(\mathcal{K}) = 0$ if and only if $\mathcal{K}$ is consistent.

Further important postulates introduced in [10] are *monotony*, *dominance* and *free-formula independence*, which we will define below. For that, we need some further notation.

First, a set $M \subseteq \mathcal{K}$ is called a *minimal inconsistent subset* (MIS) of $\mathcal{K}$ if $M \models \bot$ and there is no $M' \subset M$ with $M' \models \bot$. Let $\mathsf{MI}(\mathcal{K})$ be the set of all MISs of $\mathcal{K}$. Second, a formula $\alpha \in \mathcal{K}$ is called a *free formula* if $\alpha \notin \bigcup \mathsf{MI}(\mathcal{K})$. Let $\mathsf{Free}(\mathcal{K})$ be the set of all free formulas of $\mathcal{K}$.

For the remainder of this section, let $\mathcal{I}$ be an inconsistency measure, $\mathcal{K}, \mathcal{K}' \in \mathbb{K}$, and $\alpha, \beta \in \mathcal{L}(\mathsf{At})$. Then, the basic postulates from [10] are defined as follows.

Monotony (**MO**) If $\mathcal{K} \subseteq \mathcal{K}'$ then $\mathcal{I}(\mathcal{K}) \leq \mathcal{I}(\mathcal{K}')$.
Free-formula independence (**IN**) If $\alpha \in \mathsf{Free}(\mathcal{K})$ then
$\quad \mathcal{I}(\mathcal{K}) = \mathcal{I}(\mathcal{K} \setminus \{\alpha\})$.
Dominance (**DO**) If $\alpha \not\models \bot$ and $\alpha \models \beta$ then $\mathcal{I}(\mathcal{K} \cup \{\alpha\}) \geq \mathcal{I}(\mathcal{K} \cup \{\beta\})$.

MO states that adding formulas to the knowledge base cannot decrease the inconsistency value. IN means that removing free formulas from the knowledge base does not change the inconsistency value. DO consists of several cases, depending on the presence or absence of α or β in $\mathcal{K}$: the idea is that substituting a consistent formula α by a weaker formula β cannot increase the inconsistency.

Numerous inconsistency measures have been proposed (see [23] for a survey), many of which differ in regard to their compliance w.r.t. the introduced postulates. In this work, we will consider six measures as defined below. In order to define the *contension measure* $\mathcal{I}_c$ [7] we need some additional background on Priest's three-valued semantics [19]. A three-valued interpretation is a function $\nu : \mathsf{At} \to \{0, 1, \mathrm{B}\}$, which assigns to every atom either 0, 1 or B, where 0 and 1 correspond to *false* and *true*, respectively, and B (standing for *both*) denotes a conflict. Assuming the *truth order* $\prec_T$ with $0 \prec_T \mathrm{B} \prec_T 1$, the function ν can be extended to arbitrary formulas as follows: $\nu(\alpha \wedge \beta) = \min_{\prec_T}(\nu(\alpha), \nu(\beta))$, $\nu(\alpha \vee \beta) = \max_{\prec_T}(\nu(\alpha), \nu(\beta))$, $\nu(\neg\alpha) = 1$ if $\nu(\alpha) = 0$, $\nu(\neg\alpha) = 0$ if $\nu(\alpha) = 1$, and $\nu(\neg\alpha) = \mathrm{B}$ if $\nu(\alpha) = \mathrm{B}$. We say that an interpretation ν satisfies a formula α, denoted by $\nu \models^3 \alpha$, iff $\nu(\alpha) = 1$ or $\nu(\alpha) = \mathrm{B}$.

We will now define the measures used in this work.

Definition 3. *Let the measures* $\mathcal{I}_d$, $\mathcal{I}_{\mathsf{MI}}$, $\mathcal{I}_p$, $\mathcal{I}_r$, $\mathcal{I}_c$, *and* $\mathcal{I}_{at}$ *be defined as follows:*

$$\mathcal{I}_d(\mathcal{K}) = \begin{cases} 1 & \text{if } \mathcal{K} \models \bot \\ 0 & \text{otherwise} \end{cases}$$

$$\mathcal{I}_{\mathsf{MI}}(\mathcal{K}) = |MI(\mathcal{K})|$$

$$\mathcal{I}_p(\mathcal{K}) = \left| \bigcup_{M \in MI(\mathcal{K})} M \right|$$

$$\mathcal{I}_r(\mathcal{K}) = \min\{|X| \mid X \subseteq \mathcal{K} \text{ and } \mathcal{K} \setminus X \not\models \bot\}$$

$$\mathcal{I}_c(\mathcal{K}) = \min\{|\nu^{-1}(B) \cap \mathsf{At}| \mid \nu \models^3 \mathcal{K}\}$$

$$\mathcal{I}_{at}(\mathcal{K}) = \left| \bigcup_{M \in MI(\mathcal{K})} \mathsf{At}(M) \right|$$

A baseline approach is the drastic inconsistency measure $\mathcal{I}_d$ [11], which only differentiates between inconsistent and consistent knowledge bases. The MI-inconsistency measure $\mathcal{I}_{\mathsf{MI}}$ [11] counts the number of minimal inconsistent subsets. A similar version is the problematic inconsistency measure $\mathcal{I}_p$ [7], which counts the number of distinct formulas appearing in any inconsistent subset. The repair measure $\mathcal{I}_r$ counts the smallest number of formulas that must be removed in order to restore consistency. The contension measure $\mathcal{I}_c$ [7] quantifies inconsistency by seeking a three-valued interpretation that assigns B to a minimal number of propositions. Finally, the $\mathcal{I}_{at}$ measure counts the number of atoms in the non-free formulas.

We conclude this section with a small example illustrating the behavior of the considered inconsistency measures.

Example 1. Consider $\mathcal{K}_3$, defined via

$$\mathcal{K}_3 = \{a, \neg a, b, \neg b \wedge c \wedge d, \neg a \vee \neg b\}$$

Then we have that

$$\mathsf{MI}(\mathcal{K}_3) = \{\{a, \neg a\}, \{b, \neg b \wedge c \wedge d\}, \{a, \neg a \vee \neg b, b\}\}$$

Thus

$$\mathcal{I}_d(\mathcal{K}_3) = 1 \qquad \mathcal{I}_{\mathsf{MI}}(\mathcal{K}_3) = 3 \qquad \mathcal{I}_p(\mathcal{K}_3) = 5$$
$$\mathcal{I}_r(\mathcal{K}_3) = 2 \qquad \mathcal{I}_c(\mathcal{K}_3) = 2 \qquad \mathcal{I}_{at}(\mathcal{K}_3) = 4$$

The main focus of study in inconsistency measurement, and the introduced measures, has been on propositional logic. In this work, our aim is to apply inconsistency measures for linear time logic, which we introduce now.

2.2 Linear Temporal Logic on Fixed Traces

In this work, we consider a specific variant of LTL$_f$ that we coin linear temporal logic on *fixed* traces (LTL$_{ff}$). We consider a linear sequence of states $t_0, \ldots, t_m$, where every t_i is the state at instant i. We assume that $m > 1$ to avoid the trivial case. Note that the difference with LTL$_f$—where interpretations can vary in their length as long as they are finite—is that we keep the length of this sequence finite and fixed across all interpretations. This variant of LTL$_f$ is introduced mainly to discuss matters of inconsistency measurement, as here, the inconsistency value is computed in regard to a comparable length for all formulas. However, the ideas presented in the next sections can be extended to LTL$_f$ [4] in a straightforward manner: In the unbounded case we can use a parameter N and then proceed as in the bounded case. This also means that m must not necessarily be known or provided a priori, as a parameter N can be selected.

The syntax of LTL$_{ff}$ is the same as the syntax of LTL and LTL$_f$ [5]. Formulas are built from a set of propositional symbols At and are closed under the Boolean connectives, the unary operator **X** (*next*), and the binary operator **U** (*until*). Formally, any formula φ of LTL$_{ff}$ is built using the grammar rule

$$\varphi ::= a \,|\, (\neg\varphi) \,|\, (\varphi_1 \wedge \varphi_2) \,|\, (\varphi_1 \vee \varphi_2) \,|\, (\mathbf{X}\varphi) \,|\, (\varphi_1 \mathbf{U} \varphi_2).$$

with $a \in$ At. Intuitively, $\mathbf{X}\varphi$ denotes that φ will hold at the next state and $(\varphi_1 \mathbf{U} \varphi_2)$ denotes that φ_1 will hold until the state when φ_2 holds. Let $d(\varphi) \in \mathbb{N}$ denote the maximal number of nested temporal operators in φ.[1]

---

[1] $d(\varphi)$ is inductively defined via $d(a) = 0$ for $a \in$ At, $d(\neg\phi) = d(\phi)$, $d(\phi_1 \wedge \phi_2) = d(\phi_1 \vee \phi_2) = \max\{d(\phi_1), d(\phi_2)\}$, $d(\mathbf{X}\phi) = 1 + d(\phi)$, and $d(\phi_1 \mathbf{U} \phi_2) = 1 + \max\{d(\phi_1), d(\phi_2)\}$.

From the basic operators, some useful abbreviations can be derived, including $\mathbf{F}\varphi$ (defined as $\top \mathbf{U}\varphi$), which denotes that φ will hold (eventually) in the future and $\mathbf{G}\varphi$ (defined as $\neg\mathbf{F}\neg\varphi$), which denotes that φ will hold for all following states. Again, let $\top$ be any tautology and $\bot$ any contradiction.

An LTL$_{ff}$-interpretation $\hat{\omega}$ w.r.t. At is a function mapping each state and proposition to 0 or 1, meaning that $\hat{\omega}(t, a) = 1$ if proposition a is assigned 1 (true) in state t.[2] Then the satisfaction of a formula ϕ by an interpretation $\hat{\omega}$, denoted by $\hat{\omega} \models \phi$, is defined via

$$\hat{\omega} \models \phi \quad \Leftrightarrow \quad \hat{\omega}, t_0 \models \phi$$

where $\hat{\omega}, t_i \models \phi$ for any interpretation $\hat{\omega}$ as above and for every $t_i \in \{t_0, ..., t_m\}$ is inductively defined as follows:

$\hat{\omega}, t_i \models a$ iff $\hat{\omega}(t_i, a) = 1$ for $a \in$ At

$\hat{\omega}, t_i \models \neg\varphi$ iff $\hat{\omega}, t_i \not\models \varphi$

$\hat{\omega}, t_i \models \varphi_1 \wedge \varphi_2$ iff $\hat{\omega}, t_i \models \varphi_1$ and $\hat{\omega}, t_i \models \varphi_2$

$\hat{\omega}, t_i \models \varphi_1 \vee \varphi_2$ iff $\hat{\omega}, t_i \models \varphi_1$ or $\hat{\omega}, t_i \models \varphi_2$

$\hat{\omega}, t_i \models \mathbf{X}\varphi$ iff $i < m$ and $\hat{\omega}, t_{i+1} \models \varphi$

$\hat{\omega}, t_i \models \varphi_1 \mathbf{U}\varphi_2$ iff $\hat{\omega}, t_j \models \varphi_2$ for some $j \in \{i+1, ..., m\}$

and $\hat{\omega}, t_k \models \varphi_1$ for all $k \in \{i, ..., j-1\}$

An interpretation $\hat{\omega}$ satisfies a set of formulas K iff $\hat{\omega} \models \phi$ for all $\phi \in K$. A set K is consistent iff there exists $\hat{\omega}$ such that $\hat{\omega} \models K$. Define $X \models Y$ for (sets of) formulas X and Y if $\hat{\omega} \models X$ implies $\hat{\omega} \models Y$ for all $\hat{\omega}$.

2.3 Related Work and Contributions

This work is related to consistency- and model checking in declarative process specifications, see e.g. [9,17,21]. In particular, our approach extends recent works [2,3,17,20] on the identification of inconsistent sets in declarative process specifications by allowing to look "into" those sets and leverage inconsistency resolution with quantitative insights. For example, existing resolution approaches mainly try to minimize the *number* of deleted formulas [2,3,14]. This however completely leaves aside the semantics of those formulas or their impact on any corresponding process. Given this motivation, it is useful to consider also the degree to which certain formulas affect the following behavior; which is why we propose time sensitive inconsistency measures.

This paper is related to [6] which presents several, what we call time sensitive, inconsistency measures for branching time logics (BTL). However, in this work we are able to avoid the complicated overload of branching time as the process specifications are provided in linear time logic. Using branching time logic adds a

---

[2] Recall that we assume time of a fixed length $t_0, ..., t_m$ and interpretations only vary in what is true at each state.

layer of complexity that is unnecessary when dealing with a linear time situation. Just to take one example, consider the set $\{\mathbf{X}a, \mathbf{X}\neg a\}$. In linear time logic this gives one inconsistency at the next state. But in the case of branching time logic what does $\mathbf{X}$ mean? There may be many "next" states. If $\mathbf{X}$ means "some next state" then the set is consistent because a and $\neg a$ may hold in different next states. If $\mathbf{X}$ means "all next states" then it is inconsistent but how inconsistent depends on the number of next states. We avoid such issues by dealing only with linear temporal logic. Note also that BTL takes a different view on time than LTL_f as studied in this paper and is therefore expressively incomparable (cf. [24]).

3   Inconsistency Measurement in LTL_ff

In this section, we address the issue of measuring inconsistency in LTL_ff. As we will show, existing inconsistency measures cannot provide meaningful insights when dealing with temporal logic. Therefore, we develop a novel paraconsistent semantics as a framework for handling inconsistency and propose two concrete inconsistency measures for LTL_ff.

3.1   Motivation for Inconsistency Measures for LTL_ff

We recall the sets of LTL_ff formulas $\mathcal{K}_1$ and $\mathcal{K}_2$:

$$\mathcal{K}_1 = \{\mathbf{X}a, \mathbf{X}\neg a\} \qquad\qquad \mathcal{K}_2 = \{\mathbf{G}a, \mathbf{G}\neg a\}$$

The knowledge base $\mathcal{K}_1$ states that a is both true and false in the next state while $\mathcal{K}_2$ states that a is both true and false in all future states. Obviously, both knowledge bases are inconsistent. Yet, the inconsistencies are different in regard to the number of states they affect. For $\mathcal{K}_1$ the number is 1 and for $\mathcal{K}_2$ the number is $m > 1$. It would therefore be desirable for an inconsistency measure to take this information into account and assign $\mathcal{K}_2$ a larger inconsistency value.

In order to capture LTL_ff by the inconsistency measurement framework of Sect. 2.1, from now on a knowledge base $\mathcal{K}$ (Definition 1) will be a finite set of LTL_ff formulas and $\mathbb{K}$ is the set of all LTL_ff knowledge bases. So we can apply the inconsistency measures for $\mathcal{K}_1$ and $\mathcal{K}_2$ in a straightforward manner.

Example 2. Consider $\mathcal{K}_1$ and $\mathcal{K}_2$. Then we have that

$$\begin{aligned}
\mathcal{I}_d(\mathcal{K}_1) &= 1 & \mathcal{I}_d(\mathcal{K}_2) &= 1 \\
\mathcal{I}_{\mathsf{MI}}(\mathcal{K}_1) &= 1 & \mathcal{I}_{\mathsf{MI}}(\mathcal{K}_2) &= 1 \\
\mathcal{I}_p(\mathcal{K}_1) &= 2 & \mathcal{I}_p(\mathcal{K}_2) &= 2 \\
\mathcal{I}_r(\mathcal{K}_1) &= 1 & \mathcal{I}_r(\mathcal{K}_2) &= 1 \\
\mathcal{I}_c(\mathcal{K}_1) &= 1 & \mathcal{I}_c(\mathcal{K}_2) &= 1 \\
\mathcal{I}_{at}(\mathcal{K}_1) &= 1 &&
\end{aligned}$$

Note that all six inconsistency measures give identical values for $\mathcal{K}_1$ and $\mathcal{K}_2$, because they, or for that matter, any other propositional logic inconsistency measure, cannot distinguish between **X** and **G**. But intuitively $\mathcal{K}_2$ is more inconsistent than $\mathcal{K}_1$ because the inconsistency persists through all future states in $\mathcal{K}_2$ as opposed to the single state in $\mathcal{K}_1$. Thus, we believe that a proper inconsistency measure for LTL$_{ff}$ should distinguish between these operators. Therefore, we propose a new rationality postulate.

Time Sensitivity (TS) For all formulas φ of propositional logic,
$$\mathcal{I}(\{\mathbf{G}\varphi, \mathbf{G}\neg\varphi\}) > \mathcal{I}(\{\mathbf{X}\varphi, \mathbf{X}\neg\varphi\}).$$

In other words, the number of affected states should be reflected in the inconsistency value, i.e., inconsistency measures for LTL$_{ff}$ should be time sensitive.

Proposition 1. $\mathcal{I}_d, \mathcal{I}_{\mathsf{MI}}, \mathcal{I}_p, \mathcal{I}_r, \mathcal{I}_c, \mathcal{I}_{at}$ *violate* TS.

Following Proposition 1, the existing measures that we have from propositional logic cannot capture the desired behavior. Therefore, we introduce a novel approach to measure inconsistency in LTL$_{ff}$.

3.2   A Paraconsistent Semantics for LTL$_{ff}$

Our first contribution towards measuring inconsistency in LTL$_{ff}$ is to define an LTL$_{ff}$-variant of the three-valued semantics of [19]. By doing so, we not only develop a means to neatly express inconsistency measures for LTL$_{ff}$, but also define a general applicable paraconsistent semantics for LTL$_{ff}$.

A three-valued interpretation $\hat{\nu}$ for LTL$_{ff}$ is a function mapping each state and proposition to 0, 1 or B, that is, $\hat{\nu} : \{t_0, t_1, \ldots t_m\} \times \mathsf{At} \to \{0, 1, \mathsf{B}\}$ where as before 0 and 1 correspond to the classic logical false and true, respectively, and B (standing for *both*) denotes a conflict. We then assign

$$\hat{\nu}(\phi) = \hat{\nu}(t_0, \phi)$$

where $\hat{\nu}(t_i, \phi)$, for any interpretation $\hat{\nu}$ as above and state $t_i \in \{t_0, \ldots, t_m\}$, is inductively defined as follows:

$$\hat{\nu}(t_i, a) = \hat{\nu}(t_i, a) \text{ for } a \in \mathsf{At}$$

$$\hat{\nu}(t_i, \neg\phi) = \begin{cases} 1 \text{ if } \hat{\nu}(t_i, \phi) = 0 \\ 0 \text{ if } \hat{\nu}(t_i, \phi) = 1 \\ \mathsf{B} \text{ if } \hat{\nu}(t_i, \phi) = \mathsf{B} \end{cases}$$

$$\hat{\nu}(t_i, \varphi_1 \wedge \varphi_2) = \begin{cases} 1 \text{ if } \hat{\nu}(t_i, \varphi_1) = \hat{\nu}(t_i, \varphi_2) = 1 \\ 0 \text{ if } \hat{\nu}(t_i, \varphi_1) = 0 \text{ or } \hat{\nu}(t_i, \varphi_2) = 0 \\ \mathsf{B} \text{ otherwise} \end{cases}$$

$$\hat{\nu}(t_i, \varphi_1 \vee \varphi_2) = \begin{cases} 1 \text{ if } \hat{\nu}(t_i, \varphi_1) = 1 \text{ or } \hat{\nu}(t_i, \varphi_2) = 1 \\ 0 \text{ if } \hat{\nu}(t_i, \varphi_1) = \hat{\nu}(t_i, \varphi_2) = 0 \\ \mathsf{B} \text{ otherwise} \end{cases}$$

$$\hat{\nu}(t_i, \mathbf{X}\varphi) = \begin{cases} \hat{\nu}(t_{i+1}, \varphi) & \text{if } i < m \\ 0 & \text{otherwise} \end{cases}$$

$$\hat{\nu}(t_i, \varphi_1 \mathbf{U}\varphi_2) = \begin{cases} 1 & \text{if there is } j \in \{i+1, \ldots, m\} \text{ with} \\ & \hat{\nu}(t_j, \varphi_2) = \hat{\nu}(t_i, \varphi_1) = \ldots \\ & = \hat{\nu}(t_{j-1}, \varphi_1) = 1 \\ \mathrm{B} & \text{if there is } j \in \{i+1, \ldots, m\} \text{ with} \\ & \{\hat{\nu}(t_j, \varphi_2), \hat{\nu}(t_i, \varphi_1), \ldots, \\ & \hat{\nu}(t_{j-1}, \varphi_1)\} = \{1, \mathrm{B}\} \\ 0 & \text{otherwise} \end{cases}$$

Some comments on the above definition are in order. First, note that the evaluation of the classical Boolean connectives is the same as for propositional three-valued semantics (see Sect. 2.1). Furthermore, the evaluation of $\mathbf{X}\phi$ is simply the truth value of ϕ at the next state, or, if there is no next state, 0 (as for the classical semantics of LTL$_{ff}$). The main new feature, however, is the three-valued evaluation of a formula of the form $\varphi_1 \mathbf{U}\varphi_2$. This formula evaluates to 1 as in the classical case, i.e., if ϕ_2 evaluates to 1 in some future state and ϕ_1 evaluates to 1 in between. We evaluate $\varphi_1 \mathbf{U}\varphi_2$ to B if ϕ_2 evaluates to 1 or B in some future state and ϕ_1 evaluates to 1 or B in between (and at least one of these evaluations must be to B). Finally, $\varphi_1 \mathbf{U}\varphi_2$ evaluates to 0 otherwise, i.e., if either ϕ_2 always evaluates to 0 in the future or in-between φ_1 evaluates at least once to 0.

A three-valued LTL$_{ff}$ interpretation $\hat{\nu}$ satisfies a formula ϕ, denoted by $\hat{\nu} \models^3 \phi$, iff $\hat{\nu}(\phi, t_0) \in \{1, \mathrm{B}\}$. A three-valued interpretation $\hat{\nu}$ satisfies a set of formulas $\mathcal{K}$ iff $\hat{\nu} \models^3 \phi$ for all $\phi \in \mathcal{K}$.

Example 3. Let $\mathsf{At} = \{a, b\}$ and assume $m = 2$. Consider the knowledge base $\mathcal{K}$ defined via

$$\mathcal{K} = \{\mathbf{X}\neg a, a\mathbf{U}b\}$$

and the three-valued interpretation $\hat{\nu}$ defined via

$$\hat{\nu}(t_0, a) = 1 \qquad\qquad \hat{\nu}(t_0, b) = 0$$
$$\hat{\nu}(t_1, a) = \mathrm{B} \qquad\qquad \hat{\nu}(t_1, b) = 0$$
$$\hat{\nu}(t_2, a) = 0 \qquad\qquad \hat{\nu}(t_2, b) = 1$$

Then we have $\hat{\nu}(t_0, a\mathbf{U}b) = \mathrm{B}$ as b evaluates to 1 in t_2 and a evaluates to B in t_1. Moreover, we have $\hat{\nu}(t_0, \mathbf{X}\neg a) = \mathrm{B}$ and therefore $\hat{\nu} \models^3 \mathcal{K}$.

Define $X \models^3 Y$ for formulas X and Y if $\hat{\nu} \models X$ implies $\hat{\nu} \models Y$ for all $\hat{\nu}$.

In the propositional logic case, $\models^3$ is a faithful extension of $\models$, meaning that $\omega \models \phi$ if and only if $\omega \models^3 \phi$ for every two-valued interpretation ω and every ϕ. Our LTL$_{ff}$ extension of the three-valued semantics enjoys the same property (note that every two-valued interpretation is a also a three-valued interpretation that does not use the value B).

Proposition 2. *For every (two-valued) LTL$_{ff}$ interpretation $\hat{\omega}$ and LTL$_{ff}$ formula ϕ, $\hat{\omega} \models \phi$ if and only if $\hat{\omega} \models^3 \phi$.*

The three-valued semantics of [19] has another nice property in propositional logic, namely the non-existence of inconsistency: every propositional formula is trivially satisfiable by the interpretation that assigns B to all propositions. In general, an LTL_{ff} formula may become unsatisfiable w.r.t. to the three-valued semantics if it affects a state "beyond" t_m. However, for other formulas we obtain the following result regarding universal satisfiability.

Proposition 3. *For any LTL_{ff} formula ϕ with $d(\phi) \leq m$ there is $\hat{\nu}$ with $\hat{\nu} \models^3 \phi$.*

The semantics presented in this section allows for inconsistency-tolerant reasoning in LTL_{ff} (and it can straightforwardly be adapted for LTL_f and LTL). This provides a useful tool for the usual application scenarios of temporal logics, such as model checking and verification. While it may be worthwhile to investigate this aspect in more depth, in the remainder of this work we will focus on the application of this semantics for inconsistency measurement and postpone that endeavour to future work.

3.3 Time Sensitive Inconsistency Measures for LTL_{ff}

We will now exploit our three-valued semantics for LTL_{ff} to define some new inconsistency measures. We do this similarly as for propositional logic by assessing the amount of usage of the paraconsistent truth value B in models of an LTL_{ff} knowledge base $\mathcal{K}$ but refine it by two different levels of granularity. This yields two new inconsistency measures.

Our first approach measures the number of states affected by inconsistency. For any three-valued interpretation $\hat{\nu}$, define

$$\mathsf{AffectedStates}(\hat{\nu}) = \{t \mid \exists a : \hat{\nu}(t,a) = B\}$$

In other words, $\mathsf{AffectedStates}(\hat{\nu})$ is the set of states where $\hat{\nu}$ assigns B to at least one proposition. We can define an inconsistency measure by considering those 3-valued models of the knowledge base that affect the minimal number of states.

Definition 4 (LTL time measure). *Let $\mathcal{K}$ be a set of formulas. Then, the LTL time measure is defined via*

$$\mathcal{I}_d^{LTL}(\mathcal{K}) = \min_{\hat{\nu} \models^3 \mathcal{K}} |\mathsf{AffectedStates}(\hat{\nu})|$$

if there is $\hat{\nu}$ with $\hat{\nu} \models^3 \mathcal{K}$ and $\mathcal{I}_d^{LTL}(\mathcal{K}) = \infty$ otherwise.

This measure counts the number of states for which the knowledge base is inconsistent. It is, in fact, the extension of the drastic measure, $\mathcal{I}_d$, in that for each state it adds 1 if there is an inconsistency and 0 otherwise. This measure can be used to distinguish the knowledge bases $\mathcal{K}_1$ and $\mathcal{K}_2$, i.e., it is time sensitive.

Example 4. We recall the knowledge bases $\mathcal{K}_1 = \{\mathbf{X}a, \mathbf{X}\neg a\}$ and $\mathcal{K}_2 = \{\mathbf{G}a, \mathbf{G}\neg a\}$. Then we have

$$\mathcal{I}_d^{LTL}(\mathcal{K}_1) = 1 \qquad\qquad \mathcal{I}_d^{LTL}(\mathcal{K}_2) = m$$

As an example where there is no $\hat{\nu}$ s.t. $\hat{\nu} \models^3 \mathcal{K}$, consider the formula $\mathbf{XXX}a$. This formula cannot be satisfied for $m = 2$, so $\mathcal{I}_d^{LTL}$ would return ∞ here.

Example 4 shows that the proposed measure $\mathcal{I}_d^{LTL}$ can already provide meaningful insights for measuring inconsistency in LTL. But a potential limitation is that it can only distinguish inconsistency in individual states in a binary manner. For example, $\mathcal{I}_d^{LTL}$ cannot distinguish the knowledge base $\mathcal{K}_4 = \{\mathbf{X}a, \mathbf{X}\neg a, \mathbf{X}b, \mathbf{X}\neg b\}$ from $\mathcal{K}_1$ because all inconsistencies occur at one state, namely t_1. For this reason we believe it is useful to be able to look inside states for inconsistency. In order to do so, given a three-valued interpretation $\hat{\nu}$, define

$$\mathsf{Conflictbase}(\hat{\nu}) = \{(t, a) \mid \hat{\nu}(t, a) = \mathrm{B}\}$$

Then, define the LTL contension measure as follows.

Definition 5 (LTL contension measure). *Let $\mathcal{K}$ be a set of formulas and*

$$\mathcal{I}_c^{LTL}(\mathcal{K}) = \min_{\hat{\nu} \models^3 \mathcal{K}} |\mathsf{Conflictbase}(\hat{\nu})|$$

if there is $\hat{\nu}$ with $\hat{\nu} \models^3 \mathcal{K}$ and $\mathcal{I}_c^{LTL}(\mathcal{K}) = \infty$ otherwise.

$\mathcal{I}_c^{LTL}$ seeks an interpretation that assigns B to a minimal number of propositions individually over all the states and uses this number for the inconsistency measure. This is an extension of $\mathcal{I}_d^{LTL}$, and for that matter, of $\mathcal{I}_c$ as it calculates $\mathcal{I}_c^{LTL}$ for each state and sums the numbers obtained this way.

Example 5. We recall the knowledge bases $\mathcal{K}_1 = \{\mathbf{X}a, \mathbf{X}\neg a\}$, $\mathcal{K}_4 = \{\mathbf{X}a, \mathbf{X}\neg a, \mathbf{X}b, \mathbf{X}\neg b\}$, and consider $\mathcal{K}_5 = \{\mathbf{G}a, \mathbf{G}\neg a, \mathbf{G}b, \mathbf{G}\neg b\}$. If $m = 3$, then we have

$$\mathcal{I}_d^{LTL}(\mathcal{K}_1) = 1 \qquad \mathcal{I}_d^{LTL}(\mathcal{K}_4) = 1 \qquad \mathcal{I}_d^{LTL}(\mathcal{K}_5) = 3$$
$$\mathcal{I}_c^{LTL}(\mathcal{K}_1) = 1 \qquad \mathcal{I}_c^{LTL}(\mathcal{K}_4) = 2 \qquad \mathcal{I}_c^{LTL}(\mathcal{K}_5) = 6$$

As can be seen in Example 5, the two inconsistency measures proposed in this work can, contrary to previously existing measures, be used to provide meaningful insights into inconsistency in linear temporal logic, i.e., they are in fact time sensitive. As the two measures have a different granularity in regard to time, selecting which of the two to use depends on the intended use case.

Intuitively, it would be possible to devise further time-sensitive inconsistency measures for LTL$_{ff}$. We will however leave this discussion for future work. Importantly, the aim of this paper is to show that traditional inconsistency measures cannot be plausibly applied to temporal logics, and to present means for time sensitive inconsistency measurement. In this regard, the measures proposed in this work can be used as a baseline for measuring inconsistency in LTL. Also, they (broadly) satisfy other desirable properties and can therefore be seen as strictly better (w.r.t. the considered postulates) than their propositional logic "counterpart", i.e., $\mathcal{I}_d$ for $\mathcal{I}_d^{LTL}$, respectively $\mathcal{I}_c$ for $\mathcal{I}_c^{LTL}$. The results of this section are summarized in Table 1. Proofs can be found online.

Note that only the measures we introduced satisfy TS. Note also that $\mathcal{I}_c^{LTL}$ does not satisfy IN due to the problem of iceberg inconsistencies, cf. the provided proofs.

Table 1. Compliance of inconsistency measures with rationality postulates.

$\mathcal{I}$	CO	MO	IN	DO	TS
$\mathcal{I}_d$	✓	✓	✓	✓	✗
$\mathcal{I}_{\text{MI}}$	✓	✓	✓	✗	✗
$\mathcal{I}_p$	✓	✓	✓	✗	✗
$\mathcal{I}_r$	✓	✓	✓	✗	✗
$\mathcal{I}_c$	✓	✓	✗	✓	✗
$\mathcal{I}_{at}$	✓	✗	✗	✗	✗
$\mathcal{I}_d^{LTL}$	✓	✓	✓	✓	✓
$\mathcal{I}_c^{LTL}$	✓	✓	✗	✓	✓

4   Application to Declarative Process Models

A common application scenario for LTL_f is that of declarative process models [16], which are sets of (LTL_f-based) constraints. For such declarative process models, the issue of inconsistency is equally as problematic, as any inconsistencies between the constraints make the declarative process model unsatisfiable.

There have been a number of works addressing the issue of inconsistency in declarative process models [2,3,14]. However, those works mainly look at whether a process model is inconsistent at all (in a binary manner), or try to identify sets of inconsistent constraints. Those works can however not look "into" those sets or assess their severity. For this use case, our proposed approach can be extended to declarative process models as follows.

4.1   Inconsistency Measurement in Declarative Process Models

A declarative process model consists of a set of constraints. Typically, these constraints are constructed using predefined templates, i. e., predicates, that are specified relative to a set of propositions (e. g., company activities).

Definition 6 (Declarative Process Model). *A declarative process model is a tuple $M = (A, T, C)$, where A is a set of propositions, T is a set of constraint types, and C is the set of constraints, which instantiate the template elements in T with activities in A.*[3]

In this work, we consider the declarative modelling language Declare [16], which offers a set of "standard" templates. We will use a selection of templates shown in Table 2. We refer the reader to [3] for an overview of other Declare template types and corresponding semantics.

---

[3] For readability, we will denote declarative process models as a set of constraints (**C**).

Table 2. LTL$_{ff}$ semantics for a selection of declare templates.

Template	LTL$_{ff}$ Semantics
Init(a)	a
End(a)	$\mathbf{G}(a \vee \mathbf{F}a)$
Response(a,b)	$\mathbf{G}(a \rightarrow \mathbf{F}b)$
NotResponse(a,b)	$\mathbf{G}(a \rightarrow \neg\mathbf{F}b)$
ChainResponse(a,b)	$\mathbf{G}(a \rightarrow \mathbf{X}b)$
NotChainResponse(a,b)	$\mathbf{G}(a \rightarrow \neg\mathbf{X}b)$
AtLeast(a,n)	$\mathbf{F}(a \wedge \mathbf{X}(\text{atLeast}(a, n-1))), \text{atLeast}(a,1) = a \vee \mathbf{F}(a)$
AtMost(a,n)	$\mathbf{G}(\neg a \vee \mathbf{X}(\text{atMost}(a, n-1))), \text{atMost}(a,0) = \mathbf{G}(\neg a)$

By rewriting the constraints of a declarative process model into LTL$_{ff}$ formulas, our approach for measuring inconsistency in LTL$_{ff}$ can be applied to Declare in a straightforward manner.

Example 6. Consider the sets of constraints C_a and C_b, defined via

$$C_a = \{\text{INIT}(a), \text{RESPONSE}(a,b), \text{NOTRESPONSE}(a,b)\}$$
$$(\Leftrightarrow \{a, \mathbf{G}(a \rightarrow \mathbf{F}b), \mathbf{G}(a \rightarrow \neg\mathbf{F}b)\})$$
$$C_b = \{\text{INIT}(a), \text{RESPONSE}(a,b), \text{NOTRESPONSE}(a,b),$$
$$\text{RESPONSE}(a,c), \text{NOTRESPONSE}(a,c)\}$$
$$(\Leftrightarrow \{a, \mathbf{G}(a \rightarrow \mathbf{F}b), \mathbf{G}(a \rightarrow \neg\mathbf{F}b), \mathbf{G}(a \rightarrow \mathbf{F}c), \mathbf{G}(a \rightarrow \neg\mathbf{F}c)\})$$

then we have that $\mathcal{I}_c^{LTL}(C_a) = 1$ and $\mathcal{I}_c^{LTL}(C_b) = 2$.

Due to the recursive definition of some "existence" constraints (cf. Table 2), note that also inconsistencies concerned with cardinalities can be assessed correctly.

Example 7. Consider $C_c = \{\text{ATMOST}(a,1), \text{ATLEAST}(a,2)\}$ and $C_d = \{\text{ATMOST}(a,1), \text{ATLEAST}(a,100)\}$, then $\mathcal{I}_d^{LTL}(C_c) < \mathcal{I}_d^{LTL}(C_d)$.

As a border case, note that any inconsistency referring to a point in time beyond the assumed sequence of states will return a value of ∞ per definition, as we cannot assess any error that leaves the boundaries of our logical framework.

Example 8. Let $C_e = \{\text{END}(a), \text{CHAINRESPONSE}(a,b)\}$, then $\mathcal{I}_d^{LTL}(C_e) = \infty$.

These examples show that our approach can provide detailed insights on the severity of inconsistency in declarative process models. Such insights can prove useful for prioritizing or re-modelling different issues of the process specification. In this context, it seems intuitive that conflicts affecting only the next state ($\mathbf{X}$) should be considered as less severe than conflicts affecting multiple following states ($\mathbf{G}$), i.e., for any LTL$_{ff}$ formula φ, $\mathcal{I}(\{\mathbf{G}\varphi, \mathbf{G}\neg\varphi\}) > \mathcal{I}(\{\mathbf{X}\varphi, \mathbf{X}\neg\varphi\})$. In this regard, there are still open questions on how to distinguish the operators $\mathbf{X}$ and $\mathbf{F}$, in particular: for an LTL$_{ff}$ formula φ, what is the relation between $\mathcal{I}(\{\mathbf{X}\varphi, \neg\mathbf{X}\varphi\})$ and $\mathcal{I}(\{\mathbf{F}\varphi, \neg\mathbf{F}\varphi\})$? We address this question in the following.

C. Corea et al.

4.2 On Potentially Inconsistent States

Consider the following sets of constraints C_m and C_n, defined via

$C_m =$
$\{\text{INIT}(a) \Leftrightarrow a,$
$\text{RESPONSE}(a,b) \Leftrightarrow \mathbf{G}(a \rightarrow \mathbf{F}b),$
$\text{NOTRESPONSE}(a,b) \Leftrightarrow \mathbf{G}(a \rightarrow \neg\mathbf{F}b)\}$

$C_n =$
$\{\text{INIT}(a),$
$\text{CHAINRESPONSE}(a,b) \Leftrightarrow \mathbf{G}(a \rightarrow \mathbf{X}b),$
$\text{NOTCHAINRESPONSE}(a,b) \Leftrightarrow \mathbf{G}(a \rightarrow \neg\mathbf{X}b)\}$

Both sets are inconsistent, as they demand that b should and should not follow. However, the point in time at which the actual inconsistency can occur is different. Naturally, one question arises: which inconsistency is more severe? Or are they equally severe? We encourage the reader to come up with an own answer to this question at this point before we continue with our view on this matter.

Using the measures introduced in this work, the absolute number of affected states is 1 in both cases. So regarding the minimal number of affected states, the inconsistencies are equally severe. However, the certainty of where the inconsistency can occur at is clearly different, as visualized in Fig. 1.

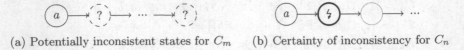

(a) Potentially inconsistent states for C_m (b) Certainty of inconsistency for C_n

Fig. 1. Visualization of the (un)certainty of where the inconsistency may occur for C_m and C_n.

In C_m, there are m different possible states to which a minimal interpretation could assign the truth value B to the proposition b, whereas the inconsistency can only occur in exactly 1 state for C_n. This could entail different severities for the inconsistencies, depending on the viewpoint:

Consider a running process which is in state t_0. For C_m, it is unclear when the inconsistency will occur. For C_n, it is directly known that the next state is inconsistent. Recovery mechanisms for such cases are well known [13], e.g., it would be possible to just skip the next state and continue with a consistent process. This is not possible for C_m without skipping all following states until the end of the process. So one might argue that the inconsistency in C_m is more severe. However, for C_n, this also means there is in fact no possible continuation as the process is in a dead-end state, thus, C_n needs to be attended to more urgently (So one might as well argue that the inconsistency in C_n is more severe).

In the field of inconsistency measurement, the dominance property states that substituting a consistent formula by a weaker formula cannot increase the inconsistency value [10]. However, when moving from C_m to C_n or vice-versa, we both replace one constraint with a stronger one and the other with a weaker one (every CHAINRESPONSE is also a RESPONSE but every NOTRESPONSE is a NOTCHAINRESPONSE). So the dominance property is not applicable here and the question remains which inconsistency is more severe. In this work, we will not give a definitive answer to this question and leave this discussion for future work. However, based on the two possible views given above, we will argue that they are, in fact, different. It would therefore be desirable to be able to distinguish the inconsistency in C_m and C_n. Here, the introduced contension concept can be adapted to quantify the certainty of *when* the inconsistency will occur.

The introduced measures quantify inconsistency by seeking an interpretation that assigns B to a minimal number of states. We denote the set of all such interpretations that assign B to a minimal number of states (at least to one) as

$$\hat{V}_{min}^{B>0}(\mathcal{K}) = \{\hat{\nu} \models^3 \mathcal{K} : |\mathsf{AffectedStates}(\hat{\nu})| > 0 \wedge |\mathsf{AffectedStates}(\hat{\nu})| = \mathcal{I}_d^{LTL}(\mathcal{K})\}$$

Every such (minimal) interpretation also encodes which exact states are affected by the inconsistency. For C_n, only one state is necessarily affected (cf. Fig 1 (b)), thus, there exists only one minimal interpretation. For C_m, there are m different interpretations that are all equally minimal in terms of how many states are affected. So the number of minimal interpretations relates to the number of distinct (sets of) states that can potentially be affected.

Definition 7 (Number of Minimal Interpretations). *Let $\mathcal{K}$ be a set of formulas. Then, define the number of minimal interpretations via*

$$\#minInterpretations(\mathcal{K}) = |\hat{V}_{min}^{B>0}(\mathcal{K})|$$

Example 9. We recall C_m and C_n. Then we have that $\#minInterpretations(C_m) = m$ and $\#minInterpretations(C_n) = 1$ as expected (cf. the above discussion)

Importantly, the function $\#minInterpretations$ is not an inconsistency measure, i.e., a higher value does not indicate a higher degree of inconsistency. It therefore also does not matter where the inconsistency in C_m eventually triggers. The value merely expresses the "certainty" of knowing where the conflict can occur at. The semantics of which is worse depends on the use case.

5   Computational Complexity

We conclude with an investigation of computational complexity in measuring inconsistency in LTL$_{ff}$. We assume familiarity with computational complexity, see [15] for an introduction. Proofs can be found online.

Note that deciding satisfiability is PSPACE-complete for LTL$_f$ [5] and also intractable for many variants of LTL$_f$ [4]. For our variant LTL$_{ff}$, as m is fixed, we get NP-completeness (think for example of a non-deterministic algorithm that guesses $\hat{\omega}$ and verifies (in polynomial time) that $\hat{\omega} \models \phi$).

Theorem 1. *Deciding whether a formula ϕ is satisfiable in LTL$_{ff}$ is NP-complete.*

If the parameter m is given in unary, the complexity result holds as it is. However, if m is given in binary then the complexity will likely increase (in the membership proof, we need to guess an interpretation and if m is given in binary, that interpretation may be exponential in the size of the input).

We continue with an investigation of the computational complexity of measuring inconsistency in LTL_{ff}. For this, let $\mathbb{L}$ denote the set of all LTL_{ff} knowledge bases. Following [23], we consider the following computational problems:

EXACT$_{\mathcal{I}}$ **Input:** $\mathcal{K} \in \mathbb{L}$, $x \in \mathbb{R}_{\geq 0}^{\infty}$
         Output: TRUE iff $\mathcal{I}(\mathcal{K}) = x$

UPPER$_{\mathcal{I}}$ **Input:** $\mathcal{K} \in \mathbb{L}$, $x \in \mathbb{R}_{\geq 0}^{\infty}$
         Output: TRUE iff $\mathcal{I}(\mathcal{K}) \leq x$

LOWER$_{\mathcal{I}}$ **Input:** $\mathcal{K} \in \mathbb{L}$, $x \in \mathbb{R}_{\geq 0}^{\infty} \setminus \{0\}$
         Output: TRUE iff $\mathcal{I}(\mathcal{K}) \geq x$

VALUE$_{\mathcal{I}}$ **Input:** $\mathcal{K} \in \mathbb{L}$
         Output: The value of $\mathcal{I}(\mathcal{K})$

For UPPER$_{\mathcal{I}}$, the same general non-deterministic algorithm can be applied.

Theorem 2. UPPER$_{\mathcal{I}_d^{LTL}}$ and UPPER$_{\mathcal{I}_c^{LTL}}$ are NP-complete.

Using the results in [23] we also get the following results for the other problems.

Corollary 1. LOWER$_{\mathcal{I}_d^{LTL}}$ and LOWER$_{\mathcal{I}_c^{LTL}}$ are coNP-complete. EXACT$_{\mathcal{I}_d^{LTL}}$ and EXACT$_{\mathcal{I}_c^{LTL}}$ are in DP. VALUE$_{\mathcal{I}_d^{LTL}}$ and VALUE$_{\mathcal{I}_c^{LTL}}$ are in FP$^{NP[\log n]}$.

In regard to the algorithmic implementation of our approach, a general approach of SAT encodings can be used. Corollary 1 gives a straightforward implementation for an algorithm to compute the measures by combining binary search with iterative calls to a SAT solver using an encoding of the problem Upper (see proof of Corollary 1). This encoding would be based on a SAT encoding for LTL_{ff} satisfiability, which is straightforward.

6   Conclusion

In this work, we have presented an approach for measuring the severity of inconsistencies in declarative process specifications, in particular those based on linear temporal logic. In this regard, we introduced a paraconsistent semantics for LTL_{ff} and developed two inconsistency measures. This provides useful insights for debugging or re-modelling declarative specifications, e.g., by allowing to compare or prioritize different inconsistencies. Here, our approach extends recent works [2,3,20] on the identification of inconsistent sets in declarative process specifications by allowing a look "into" those sets.

In future work, we aim to investigate the application of our approach to other languages such as GSM or DCR. Note that this is however not trivial, as the process models there might not be represented as orthogonal formulas. As a further limitation of our work, the current approach treats time as discrete time steps where any number of activities (within the bounds of the constraints) are allowed to occur at the same time. Real processes may however contain activities that take real time and may not be parallelizable because of resource constraints. As a result, a logically equivalent inconsistency may weigh more than another. In future work, we aim to address this issue with data-aware versions of LTL$_{ff}$.

References

1. Cecconi, A., De Giacomo, G., Di Ciccio, C., Maggi, F.M., Mendling, J.: A temporal logic-based measurement framework for process mining. In: Proceedings of the 2nd ICPM, pp. 113–120. IEEE (2020)
2. Corea, C., Nagel, S., Mendling, J., Delfmann, P.: Interactive and minimal repair of declarative process models. In: Polyvyanyy, A., Wynn, M.T., Van Looy, A., Reichert, M. (eds.) BPM 2021. LNBIP, vol. 427, pp. 3–19. Springer, Cham (2021). https://doi.org/10.1007/978-3-030-85440-9_1
3. Di Ciccio, C., Maggi, F.M., Montali, M., Mendling, J.: Resolving inconsistencies and redundancies in declarative process models. Inf. Syst. **64**, 425–446 (2017)
4. Fionda, V., Greco, G.: The complexity of LTL on finite traces: hard and easy fragments. In: Proceedings of the 30th AAAI Conference on AI, Phoenix, pp. 971–977. AAAI (2016)
5. Giacomo, G.D., Vardi, M.Y.: Linear temporal logic and linear dynamic logic on finite traces. In: Proceedings of the 23rd IJCAI, Beijing, pp. 854–860. AAAI (2013)
6. Grant, J.: Measuring inconsistency in some branching time logics. J. Appl. Non-Classic. Logics **31**, 85–107 (2021)
7. Grant, J., Hunter, A.: Measuring consistency gain and information loss in stepwise inconsistency resolution. In: Liu, W. (ed.) ECSQARU 2011. LNCS (LNAI), vol. 6717, pp. 362–373. Springer, Heidelberg (2011). https://doi.org/10.1007/978-3-642-22152-1_31
8. Grant, J., Martinez, M.V.: Measuring Inconsistency in Information. College Publication (2018)
9. Hildebrandt, T., Mukkamala, R.R., Slaats, T., Zanitti, F.: Contracts for cross-organizational workflows as timed dynamic condition response graphs. J. Logic Algebr. Program. **82**(5–7), 164–185 (2013)
10. Hunter, A., et al.: Shapley inconsistency values. KR **6**, 249–259 (2006)
11. Hunter, A., et al.: Measuring inconsistency through minimal inconsistent sets. KR **8**, 358–366 (2008)
12. Knight, K.: Measuring inconsistency. J. Philos. Logic **31**(1), 77–98 (2002)
13. Ly, L.T., Maggi, F.M., Montali, M., Rinderle-Ma, S., van der Aalst, W.M.: A framework for the systematic comparison and evaluation of compliance monitoring approaches. In: 17th IEEE EDOC, Vancouver, pp. 7–16. IEEE (2013)
14. Maggi, F.M., Westergaard, M., Montali, M., van der Aalst, W.M.P.: Runtime verification of LTL-based declarative process models. In: Khurshid, S., Sen, K. (eds.) RV 2011. Runtime verification of ltl-based declarative process models, vol. 7186, pp. 131–146. Springer, Heidelberg (2012). https://doi.org/10.1007/978-3-642-29860-8_11
15. Papadimitriou, C.: Computational Complexity. Addison-Wesley, Boston (1994)

16. Pesic, M., Schonenberg, H., Van der Aalst, W.M.: Declare: full support for loosely-structured processes. In: 11th IEEE EDOC, Annapolis, pp. 287–287. IEEE (2007)

17. Pill, I., Quaritsch, T.: Behavioral diagnosis of LTL specifications at operator level. In: 23rd International Joint Conference on Artificial Intelligence. Citeseer (2013)

18. Pnueli, A.: The temporal logic of programs. In: 18th Symposium on Foundations of Computer Science, Rhode Island, pp. 46–57. IEEE Computer Society (1977)

19. Priest, G.: Logic of Paradox. J. Philos. Logic **8**, 219–241 (1979)

20. Roveri, M., Di Ciccio, C., Di Francescomarino, C., Ghidini, C.: Computing unsatisfiable cores for LTLf specifications (Preprint). arXiv (2022)

21. Solomakhin, D., Montali, M., Tessaris, S., De Masellis, R.: Verification of artifact-centric systems: decidability and modeling issues. In: Basu, S., Pautasso, C., Zhang, L., Fu, X. (eds.) ICSOC 2013. LNCS, vol. 8274, pp. 252–266. Springer, Heidelberg (2013). https://doi.org/10.1007/978-3-642-45005-1_18

22. Thimm, M.: Inconsistency measurement. In: Ben Amor, N., Quost, B., Theobald, M. (eds.) SUM 2019. LNCS (LNAI), vol. 11940, pp. 9–23. Springer, Cham (2019). https://doi.org/10.1007/978-3-030-35514-2_2

23. Thimm, M., Wallner, J.P.: On the complexity of inconsistency measuring. AI. **275**, 411–456 (2019)

24. Vardi, M.Y.: Branching vs. linear time: final showdown. In: Margaria, T., Yi, W. (eds.) TACAS 2001. LNCS, vol. 2031, pp. 1–22. Springer, Heidelberg (2001). https://doi.org/10.1007/3-540-45319-9_1

Understanding and Decomposing Control-Flow Loops in Business Process Models

Thomas M. Prinz[1] (ID), Yongsun Choi[2(✉)] (ID), and N. Long Ha[3] (ID)

[1] Course Evaluation Service, Friedrich Schiller University Jena, Jena, Germany
Thomas.Prinz@uni-jena.de
[2] Department of Industrial and Management Engineering, Inje University,
Gimhae, Republic of Korea
yschoi@inje.edu
[3] Faculty of Economic Information Systems, University of Economics,
Hue University, Hue, Vietnam
hnlong@hueuni.edu.vn

Abstract. Business process models are usually described in a visual notation and reflect actual processes in systems. As a result, process models often are unstructured and cyclic. Unfortunately, unstructured and cyclic models are difficult to analyze and execute as research shows. Unstructuredness could be overcome using the existing studies, however, the analysis of cyclic models is still an open research problem. For this reason, this paper presents a decomposition of cyclic process models into sets of acyclic models. Together with a simple execution semantics for the acyclic models, the semantics of the decomposed model coincides with the original model if soundness is assumed. The decomposition can be achieved in a quadratic runtime complexity and gives the possibility to apply many existing analysis methodologies for acyclic process models. A short evaluation shows the feasibility of the approach.

Keywords: Process models · Loops · Decomposition · Soundness

1 Introduction

Business Process Management (BPM) examines how different operations (tasks) interact to achieve business goals [7]. This interaction is usually recorded in the form of *business process models* in special notations (like the *Business Process Model and Notation* (BPMN) [19]). Most of these process notations are visual (e.g., BPMN and Event-driven Process Chains (EPC) [14]) and the process models result from actual operational processes in systems and organizations. Therefore, such models may be unstructured and contain non-explicit *loops*, i.e., loops that result from the control-flow rather than from specific looping tasks.

The third author's work was done during his PhD program at Inje University.

C. Di Ciccio et al. (Eds.): BPM 2022, LNCS 13420, pp. 307–323, 2022.
https://doi.org/10.1007/978-3-031-16103-2_21

Unstructuredness and loops make the execution and analysis of process models difficult [2,3,8,16,17,20,22,23,30,31]. Polyvyanyy summarizes that well-structured process models are more comprehensible for humans, are more likely to contain fewer errors, and, therefore, improve their quality [20]. Arbitrary loops tend to increase the probability of errors in process models [17] and prevent the structuring of process models or at least increase the effort [20]. The *(Refined) Process Structure Tree* (RPST) describes a hierarchy of single entry and single exit (SESE) structures [2,23,30,31] that is often used to find independent structures in unstructured process models [31] and to speed up analysis [6,11], but do not help to solve the problem of *unstructured and cyclic* process components [3]. It appears that loops make the execution and analysis of process models particularly difficult. There are many efficient and simple approaches to execute and analyze processes without loops (*acyclic* processes). For example, inclusive converging gateways (OR-joins) [32] and *Dead-Path Elimination* (DPE) for WS-BPEL [18] are easy to apply and understand for acyclic process models. *Soundness* analyses [10,26] and the derivation of *behavioral relations* [11] are examples of how analyses in acyclic processes are efficient and simple. However, most of these approaches cannot be applied to cyclic processes or become more complicated to understand, implement, and prove.

Due to a large number of known state-of-the-art approaches and simpler execution semantics (e.g., for OR-joins) for acyclic process models, it would be beneficial for research and practice if there would be a transformation from a cyclic model to an acyclic model with the same execution behavior. One possibility is to transfer a process model into an RPST [2,23,30,31]. Since the RPST is a tree, it is acyclic. However, the RPST provides less information if loops are inside inherently unstructured fragments (so-called *rigids*) [20] and can, therefore, not be used in such cases. *Untanglings* [22] and *unfoldings* [8,16,20] of process models are another way of making process models acyclic. However, they use completely different representations of the original process and are not applicable to process models with OR-joins. For this reason, there is the need for a new approach to decompose an arbitrary process model with OR-joins into a set of acyclic process models while retaining its execution behavior. This paper introduces such an approach. Our approach follows ideas of Choi et al. [3], but it is applicable to process models with OR-joins and does not need to distinguish between *natural* and *irreducible* loops. A *natural* loop can only be entered at one node, while an *irreducible* loop can be entered at multiple nodes. In many situations, irreducible loops are more difficult to handle [3]. Our approach detects all loops recursively and creates acyclic versions of them. Finally, the loops in the original process are reduced to acyclic subgraphs combined with *"looping nodes"* being available in most modeling languages. If the original model is *sound*, this paper shows that the resulting acyclic processes have the same behavior as the original process. Although unsound processes exist in practice, Van Dongen et al. state that a process model should at least fulfill soundness as correctness criterion [5]. From a quality perspective, soundness is, thus, a weak constraint.

The decomposition proposed in this paper gives the following non-final improvements for research and practice regarding the state of the art: (a) It is applicable to any cyclic process model with natural and irreducible as well as nested loops. (b) The decomposition can be applied to process models with OR-splits and OR-joins. (c) Process models with complicated loop behavior should become easier to understand for humans, because they are decomposed into smaller, acyclic models. (d) A cyclic process model only needs to be transformed once. The results are again process models that can be stored and executed or analyzed as needed. (e) The execution of process models becomes trivial even with OR-joins. Actually, the kind of converging gateway becomes useless, as all converging gateways can be replaced by OR-joins. (f) Process models can be executed on any execution engine that supports acyclic processes with (sub)process calls. (g) Structuring and analyzing acyclic process models is easier and, above all, more efficient than with cyclic process models as explained before.

In summary, the contributions of this paper are: (1) Study of loops and how they can be generalized. (2) Decomposition of cyclic process models into sets of acyclic models in general process models. (2) Introduction of a semantics that executes resulting acyclic processes in the correct order to achieve the business goals of the original model (same execution semantics). (3) A short evaluation of loop decomposition regarding complexity and feasibility as well as two example applications: OR-join semantics and soundness analysis.

The rest of this paper is structured as follows: Sect. 2 introduces necessary definitions of process models, loops, and their semantics. This section is followed by Sect. 3 on related work. In Sect. 4, we describe our findings on loop structures in process models followed by a detailed description of their decomposition in Sect. 5. Section 6 describes how a decomposed process model can be executed. The complete decomposition algorithm is presented in Sect. 7. An evaluation of our loop decomposition as well as two example applications are shown in Sect. 8. Finally, Sect. 9 gives some suggestions for future research directions.

2   Preliminaries

The following section introduces notions being important to understand this work, in particular, *workflow graphs*, *loops*, *token games*, and *soundness*.

2.1   Workflow Graphs and Loops

In order to abstract process models from rich representations such as BPMN [19] and EPC [14], this paper uses *workflow graphs* [26,32]:

Definition 1 (Workflow Graph). A workflow graph $WFG = (N, E, \lambda, L)$ is a connected, directed graph (N, E). N is a set of nodes and $E \subseteq N \times N$ is a set of edges which defines the execution order between nodes. L is a set of labels {*Start, Task, AND, OR, XOR, End*} and $\lambda\colon N \mapsto L$ is a total mapping that assigns to each node a label. Depending on its label, a node has different properties:

- Nodes with label *Start* (the *start nodes*) have no incoming but exactly one outgoing edge. Nodes with label *End* (the *end nodes*) have exactly one incoming but no outgoing edge. There is at least one start and one end node.

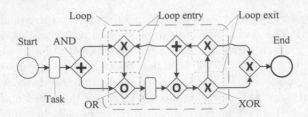

Fig. 1. A workflow graph with an irreducible loop with its loop entries and exits.

- Each node lies on a path from a start to an end node.
- Nodes with label *Task* have exactly one incoming and one outgoing edge.
- All other nodes have labels *AND*, *OR*, and *XOR*. They are divided into *split* and *join nodes*. Split nodes (i.e., AND-split, OR-split, and XOR-split) have exactly one incoming and at least two outgoing edges. Join nodes (i.e., AND-join, OR-join, and XOR-join) have at least two incoming edges and exactly one outgoing edge.

For the visualization of workflow graphs, we use notions of BPMN [19] as shown in the workflow graph in Fig. 1. Start and end nodes are represented as circles, with end nodes having thicker lines. Tasks are represented as rounded rectangles. Split and join nodes are depicted with diamonds. AND-splits and AND-joins have + signs, XOR-splits and XOR-joins have a cross, and OR-splits and OR-joins have circles in their diamonds. Later in this paper, we include special *loop nodes* (represented as rectangles with a circular arrow) in workflow graphs, informally extending Definition 1. They have at least one incoming edge and at least one outgoing edge. Details are explained during their introduction.

Loops in workflow graphs are special subgraphs:

Definition 2 (Loop/Cycle). Let $WFG = (N, E, \lambda, L)$ be a workflow graph.

A *loop* is a strongly connected component (SCC) [4] $\mathcal{L} = (N_{\mathcal{L}}, E_{\mathcal{L}})$ of WFG containing at least two nodes, i.e., each node of $\mathcal{L}$ has a path to each other node of $\mathcal{L}$ [1]. If WFG contains any loop, it is called *cyclic*. Otherwise, it is *acyclic*.

Usually, loops in workflow graphs contain at least one task node. It is important to note that a SCC is *maximal* by definition [4]. Consequently, our loops do *not* contain subloops in the classical sense. However, the approach presented in this paper later shows that *it identifies loops within loops if necessary*. A similar approach was taken by Steensgaard [28]. He also generalized the definitions of *loop entries* and *exits*:

Definition 3 (Loop Entries and Exits). Let WFG be a cyclic workflow graph with $\mathcal{L} = (N_{\mathcal{L}}, E_{\mathcal{L}})$ is one of its loops.

Loop Entry. All nodes of $\mathcal{L}$ that have at least one incoming edge from outside $\mathcal{L}$ are called *loop entries* and these incoming edges are *loop-entry edges*.

Loop Exit. All nodes of $\mathcal{L}$ that have at least one outgoing edge to a node outside the loop are called *loop exits* and these outgoing edges are *loop-exit edges*.

Remark 1. Each loop entry of a loop in a workflow graph is of course a join node and each loop exit of a loop is of course a split node.

Figure 1 shows a loop (gray large dashed rounded rectangle) in our example workflow graph and its loop entries (small gray rounded rectangles on the left) and loop exits (small gray rounded rectangles on the right).

Depending on the number of loop entries, two types of loops are distinguished in loop research: *natural* and *irreducible loops*. Although natural loops are more intuitive, there is no need to distinguish between the two in this paper. Finally, it is also important to note that loop-exit edges are not part of any loop:

Corollary 1. Since a loop is a maximal SCC by Definition 2, loop-exit edges cannot be part of any loop.

2.2 Semantics

The semantics of cyclic workflow graphs with OR-joins is not trivial [32]. The reason is that situations can arise where two OR-joins mutually wait for each other [15, 32]. In this paper, we refer to the semantics of our previous work [24] that is complete for *sound* workflow graphs. However, *we emphasize that an additional OR-join semantics for cyclic workflow graphs is not needed after decomposition and is not required in detail in the proofs.* Therefore, it is only used here for the sake of completeness. We use a *token game* semantics in the following describing state transitions in a workflow graph $WFG = (N, E, \lambda, L)$.

A *state S* of *WFG* is a total mapping from the set of edges E to the set of natural numbers, $S: E \mapsto \mathbb{N}_0$. It describes the number of *tokens* on each edge, e.g., $S(e) = 1$ means that edge e in state S carries 1 token. An *initial state* of *WFG* is a state in only one outgoing edge of exactly one start node has a token. Every other edge has 0 tokens.

A node n of *WFG* is *waiting* in a state S if at least one incoming edge of n has a token. If n is neither an AND-join nor an OR-join, n is *enabled* if it is waiting in S. If n is an AND-join and all incoming edges of n carry a token in S, n is *enabled* in S. If n is an OR-join, then it has a waiting area $\omega(n)$ that contains all edges where n must wait for their tokens (for more details, please take a look on Prinz and Amme [24]). n is enabled in S, if it is waiting in S and no token is in n's waiting area except on n's incoming edges. If there is an enabled node n in S, then S can change into a state S' by executing n, written $S \xrightarrow{n} S'$. The resulting state S' is based on S with the following modifications: (1) Each incoming edge *in* of n with at least one token loses a token in S', except if n is an XOR-join, then only one incoming edge loses a token. (2) The number of tokens on n's outgoing edges depends on n's type. If n is an OR-split, then a non-empty set of outgoing edges of n gets an extra token in S'. If n is an XOR-split, then exactly one outgoing edge of n gets an extra token in S'. Otherwise, each outgoing edge of n gets an extra token in S'.

The execution of a workflow graph starts with an initial state, and is executed node by node, resulting in a chain of node executions and state transitions.

A state S' is *directly reachable* from a state S, depicted $S \to S'$, if there is a possible state transition $S \overset{n}{\to} S'$, i.e., node $n \in N$ is executed in state S. S' is *reachable* from a state S, depicted $S \to^* S'$, if there is a sequence/chain of directly reachable states $S_1 \to S_2 \to \ldots \to S_k$, $k \geq 1$, with $S_1 = S$ and $S_k = S'$.

2.3 Soundness

Two types of structural conflicts can occur in workflow graphs, namely *deadlocks* and *lacks of synchronization* [9]. A deadlock arises from an initial state when an AND-join or OR-join is waiting in a reachable state S, but is never enabled in reachable states from S [9]. A lack of synchronization results from an initial state when an edge carries more than one token in a reachable state. A workflow graph is said to be *sound* if it has neither a deadlock nor a lack of synchronization [9]. This soundness is defined on workflow graphs with OR-joins. In our previous work, we showed that no other semantics allows running more workflow graphs sound than the one used here [24]. In the context of the present work, we assume that each XOR- and OR-split independently decides after their execution which outgoing edges receive an additional token.

3 Related Work

Finding and restructuring loops (cycles) in graphs has a long tradition in compiler theory. Tarjan defined an efficient algorithm to find cycles (SCCs) in arbitrary graphs [29]. Since one main application of finding cycles in compiler theory is optimization [12], further algorithms arose to detect in particular nested loops, i.e., loops within loops. The representation of loops within loops is called a *loop nesting forest*. Depending on the applied algorithm, different kinds of loop nesting forests can be derived from the same graph. Prominent examples are the Sreedhar-Gao-Lee [27], Steensgaard [28], and the Havlak [12] forests. As the approach in this paper decomposes loops, it leads to a loop nesting forest that may differ from those in the literature. Depending on the loop nesting forest, optimizations can be applied worse or better. Exactly what the effects are, should be investigated in the future and is beyond the scope of this paper.

The idea of finding independent structures in unstructured graphs is to improve analyses and visualization [20]. In BPM, a prominent approach is the decomposition of a graph into SESE components resulting in a tree, the RPST, that hierarchically orders those components [2,23,30,31]. Each component can be analyzed independently and, therefore, in parallel.

As explained in the introduction, loops make the analysis of process models difficult [3]. SESE decompositions do not help to improve the analysis of unstructured components with loops. Polyvyanyy et al. [21] use SESE decomposition (in the form of SPQR-trees after recognizing triconnected components) to show how sound, unstructured process models can be restructured into structured process models. Their structuring is based on *quasi block-structured* process models. A process model is quasi block-structured if its RPST does not contain a so-called *rigid* component. Loops can then occur as *Loop Case* components with single entries and single exits. Polyvyanyy et al. argue for such a loop case component that the entry and exit must be XOR gateways. This is consistent with our findings later in this paper. However, our loops can occur in any kind of component of a process model—our approach, therefore, generalizes the findings of Polyvyanyy et al. Furthermore, our process models can also contain OR-gateways. Finally, there are loops that cannot be structured and, therefore, cannot be analyzed. For this reason, Choi et al. [3] proposed a decomposition of workflow graphs into acyclic components so that loops no longer exist for soundness analysis. This decomposition separates each loop in the graph into a *forward* and *backward* *flow*. In natural loops, the forward flow contains the loop nodes on the paths between the loop entry and its exits; the backward flow contains all other nodes and some nodes overlapped with the forward flow. This approach is well suited for testing soundness of workflow graphs with natural loops. Irreducible loops, however, are not decomposed. They are instantiated each into multiple distinct loops depending on all combinations of how their loop entries may have tokens. When an instantiated loop has parallel loop entries, it is transformed into a natural loop. Our approach does not have to differentiate between natural and irreducible loops and decomposes both in a similar way. Moreover, it can be applied to processes with OR-splits and -joins, to which the approach of Choi et al. is not applicable.

Untanglings represent final extracts of executions of a process model [22]. They are similar to *instance subgraphs* [26], but work for cyclic process models free of lacks of synchronization. Although untanglings are acyclic versions of process models, they describe the processes within a different representation. One reason for this is that they have been used as index for querying process models within repositories. Our decomposition, in contrast, converts cyclic models into acyclic models within the same representation (e.g., BPMN into BPMN).

Another approach to the study of cyclic process models is *unfolding* [8, 16, 20]. Unfolding extends the process model (as Petri net) so that the resulting Petri net represents the same behavior as before, but without loops. Since unfoldings due to loops can be large or infinite, the resulting Petri net covers only a so-called *prefix* of the process, which represents the behavior of the entire process. Although this approach is well suited for analysis, it cannot be reused later, e.g., for execution, because it only covers an extract of the process model. In contrast, our decomposition uses (re)calls of subprocesses in loop nodes to cover the complete semantics of the original process model more naturally. In addition, since unfoldings use Petri nets, processes with OR-gateways cannot be handled.

4 Understanding Loops in Business Process Models

Loops are a classic term from programming language theory and control-flow graphs. Natural loops have a *header*, a *body*, and a *condition* [28]. The header is the entry to the loop, the body contains the code that may be executed several times, and the condition checks whether the body should be iterated or exited, i.e., a condition is a loop exit. Since business process models are very similar to control-flow graphs, it seems natural to use these terms for them as well.

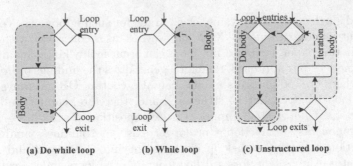

Fig. 2. Different types of loops with their loop entries, bodies, and loop exits.

4.1   Types of Loops

Classical loops can be divided into two types: *do-while* (repeat-until) and *while* loops [1]. *For* and *for-each* loops are interpreted in this context as special forms of while loops. In a do-while loop, the body of the loop is executed at least once. After the body has been executed, the loop exit (condition) is reached and it is checked whether the body is executed again. A do-while loop results in a graph as shown in Fig. 2(a). In contrast, the body of a while loop is only executed if the loop iterates at least once. Figure 2(b) shows a graph of a while loop.

Do-while and while loop structures in control-flow graphs usually result from loop structures in programming languages (e.g., do { ... } while(cond);). However, loops in control-flow graphs can also be unstructured, especially if *goto* statements are present [12]. On closer inspection, an unstructured loop may contain a subgraph that can be executed even though the loop does not iterate—as in a do-while loop. We call this subgraph *do-body*. And there can be a subgraph that can only be executed if the loop iterates at least once—as in a while loop. We call this subgraph *iteration-body*. Do- and iteration-bodies can be very complex, with subloops that may not pass the loop exits or entries.

A general unstructured loop can have multiple loop entries and exits. Loop entries are important for the first execution of the loop, but during iteration, they are negligible because the loop exits decide whether the loop iterates or not. For this reason, the do-body is part of the iteration-body, cf. Fig. 2*(c)*. In abstract terms, the do-body is the subgraph that starts at the loop entries and ends at its exits. It can be interpreted as the initialization of the loop or as a subgraph that must at least be executed even if the loop does not iterate. Iteration-bodies contain all nodes and edges of the loop, but can be split so that they start at the loop exits and end at them—like an unrolling of the loop.

4.2   Loops in Process Models

Process models as workflow graphs can contain all kinds of loops: do-while, while, and unstructured loops. Since workflow graphs can be designed unstructured in a graphical editor, unstructured loops with multiple entries and exits are not uncommon. To make matters worse, workflow graphs have explicit parallelism. Therefore, loop headers and nodes in loops can be executed in parallel. These circumstances make the analysis of process models particularly difficult.

From a business and intuitive point of view: When a token leaves a loop, the loop should be finished. Or in words of token game semantics: If a token is on a loop-exit edge, no other edge of the loop should carry a token anymore. Fortunately, this corresponds to the soundness property:

Theorem 1 (Single Token Loop Exits). Let *WFG* be a cyclic workflow graph with one of its loops $\mathcal{L} = (N_{\mathcal{L}}, E_{\mathcal{L}})$.

If *WFG* is sound and a loop-exit edge of $\mathcal{L}$ has a token in a state S, then the following both statements are valid:

(1) No edge in $\mathcal{L}$ has a token in S.
(2) No other loop-exit edge has a token in S.

Proof. We assume that *WFG* is sound and within a state S while a loop-exit edge *ex* of loop $\mathcal{L}$ carries a token.

To (1): *No edge in $\mathcal{L}$ has a token in S.* Assume there is an edge e of $\mathcal{L}$ that has a token or gets a token in a subsequent state S', $S \rightarrow^* S'$, while *ex* keeps its token. Since $\mathcal{L}$ is by Definition 2 a SCC, each node (and edge) within $\mathcal{L}$ has a path to each other node (and edge) in $\mathcal{L}$—and, therefore, also to *ex*. Let us take a path $P_{e \rightarrow ex}$ from e to *ex* in $\mathcal{L}$. Since each node of *WFG* has a path to at least one end node (cf. Definition 1), there is also a path $P_{ex \rightarrow end}$ from *ex* to a single incoming edge *end* of an end node. $P_{e \rightarrow ex}$ and $P_{ex \rightarrow end}$ are disjoint except of *ex* by Corollary 1. The combination of $P_{e \rightarrow ex}$ and $P_{ex \rightarrow end}$ results in the path $P_{e \rightarrow end}$. We now treat each XOR- and OR-split in each subsequent state of S' to always put a token on an edge on $P_{e \rightarrow end}$ (if it has any). In visual words, the tokens on *ex* and e follow finally the same path. Since *WFG* is sound and has no deadlock, there is a reachable state from S' where *end* carries at least two tokens. A lack of synchronization is reachable and, therefore, *WFG* is unsound. ⨍

To (2): *No other loop-exit edge has a token in S.* Assume there would be another loop-exit edge *ex'* of $\mathcal{L}$ that carries a token in S. There are two possibilities without loss of generality: (a) *ex'* got a token some states after *ex* or (b) *ex* and *ex'* got the tokens within the same state transition.

To (a): *ex' got a token some states after ex .* In other words, there was a previous state, in which *ex* and another edge in $\mathcal{L}$ have tokens. (1) shows that this is a contradiction to the soundness of *WFG*. ⨍

To (b): *ex and ex' got the tokens within the same state transition.* In each state transition, only one node is executed. Therefore, edges *ex* and *ex'* must have the same source node n. This node n has to be an AND- or OR-split since its execution results in more than one token. n—in contrast to *ex* and *ex'*—is part of $\mathcal{L}$. As a consequence, n must have at least one outgoing edge *in* within $\mathcal{L}$. Therefore, an execution of n can result in a state where *ex* and *in* both have tokens. This is again (1). ⨍

Corollary 2. *Resulting from Theorem 1 and its proof (2) (b), each loop exit in a sound workflow graph is an XOR-split.*

Both—that every loop exit should be an XOR-split and that a loop is finished when a token leaves the loop—are intuitive but important properties of a loop in sound workflow graphs. It follows from these properties that any parallelism within a loop is synchronized before any loop exit is reached. More specifically, (1) do-bodies synchronize multiple tokens at loop entries into a single one when any loop exit is reached. (2) Each iteration of iteration-bodies starts and ends with a single token, i.e., each parallelism in iteration-bodies starts and ends in one iteration. And another important point: (3) No iteration of the iteration-body interacts with a previous iteration or with the do-body. Finally, (4) the iteration-body of a complex loop structure in a workflow graph can be replaced by a single node, making the graph acyclic. Knowing loops from control-flow graphs running not in parallel, all these facts are intuitively correct. But Theorem 1 shows that the same is true for loops in sound workflow graphs.

5   Loop Decomposition

Theorem 1 gives us the legitimacy to split a loop into two (sub) workflow graphs, one representing the do-body and the other the iteration-body. Then, the execution of a loop is done in two steps: The execution of its do-body and, subsequently, when the loop iterates, the multiple executions of the iteration-body. The separation is similar to converting a do-while loop into a while loop in programming languages where the do-body and iteration-body are the same, i.e., do {<body>} while(cond); is transferred to <body> while(cond) {<body>}. In other words, by separating the do-body from the iteration-body, an abstract do-while loop is transferred to an abstract while loop, very simplified as <do-body> while(cond) {<iteration-body>}.

The separation of the do- and iteration-body can generally be achieved by cutting the loop by at least one loop exit. In doing so, the do-body should contain most nodes and edges between the loop entries and the loop exits, so that execution enters the iteration-body only when required. Removing one *inner-loop* outgoing edge of a loop exit is sufficient to cut the initial loop structure. However, the more inner-loop outgoing edges of loop exits are removed, the more looping structures are destroyed. A do-body can be now defined as follows:

Definition 4 (Do-Body). Let there be a workflow graph with a loop $\mathcal{L}$.

The *do-body* of loop $\mathcal{L}$ is a maximal connected subgraph of $\mathcal{L}$ containing all loop entries and loop exits but a minimum number of *inner-loop* outgoing edges of loop exits.

Do-bodies can be identified by a variant of depth-first search. Their definition leads to the definition of iteration-bodies and cutoff edges:

Definition 5 (Iteration-Body and Cutoff Edges). Let there be a workflow graph with one of its loops $\mathcal{L}$ and $\mathcal{L}$'s do-body Do.

Every *inner-loop* outgoing edge of a loop exit of $\mathcal{L}$ that does not lie in *Do* is a *cutoff edge*. The *iteration-body* of $\mathcal{L}$ is the connected subgraph of $\mathcal{L}$ that contains all nodes and edges of $\mathcal{L}$ without the cutoff edges.

Because of the exclusivity of loop exits and, therefore, of cutoff edges, we can reduce each loop in a workflow graph to its do-body. To maintain loop behaviors, new *loop nodes* are inserted representing the iteration bodies of loops (for this case, Definition 1 is extended as mentioned in Sect. 2). Loop nodes may have multiple incoming and multiple outgoing edges and have an exclusive behavior according to Theorem 1. They are executed if one of their cutoff edges is taken, i.e., the cutoff edges are redirected to the loop nodes. After executing the loop nodes (their iteration-bodies), the execution can return to the previous loop-exit edges in the surrounding workflow graph. We call the modified workflow graph *reduced*. The iteration-body is extended with start and end nodes to a workflow graph:

Definition 6 (Workflow Graph Extensions). Let there be a workflow graph *WFG* with one of its loops $\mathcal{L}$.

The *$\mathcal{L}$-loop-reduced WFG* is a modified *WFG* regarding $\mathcal{L}$:
  - each node and edge being part of $\mathcal{L}$ but not of $\mathcal{L}$'s do-body is removed from *WFG* (i.e., at least the cutoff edges),
  - a *loop node* is inserted representing the iteration-body of loop $\mathcal{L}$,
  - the cutoff edges of $\mathcal{L}$ are redirected to the new loop node,
  - for each loop-exit edge, an outgoing edge from the loop node to the target of the loop-exit edge is introduced,
  - each join node with one incoming edge is replaced by a single edge, and
  - if non-join node has multiple incoming edges, these edges are combined with a new XOR-join in front of it.

The *extended iteration-body* $It(\mathcal{L})$ is $\mathcal{L}$'s iteration-body as a workflow graph:
  - for each cutoff edge, two edges, one from a separate start node to its original target node and the other from the corresponding loop exit to a separate end node, are introduced,
  - for each loop-exit edge, a corresponding edge to a separate end node is introduced, and
  - each join node with only one incoming edge is replaced by a single edge.

Although iteration-bodies and extended iteration-bodies are different, we sometimes use both terms as synonyms when the context is clear.

In summary, three types of edges are redirected or added to modify and extend the original workflow graph and the iteration-bodies: loop-entry edges, loop-exit edges, and cutoff edges, each associated with a start, end, or loop node. Each cutoff edge appears twice in the extended iteration-body, once in connection with a start and once in connection with an end node.

Figure 3 *(a)* shows our example workflow graph from Fig. 1 with its loop (large gray rounded rectangle). The entries of the loop are A and B, its exits F and G. The do-body contains these loop entries and exits and the subgraph between

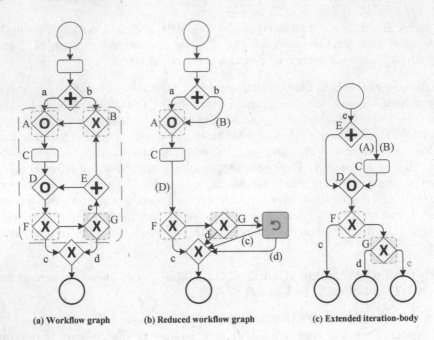

(a) Workflow graph   (b) Reduced workflow graph   (c) Extended iteration-body

Fig. 3. Decomposition of the example workflow graph of Fig. 1 *a)* into its reduced workflow graph with a loop node *b)* and extended iteration-body *c)*.

them: A, B, C, D, F, and G. This do-body is maximal including all loop entries and exits and minimal in terms of the number of outgoing edges of loop exits. Therefore, e is a cutoff edge. The reduced version of the original workflow graph is illustrated in Fig. 3(b). The reader can verify that it contains all nodes of the do-body (except B and D being replaced by edges (B) and (D)). The loop's iteration-body contains all nodes and edges of the loop, but it is truncated at the cutoff edges, e in this example, and join nodes A and B are merged into single edges (A) and (B). The extended iteration-body of Fig. 3(c) is the result. The graph has been arranged for clarity. We can see that the possible subloop of nodes D, F, G, and E in the original workflow graph is actually a parallel pattern and not a subloop anymore in the iteration-body. In Fig. 3(b) and (c) all edge-replaced nodes are either XOR- or OR-joins. Edge-replaced nodes can only be join nodes by Definition 6. If they could be AND-joins, the workflow graph cannot be sound following Theorem 1 and Choi et al. [3]:

Corollary 3. *Resulting from Definition 6, Theorem 1 and its proof, each join node being an edge in the reduced workflow graph or in $It(\mathcal{L})$ of a loop $\mathcal{L}$ of a sound workflow graph WFG is an XOR- or OR-join in WFG.*

Definition 6 describes a decomposition of a workflow graph into two workflow graphs regarding a single loop. In fact, both resulting workflow graphs can be cyclic again. However, since the original loop is cut on at least one outgoing edge of a loop exit, the remaining loops in both graphs are always smaller than the original loop. It is possible to identify these loops within the reduced workflow graph and iteration-body by using the same decomposition as described. In other words, we can recursively decompose each loop within a workflow graph, as Steensgaard [28] did. Eventually, this recursion terminates as the loops in each recursion become smaller and smaller.

6   Loop Execution

In the reduced workflow graph, a previously iteration-body of the loop is represented by a single node, the loop node (see Fig. 3(b)). The semantics of those is exclusive following Theorem 1. In other words, the special loop node is semantically a combination of an XOR-join and an XOR-split. If each loop of a workflow graph is recursively reduced to its do-body and a loop node, the workflow graph is *acyclic* by definition. Therefore, a sound, cyclic workflow graph has been transformed into an acyclic one.

The semantics of do-bodies of loops is already represented in the reduced workflow graph. To maintain the semantics of loop iterations, their extended iteration-bodies must be included. Their inclusion occurs when a corresponding loop node is enabled. When a loop node is enabled, the extended iteration-body is instantiated with a token on the corresponding start edge (e.g., the outgoing edge e of the start node in Fig. 3(c)). The extended iteration-body is executed until a token reaches the incoming edge of one of its end nodes. According to Theorem 1, only this incoming edge has a token in the entire iteration-body. In other words, there is always only one execution instance of the iteration-body. If the incoming edge of the end node matches the outgoing edge of one of its start nodes (e.g., the incoming edge e of the right end node in Fig. 3(c)), the iteration-body is terminated and instantiated again with a token on this edge (above edge e in Fig. 3(c)). If the incoming edge is a loop-exit edge (e.g., c and d in Fig. 3(c)), the loop node finishes its execution with a token on this edge.

If the original workflow graph is sound, then the execution semantics described above coincides with the behavior of the original workflow graph. But its acyclic workflow graphs are much easier to validate, verify, and execute. They should be executable on a process engine without much effort.

7   Algorithm

The combination of the recursive loop decomposition described above and the reduction of each loop and the insertion of loop nodes describes a complete decomposition algorithm from a cyclic workflow graph to a set of acyclic ones, summarized by Algorithm 1. The algorithm defines the function *decompose*, which first searches all loops within the given workflow graph, e.g., by Tarjan's algorithm for finding SCCs [29]. If there is no loop in the workflow graph, the function is finished and returns this graph. Otherwise, it initializes the set of acyclic workflow graphs. Subsequently, for each loop, the workflow graph is reduced and a loop node is inserted in the input workflow graph. The extended

Algorithm 1. Decomposition of a cyclic workflow graph WFG into a set of acyclic workflow graphs.

```
1: function DECOMPOSE(WFG)
2:     Find set of loops £ in WFG.
3:     if £ = ∅ then
4:         return {WFG}
5:     acyclic ← ∅
6:     for all L ∈ £ do
7:         Reduce L in WFG to its do-body and include a loop node.
8:         Derive the extended iteration-body workflow graph It(L) from L.
9:         acyclic ← acyclic ∪ DECOMPOSE(It(L))
10:    acyclic ← acyclic ∪ DECOMPOSE(WFG)
11:    return acyclic
```

iteration-body is derived and applied recursively to the *decompose* function. Since the input workflow graph has changed, *decompose* is applied again to it to find further (nested) loops. Finally, the result set of acyclic workflow graphs is returned at the end of the function.

Finding all loops in a workflow graph can be achieved in linear time, $O(N+E)$, using Tarjan's algorithm for finding SCCs [29]. Reducing the workflow graph and deriving the extended iteration-bodies can also be done in linear time mainly by using a variant of depth-first search. In the worst case, the number of (nested) loops in a workflow graph is as large as the number of edges E, i.e., the function *decompose* and the for-each-loop are both called E times at maximum. Overall, the asymptotic runtime complexity is $O(EN + E^2)$ in the worst case.

8   Evaluation and Example Applications

We have implemented our loop decomposition as a plugin for the research tool *Mojo*[1]. As input files, we used the PNML version of a library of real process models[2] from *IBM WebSphere Business Modeler* [9]. The models were originally in an XML format. The library contains 1,368 process models. 178 (approx. 13%) process models are cyclic—169 have a single and 9 have two independent SCCs (loops) by Definition 2. Therefore, there are in total 187 loops in all process models. During our decomposition, only 28 of 187 loops contain nested loops— 23 contain a single nested loop and 5 contain two independent nested loops. There is no process model with a nesting depth greater than 1.

After loop decomposition, the process models are reduced to about 94% (SD 0.27) of their original size. The worst case complexity of the algorithm is quadratic, however, it appears to be linear in practice, as the number of edges visited during the decomposition does not increase quadratically. This

---

[1] https://github.com/guybrushPrince/mojo.plan.loop, last visited March 2022.

[2] https://web.archive.org/web/20131208132841/http://service-technology.org/public
ations/fahlandfjklvw_2009_bpm, last visited March 2022.

seems to depend on the nature of the process model. Without the elimination of background processes and running on a standard computer, a loop decomposition takes on average less than 1 ms (SD 0.9).

Loop decomposition has a direct effect on the execution semantics of OR-joins. The definition of the semantics for OR-joins is intuitive for acyclic but difficult for cyclic process models [24,32]. Most execution semantics in research have, finally, a problem with so-called *vicious circles*—loops in which two OR-joins are mutually waiting for each other [15]. Such vicious circles are no longer possible after decomposition into acyclic process models. Therefore, understanding the semantics of each OR-join in the decomposed models should be intuitive. Moreover, each join node can be replaced by an OR-join (or an undifferentiated join), while retaining the semantics.

Although soundness is a precondition of loop decomposition, it can be used to check soundness itself. This is the first method (to the authors' knowledge) that can be used to check soundness of cyclic process models with OR-joins. We just outline the idea without proof in the following: First, loop decomposition is applied on the process model resulting in a set of acyclic models. Starting from the acyclic version of the original process model, we can apply an arbitrary soundness verification algorithm for acyclic models with OR-splits and OR-joins (e.g., Favre and Völzer [10] or Prinz and Amme [25]). If the process is unsound, its original process is unsound too. This is valid since, otherwise, a loop could not be reduced and replaced with a loop node, being a violation against Theorem 1. Second, all of the process model's extended iteration-bodies can be simplified so that they have a single start and end node since all their start and end nodes are exclusive. Subsequently, all iteration-bodies are checked against soundness and whether Corollary 3 holds. For all sound (sub) models, simplifications and analyses can be applied iteratively. If the process model contains any unsound iteration-body, the model is naturally unsound.

9   Conclusion

In this paper, we have presented an algorithm that decomposes cyclic process models in the form of workflow graphs into sets of acyclic workflow graphs. The approach is based on a general view of loops and a simple execution semantics. The execution semantics of the acyclic workflow graphs coincides when the original workflow graph is sound. A short evaluation shows that loop decomposition is feasible in practice. Some advantages of decomposition were discussed using OR-join semantics and soundness analysis.

The BPM community should benefit from this approach. Many analyses in BPM are limited to acyclic process models and can now be applied to cyclic models. In addition, execution engines of processes can be simplified, especially regarding the semantics of OR-joins. In general and in terms of compiler theory, our presented algorithm leads to a loop nesting forest that may differ from those in the literature. The differences arise because each loop is transformed into a while-loop-like form. In the future, it will be interesting to study its impact.

For future work, we plan to incorporate algorithms and analyses from acyclic process models to our acyclic decomposition. One application could be the first detailed and complete soundness approach for cyclic workflow graphs with OR-joins. Another useful application is to combine the RPST with our loop decomposition. This should facilitate divide-and-conquer approaches for process analysis, especially for unstructured components.

References

1. Aho, A.V., Sethi, R., Ullman, J.D.: Compilers: Principles, Techniques, and Tools. Addison-Wesley, Boston (1986)
2. Choi, Y., Ha, N.L., Kongsuwan, P., Han, K.H.: An alternative method for refined process structure trees (RPST). Bus. Process. Manag. J. **26**(2), 613–629 (2020). https://doi.org/10.1108/BPMJ-11-2018-0319
3. Choi, Y., Kongsuwan, P., Joo, C.M., Zhao, J.L.: Stepwise structural verification of cyclic workflow models with acyclic decomposition and reduction of loops. Data Knowl. Eng. **95**, 39–65 (2015). https://doi.org/10.1016/j.datak.2014.11.003
4. Cormen, T.H., Leiserson, C.E., Rivest, R.L., Stein, C.: Introduction to Algorithms, 3rd edn. MIT Press, Cambridge (2009)
5. van Dongen, B.F., Mendling, J., van der Aalst, W.M.P.: Structural patterns for soundness of business process models. In: Tenth IEEE International Enterprise Distributed Object Computing Conference (EDOC 2006), 16–20 October 2006, Hong Kong, China, pp. 116–128. IEEE Computer Society (2006). https://doi.org/10.1109/EDOC.2006.56
6. Dumas, M., García-Bañuelosa, L., La Rosa, M., Ubaa, R.: Fast detection of exact clones in business process model repositories. Inf. Syst. **38**(4), 619–633 (2013). https://doi.org/10.1016/j.is.2012.07.002
7. Dumas, M., La Rosa, M., Mendling, J., Reijers, H.A.: Fundamentals of Business Process Management. Springer, Heidelberg (2013). https://doi.org/10.1007/978-3-642-33143-5
8. Esparza, J., Römer, S., Vogler, W.: An improvement of mcmillan's unfolding algorithm. Formal Methods Syst. Des. **20**(3), 285–310 (2002). https://doi.org/10.1023/A:1014746130920
9. Fahland, D., Favre, C., Koehler, J., Lohmann, N., Völzer, H., Wolf, K.: Analysis on demand: instantaneous soundness checking of industrial business process models. Data Knowl. Eng. **70**(5), 448–466 (2011)
10. Favre, C., Völzer, H.: Symbolic execution of acyclic workflow graphs. In: Hull et al. [13], pp. 260–275. https://doi.org/10.1007/978-3-642-15618-2_19
11. Ha, N.L., Prinz, T.M.: Partitioning behavioral retrieval: an efficient computational approach with transitive rules. IEEE Access **9**, 112043–112056 (2021)
12. Havlak, P.: Nesting of reducible and irreducible loops. ACM Trans. Program. Lang. Syst. **19**(4), 557–567 (1997). https://doi.org/10.1145/262004.262005
13. Hull, R., Mendling, J., Tai, S. (eds.): Business Process Management. LNCS, vol. 6336. Springer, Heidelberg (2010). https://doi.org/10.1007/978-3-642-15618-2
14. Keller, G., Scheer, A.W., Nüttgens, M.: Semantische Prozeßmodellierung auf der Grundlage "Ereignisgesteuerter Prozeßketten (EPK)". Inst. für Wirtschaftsinformatik (1992)
15. Kindler, E.: On the semantics of EPCs: resolving the vicious circle. Data Knowl. Eng. **56**(1), 23–40 (2006)

16. McMillan, K.L.: Using unfoldings to avoid the state explosion problem in the verification of asynchronous circuits. In: von Bochmann, G., Probst, D.K. (eds.) CAV 1992. LNCS, vol. 663, pp. 164–177. Springer, Heidelberg (1993). https://doi.org/10.1007/3-540-56496-9_14

17. Mendling, J., Neumann, G., van der Aalst, W.: Understanding the occurrence of errors in process models based on metrics. In: Meersman, R., Tari, Z. (eds.) OTM 2007. LNCS, vol. 4803, pp. 113–130. Springer, Heidelberg (2007). https://doi.org/10.1007/978-3-540-76848-7_9

18. OASIS: Web Services Business Process Execution Language Version 2.0, April 2007. http://docs.oasis-open.org/wsbpel/2.0/OS/wsbpel-v2.0-OS.pdf

19. Object Management Group (OMG): Business Process Model and Notation (BPMN) Version 2.0. formal/2011-01-03, January 2011. http://www.omg.org/spec/BPMN/2.0

20. Polyvyanyy, A.: Structuring process models. Ph.D. thesis, University of Potsdam (2012)

21. Polyvyanyy, A., García-Bañuelos, L., Weske, M.: Unveiling hidden unstructured regions in process models. In: Meersman, R., Dillon, T., Herrero, P. (eds.) OTM 2009. LNCS, vol. 5870, pp. 340–356. Springer, Heidelberg (2009). https://doi.org/10.1007/978-3-642-05148-7_23

22. Polyvyanyy, A., La Rosa, M., ter Hofstede, A.H.M.: Indexing and efficient instance-based retrieval of process models using untanglings. In: Jarke, M., et al. (eds.) CAiSE 2014. LNCS, vol. 8484, pp. 439–456. Springer, Cham (2014). https://doi.org/10.1007/978-3-319-07881-6_30

23. Polyvyanyy, A., Vanhatalo, J., Völzer, H.: Simplified computation and generalization of the refined process structure tree. In: Bravetti, M., Bultan, T. (eds.) WS-FM 2010. LNCS, vol. 6551, pp. 25–41. Springer, Heidelberg (2011). https://doi.org/10.1007/978-3-642-19589-1_2

24. Prinz, T.M., Amme, W.: A complete and the most liberal semantics for converging OR gateways in sound processes. Complex Syst. Informatics Model. Q. **4**, 32–49 (2015). https://doi.org/10.7250/csimq.2015-4.03

25. Prinz, T.M., Amme, W.: Control-flow-based methods to support the development of sound workflows. Complex Syst. Informatics Model. Q. **27**, 1–44 (2021)

26. Sadiq, W., Orlowska, M.E.: Analyzing process models using graph reduction techniques. Inf. Syst. **25**(2), 117–134 (2000)

27. Sreedhar, V.C., Gao, G.R., Lee, Y.: Identifying loops using DJ graphs. ACM Trans. Program. Lang. Syst. **18**(6), 649–658 (1996)

28. Steensgaard, B.: Sequentializing program dependence graphs for irreducible programs. Technical report, Microsoft Research (1993)

29. Tarjan, R.E.: Depth-first search and linear graph algorithms. SIAM J. Comput. **1**(2), 146–160 (1972). https://doi.org/10.1137/0201010

30. Vanhatalo, J., Völzer, H., Koehler, J.: The refined process structure tree. Data Knowl. Eng. **68**(9), 793–818 (2009). https://doi.org/10.1016/j.datak.2009.02.015

31. Vanhatalo, J., Völzer, H., Leymann, F.: Faster and more focused control-flow analysis for business process models through SESE decomposition. In: Krämer, B.J., Lin, K.-J., Narasimhan, P. (eds.) ICSOC 2007. LNCS, vol. 4749, pp. 43–55. Springer, Heidelberg (2007). https://doi.org/10.1007/978-3-540-74974-5_4

32. Völzer, H.: A new semantics for the inclusive converging gateway in safe processes. In: Hull et al. [13], pp. 294–309. https://doi.org/10.1007/978-3-642-15618-2_21

Reasoning on Labelled Petri Nets and Their Dynamics in a Stochastic Setting

Sander J. J. Leemans[1]([⊠])[iD], Fabrizio Maria Maggi[2][iD], and Marco Montali[2][iD]

[1] RWTH Aachen, Aachen, Germany
s.leemans@qut.edu.au
[2] Free University of Bozen-Bolzano, Bolzano, Italy
{maggi,montali}@inf.unibz.it

Abstract. Interest in stochastic models for business processes has been revived in a recent series of studies on uncertainty in process models and event logs, with corresponding process mining techniques. In this context, variants of stochastic labelled Petri nets, that is with duplicate labels and silent transitions, have been employed as a reference model. Reasoning on the stochastic, finite-length behaviours induced by such nets is consequently central to solve a variety of model-driven and data-driven analysis tasks, but this is challenging due to the interplay of uncertainty and the potentially infinitely traces (including silent transitions) induced by the net. This explains why reasoning has been conducted in an approximated way, or by imposing restrictions on the model. The goal of this paper is to provide a deeper understanding of such nets, showing how reasoning can be properly conducted by leveraging solid techniques from qualitative model checking of Markov chains, paired with automata-based techniques to suitably handle silent transitions. We exploit this connection to solve three central problems: computing the probability of reaching a particular final marking; computing the probability of a trace or that a temporal property, specified as a finite-state automaton, is satisfied by the net; checking whether the net stochastically conforms to a probabilistic Declare model. The different techniques have all been implemented in a proof-of-concept prototype.

Keywords: Stochastic Petri nets · Stochastic process mining · Qualitative verification · Markov chains

1 Introduction

In process mining, recorded organisational process data is leveraged to gain insights into business processes by means of analysis techniques. Process mining techniques have traditionally taken the frequency and timing of observed behaviour into account implicitly: depending on the type of analysis performed, the most-occurring happy paths vs. little-occurring deviations, as well as quick vs. slow performing activities, might be of interest and is essential for quantifiable insights. For instance, a quality control process with 30% failed checks is a considerably different process than a process with 2% failed checks, even though the control flow would be equivalent. Explicitly modelling the stochastic perspective of process models – how likely each behaviour is – may assist

© Springer Nature Switzerland AG 2022
C. Di Ciccio et al. (Eds.): BPM 2022, LNCS 13420, pp. 324–342, 2022.
https://doi.org/10.1007/978-3-031-16103-2_22

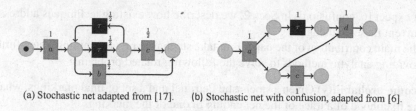

(a) Stochastic net adapted from [17].   (b) Stochastic net with confusion, adapted from [6].

Fig. 1. Two examples of labelled stochastic Petri nets.

in obtaining quantifiable insights, and in ensuring the quality of simulation, prediction and recommendation. Recent work includes discovery techniques to automatically discover stochastic process models [5,22].

Stochastic process models (such as generalised stochastic Petri nets [19]) provide this information explicitly by indicating the likelihood and timing of steps in the process and, indirectly, of process behaviour (traces). Not surprisingly, interest in these stochastic models has been revived in the context of process mining, with a series of recent studies focused on: (1) discovery of stochastic process models that indicate the likelihood of behaviour [5,22]; (2) repair of process models [21]; (3) conformance checking, either comparing the stochastic behaviour of a log with that expected by a stochastic process models to gain insights from their differences [15,16], or using the likelihood of model traces when aligning observed traces with a reference model [2].

When attacking these problems, it becomes essential to reason on the stochastic behaviour captured by the process, for example to determine the likelihood of model traces. Traditional techniques relying on the connection between stochastic Petri nets and Markov chains [19,20] cannot be readily applied to this setting, due to key conceptual mismatches related to the usage of stochastic Petri nets to represent business processes. First, transitions in the net must be labelled with corresponding (names of) activities in the processes, possibly using the same label for multiple transitions. Second, silent transitions should be supported, to represent control-flow structures in the process (such as gateways) that do not correspond to any visible activity. Third, when analysing the dynamics of the net the focus is not on infinite, recurring behaviour, but on finite traces representing the possible executions of process instances, moving a case object from the initial to a final state without considering which silent steps have been taken in between.

Supporting all these modelling requirements makes it difficult to actually reason on the traces supported by these nets and their probabilities. To see this, consider the labelled stochastic Petri net shown in Fig. 1(a) (we will introduce these nets formally in Sect. 3). This model has two traces, however computing the likelihood of the traces may be counterintuitive: the likelihood of the trace of a followed by b is $\frac{2}{3}$ [17]. The challenge here stems from the loop of silent transitions, which "favours" b over c.

Another example of potentially counterintuitive likelihoods of traces is shown in Fig. 1(b). In this net, the likelihood of a followed by c is $\frac{3}{4}$. The challenge in this example is again the silent transition, which is used here in a semi-concurrent context: the transition c is mutually exclusive with transition d, but as c is part of two runs (that is, executed before or after the silent transition), its probability is higher than one might expect [6, confusion]. In Sect. 2, we describe how existing techniques address or circumvent these challenges.

The main contribution of the paper is to take stochastic process mining a step further by providing analytic methods to solve the following related problems:

(Outcome probability) Given a stochastic Petri net and a set of final markings, what is the likelihood of a trace of the net ending in one of the markings?

(Verification) Given a temporal property captured by a finite-state automaton (e.g., an LTL_f formula or a Declare model), what is the probability that the net generates a trace satisfying the property? A special case of this problem is calculating the probability of a trace.

(Stochastic model conformance) Given a set of probabilistic temporal constraints [18] and a labelled stochastic Petri net, does the net conform to the set by comparing their stochastic behaviours?

We address these problems by transforming them into problems that can be solved using well-established techniques. In particular, we build on the connection [19] between stochastic Petri nets and Markov chains [11,12], paired with automata-based techniques to handle their qualitative verification against temporal properties [1, Ch.10]. We use the former to compute the probability of reaching a target marking, and the latter to handle silent transitions, and to isolate the behaviour induced by the net that satisfy a property of interest.

The methods have been implemented as part of the ProM framework [10].

2   Related Work

Stochastic process-based models have been studied extensively in literature. In the context of this work, we are interested in formal, Petri net-based stochastic models that are at the basis of the recent series of approaches in stochastic process discovery [5,22] and conformance checking [3,15,16]. Such approaches all refer to the model of (generalised) stochastic Petri nets, or fragments thereof. A first version of this model was proposed in [20], extending Petri nets by assigning exponentially distributed firing *rates* to transitions. This was extended in [19] by distinguishing timed (as in [20]) and immediate transitions. Immediate transitions have priority over timed ones, and have *weights* to define their relative likelihood. As these two types of transitions, abstracting from time, behave homogeneously, we may capture the stochastic behaviour of the net through a discrete-time Markov chain [19].

Several variants of stochastic Petri nets have been investigated starting from the seminal work in [19]. These variants differ from each other depending on the features they support (e.g., arbiters to resolve non-determinism, immediate vs timed transitions) and the way they express probabilities. Such nets may aid modellers in expressing certain constructs. An orthogonal, important dimension is to ensure that probabilities and concurrency interact properly. This can be achieved through good modelling principles [6,19] or automated techniques [4].

Contrasting these formal models with recent works in stochastic process mining, key differences exist. Traditional stochastic nets do not support transition labels nor silent transitions, and put emphasis on recurring, infinite executions and the so-called steady-state analysis, focused on calculating the probability that an execution is currently placed in a given state. This is done by constructing a discrete-time Markov chain that characterises the stochastic behaviour of the net [19,20]. Finding the probability of a finite-length trace in such nets is trivial, as every trace corresponds to a single path. However, no transition labels or silent steps are supported, which limits their usefulness for process mining due to the omnipresence of such transitions in process models. On the other hand, when these features are incorporated in stochastic Petri nets, which is precisely what we target in this paper, computing the probability of a trace cannot be approached directly anymore, as infinitely many paths would in principle need to be inspected. At the same time, in business processes we are interested in behaviour at the trace level rather than at the process level – that is, traces have a finite length and are in principle independent – thus the large body of work on steady-state-based analyses does not apply for our purposes. This explains why reasoning on the stochastic behaviour of such extended nets has been conducted in an approximated way [15,16], or by imposing restrictions on the model [3].

To bridge this gap, in this paper we take the most basic stochastic Petri nets: we do not consider time or priority, but we add (duplicate) labels and silent transitions. Importantly, *our results seamlessly carry over bounded, generalised stochastic Petri nets*, thanks to the fact that incorporating priorities in bounded nets is harmless, and that timed and immediate transitions are homogeneous from the stochastic point of view. To the best of our knowledge, outside of recent work using stochastic Petri nets with silent transitions [3,15,17], such nets have not been defined or studied before.

While intuitively stochastic conformance checking techniques need to obtain the probability of a given trace in a stochastic process model (for instance, [17] explicitly obtains this probability to compute a distance measure between a log and a stochastic process model), some stochastic conformance checking techniques avoid computing the probability for a single trace, for instance by playing out the model to obtain a sample of executions [15], or by assuming that the model is deterministic [16]. The results presented in this paper therefore enable the practical application of [17], and may enable further stochastic conformance checking techniques and, consequently, new types of analysis.

Silent steps have been studied in the context of automata. For instance, in [13] an ad-hoc method is described to iteratively remove all silent steps from a stochastic automaton. Due to concurrency and confusion (see for instance Fig. 1(b)), such techniques are not directly applicable to stochastic Petri nets. A result of this paper is that silent steps can be handled directly, without the need for ad-hoc techniques.

3   Stochastic Petri Net-Based Processes

We first provide some brief preliminaries on multisets. A multiset a over a set U (which defines the support of the multiset) is a function $a : U \to \mathbb{N}$, where for $u \in U$, $a(u)$ indicates the multiplicity (i.e., the number of occurrences) of u. Given two multisets a and b over U, we write:

- $a + b$ for the *union* of a and b, defined as the multiset that assigns to each $u \in U$ multiplicity $a(u) + b(u)$;
- $a \leq b$ if for every $u \in U$, we have $a(u) \leq b(u)$;
- assuming $a \leq b$, $b - a$ for the *difference* of b and a, defined as the multiset that assigns to each $u \in U$ multiplicity $b(u) - a(u)$.

The set of all multisets over U is defined as $\mathbb{M}(U)$. Multiset a is explicitly represented placing inside squared brackets $[\ldots]$ each element u with non-zero multiplicity, using notation $u^{a(u)}$.

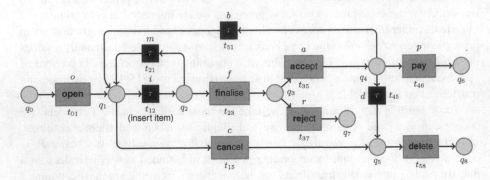

Fig. 2. Stochastic net of an order-to-cash process. Weights are presented symbolically. Transition t_{12} captures a task that cannot be logged, and so is modelled as silent.

3.1 Labelled Petri Nets

As underlying control-flow structure to specify work processes, we consider **Petri nets** that are **labelled** with (atomic) tasks. As customary in Business Process Management, the same task may be used to **label multiple** transitions in the process, and among all labels, we include a special label to indicate **silent steps**, which are internal execution steps of the process that are not explicitly exposed to the external environment (and are thus not recorded in event logs). To capture such labels, we assume a finite set Σ to denote task (names), and a special label $\tau \notin \Sigma$ to indicate a silent step. We also use $\overline{\Sigma} = \Sigma \cup \{\tau\}$ to denote the extended set containing task names and the silent label.

Definition 1 (Labelled Petri net). *A labelled Petri net N is a tuple $\langle Q, T, F, \ell \rangle$, where: (i) Q is a finite set of* places; *(ii) T is a finite set of* transitions, *disjoint from Q (i.e., $Q \cap T = \emptyset$); (iii) $F \subseteq (Q \times T) \cup (T \times Q)$ is a* flow *relation connecting places to transitions and transitions to places; (iv) $\ell : T \to \overline{\Sigma}$ is a* labelling *function mapping each transition $t \in T$ to a corresponding label $\ell(t)$ that is either a task name from Σ or the silent label τ.* ◁

In the paper, we adopt a dot notation to extract the component of interest from a net, that is, given a net N, its places are denoted by $N.Q$, etc. We will adopt the same notational convention for the other definitions as well. Given a net N and an element

$x \in N.Q \cup N.T$, the *preset* and *post-set* of x are respectively defined by $^{\bullet}x = \{y \mid \langle y, x \rangle \in F\}$ and $x^{\bullet} = \{y \mid \langle x, y \rangle \in F\}$. If x is a transition, then its pre- and post-set respectively denote its input and output places.

Figure 2 shows a labelled Petri net where silent transitions are either used to capture control-flow structures (t_{41} for looping, and t_{45} for rerouting), or tasks that cannot be logged (t_{12}, which represents a non-loggable task for inserting an item). Silent transitions may result from modelling skips, loopbacks, or to start and join concurrent branches, however can also be used to represent processes with non-loggable tasks (Fig. 2 even contains a loop of silent transitions). Also, Petri net discovery algorithms may produce nets containing silent loops. This motivates why we study such nets.

An execution state of a net is described by a marking, which is a multiset of places. A transition is enabled in a marking if its input places contain at least one token each. Firing an enabled transition produces a new marking where one token per input place is consumed, and each output place gets one token more.

Definition 2 (Marking). *A marking m of a net N is a multiset over the places of N, mapping each place $q \in N.Q$ to the number $m(q)$ of tokens on q. Given a marking m of N, and a transition $t \in N.T$, we say that:*

- *t is enabled in m, written $m[t\rangle_N$, if $^{\bullet}t \leq m$;*
- *$E_N(m)$ is the set of enabled transitions in a marking m.*
- *assuming $m[t\rangle_N$, t fires in m for N producing a new marking m' of N, written $m[t\rangle_N m'$, if $m' = (m - {}^{\bullet}t) + t^{\bullet}$;*   ◁

The next definitions are essential to capture that we are interested in finite-trace executions over nets. Specifically, as customary in BPM, each execution represents the evolution of a process instance from the initial state to a final state.

Definition 3 (Execution). *An execution of a net N from a marking m_s to a marking m_f of N is a (possibly empty) finite sequence $t_0, \ldots, t_n$ of transitions in $N.T$ such that there exist markings $m_0, \ldots, m_{n+1}$ of N with (i) $m_0 = m_s$, (ii) $m_{n+1} = m_f$, (iii) for every $i \in \{0, \ldots, n\}$ we have $m_i[t_i\rangle_N m_{i+1}$.*   ◁

Definition 4 (Deadlock, livelock). *A marking m of a net N is a:*

- *deadlock if there is no transition enabled: $E_N(m) = \emptyset$;*
- *livelock if there is no execution of N from m to a deadlock marking.*

Definition 5 (Petri net-based process, runs). *A Petri net-based process (PNP) is a triple $\langle N, m_0, M_f \rangle$, where: (i) N is a net; (ii) m_0 is a marking of N denoting the initial state; (iii) M_f is a finite set of deadlock markings of N denoting its possible final states. An execution (Definition 3) starting in m_0 and ending in an $m' \in M_f$ is a run. A PNP $\mathcal{N}$ is:*

- *deadlock-free if the only reachable deadlock markings are from $\mathcal{N}.M_f$;*
- *livelock-free if $RG(\mathcal{N})$ does not contain any livelock marking.*   ◁

A single transition in a PNP without an input place makes the net unable to reach a final marking, and thus the PNP has no runs. This is a special case of an unavoidable livelock.

By fixing an initial marking and a set of final markings, we define a process:

Restricting final states to deadlocks markings is without loss of generality: one can take a non-deadlock marking and turn it into a deadlock one by introducing a new silent transition pointing to a dedicated, exclusive "final" deadlock place.

Remark 1. There are two types of execution of a PNP $\mathcal{N}$ that, starting from its initial state, cannot be extended into proper runs:

- Executions ending in a deadlock marking not being a final marking in $\mathcal{N}$;
- Executions in a livelock from which no deadlock marking can be reached.   ◁

A *trace* is a sequence $\sigma = e_0, \ldots, e_n \in \Sigma^*$ of *events* over Σ, where, for simplicity, each event e_i indicates the execution of a task by means of the firing of a transition. A trace is a *model trace* for a PNP if it is produced by one of its runs, considering only the visible labels of the transitions contained in the run.

Definition 6 (Model trace). *A trace σ is a model trace of PNP $\mathcal{N}$ if there exists a run $\eta = t_0, \ldots, t_m$ of $\mathcal{N}.N$ whose corresponding sequence of labels $\mathcal{N}.N.\ell(t_0), \ldots, \mathcal{N}.N.\ell(t_m)$ coincides with σ once all τ elements are removed. In this case, we say that η induces σ.*   ◁

A model trace σ of PNP $\mathcal{N}$ may be induced by multiple, possibly infinitely many, runs. The set of runs of $\mathcal{N}$ inducing σ is denoted by $runs_{\mathcal{N}}(\sigma)$.

The execution semantics of a PNP can be described through a reachability graph, namely a (possibly infinite-state) labelled transition system whose states correspond to reachable markings, and whose transitions match transition firings of the PNP.

Definition 7 (Labelled transition system). *A labelled transition system is a tuple $\langle S, s_0, S_f, \varrho \rangle$ where: (i) S is a (possibly infinite) set of states; (ii) $s_0 \in S$ is the initial state; (iii) $S_f \subseteq S$ is the set of accepting states; (iv) $\varrho \subseteq S \times \overline{\Sigma} \times S$ is a $\overline{\Sigma}$-labelled transition relation. A run is a finite sequence of transitions leading from s_0 to one of the states in S_f in agreement with ϱ.*   ◁

Due to our requirement that all final markings are deadlock markings, accepting states have no outgoing transitions either.

Definition 8 (Reachability graph). *The reachability graph $RG(\mathcal{N})$ of a PNP $\mathcal{N}$ is a labelled transition system $\langle S, s_0, S_f, \varrho \rangle$ whose components are defined by mutual induction as the minimal sets satisfying the following conditions:*

1. *$s_0 = m_0 \in S$;*
2. *for every state $m \in S$, every transition $t \in T$, and every marking $m' \in \mathbb{M}(Q)$, if $m[t\rangle_N m'$ we have that (a) $m' \in S$; (b) if $m' \in \mathcal{N}.M_f$, then $m' \in S_f$; (c) $\langle m, \ell(t), m' \rangle \in \varrho$.*   ◁

The runs of $RG(\mathcal{N})$ capture all and only the runs of $\mathcal{N}$. It will be useful later to refer to outgoing transitions from a given state s. We do so with notation $succ_{RG(\mathcal{N})}(s)$.

We close this part by defining some key, standard properties of PNPs. In particular, we fix the last control-flow feature of our model, namely the fact that we focus on **bounded** processes.

Definition 9 (Bounded PNP). *A PNP $\mathcal{N}$ is* bounded *if there exists a number k such that, for every reachable marking $m \in RG(\mathcal{N}).S$ and every place $q \in \mathcal{N}.N.Q$, we have $m(p) \leq k$.*   ◁

A key property of bounded PNPs is that they induce a reachability graph that has finitely many states. Boundedness is a standard property assumed when capturing business processes; verifying boundedness is decidable [9] and well-known techniques exist. In the remainder of this paper, we assume bounded PNPs.

3.2 Stochastic Behaviour

We now extend PNPs with stochastic behaviour, by incorporating **stochastic decision making to determine which enabled transition to fire**. Technically, this is done by adding a weight to each transition in a PNP [19]. The probability of firing an enabled transition is the fraction of the weight of the transition compared to the sum of the weights of all enabled transitions.

A *stochastic PNP* is then a PNP of which the transitions similarly have a weight.

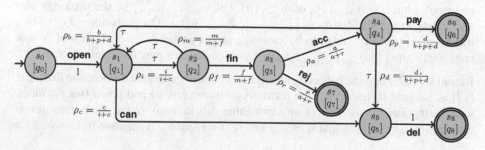

Fig. 3. Stochastic reachability graph of the order-to-cash bounded stochastic PNP. States are named. The initial state is shown with a small incoming edge. Final states have a double contour.

Definition 10 (Stochastic Petri net). *A stochastic Petri net N is a tuple $\langle Q, T, F, \ell, w \rangle$, where $\langle Q, T, F, \ell \rangle$ is a labelled Petri net, and $W \colon N.T \to \mathbb{R}^+$ is a weight function assigning a positive weight to each transition in N. Given a marking m of N, and an enabled transition $t \in E_N(m)$, the firing probability of t in m, $\mathbb{P}_{m,N}(t)$, is $\frac{N.w(t)}{\sum_{t' \in E_N(m)} N.w(t')}$ [19].*   ◁

Please note that stochastic Petri nets inherit the concurrency properties of Petri nets: execution is atomic, and "true" concurrency needs to be added on top. The stochastic perspective may still matter in concurrency, for instance in resource-constrained settings. It is easy to see that the firing probability defines a discrete probability distribution over $E_N(m)$, as $\sum_{t \in E_N(m)} \mathbb{P}_{m,N}(t) = 1$. Then, we can define the semantics of stochastic PNPs using a stochastic transition system and a stochastic reachability graph.

Definition 11 (Stochastic transition system). *A stochastic transition system is a tuple* $\langle S, s_0, S_f, \varrho, p \rangle$ *where* $\langle S, s_0, S_f, \varrho \rangle$ *is a transition system, while p is a transition probability function mapping each transition in ϱ to a corresponding probability value in* $[0, 1]$, *such that for every state $s \in S$, $\sum_{\xi \in succ_N(s)} p(\xi) = 1$.*   ◁

The reachability graph $RG(N)$ of a stochastic PNP N is hence defined as a stochastic transition system obtained as in Definition 8, defining the transition probability function as follows: for every transition $\langle m, \ell(t), m' \rangle$, its probability is set to $\mathbb{P}_{m,N.N}(t)$.

Example 1. Figure 2 shows an stochastic PNP (N_{order}) capturing an order-to-cash process. The reachability graph of N_{order} is shown in Fig. 3, where transition probabilities are calculated using the weights of the stochastic net. If, for example, we fix the weight a of transition t_{35} to 80, and the weight r of transition t_{37} to 20, we get that in marking $[q_3]$ (corresponding to the state where the order has been finalised), there is 0.2 chance that the order is rejected, and 0.8 chance that the order is accepted.

Remark 2. The probability of firing an enabled transition only depends on the current marking, thus stochastic Petri nets and stochastic PNPs are Markovian.   ◁

Consequently, to calculate the probability of a run, we may consider each choice therein as independent. In a stochastic PNP $N = \langle N, m_0, M_f \rangle$, we denote the probability of a run $\eta = t_0, \ldots, t_n$ with $\mathbb{P}_N(\eta)$. Let $m_0, \ldots, m_{n+1}$ be the markings corresponding to η; then $\mathbb{P}_N(\eta) = \prod_{i \in \{1,\ldots,n\}} \mathbb{P}_{m_{i-1},N}(t_i)$. The probability $\mathbb{P}_N(\eta)$ of a trace σ of N is in turn obtained by summing up the probabilities of all runs of N that induce σ: $\mathbb{P}_N(\sigma) = \sum_{\eta \in runs_N(\sigma)} \mathbb{P}_N(\eta)$. Notice that this may be an infinite sum.

Remark 3. Our approach directly lifts to full, bounded generalised stochastic Petri nets [19] as follows: *(i)* transitions are partitioned into *immediate* and *timed* (for the latter, interpreting weights as *rates* of an exponential distribution); *(ii)* a timed transition is enabled if it is so in the usual sense, and *there is no enabled immediate transition.*   ◁

4   Outcome Probability

In this section, we tackle a first, fundamental problem: *computing outcome probabilities*, that is, computing what the probability is that a process instance of the bounded stochastic PNP of interest evolves from the initial to one among a desired set of final states (representing the desired outcomes). For example, we may be interested in knowing the probability that the bounded stochastic PNP N_{order} of our running example (Fig. 2) evolves an order from opening to payment.

Technically, given a bounded stochastic PNP N, we borrow the standard notion of conditional probability and indicate the probability that N evolves marking m into some marking from a set M as $\mathbb{P}_N(M|m_1)$. Formally, this corresponds to the sum of the probabilities of all executions of N from m_1 to some marking in M (in the sense of Definition 3). This leads us to the formulation of the OUTCOME-PROB(N, F) problem:

Input: Bounded stochastic PNP $\mathcal{N}$, set $F \subseteq \mathcal{N}.M_f$ of desired final states;
Output: Probability value $\mathbb{P}_{\mathcal{N}}(F|\mathcal{N}.m_0) = \sum_{\eta \text{ run of } \mathcal{N} \text{ ending in } m \in F} \mathbb{P}_{\mathcal{N}}(\eta)$.

Notice that the same problem can also get, as input, a stochastic transition system in place of a bounded stochastic PNP.

OUTCOME-PROB cannot be solved exactly through an enumeration of runs, as there may be infinitely many. It can be approximated by fixing a maximum threshold either on the length of runs [15], or on their minimum probability [2]. To obtain an exact answer, we build on the connection between bounded stochastic PNPs and discrete-time Markov chains [11], lifting [19] to our setting.[1]

Remark 4. The reachability graph $RG(\mathcal{N}) = \langle S, s_0, S_f, \varrho, p \rangle$ of a bounded stochastic PNP $\mathcal{N}$ can be seen as a discrete-time Markov chain C where: *(i)* S is the finite set of states of C, with s_0 the initial state; *(ii)* S_f are the absorption/exit states of C; *(iii)* ϱ and p define the transition matrix of C, where the entry for a pair $s_1, s_2 \in S$ gets value $p(s, l, s')$ for some label $l \in \overline{\Sigma}$ if $\langle s, l, s' \rangle \in$, 0 otherwise. ◁

We exploit this, noticing that the OUTCOME-PROB problem corresponds to the problem of calculating *exit distributions* in a discrete-time Markov chain [11] (also called the problem of calculating *absorption/hit probabilities* [12]). To analytically solve the problem, we take OUTCOME-PROB$(\mathcal{N}, F)$ and create a system of equations, starting from the reachability graph $RG(\mathcal{N})$. Specifically, each state s of $RG(\mathcal{N}).S$ corresponds to a state variable x_{s_i} denoting the probability $\mathbb{P}_{\mathcal{N}}(F|s)$ of reaching one of the states in F from s; hence $x_{RG(\mathcal{N}).s_0}$ represents the solution of the problem. Then, each equation defines the value of one of the state variables x_s as follows:

Base case if s has no successor states (i.e., is a deadlock marking), then $x_{s_i} = 1$ if s corresponds to a final marking, otherwise $x_{s_i} = 0$;
Inductive case if s has successors, its variable is equal to sum of the state variables of its successor states, weighted by the transition probability to move to that successor.

Formally, OUTCOME-PROB$(\mathcal{N}, F)$ with $RG(\mathcal{N}) = \langle S, s_0, S_f, \varrho, p \rangle$ gets encoded into the following linear optimisation problem $\mathcal{E}_{\mathcal{N}}^F$:
    Return x_{s_0} from the minimal non-negative solution of

$$x_{s_i} = 1 \qquad \text{for each } s_i \in F \qquad (1)$$
$$x_{s_j} = 0 \qquad \text{for each } s_j \in S \setminus F \text{ s.t. } |succ_{RG(\mathcal{N})}(s_j)| = 0 \qquad (2)$$
$$x_{s_k} = \sum_{\langle s_k, l, s'_k \rangle \in succ_{RG(\mathcal{N})}(s_k)} p(\langle s_k, l, s'_k \rangle) \cdot x_{s'_k} \quad \text{for each } s_k \in S \text{ s.t. } |succ_{RG(\mathcal{N})}(s_k)| > 0 \quad (3)$$

By recalling that states of $RG(\mathcal{N})$ are markings of $\mathcal{N}$, the schema (1) of equations deals with final (deadlock) states, that in (1) with non-final deadlock states, and that in (1) with non-final, non-deadlock states.

---

[1] In case of generalised stochastic Petri nets, the resulting discrete-time Markov chain is the so-called *embedded/jump* chain obtained from the continuous-time Markov chain capturing the execution semantics of the net [19,20].

$\mathcal{E}_{\mathcal{N}}^{F}$ has always at least a solution. However, it may be indeterminate and thus admit infinitely many ones, requiring in that case to pick the least committing (i.e., minimal non-negative) solution. The latter case happens when $\mathcal{N}$ contains livelock markings. This is illustrated in the following examples.

Example 2. Consider bounded stochastic PNP $\mathcal{N}_{\text{order}}$ (Fig. 2). We want to solve the problem OUTCOME-PROB($\mathcal{N}_{\text{order}}, [q_6]$), to compute the probability that a created order eventually completes the process by being paid. To do so, we solve $\mathcal{E}_{\mathcal{N}_{\text{order}}}^{[q_6]}$ by encoding the reachability graph of Fig. 3 into:

$$x_{s8} = 0 \qquad x_{s5} = x_{s8} \qquad\qquad x_{s2} = \rho_m x_{s1} + \rho_f x_{s3}$$
$$x_{s7} = 0 \qquad x_{s4} = \rho_b x_{s1} + \rho_d x_{s5} + \rho_p x_{s6} \qquad x_{s1} = \rho_i x_{s2} + \rho_c x_{s5}$$
$$x_{s6} = 1 \qquad x_{s3} = \rho_a x_{s4} + \rho_r x_{s7} \qquad\qquad x_{s0} = x_{s1}$$

This yields $x_{s0} = \frac{\rho_i \rho_f \rho_a \rho_p x_{s6} + \rho_i \rho_f \rho_r x_{s7} + (\rho_i \rho_f \rho_a \rho_d + \rho_c) x_{s8}}{1 - \rho_i \rho_m - \rho_i \rho_f \rho_a \rho_b} = \frac{\rho_i \rho_f \rho_a \rho_p}{1 - \rho_i \rho_m - \rho_i \rho_f \rho_a \rho_b}$, which is the only solution. If we assume that the weights of $\mathcal{N}_{\text{order}}$ are all equal, the probability distributions for choosing the next transition are all uniform, leading to $\rho_i = \rho_f = \rho_m = \rho_a = \frac{1}{2}$ and $\rho_p = \rho_b = \frac{1}{3}$, and, in turn, that the probability of completing the process by paying the order is $x_{s0} = \frac{1}{17} \sim 0.06$.

With an analogous approach, we can prove that the probability that an order gets deleted is $\frac{13}{17}$, and the one that an order gets rejected is $\frac{3}{17}$. Notice that the sum of all such probabilities is, as expected, 1, that is, every order gets paid, deleted or rejected. ◁

Example 3. Consider the bounded stochastic PNP $\mathcal{N}_{\text{live}}$ in Fig. 4. To compute the outcome probability of its single final state, we solve $\mathcal{E}_{\mathcal{N}_{\text{live}}}^{[q_1]}$ by encoding the reachability graph of Fig. 4(b) into:

$$x_{s0} = \rho_a x_{s1} + \rho_b x_{s2} \qquad x_{s1} = 1 \qquad x_{s2} = x_{s3} \qquad x_{s3} = \rho_d x_{s2} + \rho_e x_{s3}$$

We get $x_{s3} = \rho_d x_{s3} + \rho_e x_{s3} = (\rho_d + \rho_e) x_{s3} = x_{s3}$, making the system indeterminate. Its minimal non-negative solution is then the one where $x_{s3} = 0$, and in turn $x_{s0} = \rho_a$. ◁

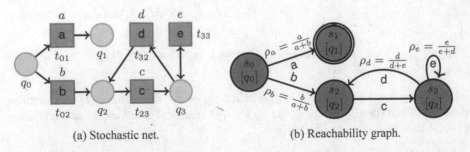

(a) Stochastic net.   (b) Reachability graph.

Fig. 4. Reachability graph (b) of a bounded stochastic PNP with net shown in (a), initial marking $[q_0]$ and final marking $[q_1]$. States s_2 and s_3 are livelock markings.

Example 3 illustrates how the technique implicitly gets rid of livelock markings, associating to them a 0 probability. This captures the essential fact that, by definition, a livelock marking can never reach any final marking. More in general, we can in fact solve OUTCOME-PROB$(\mathcal{N}, F)$ by turning the linear optimisation problem $\mathcal{E}_{\mathcal{N}}^{F}$ into the following system of equalities, which is guaranteed to have exactly one solution:

$$x_{s_i} = 1 \qquad\qquad\qquad \text{for each deadlock marking } s_i \in F \qquad (4)$$

$$x_{s_j} = 0 \qquad\qquad\qquad \text{for each deadlock marking } s_j \in S \setminus F \quad (5)$$

$$x_{s_k} = 0 \qquad\qquad\qquad \text{for each livelock marking } s_k \in S \qquad (6)$$

$$x_{s_h} = \sum_{\langle s_h, l, s'_h \rangle \in succ_{RG(\mathcal{N})}(s_h)} p(\langle s_h, l, s'_h \rangle) \cdot x_{s'_h} \quad \text{for each remaining marking } s_h \in S \quad (7)$$

Recall that checking whether a marking s is livelock can be done over $RG(\mathcal{N})$ by checking (non-)reachability of some deadlock marking in $RG(\mathcal{N})$ from s. This check does not involve probabilities at all, but extends to probabilistic settings as per Definition 10, all transitions have a non-zero weight.

5 Qualitative Verification and Trace Probability

We now further leverage the connection between bounded stochastic PNPs and discrete-time Markov chains (cf. Remark 4), to deal with the verification of (qualitative, i.e., non-probabilistic) temporal/dynamic properties over bounded stochastic PNPs. This amounts to compute the probability that a run of the PNP indeed satisfies the property of interest. We rely on [1, Ch. 10] and employ automata-theoretic techniques coupled with the computation of outcome probabilities to solve the problem. We then show how this technique also solves another, related problem: that of computing trace probabilities.

5.1 Verification of Temporal Properties

Properties of interest intensionally describe a (possibly infinite) set of desired finite-length traces that may be induced by runs of the stochastic PNP under scrutiny. Such traces are defined over the task names in Σ (*without* τ). We opt for a very general formalism to describe such properties: (deterministic) finite-state automata.

Definition 12 (DFA, acceptance, language). *A deterministic finite-state automaton (DFA) over L is a tuple $A = \langle L, S, s_0, S_f, \delta \rangle$, where: (i) L is a finite alphabet of symbols; (ii) S is a finite set of states, with $s_0 \in S$ the initial state and $S_f \subseteq S$ the set of final states; (iii) $\delta : S \times L \to S$ is a transition transition function that, given a state $s \in S$ and a label $l \in L$, returns the successor state $\delta(s, l)$. A accepts a trace $\sigma = l_0, \ldots, l_n$ over $L^\star$ if there exists a sequence of states $s_0, \ldots, s_{n+1}$ starting from the initial state and such that: (i) $s_{n+1} \in S_f$, and (ii) for every $i \in \{0, \ldots, n\}$, we have $s_{i+1} = \delta(s_i, l_i)$. The language $\mathcal{L}(A)$ of A is the set of all traces accepted by A.* ◁

This accounts for non-deterministic automata (NFAs), as each NFA can be encoded into a corresponding DFA. Also, it makes our approach directly operational for other property specification languages, as long as they can get encoded into DFAs. This holds, e.g., when for regular expressions, $\text{LTL}_f/\text{LDL}_f$ temporal formulae over finite traces [8], and Declare possibly extended with meta-constraints [7].

In this setting, verification takes as input a bounded stochastic PNP $\mathcal{N}$ and an automaton A whose transitions are labelled by task names, and returns the probability that $\mathcal{N}$ generates a model trace that belongs to the language of A. Technically, we define the VERIFY-PROB$(\mathcal{N}, A)$ problem as follows:

Input: Bounded stochastic PNP $\mathcal{N}$, DFA A over Σ;
Output: Probability value equal to $\sum_{\sigma \text{ model trace of } \mathcal{N} \ s.t. \ \sigma \in \mathcal{L}(A)} \mathbb{P}_{\mathcal{N}}(\sigma)$.

To solve the problem, we need to account for three different aspects:

1. deal with the mismatch between runs over $\mathcal{N}$ and traces of A;
2. single out all and only those model traces of $\mathcal{N}$ that are also traces of A;
3. compute the collective probability of all such traces.

We tackle these three aspects with corresponding three steps.

Automaton with Silent Transitions. Definition 6 indicates that the set of runs inducing a trace consists of all those runs that insert an arbitrary number of τs before and after each event in the trace. For a trace $\sigma = a_0, \ldots, a_n$, this set corresponds to the language of the regular expression $\tau^*; a_0; \tau^*; \ldots; \tau^*; a_n; \tau^*$. We then take the input automaton A over Σ and turn it into a corresponding automaton $\bar{A}$ over $\overline{\Sigma}$ whose language $\mathcal{L}(\bar{A})$ corresponds to all and only the possible runs that induce the traces of $\mathcal{L}(A)$. This is done by simply expanding it with τ-labelled self-loops connecting every state to itself.

Definition 13 (Run DFA). *Given a DFA $A = \langle \Sigma, S, s_0, S_f, \delta \rangle$ over Σ, its run DFA $\bar{A}$ is a DFA over $\overline{\Sigma}$ defined as $\langle \overline{\Sigma}, S, s_0, S_f, \delta' \rangle$ with identical states (including the initial and final ones), and where $\delta' = \delta \cup \{\langle s, \tau \rangle \to s \mid s \in S\}$.* ◁

Product Stochastic Transition System. We now consider $RG(\mathcal{N})$ and $\bar{A}$. Since they are both run-generating devices, we can obtain an intensional representation of all the runs of $\mathcal{N}$ by constructing a *product* stochastic transition system generates all and only runs that are common to $\mathcal{N}$ and $\bar{A}$, which in turn are the runs of $\mathcal{N}$ that induce traces of A. This can be done by the usual product automaton construction, with the only difference that we need to retain the stochastic information coming from $\mathcal{N}$. This is straightforward, as $\bar{A}$ is qualitative, i.e., does not contain probabilities.

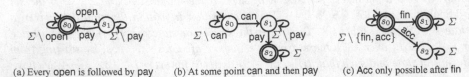

(a) Every open is followed by pay   (b) At some point can and then pay   (c) Acc only possible after fin

Fig. 5. DFAs of three properties for the order-to-cash example. A single edge labelled by a set L of task names describes a set of edges, each labelled by a task name from L.

Definition 14 (Product system). *Let $\mathcal{N}$ by a bounded stochastic PNP with $RG(\mathcal{N}) = \langle S_1, s_0^1, S_f^1, \varrho_1, p_1 \rangle$, and $\bar{A} = \langle \overline{\Sigma}, S_2, s_0^2, S_f^2, \delta_2 \rangle$ a DFA over $\overline{\Sigma}$. The product system $\Upsilon_{\mathcal{N}}^{\bar{A}}$ of $\mathcal{N}$ and $\bar{A}$ is a stochastic transition system $\langle S, s_0, S_f, \varrho, p \rangle$ whose states are pairs of states from $S_1 \times S_2$, and whose components are defined by mutual induction as the sets satisfying the following conditions:*

1. *$s_0 = \langle s_0^1, s_0^2 \rangle \in S$;*
2. *for every state $\langle s_1, s_2 \rangle \in S$ and every label $l \in \overline{\Sigma}$ such that (i) $\langle s_1, l, s_1' \rangle \in \varrho_1$ for some $s_1' \in S^1$, and (ii) $\delta_2(s_2, l) = s_2'$ for some $s_2' \in S^2$, by fixing $s' = \langle s_1', s_2' \rangle$ we have: (a) $s' \in S$, (b) $\langle s, l, s' \rangle \in \varrho$, (c) $p(\langle s, l, s' \rangle) = p_1(s_1')$, (d)if $s_1' \in S_f^2$ and $s_2' \in S_f^2$, then $s' \in S_f$.*   ◁

The so-defined product system is not a complete stochastic transition system: there may be states whose successor probabilities do not add up to one. It can be made complete by adding a fresh non-final sink state and transitions pointing from such incomplete states to the fresh one, each decorated with the probability value needed to reach 1, and labelled with whatever label from $\overline{\Sigma}$. This completion is not essential for the consequent computation (as the state variable for such a sink state would be equal to 0).

Verification as Outcome Probability Computation. We are now ready to bring everything together, exploiting the notions of run DFA and product system to show how the VERIFY-PROB problem can be reduced to the OUTCOME-PROB, invoked on $\Upsilon_{\mathcal{N}}^{\bar{A}}$ considering all its final states.

Theorem 1. *For every bounded stochastic PNP $\mathcal{N}$ and DFA A, we have that* VERIFY-PROB$(\mathcal{N}, A) = $ OUTCOME-PROB$(\Upsilon_{\mathcal{N}}^{\bar{A}}, \Upsilon_{\mathcal{N}}^{\bar{A}}.S_f)$.   ◁

Proof. Considering that the definition of VERIFY-PROB, and that the probability of a model trace of $\mathcal{N}$ is the sum of the probabilities of the runs of $\mathcal{N}$ inducing that trace, we have VERIFY-PROB$(\mathcal{N}, A) = \sum_{\sigma \text{ model trace of } \mathcal{N} \text{ s.t. } \sigma \in \mathcal{L}(A)} \mathbb{P}_{\mathcal{N}}(\sigma) = \sum_{\eta \text{ run of } \mathcal{N} \text{ inducing } \sigma \text{ s.t. } \sigma \in \mathcal{L}(A)} \mathbb{P}_{\mathcal{N}}(\eta)$. By Definitions 13 and 14, we have that the set of runs of $\mathcal{N}$ inducing traces in $\mathcal{L}(A)$ coincides with the set of runs of $\Upsilon_{\mathcal{N}}^{\bar{A}}$. This, together with the definition of OUTCOME-PROB, yields: $\sum_{\eta \text{ run of } \mathcal{N} \text{ inducing } \sigma \text{ s.t. } \sigma \in \mathcal{L}(A)} \mathbb{P}_{\mathcal{N}}(\eta) = \sum_{\eta \text{ run of } \Upsilon_{\mathcal{N}}^{\bar{A}}} = $ OUTCOME-PROB$(\Upsilon_{\mathcal{N}}^{\bar{A}}, \Upsilon_{\mathcal{N}}^{\bar{A}}.S_f)$.   ⊣

Example 4. Figure 5 shows three properties of interest for $\mathcal{N}_{\text{order}}$. Solving VERIFY-PROB for them gives: (1) The probability that $\mathcal{N}_{\text{order}}$ verifies the property of Fig. 5(a) coincides with the solution of OUTCOME-PROB when asking the probability that an order gets paid (cf. Example 2). (2) The probability that $\mathcal{N}_{\text{order}}$ verifies the property of Fig. 5(b) is 0, as there is no run of $\mathcal{N}_{\text{order}}$ where an order is first cancelled and then paid. (3) The probability that $\mathcal{N}_{\text{order}}$ verifies the property of Fig. 5(c) coincides with that of a run of $\mathcal{N}_{\text{order}}$ reaching completion, as each run either does not finalise the order, or does so before possibly accepting it. This probability is actually 1 (as the process does not contain livelocks nor non-final deadlocks).   ◁

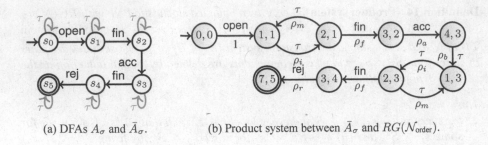

(a) DFAs A_σ and $\bar{A}_\sigma$.   (b) Product system between $\bar{A}_\sigma$ and $RG(\mathcal{N}_{\text{order}})$.

Fig. 6. DFAs for a trace and product system with the reachability graph of Fig. 3.

5.2 Computing Trace Probabilities

A key problem in stochastic conformance checking [2, 15] is that of computing the probability of a trace in a stochastic Petri net. We cast this problem in our setting as the TRACE-PROB problem, where TRACE-PROB$(\mathcal{N}, \sigma)$ is defined as:

Input: Bounded stochastic PNP $\mathcal{N}$, trace σ over Σ;
Output: Probability $\mathbb{P}_\mathcal{N}(\sigma)$.

This problem is clearly subsumed by the VERIFY-PROB problem described before. In fact, we can simply solve it by constructing a so-called trace automaton that trivially encodes σ as a DFA that only accepts that trace, then invoking VERIFY-PROB on it.

Definition 15 (Trace DFA). *Given a trace* $\sigma = a_0, \ldots, a_n$ *over* Σ, *its trace DFA* A_σ *is the DFA* $\langle \Sigma, S, s_0, S_f, \delta \rangle$ *over* Σ *such that: (a)* $S = \{s_0, \ldots, s_{n+1}\}$ *contains* $n + 1$ *states; (b)* $S_f = \{s_{n+1}\}$; *(c) for every* $i \in \{0, \ldots, n\}$, $\delta(s_i, a_i) = s_{i+1}$ *(and nothing else is in* δ).

Theorem 2. *For every bounded stochastic PNP* $\mathcal{N}$ *and every trace* σ *over* Σ^*, *we have that* TRACE-PROB$(\mathcal{N}, \sigma) = $ OUTCOME-PROB$(\mathcal{N}, A_\sigma)$.   ◁

Proof. Direct from the definition of the problems, noticing that $\mathcal{L}(A_\sigma) = \{\sigma\}$.   ⊣

Example 5. We compute the probability that $\mathcal{N}_{\text{order}}$ generates trace $\sigma = $ open, fin, acc, fin, rej, where an order is filled, finalised, accepted, then modified, finalised again, and this second time rejected. Following the described technique, we first transform σ into its trace DFA A_σ, and then further into its run DFA $\bar{A}_\sigma$. This is shown in Fig. 6(a). We then compute the product system $\Upsilon^{\bar{A}}_{RG(\mathcal{N}_{\text{order}})}$ of $\bar{A}_\sigma$ and $RG(\mathcal{N}_{\text{order}})$ (shown in Fig. 3), obtaining Fig. 6(b) (notice how silent transitions unfold in this transition system). Finally, we construct $\mathcal{E}^{\langle 7,5 \rangle}_{\Upsilon^{\bar{A}}_{RG(\mathcal{N}_{\text{order}})}}$ getting $x_{00} = \frac{\rho_i \rho_f \rho_a \rho_b \rho_i \rho_f \rho_r}{(1 - \rho_i \rho_m)^2}$, which yields the solution to the TRACE-PROB$(\mathcal{N}_{\text{order}}, \sigma)$ problem.   ◁

6 Stochastic Conformance with Probabilistic Declare

We now employ the verification machinery from Sect. 5 to check how the probabilistic behaviour encoded in a stochastic PNP relates to that declaratively specified using Prob-Declare [18]. We start with a gentle introduction to ProbDeclare, then showing how we can check whether a bounded stochastic PNP conforms to a ProbDeclare model.

6.1 Probabilistic Declare

Declare is a constraint-based process modelling language based on LTL_f. A model comes with a set of LTL_f constraint, and their conjunction must be respected by the process. This imposes a *crisp* interpretation of constraints: a trace satisfies a Declare model if it satisfies *every* constraints contained therein. This crisp semantics was relaxed in [18]: there, each constraint comes with a probability condition indicating the allowed probabilities for which a satisfying trace should be generated by the process. The semantics is formally defined using stochastic languages, and is therefore compatible with that of stochastic PNPs. We recall the necessary definitions.

Definition 16 (Probabilistic constraint). *A probabilistic constraint is a triple* $\langle \varphi, \bowtie , p \rangle$, *where: (i)* φ *is an LTL_f formula over Σ representing the* constraint formula; *(ii)* $\bowtie \in \{=, \neq, \leq, \geq, <, >\}$ *is the* constraint probability operator; *(iii) p is a rational value in $[0, 1]$ representing the* constraint probability.   ◁

Definition 17 (ProbDeclare). *A ProbDeclare model is a triple* $\langle \Sigma, \mathcal{C}, \mathcal{P} \rangle$, *where $\mathcal{C}$ is a finite set of LTL_f formulae called* crisp constraints, *while $\mathcal{P}$ is a set of (genuinely)* probabilistic constraints.   ◁

Since each constraint in $\mathcal{P}$ can be satisfied or violated, a ProbDeclare model induces $2^{|\mathcal{P}|}$ scenarios, each associated to a corresponding LTL_f formula.

Definition 18 (Scenario). *A scenario for a ProbDeclare model* $\mathcal{D} = \langle \Sigma, \mathcal{C}, \mathcal{P} \rangle$ *is a total boolean function* $\mathcal{S} : \mathcal{P} \rightarrow \{0, 1\}$ *indicating which probabilistic constraints are satisfied, and which violated. The set of scenarios is denoted by* $\mathbb{S}_{\mathcal{D}}$.   ◁

Scenarios come with two dimensions, induced by the crisp and probabilistic constraints: a temporal dimension indicating which traces belong to which scenarios, and a probabilistic dimension indicating how likely it is that a trace belongs to a scenario.

Definition 19 (Characteristic formula). *The characteristic formula* $\Phi_{\mathcal{S}}$ *of a scenario* $\mathcal{S}$ *for ProbDeclare model* $\langle \Sigma, \mathcal{C}, \mathcal{P} \rangle$ *is the LTL_f formula* $\bigwedge_{\varphi_i \in \mathcal{C}} \varphi_i \wedge \bigwedge_{\langle \varphi_j, \bowtie_j, p_j \rangle \in \mathcal{P}, \ \mathcal{S}(\langle \varphi_j, \bowtie_j, p_j \rangle) = 1} \varphi_j \wedge \bigwedge_{\langle \varphi_k, \bowtie_k, p_k \rangle \in \mathcal{P}, \ \mathcal{S}(\langle \varphi_k, \bowtie_k, p_k \rangle) = 0} \neg \varphi_k$. *Scenario* $\mathcal{S}$ *is consistent if $\Phi_{\mathcal{S}}$ is satisfiable (i.e., has at least one satisfying trace).*   ◁

Definition 20 (Valid scenario distribution). *A valid scenario distribution over scenarios* $\mathbb{S}_{\mathcal{D}}$ *of a ProbDeclare model* $\mathcal{D} = \langle \Sigma, \mathcal{C}, \mathcal{P} \rangle$ *is a probability distribution* $\mathbb{P}_{\mathcal{D}} : \mathbb{S}_{\mathcal{D}} \rightarrow [0, 1]$ *such that: (a) for every scenario $\mathcal{S} \in \mathbb{S}_{\mathcal{D}}$, if $\mathcal{S}$ is not consistent then $\mathbb{P}_{\mathcal{D}}(\mathcal{S}) = 0$; (b) For every probabilistic constraint $\langle \varphi, \bowtie, p \rangle \in \mathcal{P}$, we have* $\sum_{\mathcal{S} \in \mathbb{S}_{\mathcal{D}} \ s.t. \ \mathcal{S}(\langle \varphi, \bowtie, p \rangle) = 1} \mathbb{P}_{\mathcal{D}}(\mathcal{S}) \bowtie p$.   ◁

Example 6. Consider the order-to-cash ProbDeclare model $\mathcal{D}_{order}$, with one crisp *not coexistence* constraint C_1 indicating that pay and rej cannot be both in a trace, and two *response* probabilistic constraints C_2 and C_3, indicating that open must be eventually followed by pay with a probability of $\geq \frac{1}{20}$, and that open must be eventually followed by rej with a probability of $\leq \frac{1}{4}$. Of the four scenarios over C_2 and C_3, one is that both C_2 and C_3 are satisfied and is inconsistent as it clashes with the crisp constraint C_1. ◁

6.2 Checking Stochastic Conformance

In [18], it is shown how a system of inequalities can be constructed so as to compute possible valid scenario distributions in accordance with Definition 20. Here we are just interested in using that system to *check* whether a probability distribution over scenarios is indeed valid, and thus we can keep this system of inequalities as a black box.

In a non-stochastic setting, one can check whether a bounded Petri net satisfies a Declare model by verifying that every trace it generates belongs to the language accepted by the model. In our stochastic setting, we define a stochastic variant of this problem. We start by showing that a bounded stochastic PNP induces a probability distribution over scenarios of a ProbDeclare model, obtained by collecting, scenario by scenario, the probability of all PNP traces that satisfy the characteristic formula of that scenario.

Definition 21 (Induced scenario distribution). *Let $\mathcal{N}$ be a bounded stochastic PNP, and $\mathcal{D}$ a ProbDeclare model. The* scenario distribution $\mathbb{P}_{\mathcal{D}}^{\mathcal{N}}$ *induced by $\mathcal{N}$ over scenarios* $\mathbb{S}_{\mathcal{D}}$ *is the probability distribution defined as follows: for every $\mathcal{S} \in \mathbb{S}_{\mathcal{D}}$, we have that* $\mathbb{P}_{\mathcal{D}}^{\mathcal{N}}(\mathcal{S}) = \sum_{\sigma \text{ trace of } \mathcal{N} \text{ s.t. } \sigma \text{ satisfies } \Phi_{\mathcal{S}}} \mathbb{P}_{\mathcal{N}}(\sigma)$. ◁

We then define the stochastic conformance problem S-CONFORM$(\mathcal{N}, \mathcal{D})$ as:

Input: bounded stochastic PNP $\mathcal{N}$, ProbDeclare model $\mathcal{D}$;
Output: Whether $\mathbb{P}_{\mathcal{D}}^{\mathcal{N}}$ is valid (in the sense of Definition 20).

We address the problem through iterated invocations of the VERIFY-PROB problem, one per scenario. Specifically, for each scenario $\mathcal{S} \in \mathbb{S}_{\mathcal{D}}$: (1) Construct DFA $A_{\mathcal{S}}$ for characteristic formula $\Phi_{\mathcal{S}}$ with standard techniques [7, 18]; (2) Get the probability value $p = $ VERIFY-PROB$(\mathcal{N}, A_{\mathcal{S}})$; (3) Check whether p is valid for $\mathcal{S}$ using the system of inequalities for Definition 20; (4) If this is the case, proceed with the next scenario, otherwise return No; and (5) If all scenarios have been checked, return Yes.

Example 7. Consider the bounded stochastic PNP $\mathcal{N}_{order}$ (assuming equal weights for all transitions) and the ProbDeclare model $\mathcal{D}_{order}$. The only scenario where C_2 holds is the one where C_2 is satisfied while C_3 is violated. The probability induced by $\mathcal{N}_{order}$ for this scenario corresponds to the outcome probability for $\mathcal{N}_{order}$ to finish with a payment. As discussed in Example 2, this is $\frac{1}{17}$, which is indeed $\geq \frac{1}{20}$. The only scenario where C_3 holds is the one where C_2 is violated while C_2 is satisfied. The probability induced by $\mathcal{N}_{order}$ for this scenario corresponds to the outcome probability for $\mathcal{N}_{order}$ to finish with a rejection. As discussed in Example 2, this is $\frac{3}{17}$, which is indeed $\leq \frac{1}{4}$. This witnesses that $\mathcal{N}_{order}$ stochastically conforms to $\mathcal{D}_{order}$. ◁

In case S-CONFORM is negative, the standard Earth Mover's Distance can be used to measure the deviation between the scenario distribution induced by $\mathcal{N}$ and the closed valid scenario distribution for $\mathcal{D}$. This realises a form of *stochastic delta analysis*.

7 Conclusion

We have provided formal methods and algorithmic techniques solving three key problems concerning reasoning on labelled Petri nets and their executions in a stochastic setting: outcome probability, verification, and stochastic model conformance. All techniques are implemented in the *StochasticLabelledPetriNets* plug-in of ProM. For solving systems of inequalities, we use a Java LP solver (LPSolve). Our approach lazily handles silent transitions when combining the reachability graph of the net with the automaton of a temporal property of interest.

A natural extension of this work is to incorporate our techniques into stochastic process mining pipelines, validating the resulting framework experimentally either using our own implementation or by invoking probabilistic model checkers [14]. We also want to study if and how our results transfer to richer settings, such as stochastic nets that map to Markov decision processes, as well as non-Markovian nets.

Acknowledgement. Marco Montali is partially supported by the Italian PRIN project PIN-POINT and the UNIBZ projects ADAPTERS and SMART-APP.

References

1. Baier, C., Katoen, J.: Principles of Model Checking. MIT Press, Cambridge (2008)
2. Bergami, G., Maggi, F.M., Montali, M., Peñaloza, R.: Probabilistic trace alignment. In: ICPM, pp. 9–16. IEEE (2021)
3. Bergami, G., Maggi, F.M., Montali, M., Peñaloza, R.: A tool for computing probabilistic trace alignments. In: Nurcan, S., Korthaus, A. (eds.) CAiSE 2021. LNBIP, vol. 424, pp. 118–126. Springer, Cham (2021). https://doi.org/10.1007/978-3-030-79108-7_14
4. Bruni, R., Melgratti, H.C., Montanari, U.: Concurrency and probability: removing confusion, compositionally. Log. Methods Comput. Sci. **15**(4) (2019)
5. Burke, A., Leemans, S., Wynn, M.: Stochastic process discovery by weight estimation. In: PQMI (2020)
6. Chiola, G., Marsan, M.A., Balbo, G., Conte, G.: Generalized stochastic petri nets: a definition at the net level and its implications. IEEE Trans. Soft. Eng. **19**(2), 89–107 (1993)
7. De Giacomo, G., De Masellis, R., Maggi, F.M., Montali, M.: Monitoring constraints and metaconstraints with temporal logics on finite traces. ACM TOSEM (2022)
8. De Giacomo, G., Vardi, M.Y.: Linear temporal logic and linear dynamic logic on finite traces. In: Proceedings IJCAI. AAAI Press (2013)
9. Desel, J., Reisig, W.: Place/transition petri nets. In: Reisig, W., Rozenberg, G. (eds.) ACPN 1996. LNCS, vol. 1491, pp. 122–173. Springer, Heidelberg (1998). https://doi.org/10.1007/3-540-65306-6_15
10. van Dongen, B.F., de Medeiros, A.K.A., Verbeek, H.M.W., Weijters, A.J.M.M., van der Aalst, W.M.P.: The ProM framework: a new era in process mining tool support. In: Ciardo, G., Darondeau, P. (eds.) ICATPN 2005. LNCS, vol. 3536, pp. 444–454. Springer, Heidelberg (2005). https://doi.org/10.1007/11494744_25

11. Durrett, R.: Essentials of Stochastic Processes. STS, 2nd edn. Springer, New York (2012). https://doi.org/10.1007/978-1-4614-3615-7
12. Fewster, R.: Stochastic Processes. Course Notes Stas 325, University of Auckland (2008)
13. Hanneforth, T., De La Higuera, C.: Epsilon-removal by loop reduction for finite-state automata over complete semirings. Studia Grammatica **72**, 297–312 (2010)
14. Kwiatkowska, M., Norman, G., Parker, D.: PRISM 4.0: verification of probabilistic real-time systems. In: Gopalakrishnan, G., Qadeer, S. (eds.) CAV 2011. LNCS, vol. 6806, pp. 585–591. Springer, Heidelberg (2011). https://doi.org/10.1007/978-3-642-22110-1_47
15. Leemans, S.J.J., van der Aalst, W.M.P., Brockhoff, T., Polyvyanyy, A.: Stochastic process mining: earth movers' stochastic conformance. Inf. Syst. **102**, 101724 (2021)
16. Leemans, S.J.J., Polyvyanyy, A.: Stochastic-aware conformance checking: an entropy-based approach. In: Dustdar, S., Yu, E., Salinesi, C., Rieu, D., Pant, V. (eds.) CAiSE 2020. LNCS, vol. 12127, pp. 217–233. Springer, Cham (2020). https://doi.org/10.1007/978-3-030-49435-3_14
17. Leemans, S.J.J., Syring, A.F., van der Aalst, W.M.P.: Earth movers' stochastic conformance checking. In: Proceedings of the BPM Forum, vol. 360, pp. 127–143 (2019)
18. Maggi, F.M., Montali, M., Peñaloza, R., Alman, A.: Extending temporal business constraints with uncertainty. In: Fahland, D., Ghidini, C., Becker, J., Dumas, M. (eds.) BPM 2020. LNCS, vol. 12168, pp. 35–54. Springer, Cham (2020). https://doi.org/10.1007/978-3-030-58666-9_3
19. Marsan, M.A., Conte, G., Balbo, G.: A class of generalized stochastic petri nets for the performance evaluation of multiprocessor systems. ACM TOCS **2**(2), 93–122 (1984)
20. Molloy, M.K.: Performance analysis using stochastic petri nets. IEEE Trans. Comput. **31**, 913–917 (1982)
21. Rogge-Solti, A., Mans, R.S., van der Aalst, W.M.P., Weske, M.: Repairing event logs using timed process models. In: Demey, Y.T., Panetto, H. (eds.) OTM 2013. LNCS, vol. 8186, pp. 705–708. Springer, Heidelberg (2013). https://doi.org/10.1007/978-3-642-41033-8_89
22. Rogge-Solti, A., van der Aalst, W.M.P., Weske, M.: Discovering stochastic petri nets with arbitrary delay distributions from event logs. In: BPMW 2013, pp. 15–27 (2013)

Incentive Alignment Through Secure Computations

Frederik Haagensen[ID] and Søren Debois[✉][ID]

IT University of Copenhagen, Copenhagen, Denmark
{haag,debois}@itu.dk

Abstract. We present a game-theoretic approach to analyzing the incentive structure in formal models of inter-organizational businesses processes. In such processes, the choices of each participants influence the outcome of others. A potential participant may be torn between the prospect of a highly preferable outcome on the one hand (e.g., a bonus on timely delivery), and the possibility that another player may make a choice (e.g., reallocation of the fast trucks) which renders that outcome impossible to achieve. We propose (a) an analysis which given the preferences of participants determines if the collaboration is at all meaningful; (b) an algorithm for modifying the incentive structure of such a process using both fines and outcome re-distribution to increase the benefit for all participants; and (c) a practical way of computing this algorithm while *concealing* the preferences of the collaborators for each other using secure multi-party computation.

Keywords: Game theory · Multi-party computations · MPC

1 Introduction

It is in the nature of businesses and organisations to collaborate, out of necessity and self-interest. Collaborations hold both the promise of great rewards, and the pitfalls of great losses. A collaborating organisation necessarily assumes some risk by relying on the actions of other parties. When such collaborations are structured enough, we model their interactions in formal notation, such as Collaboration Diagrams in BPMN [15] or distributed Dynamic Condition Response (DCR) Graphs [9]. Until recently, such formal models emphasised the sequencing of interactions between collaborators, disregarding the promises and pitfalls of engaging with others. However, with the advent of blockchain technology, there has been a surge of interest in the question of whether and how much an organisation must necessarily trust its collaborators, and to what degree that trust can be replaced by computer systems, e.g. [3,6,12,13].

This paper takes a game-theoretic approach to the same question: Rather than trying to enforce that collaborators follow protocol using blockchain technologies, we rely on their self-interest to deliver a particular outcome.

In practice, we achieve this by supplementing formal models of collaborations with profits ("utility") earned from their execution. Such models are dynamic

© Springer Nature Switzerland AG 2022
C. Di Ciccio et al. (Eds.): BPM 2022, LNCS 13420, pp. 343–360, 2022.
https://doi.org/10.1007/978-3-031-16103-2_23

344   F. Haagensen and S. Debois

games in the game theoretical sense—an idea recently investigated by Heindel and Weber [8]. Games can be analyzed to predict the behavior of rational players, answering the question "would this collaboration be beneficial for me?" assuming rational collaborators. We present a formal means of answering this question for collaborations modelled as finite-state machines/extensive form games in Sect. 3. We leave the generalisation to Petri Nets/BPMN as future work: concurrency does not appear to be central to incentive alignment.

Once we are able to compute such outcomes, we can strategically transfer utility (e.g., by shifting prices in contracts) in such a way that the new rational outcome makes all collaborators better off. We present an algorithm achieving such mechanical incentive alignment in Sect. 4.

However, an individual organization likely knows only about its own and shared parts of the process, not the internals of other parts, nor, crucially, the utility of outcomes to *other* collaborators. Moreover, the latter are almost certainly confidential: A shipping company generally does not want its profit on a particular delivery known to its customers. We therefore present in Sect. 5 a conceptual implementation of the incentive alignment in terms of Secure Multi-party computation (MPC) [4,7], which enables computing the best outcome and necessary utility transfer *without* the collaborators having to disclose their utilities (profits) to each other. We realise this conceptual implementation into an MP-SPDZ [10] based actual implementation in Sect. 6.

In summary, our contributions are as follows.

- We present a formal model of collaborative business processes as extensive form games (Sect. 3).
- We present algorithms computing the expected outcome for rational collaborators with known utilities (Algorithm 1), and for aligning incentives by utility transfer (Algorithm 2).
- We present a proof-of-concept multi-party computation implementation of these algorithms to avoid collaborators needing to disclose utilities such that the businesses do not have to reveal private information (Sect. 6).

Related Work. Heindel and Weber pioneered re-casting collaborative business process as games [8]. They used this formalisation to relate soundness of workflow nets to alignment of incentives. We relax their assumption of known utility functions to consider cases where a contract may not have been signed yet, or the utility functions for some other reasons are unknown; our collaborations are modelled as finite-automata rather than Petri nets; we treat presently only non-repeated games, and we model utilities only on outcomes, not on internal choices. (For non-repeating games, those two are equivalent.)

The present work also relates to the research area of trust in business processes. Rosemann [18] gives a four-stage model for how to build trust in a business process. The solution presented in this paper can be used in tandem with these stages and alleviate uncertainties and vulnerabilities in the process.

Finally, the present application of MPC technology (and the finite-state process model) is inspired by [7]. The aim in that work is not to compute properties of a potential collaboration before it begins, but rather to keep the state of each participants workflow secret from the others to the extent possible *during* the collaboration. However, the application is morally the same if one considers our proposed incentive-alignment algorithms as the beginning of the collaboration.

2   A Running Example

Here is a simple collaborative process presented as a game.

Example 1. Imagine a production company in the greater Copenhagen area engaging with a shipping company to have a very large turbine shipped from Berlin for use in a factory. The turbine is expensive and delicate, but bulky and unwieldy. The production company does not know exactly when the shipping needs take place, except within the coming 18 months. For this reason, the shipping company needs to keep its options open, and so the parties agree to this process:

- The production company initiates shipping, choosing whether the delivery should be by helicopter airlift or by specialised truck.
- If airlifted, the shipping company chooses whether to deliver to the Copenhagen CPH or the Malmö MMX airport.
- If delivered by truck, the shipping company chooses whether to deliver in the following weekend, mon-wed or thu-fri.
- If delivery is in the weekend, the production company chooses whether to take delivery saturday or sunday

The two parties have vastly different preferences for these outcomes: For the production company, taking delivery in airports is possible but requires additional in-house shipping. Truck delivery is better, with mon-wed delivery the best compromise between disruption to normal operations and the cost of calling in installation specialists outside normal working hours. However, thu-fri delivery will for complicated reasons incur problems with the local union, and should be avoided if at all possible.

For the shipping company, which has better agreements with its unions, saturday delivery is actually more cost-effective (because bulk of the actual transportation happens in the preceding work days), followed by thu-fri, airlift to the cheap Malmö airport, mon-wed and the expensive Copenhagen airport.

We can model this collaborative process using BPMN as seen in the top of Fig. 1. This model could be extended with extra activities to handle day or location delivery specific details which may run in parallel. Focusing on only the choices of the process we can extract a subprocess which can be represented as a tree, where internal nodes represent choices, leaf nodes outcomes; and where we annotate the internal nodes with the party making the choice, and the outcomes with the value (utility) of that outcome for either player. In this case, refining the above preferences to numerical values, we get the tree in the bottom of Fig. 1. Here P is the production company and S is the shipping company; choices and utilities are color-coded accordingly.

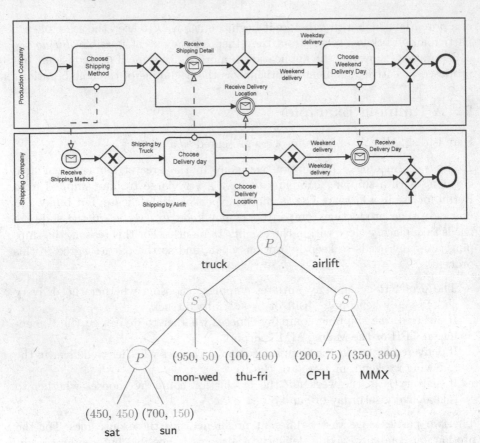

Fig. 1. The production-shipper BMPN (top) and extensive form game (bottom)

The four activities and their relationship from the BPMN are represented by the four decision nodes in the tree, and the six outcomes and utility values could be encoded in the BPMN using state variables set by the choice activities, or alternatively as unique combinations of performed activities.

It is important to note that the present work considers only this mutually observable and synchronized subprocess of the collaborative process model, i.e., the interface. Other internal activities by the production company or the shipping company leading to the decisions, or preparing for delivery, may exist, but are unnecessary for the purpose of analyzing the decisions. In essence, these details can be abstracted away in the utility values, which then allows for game theoretical analysis of which choices to make. When we give our formal model below, it is important to keep in mind that only this interface part of the collaborative process need to conform to our particular assumptions, *not* the process as a whole. We return to this point after the formalization below.

3   Games

We proceed to formalise the ideas of this example. Following [8], which extended BPMN and Petri nets with utility values and roles at transitions in the model, we extend deterministic finite state automata with a set of players, i.e. the businesses, utility functions for giving value to outcomes in the process, and a partitioning function to describe who acts when. We call these *automata deterministic finite-state game automata* (DFGA); they correspond to the first step in the four-stage model [18] to identify moments of trust in the process.

A detailed description of all the game theoretical concepts used is outside the scope of this paper. The curious reader should refer to [5] for an in-depth dive into game theory, or to [14] for a more computational oriented overview.

Definition 2. *A **deterministic finite-state game automaton** is a tuple* $A = (\Sigma, Q, q_0, \delta, I, \vec{u}, \pi)$, *where:*

- Σ *is a set of symbols, i.e. an alphabet, representing actions by a business*
- Q *is a finite set of states in the combined business processes*
- q_0 *is an initial state of Q; i.e. the beginning of the process*
- δ *is a transition function $Q \times \Sigma \to Q$ taking the process from one state to the next through some action*
- I *is a set of k players, i.e. the businesses*
- $\vec{u} = (u_1, ..., u_k)$ *is a player indexed vector of utility functions $\mathcal{L}(A) \to \mathbb{R}$ where given a complete trace of an execution, u_i assigns some real valued number as the payoff to player $p_i \in I$.*
- π *is a partitioning function $\mathsf{pref}(\mathcal{L}(A)) \to I$ which specifies the next party to continue the execution.*

Notation. For some DFGA A, and a fixed universe of symbols Σ, let $\mathcal{L}(A)$ denote the language of A, i.e. a set of traces, which are tuples of symbols from Σ, representing the history of a finished (complete) execution of a process, $\mathsf{pref}(\mathcal{L}(A))$ the set of trace prefixes of traces in $\mathcal{L}(A)$, and for a trace $\tau \in \mathsf{pref}(\mathcal{L}(A))$ and a symbol $\alpha \in \Sigma$ let $\tau\alpha$ be the trace obtained by extending the τ with α, that is, a continuation of the history. Note that $\mathcal{L}(A)$ and $\mathsf{pref}(\mathcal{L}(A))$ may be distinct, since while it may possible that at any point a business may stop cooperating, the other business may still have more work to be done to deal with this situation, like filing papers, or reallocating resources.

Utility functions represent the preferences over outcomes, which can be influenced by any factor imaginable, in general, too many to properly and accurately describe as a single number. We assume the utility function can be described using just profits, or losses, gained from executing the process. For publicly traded companies with a fiduciary duty this assumption seems well-founded.

Heindel and Weber assign utility gains and losses to transitions [8], whereas we require only the total value for a completed execution of the process. We are assuming the games do not loop, and therefore eventually terminate, and that players can not stop at an arbitrary point in the process, but only in states where they have such action available. In such situations there is no difference between knowing the final (aggregate) utility, or individual values at each transition since, in either case, we assume players will choose the set of transitions which optimizes their total utility. Games where players can stop arbitrarily can be modeled by including such action at every node in the process.

We assume the games do not loop, and are finite, and while many problems in process modelling are trivial in the finite case, solving games is hard already in the finite setting when actions can be taken asynchronously [14]. To keep the model finite, we require some maximum bound on the number of iterations through cycles, such that it is certain that the process eventually completes. This is equivalent to unrolling the cycles, transforming the process to a finite DAG. We therefore leave indefinitely iterative processes as future work.

The model is deterministic since we assume actions can be chosen deterministically even though the actual outcome may depend on unpredictable circumstances. This also corresponds to how businesses would expect each other to act. While the preference over actions may depend on what actually happens, this can be represented in the utility function by taking the expected value of outcomes. Rosemann [18] focus on uncertainties caused by inabilities to provide the necessary service, which can be modeled using randomness, but never addresses the problem of whether incentives in the process are properly aligned which is the focus of this paper. We leave models which require non-deterministic or random transitions, as a point of future work.

Asynchronous Behaviour. The model is synchronous in the game theoretical sense, meaning that players can observe each other's actions before acting themselves. In the context of the example in Fig. 1, this means, e.g., that the shipping company can react to the production company's choice of either truck or airlift. Using terminology from game theory, we limit ourselves to processes which can be described as games of *perfect information* and leave processes which needs to be described as *imperfect information* for future work.

Example 3. Expanding on the example in Fig. 1, if the production company's preference between a delivery sat or sun was entirely dependent on an internal decision by the shipping company which could not be observed before the date was picked then the model would fail the requirement of perfect information. But even if the shipping company and production company have to perform internal work asynchronously to prepare for a sat delivery, then if the utility value for the outcome can be captured by a static value, e.g., by expected value, then we consider the game perfect information and therefore fits our proposed solution.

Connections to Process Models. DFGAs may be considered process models in their own right, albeit downright primitive ones, when compared to contemporary process models such as a BPMN [15], DECLARE [16], or DCR Graphs [2]. They are nonetheless interesting to study because, as mentioned above and in Sect. 2, to study incentives in collaborative processes using game theory, it is sufficient to consider the sub-process or projection of the collaborative process that must be observable to both players. Internal behaviour *need not necessarily* be considered when analysing incentives.

For example, imagine that the our running example stems from a larger BPMN 2.0 Choreography [15] diagram, which included asynchronous internal behaviour and activities for both companies. In this case, the game in Fig. 1 is the result of some a projection of this larger process onto only those activities that occur immediately before a message is sent between the two collaborators.

What is required to apply the present work is, in this case and in general, *either* that the process being investigated is simple enough to be represented as a DFGA, *or* that such a projection exists, that it faithfully represents the behaviour of the larger process, and that it is representable as a DFGA.

These latter conditions will obviously not be met by every process, and it is an open question how to generally define such projections. We leave this important question for future work, noting here simply that we are encouraged by existing work on formalising projections of BPMN Choreography diagrams, e.g., [17, 19].

3.1 Game Behaviour of DFGAs

We can now describe the game theoretical aspects. A game is a description of a situation with rational actors, called players, actions they can take, and rewards or losses (utility), based on the actions they choose. The players being rational means they aim to maximize their utility. We will consider *extensive form games* which are games consisting of multiple stages with one player acting at a time; choosing the next action to take. In this model the players are rewarded at the end of the game [5]. Heindel and Weber uses *stochastic games* which are games where at each stage every player chooses an action, and the next transition is then the result of all player's actions and a probability distribution. This model is better for situations with asynchronous actions which can have random outcomes [8], but did not seem appropriate for our use case as explained above.

Definition 4. *An **Extensive form game** is a tuple $G = (I, T, \pi, \vec{u})$ where I is a finite set of k (rational) player, T is a rooted (game) tree consisting of non-terminal (inner) nodes, and terminal nodes (leaves), connected by edges. π is a function which partitions the set of non-terminal nodes into a subset for each player p_i, such that in the node n, player p_i chooses an edge leading to a new node n'. Finally, $\vec{u}$ is a player indexed vector of functions from a terminal node to a utility payoff for player p_i [5, 14].*

Game Theoretic Model. In summary, our game theoretical model uses extensive form games of perfect information where the utility functions are known. Extensive form games allows us to model the dynamic aspects of the processes, perfect information allows for much faster computational analysis than imperfect information games, and known utility functions allows for better optimization of business processes than if these were unknown. The issue of how to learn the utility function of other businesses is addressed in Sect. 5.

The present work is also a good candidate for applications of Cooperative game theory [14], which allows for analysis of cooperative behavior between players when external enforcement can be assumed, such as in our case using contracts. In particular, the notion of "Share distribution function" in Definition 11, seems amenable to analysis using techniques from cooperative game theory. However, the present first steps can be taken comfortably in the setting of the arguably simpler standard game theoretical tools.

Notation. In an extensive form game G, let $\mathsf{root}(G)$ denote the non-terminal root node of T, and $\mathsf{children}(N)$ the successors of the non-terminal node N. Lastly, we lift $\mathcal{L}$ to games such that $\mathcal{L}(G)$ denotes the set of terminal nodes in G, i.e. those representing complete traces, or outcomes, of the process.

We can then view a DFGA as an extensive form game by folding out the possible executions of the automaton as a tree, assigning players to non-terminal nodes and payoffs to terminal nodes. This leads to the intuitive Definition 5:

Definition 5. *The **induced game** of a deterministic finite-state game automaton $A = (\Sigma, Q, q_0, \delta, I, \vec{u}, \pi)$ is the extensive form game given by $G = (I, T, \pi, \vec{u})$ with players I, a rooted game tree T with non-terminal nodes $\mathsf{pref}(\mathcal{L}(A))$ and terminal nodes $\mathcal{L}(A)$, the partitioning function π and utility functions $\vec{u}$. There is an edge from some non-terminal node n to a non-terminal child n' iff for some $\alpha \in \Sigma : n\alpha \in \mathsf{pref}(\mathcal{L}(A))$ and an edge to a terminal child iff $n \in \mathcal{L}(A)$.*

In the remainder of this paper we only treat such induced games, with the understanding they were induced from by some process automaton.

The behaviour of players are represented using *strategies* [5, p. 4]. A strategy σ_i for a player p_i is an assignment of which successor node to choose for each node where p_i is choosing the successor[1]. To predict the outcome of a game different *solution concepts* exists. A solution concept is a restriction on the strategies of players, to describe only those strategies we would expect the players to follow, narrowing the possible outcomes of the game to those rational players would pursue to maximize their utility [5,14].

Following convention [5,14], we will refer to the vector of strategies as σ, and the vector of strategies of every player except p_i as σ_{-i}. As notation, we will use the strategy vector σ or (σ_i, σ_{-i}) interchangeably with the outcome of the game resulting from the players playing that strategy vector. The standard solution concept for the type of game described in this paper is the *sub-game perfect Nash equilibrium* [5]. A sub-game perfect Nash equilibrium (SPNE) is an outcome where it is not in players' interest to unilaterally change strategy, and where the players are doing the best response to each others possible actions. The traditional Nash equilibrium has the problem of allowing *empty threats* in extensive form games, which are strategies that a rational player would not credibly stick to, were the other player to ignore the threat.

Definition 6. *For some extensive form game $G = (I, T, \pi, \vec{u})$, a strategy vector σ is a **Nash equilibrium** in G iff, for all $p_i \in I$, and for all strategies σ_i':*

$$u_i(\sigma_i, \sigma_{-i}) \geq u_i(\sigma_i', \sigma_{-i})$$

---

[1] From a practical perspective, if a node can never be reached based on a previous choice it need not an assignment.

*A **sub-game** is a game obtained by using a non-terminal node as the root of a new game. A strategy vector σ is a **sub-game perfect Nash equilibrium** iff σ is a Nash equilibrium in any prefix of the original game [14].*

For deterministic games, if payoffs are distinct, there exists only a single SPNE [5,14]. That is, there exists a single outcome which maximize utility given the expected behavior of the other party. For our applications, we can assume such distinctness without loss of generality as a technical convenience.

We can find the unique SPNE using backwards induction [5,14], which is the technique to analyze the behavior of players by starting at the end of the game and working backwards. At the very last choice in the game, the player to move simply chooses the terminal node available to him that gives him the greatest utility. The non-terminal nodes are thus morally replaced by the outcome picked by the player and the process can propagate up the tree, applying the same logic at each step [5,14]. This also motivates the SPNE as the most reasonable solution concept, since any strategy which would not follow this chain of logic would seem irrational. We will refer to the player strategy of using backwards induction as the backwards induction strategy (BIS) and the outcome resulting from every player using BIS, as the backwards induction outcome (BIO).

Algorithm 1. BI

Input

    $G = (I, T, \pi, \vec{u})$ ▷ An extensive form game

Output

    $N \in \mathcal{L}(G)$ ▷ BIO terminal-node

1: **function** BackwardsInduction(n)
2:    **if** n is terminal **then**
3:       **return** n
4:    **end if**
5:    $n' \leftarrow \arg\max_{c \in \text{children}(n)} u_{\pi(n)}(\text{BackwardsInduction}(c))$
6:    **return** BackwardsInduction(n')
7: **end function**
8: **return** BackwardsInduction(root(G))

Backwards induction does, however, require that the game is finite, which we addressed earlier, and that the game is *Common Knowledge*, meaning that every player knows the game (i.e. set of players, game tree, the partition function, as well as the utility function of other players), and every player knows that the other players knows the game, and that every player knows that every player knows that the other players knows the game and so on ad infinitum[2]. In addition to the game being common knowledge, we also need to assume Common Knowledge of Rationality. Common knowledge is required to reason over other player's behavior which in turn guides one's own actions.

---

[2] Technically the players are only required to have k-level knowledge of the game, where k is the depth of the game tree.

Example 7. The SPNE of the game in Fig. 1, is for the turbine to be shipped to Malmö MMX, giving the production company P 350 utility and the shipping company S 300 utility. P picks shipping by helicopter airlift, since P knows that S will ship to MMX, not CPH, yielding 350 utility to P. This is prefered to P over shipping by truck, since P knows that then S will deliver thu-fri, and will thus only get 100 utility. P knows this, since P knows that S knows that were S to deliver in the weekend instead, then P wants it delivered on sun, which would only give S 150 utility instead of 400. Backwards induction uses the assumptions of Common Knowledge when nodes are propagated up the tree.

Using Algorithm 1 on the induced game of a DFGA, businesses can determine if it is in their interest to collaborate based on what they gain from the BIO. If it is not beneficial for some party to collaborate, it may be possible to modify the incentive structure such that it is.

4   Aligning Incentives Better

The previous section described how we can use game theory to find an outcome which we expect businesses not to deviate from, since it is in nobody's interest to *unilaterally* deviate from their strategy leading to this outcome. But what if this outcome is inefficient, and everybody agree another outcome would be better?

Example 8. We saw previously that the SPNE of the game in Fig. 1 is shipping by airlift to MMX, yielding utilities (350, 300). However, shipping by truck on sat with values (450, 450) is better for both. Why did they not choose that? This is because, if P is given the choice of which day of the weekend to get the turbine delivered, then P will always prefer sun, to get 700 utility instead of sat giving only 450, meaning P cannot be trusted to cooperate, since it would be irrational, without any external force, to choose to get less utility.

The issue with the SPNE, is that it may not be Pareto efficient [5,14], meaning there might exist some other outcome which would give at least one players more utility without resulting in less utility for the rest:

Definition 9. *For all strategy vectors σ in some extensive form game G, an outcome ϕ is **Pareto dominated** if there exists a strategy vector σ' with outcome ϕ', for which $u_i(\phi') > u_i(\phi)$ for some player p_i and for all $j \neq i : u_j(\phi') \geq u_j(\phi)$. If an outcome is not Pareto dominated, it is **Pareto efficient** [5,14].*

To improve efficiency of the game, we add fines for deviating from strategies leading to a Pareto efficient outcome. We call the strategy leading to the prefered outcome the *collaborative strategy* and modify the game such that the collaborative strategy is an SPNE. These fines can be implemented in practice as clauses in a contract between the businesses, making the modifications to the game binding, ensuring that if a player deviates, they can be taken to court to pay. In the example above, P would be fined for deviating from the collaborative strategy by choosing sun when choosing the day of the weekend for delivery.

We note that such fines correspond to the second stage in the four-stage model of Rosemann [18], by adding fines for deviating we reduce perceived uncertainty in the process, by making it clear which behaviour is undesired. Furthermore, fines can serve as compensation to a business for any lost revenue which also serves as the third stage in the four-stage model to reduce vulnerability.

There may, however, be multiple Pareto efficient outcomes in a game with each player having conflicting preferences over which of these should be picked. In Fig. 1, the outcomes giving $(450, 450), (700, 150)$, and $(950, 50)$ are all Pareto efficient, where S mostly prefers shipping by truck sun, giving $(450, 450)$, and P prefers shipping by truck mon-wed, giving $(950, 50)$. In order to maximize efficiency, we pick the outcome which maximize total utility gained. This outcome is Pareto efficient, since if it was not, then there would exist another outcome which gave at least one player more utility without giving any other player less, contradicting that it maximizes total utility gained.

Definition 10. *For some game $G = (I, T, \pi, \vec{u})$, the **total utility maximizing outcome** (TUMO) is the outcome $\Phi = \arg\max_{\phi \in \mathcal{L}(G)} \sum_{i \in I} u_i(\phi)$*

The TUMO may, however, lead to less utility for one player than if they simply played BIS. To make sure it is beneficial for every player, the total utility can be split, such that players get at least as much as the BIO, and any remaining excess is then split according to some agreed upon parameter. Splitting utility is possible, because in our case, utility is just money. This new modified outcome is also still Pareto efficient following the same argument as before.

Definition 11. *Let $S = (S_1, ..., S_k)$, be a tuple of unbounded, strictly monotonically increasing, **share distribution functions**, such that for all i and all $v \geq 0 : S_i(v) \geq 0$ and $\sum_{i=1}^{k} S_i(v) = v$.*

It is then possible to redistribute the utility of the TUMO such that each player gets at least as much utility from this outcome as they do the BIO.

Definition 12. *For some game $G = (I, T, \pi, \vec{u})$, with a TUMO Φ, call $\mathcal{T}(G)$ the surplus of G and let $\mathcal{T}(G) = \sum_{i \in I} u_i(\Phi) - u_i(\mathsf{BI}(G))$*

Proposition 13. *For some game $G = (I, T, \pi, \vec{u})$, the TUMO Φ in G, the BIO $\mathcal{B} = \mathsf{BI}(G)$, and for every share distribution function tuple S where $|S| = |I|$:*

$$\sum_{i \in I} u_i(\mathcal{B}) + S_i(\mathcal{T}(G)) = \sum_{i \in I} u_i(\Phi)$$

Proof. By definition $\Phi = \arg\max_{\phi \in \mathcal{L}(G)} \sum_{i \in I} u_i(\phi)$, it follows that $\mathcal{T}(G) \geq 0$ and by definition of S, $\sum_{i \in I} S_i(v) = v$, which leads to $\sum_{i \in I} u_i(\mathcal{B}) + S_i(\mathcal{T}(G)) = \sum_{i \in I} u_i(\mathcal{B}) + u_i(\Phi) - u_i(\mathcal{B}) = \sum_{i \in I} u_i(\Phi)$ ☐

Example 14. In Fig. 1 the outcome which maximizes total utility is shipping by truck mon-wed, giving $(50, 950)$. It is not in S's interest to pick this outcome over shipping thu-fri, which gives them 400 utility instead of only 50, but if P were to pay 351 to S for shipping mon-wed then S would prefer $50 + 351 = 401$ over 400. And P would have $950 - 351 = 599$. Both parties are now better off over the BIO, and also neither should want to deviate.

To account for businesses' inabilities to predict optimal play in cases where another party deviates, we base fines on the difference between the selected outcome after redistribution and the businesses' worst case outcome.

Proposition 13 along with the fines mentioned above, leads to the **Incentive Alignment** (IA) Algorithm 2. The algorithm yields an outcome, which together with redistributions and fines is an SPNE as specified in Proposition 15.

Algorithm 2. IA

    Input

        $G = (I, T, \pi, \vec{u})$ ▷ Extensive form game of a DFGA

        $S = (S_1, ..., S_{|I|})$ ▷ Player indexed utility share distribution functions

    Output

        $\Phi \in \mathcal{L}(G)$ ▷ Utility maximizing outcome

        $D = (D_1, ..., D_{|I|})$ ▷ Player indexed list of utility to be redistributed at Φ

        $F = (F_1, ..., F_{|I|})$ ▷ Player indexed list of fines for deviating

 1: $\mathcal{B} \leftarrow \mathsf{BI}(G)$

 2: $\Phi \leftarrow \arg\max_{\phi \in \mathcal{L}(G)} \sum_{i \in I} u_i(\phi)$ ▷ In case of ties with $\mathcal{B}$, pick $\mathcal{B}$

 3: **for all** $i \in I$ **do**   ▷ Redistribute total utility and calculate fines

 4:     $D_i \leftarrow u_i(\mathcal{B}) + S_i(\mathcal{T}(G)) - u_i(\Phi)$ ▷ A positive value indicates getting paid

 5:     $F_i \leftarrow \sum_{j \in I \setminus \{i\}} u_j(\mathcal{B}) + S_j(\mathcal{T}(G)) - min_{\phi \in \mathcal{L}(G)} u_j(\phi)$

 6: **end for**

 7: **return** (Φ, D, F)

Line 1 computes the BIO $\mathcal{B}$, and line 2 the TUMO Φ. Lines 3 through 6 calculates, for each player, how much the player should pay or get paid at Φ, the redistribution value, as well as the fine for deviating. The redistribution is calculated according to Proposition 13 using S, subtracting what the player gains at Φ. The fines are calculated as specified above, as the difference between the worst possible outcome and what a player would otherwise have gotten.

Proposition 15. *For any game $G = (I, T, \pi, \vec{u})$, and share distribution function S, with $(\Phi, D, F) = \mathsf{IA}(G, S)$, the collaborative strategy vector σ in G, leading to the TUMO Φ, is an SPNE in G after fines and redistributions have been added, that is, for all i and any σ_i': $u_i(\sigma_i, \sigma_{-i}) + D_i \geq u_i(\sigma_i', \sigma_{-i}) - F_i$.*

Algorithm 2 terminates with $O(|T| \times |I| + |I|^2)$ calls to $\vec{u}$. Backwards induction can be implemented such that each node is only visited a constant number of times, and TUMO can similar be computed by a single pass through the terminal nodes, but requires computing the total utility sum which means looping through every player, leading to the upper bound $O(|T| \times |I|)$. Since $\mathcal{T}(G)$ only needs to be calculated once D can be computed by looping through the players once. Computing F requires looping through each player in the outer loop on line 3, and then also another inner loop of players on line 5, which gives an upper bound of $O(|I|^2)$ calls to $\vec{u}$. The min value of each player only needs to be computed once which can again be done in $O(|T| \times |I|)$ calls to $\vec{u}$, leading to the total mentioned $O(|T| \times |I| + |I|^2)$ calls to $\vec{u}$. For games of reasonable size, calls to $\vec{u}$ can be done in constant time using a simple lookup table.

Even though Algorithm 2 returns just a single outcome, this outcome may not always be what happens in practice. A business may not have the resources available at the time in order to fulfill their obligations, e.g. the fast delivery of goods may be impossible because everybody is already working on something else. Alternatively, an accident may happen which causes a delay. Algorithm 2 only insures that it is not in the businesses' interests to trick each other, so when everything goes as planned, everybody works towards the optimal outcome. In practice we need to make sure the fines are not too punishing and that businesses can renegotiate in case of unforseen situations, such that the businesses may deviate from the computed outcome. This requirement can already somewhat be accounted for in the choice of how to distribute excess profit and in the initial utility values picked, e.g. using expected value. Including randomness into the games to model unforseen situations is left as a point of future work.

In this section we demonstrated how to improve the efficiency of a process by playing a collaborative strategy, leading to the TUMO of the process. Using Algorithm 2, we can find the TUMO, redistributions of the surplus generated, and appropriate fines, to make it a SPNE for every player to play the collaborative strategy without anybody gaining less than if players had played the BIS instead.

5 Working Together Under Uncertainty

Section 4 demonstrated inefficiencies with SPNE and how using Algorithm 2 leads to a more efficient process for everybody by redistributing gains and adding fines for deviating from the optimal outcome. But the DFGA is usually not common knowledge, and businesses do not want to share their internal costs and profits, since if leaked, competitors may be able to take advantage of this to take over their market share. So, how can we compute the TUMO, redistributions, and fines, without leaking private information?

Without common knowledge of the players' utility functions it is not possible for the players to use the chain of logic required to compute the SPNE. In Fig. 1, without common knowledge, P cannot predict which choice S will take, so P can either choose to ship by truck, and risk getting as little as 100 utility for the chance of up to 950 utility, or choose to ship by airlift, to increase the lower bound but also be unable to get more than 350 utility. This kind of game is refered to as a game of *incomplete information* and uses different solution concepts than the SPNE to take into consideration the players' lack of knowledge [5,14]. But, because these solution concepts require assumptions that players have some knowledge of each others' utility functions, we will introduce a different approach which works when nothing is known.

The solution we propose is to use Multi-party computations (MPC), also known as secure computations. MPC uses cryptographic techniques for various computational operations such that we can compute arbitrary algorithms on player input without revealing their values to other parties, only revealing the output of the algorithm [4]. Using MPC, businesses can keep their utility functions secret during computation of Algorithm 2 and calculations made using them will likewise be kept secret. When the outcome and fines have been calculated, these alone can be revealed to all parties involved. There exists MPC protocols robust against malicious adversaries: any attempts to learn the input of another party, or modify the result of the computation, can be detected before they succeed and the protocol can thus be aborted to avoid it [4].

We assume there are no hidden internal parts of the process which influence the decisions of any party. That is, the process can be accurately represented such that everything except the utility functions of the induced game is common knowledge. Incentive alignment in cases where businesses have internal actions which are significant for the analysis of the game is left as a point of future work, however, we note Guanciale et al. [7] as an example of how to use MPC to create the combined process, without leaking one's own, for the computation alone.

What if players misrepresent their utilities? Using MPC to run Algorithm 2, based on the private input of the players, gives rise to an extension of the original game, where players first pick the utility function they want to give to the algorithm, without knowing the input of other players, and then execute the process given the output of the algorithm. Game theory can describe when players may want to lie about their utility function, and assuming rationality, if it is in a player's interest to lie, we would expect them to, unless changes are made to the model, such that nothing to can be gained by lying.

We limit ourselves to security against a *rational adversary* who has *strict incomplete information* over other player's utility function. The adversary being rational will in this context mean they will not lie about their own utility function unless it can guarantee more utility without risking losing any, and strict incomplete information meaning that nothing is known about other player's utility functions, neither values, nor relations, and this information can thus not be used to inform how to lie. We leave the point of how to extend the proposed solution to adversaries with partial information as future work.

As notation, let $G_{u_i'}$ be the game obtained by changing the utility function u_i of p_i with u_i', that is $G_{u_i'} = (I, T, \pi, \vec{u'})$ where $\vec{u'} = (u_1, ..., u_{i-1}, u_i', u_{i+1}, ..., u_k)$. Assume a fixed $G^{00} = (I, T, \pi, \vec{u})$, where u_i is the true utility function of player p_i. Let $G^{10} = G_{u_i'}$ be the game where p_i reported some function $u_i' \neq u_i$. When introducing a u_j', we let $G^{01} = G_{u_j'}^{00}$ and $G^{11} = G_{u_j'}^{10}$. Here G^{00} and G^{01} are to be understood as two different games where p_i gives their true utility function, and G^{10} and G^{11} as the same games from before except p_i lies. Figure 2 gives a visual representation of the relationships between games where the arrows are to be understood as a player reporting a different utility function.

Assume a fixed player indexed tuple of share distribution functions S, and let $(\Phi^t, D^t, F^t) = \mathsf{IA}(G^t, S)$ for each of version of the game mentioned above. Note that when a lying player p_i reports u_i' instead of their true utility function u_i, they will still gain $u_i(\phi)$ at any reached outcome ϕ. We conjecture that there is no guaranteed way to gain utility from misrepresenting one's utility function or deviating from the TUMO:

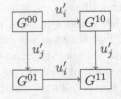

Fig. 2. Relation between games.

Conjecture 16. For all functions, which given a game tree, a partition function, and a player's utility function u_i, can compute a u_i' such that $u_i(\Phi^{00}) + D_i^{00} < u_i(\Phi^{10}) + D_i^{10}$, there exists some u_j' for some player $p_j \neq p_i$ such that $u_i(\Phi^{11}) + D_i^{11} < u_i(\Phi^{01}) + D_i^{01}$ and for all $\phi \neq \Phi^{11}$, $u_i(\phi) - F_i^{11} < u_i(\Phi^{11}) + D_i^{11}$.

6 Implementation

The previous sections describe how we can use MPC and game theory to analyze an inter-organizational business process and align incentives such that every party is better off, without leaking private information. In this section we will present a proof-of-concept prototype implementation of Algorithm 2 made using MPC techniques such that the utility values of businesses are kept private and only the outcome of the algorithm is revealed.

There are many different MPC protocols, the difference between these being the necessary assumptions they work under, the security guarantees they provide, the kinds of computations they can most efficiently compute, and whether or not communication is expensive, i.e. the latency between parties. In our case, we want to support an arbitrary number of parties, has security against malicious parties, and is efficient for computing comparisons and simple additions.

For the proof-of-concept we assume the communication is cheap, and based on this the SPDZ [1] protocol based on secret sharing, with the MASCOT [11] protocol for the offline phase seems appropriate. If communication is expensive the more appropriate protocols are based on garbled circuits such as BMR [4]. The SDPZ protocol consists of an offline and online phase. The offline phase is preprocessing to generate multiplication triplets, also known as Beaver triplets, and message authentication codes (MACs) to be used in the online phase [1,4].

To implement the prototype, we have used MP-SPDZ version 0.2.9, a framework for creating and testing MPC applications. MP-SPDZ consists of a virtual machine, a compiler which compiles (restricted) python code to bytecode for the virtual machine, and a library for privacy preserving data types and operations. MP-SPDZ supports writing MPC applications using a wide variety of existing protocols, for example the MASCOT and SPDZ combination we use [10].

The prototype consists of a Haskell program[3] for transforming a binary tree representation of the process structure, into code for the MP-SPDZ's compiler. The tree consists of information about which player is acting at which node and unique identifiers for each node, but not the utility functions of the parties. This tree is assumed common knowledge to the businesses. The generated code is then compiled with the MP-SPDZ compiler to bytecode for the protocols.

Figure 3 gives an overview of the toolchain. Boxed nodes indicate executable, dashed boxes indicate protocols used by an executable, and everything else is input or the final output. MASCOT is ran alongside SPDZ, but could in theory be ran at any point in the process. The player inputs are secret and are only needed for SPDZ, the rest is common knowledge.

---

[3] Online at https://github.com/Frehaa/mpc-games.

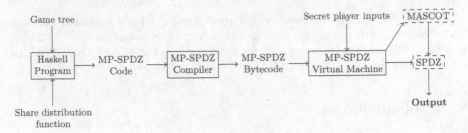

Fig. 3. Overview of toolchain

We assume without loss of generality that the game tree is binary. This can be achieved by replacing internal nodes with only a single child node with that child, and splitting any internal node with more than two children into a new node with one child as one of the choices, and another child as a new node, with the same player in control, and with the remaining choice, repeating the process if necessarily. Nodes with only a single child do not represent a meaningful decision (there is no choice), and nodes with multiple children still maintain the same choices, possibly by going through a series of the new split nodes.

Benchmarks. MPC sometimes incurs substantial overhead, so we present exploratory benchmarks in Fig. 4 to demonstrate the feasibility of the solution in practice. The benchmarks ran on an i7-10750H CPU @ 2.60 GHz with 32 GB of RAM. The benchmarks comprised randomly generated complete binary trees of specified increasing heights using the Haskell pro-

Nodes	511	1023	2047	4095
Compilation	53s	110s	216s	695s
Runtime	96s	190s	397s	774s
Total	149s	300s	613s	1469s

Fig. 4. Benchmarks

gram "generateRandomTree.hs", and player input was generated using "generate_data.py", both using the initial seed 1331. We use the default security parameter of 40 specified by MP-SPDZ. Multiplication triples are calculated in batches of 50. We used a constant 2 to 1 split and 2 players. For compilation, the -M flag is used to ensure memory accesses are performed in the correct order. Execution is done using the "mascot-party.x" virtual machine.

7   Conclusion and Future Work

We demonstrated how a combined inter-organizational business process can be analyzed using game theory, in order to determine if it is worth collaborating, or if incentives need to be better aligned. This includes a way to improve the process such that the selected outcome is maximizing effectively without businesses lossing out on profits. All done using MPC to make sure private information about a business' losses and profits are not leaked.

Besides deciding Conjecture 16, we see repeated games as an avenue for future work. Repeated games makes it possible to punish unwanted behaviour in a following game from one a player is already deviating in. Since businesses rarely collaborate only once and processes are often times repeated, the potential for lost utility in subsequent games can itself be a better deterrent than any fine.

Acknowledgements. We gratefully acknowledge helpful discussions related to MPC with Bernardo David, as well as useful feedback from reviewers. This work is supported by Independent Research Fund Denmark, grant 0136-00144B, "DISTRUST" project.

References

1. Damgård, I., Pastro, V., Smart, N., Zakarias, S.: Multiparty computation from somewhat homomorphic encryption. In: Safavi-Naini, R., Canetti, R. (eds.) CRYPTO 2012. LNCS, vol. 7417, pp. 643–662. Springer, Heidelberg (2012). https://doi.org/10.1007/978-3-642-32009-5_38
2. Debois, S., Hildebrandt, T.T., Slaats, T.: Replication, refinement & reachability: complexity in DCR graphs. Acta Informatica **55**(6), 489–520 (2018)
3. Di Ciccio, C., et al.: Blockchain support for collaborative business processes. Informatik Spektrum **42**(3), 182–190 (2019)
4. Evans, D., Kolesnikov, V., Rosulek, M.: A pragmatic introduction to secure multiparty computation. Found. Trends Privacy Secur. **2**(2–3), 70–246 (2018)
5. Fudenberg, D., Tirole, J.: Game Theory. MIT Press, Cambridge (1991)
6. García-Bañuelos, L., Ponomarev, A., Dumas, M., Weber, I.: Optimized execution of business processes on blockchain. In: BPM 2017, pp. 130–146 (2017)
7. Guanciale, R., Gurov, D., Laud, P.: Business process engineering and secure multiparty computation. Cryptol. Inf. Sec. Ser. **13**, 129–149 (2015)
8. Heindel, T., Weber, I.: Incentive alignment of business processes. In: Fahland, D., Ghidini, C., Becker, J., Dumas, M. (eds.) BPM 2020. LNCS, vol. 12168, pp. 93–110. Springer, Cham (2020). https://doi.org/10.1007/978-3-030-58666-9_6
9. Hildebrandt, T., Mukkamala, R.R., Slaats, T.: Declarative modelling and safe distribution of healthcare workflows. In: Liu, Z., Wassyng, A. (eds.) FHIES 2011. LNCS, vol. 7151, pp. 39–56. Springer, Heidelberg (2012). https://doi.org/10.1007/978-3-642-32355-3_3
10. Keller, M.: MP-SPDZ: a versatile framework for multi-party computation. In: Proceedings of the 2020 ACM SIGSAC Conference on Computer and Communications Security (2020)
11. Keller, M., Orsini, E., Scholl, P.: MASCOT: faster malicious arithmetic secure computation with oblivious transfer. In: Proceedings of the 2016 ACM SIGSAC Conference on Computer and Communications Security, CCS 2016, pp. 830–842. ACM, October 2016
12. Madsen, M.F., Gaub, M., Kirkbro, M.E., Høgnason, T., Slaats, T., Debois, S.: Collaboration among adversaries: distributed workflow execution on a blockchain. In: Symposium on Foundations and Applications of Blockchain (FAB 2018) (2018)
13. Mendling, J., Weber, I., Aalst, W.V.D., Brocke, J.V., et al.: Blockchains for business process management - challenges and opportunities. ACM Trans. Manag. Inf. Syst. **9**(1), 4:1–4:16 (2018)
14. Nisan, N., Roughgarden, T., Tardos, E., Vazirani, V.V.: Algorithmic Game Theory. Cambridge University Press, Cambridge (2007)

15. Object Management Group BPMN Technical Committee: Business Process Model and Notation, Version 2.0 (2013)
16. Pesic, M., Schonenberg, H., Van der Aalst, W.M.: Declare: full support for loosely-structured processes. In: 11th IEEE International Enterprise Distributed Object Computing Conference (EDOC 2007), pp. 287–287. IEEE (2007)
17. Poizat, P., Salaün, G.: Checking the realizability of BPMN 2.0 choreographies. In: Proceedings of the 27th ACM Symposium on Applied Computing, pp. 1927–1934 (2012)
18. Rosemann, M.: Trust-aware process design. In: Hildebrandt, T., van Dongen, B.F., Röglinger, M., Mendling, J. (eds.) BPM 2019. LNCS, vol. 11675, pp. 305–321. Springer, Cham (2019). https://doi.org/10.1007/978-3-030-26619-6_20
19. Tiezzi, F., Re, B., Polini, A., Morichetta, A., Corradini, F.: Collaboration vs. choreography conformance in BPMN. Log. Methods Comput. Sci. **16** (2020)

Business Process Simulation with Differentiated Resources: Does it Make a Difference?

Orlenys López-Pintado and Marlon Dumas$^{(\boxtimes)}$

University of Tartu, Tartu, Estonia
{orlenyslp,marlon.dumas}@ut.ee

Abstract. Business process simulation is a versatile technique to predict the impact of one or more changes on the performance of a process. Mainstream approaches in this space suffer from various limitations, some stemming from the fact that they treat resources as undifferentiated entities grouped into resource pools. These approaches assume that all resources in a pool have the same performance and share the same availability calendars. Previous studies have acknowledged these assumptions, without quantifying their impact on simulation model accuracy. This paper addresses this gap in the context of simulation models automatically discovered from event logs. The paper proposes a simulation approach and a method for discovering simulation models, wherein each resource is treated as an individual entity, with its own performance and availability calendar. An evaluation shows that simulation models with differentiated resources more closely replicate the distributions of cycle times and the work rhythm in a process than models with undifferentiated resources.

Keywords: Process simulation · Resource allocation · Process mining

1 Introduction

Business Process (BP) simulation [1] is a technique to analyze "what-if" scenarios, such as "what would be the cycle time of a process if the number of daily new cases increases by 20%?" (**S1**) or "what if two resources involved in a process become unavailable for an extended period of time?" (**S2**).

The starting point for BP simulation is a simulation model consisting of a process model enhanced with parameters capturing the available resource capacity, activity processing times, arrival rate of new cases, etc. It has been noted that existing BP simulation approaches suffer from various limitations [1,2,8]. Some of these limitations stem from incompleteness of, or inaccuracies in, the BP simulation model. These limitations are partly addressed by data-driven simulation methods [5,11], which automatically discover and calibrate simulation models from execution data (*event logs*). These methods ensure that the simulation model is better aligned with the observed reality [1,5,11]. Other limitations

© The Author(s) 2022
C. Di Ciccio et al. (Eds.): BPM 2022, LNCS 13420, pp. 361–378, 2022.
https://doi.org/10.1007/978-3-031-16103-2_24

of BP simulation approaches relate to assumptions made by the underlying BP simulator [1,8], most notably the assumption that resources are interchangeable entities. Specifically, mainstream BP simulation approaches, including data-driven ones, make the following assumptions:

A1 *Pooled resource allocation.* Each resource belongs to one resource pool (e.g., a role or group). Resource pools are disjoint. All instances of an activity are allocated to the same resource pool. For example, all instances of tasks *Check invoice* and *Schedule payment* are allocated to an *Accountant* pool.

A2 *Undifferentiated performance.* The processing time of an activity does not depend on the resource who performs it.

A3 *Undifferentiated availability.* All resources in a pool are available for work during the same time periods, e.g., Monday to Friday, 9:00–17:00.

In practice, each (human) resource has their own capabilities, performance, and availability. Previous studies have hypothesized that the above assumptions affect the accuracy of simulation models [1–3,8], but without quantifying their impact. In this setting, this paper addresses the following question: *Do assumptions A1–A3 affect the accuracy of a business process simulation model, and if so, to what extent?* The paper studies this question in the context of simulation models discovered from event logs. To address this question, the paper proposes and evaluates: (1) a business process simulation approach with differentiated resources; and (2) an automated method to discover a simulation model with differentiated resources from an event log. In the proposed approach, resources are not grouped into pools, but treated as individuals (unpooled allocation), the performance of each resource is independent of that of other resources (differentiated performance), and each resource may have its own availability calendar (differentiated availability). As a result, a simulation model can be used not only to answer what-if scenarios **S1** and **S2** above, but also scenarios such as: "what if resource R is replaced by resource R' with lower performance?" (**S3**) or "what if a resource changes their availability from full-time to part-time?" (**S4**).

The paper is structured as follows. Section 2 discusses related work. Section 3 formalizes assumptions **A1–A3** by presenting a simulation approach with undifferentiated resources. Section 4 presents a simulation approach with differentiated resources, while Sect. 5 proposes a corresponding method to discover simulation models. Section 6 empirically compares simulation models with differentiated vs. undifferentiated resources, and Sect. 7 concludes and sketches future work.

2   Related Work

Van der Aalst et al. [1,2] analyze three limitations of BP simulation approaches: unreliability of simulation models for short-term prediction, insufficient reliance on execution data to construct simulation models, and incorrect modeling of resources. The authors emphasize that resources often work part-time and that failure to capture this, leads to inaccurate simulations. In [13], the authors study the impact of workload on resource performance, i.e., to what extent resource performance varies depending on workload and the impact of this variability on simulation accuracy. Our contribution is related to these studies, but we focus on limitations that arise when resources are modeled as undifferentiated entities.

Afifi et al. [3] note that existing BP simulation approaches, including the BPSim simulation modeling standard [16], rely on role-based resource allocation, and do not support a wider range of resource allocation styles such as those identified in [15]. However, the authors do not quantify the impact of the identified limitations (e.g., role-based allocation) on concrete simulation scenarios.

Freitas & Pereira [8] reviews five BP simulation tools. They find that these tools do not allow one to define unavailability periods for individual resources. However, they do not evaluate the impact of this limitation. Some commercial simulation engines such as IBM Websphere Modeler[1] support the definition of "named resources", which can have their own timetables (differentiated availability). However, the activity processing times are defined at the level of tasks, and hence they do not support differentiated performance.

This paper studies the impact of resource differentiation on simulation models discovered from logs. Prior studies on BP simulation model discovery [5, 11, 14] assume that resources are available 24/7. In [7], the authors address this limitation by integrating a technique for discovering timetables into a simulation model discovery pipeline, assuming all resources in a pool have the same timetable.

3   Simulation Models with Undifferentiated Resources

A BP simulation model with pooled allocation and undifferentiated resources (herein, a *classic BP simulation model*) consists of a process model M (e.g., a BPMN diagram) enhanced with simulation metadata described in Definition 1.

Definition 1 (Classic BP Simulation Model). *A classic BP simulation model is a tuple $<E, A, G, F, RPools, Alloc, PT, BP, AT, AC>$, where E, A, G are respectively the sets of events, activities, and gateways of a BPMN model, F is the set of directed flow arcs of a BPMN model, and the remaining elements capture simulation parameters as follows:*

1. *RPools is a set of resource pools. Each resource pool $p \in RP$ represents a group of resources. The resource pools are disjoint, i.e., $\forall\ p_1, p_2 \in RPools$: $p_1 \cap p_2 = \emptyset$. Each resource pool is described by the following properties:*
   - SIZE(P) $\in \mathbb{N}$ *is the number of resources in the pool.*
   - AVAIL(P) *is a calendar (a set of intervals) during which every resource in p is available to perform activity instances.*
   - COST(P) *is the cost of each pool p per time unit (e.g., hour).*
2. ALLOC $: A \rightarrow RP$ *is a function mapping each activity $a \in A$ to one resource pool $p \in RPools$. A resource pool can perform many activities.*
3. *$PT : A \rightarrow \mathcal{P}(\mathbb{R}+)$ is a mapping from each activity $a \in A$ to a probability density function, modeling the the processing times of activity a.*

---

[1] https://www.ibm.com/support/pages/download-websphere-business-modeler-advanced-v70.

4. BP: $F \to [0,1]$ *is a function that maps each flow* $f \in F$ *s.t., the source of* f *is an element of* G *to a probability (a.k.a., the branching probability).*
5. AT $\in \mathcal{P}(\mathbb{R}+)$ *is a probability density function modeling the inter-arrival times between consecutive case creations.*
6. AC *is calendar (set of intervals) such that cases can only be created during an interval in* AC.

Given that in classic BP simulation models, resource pools are disjoint, they cannot capture scenarios where participants share their time across multiple pools (cf. assumption **A1** in Sect. 1). Also, since all resources in a pool have the same timetable, these models cannot capture scenarios where a pool incorporates some part-time resources and some full-time ones (assumption **A3**). Finally, in classic BP simulation models, the processing times of an activity do not depend on the resource that performs it. Hence, such models cannot capture scenarios where some resources in a pool are faster or slower than others (assumption **A2**).

When executed in a simulation engine, a (classic) BP simulation model produces an event log as per Definition 2. Herein, we call *simulated logs* those logs produced by a simulation and *real logs* those extracted from information systems.

Definition 2 (Event log). *An event log* E *is a set of events, each representing the execution of an activity instance in a process. An event* $e \in E$ *is a tuple* $e = <\alpha, r, \tau_0, \tau_s, \tau_c>$, *where* α *is the label of one activity in a business process (i.e.,* e *is an instance of the activity* α*),* r *is the resource who performed* α*,* τ_0 *is the timestamp in which the activity instance was enabled to be executed, and* τ_s, τ_c *are, respectively, the timestamps corresponding to the beginning and end of the activity instance. A trace (a.k.a., process case) is a non-empty sequence of events* $t = <e_1, e_2, ..., e_n>$, *and an event log* $L = <t_1, t_2, ..., t_m>$ *is a non-empty sequence of traces, each capturing one instance of a process (i.e., a case).*

Various performance metrics can be computed from a log, including: *waiting time* – the time-span from the moment the activity is enabled until the starting of the corresponding event; *processing time* – the time-span between beginning and end of the event; *cycle time* – the difference between the end time and start time of a case; and *resource utilization* – the ratio between the time a resource is busy executing activity instances, divided and its total availability time.

4   Simulation Models with Differentiated Resources

To lift the limitations imposed by assumptions **A1–A3** (cf. Sect. 1), we propose an approach to BP simulation with differentiated resources. In this simulation model, the notion of *resource pool* is replaced by that of *resource profile*. Like a resource pool, a resource profile models a set of resources that share the same availability calendar. However, unlike classic BP simulation models, an activity in a process model may be assigned to multiple resource profiles and the same resource profile may be shared by multiple pools. For example, in a claims handling process, there may be a resource profile for *junior claims handler*, another for *senior claims handler* and a third for *lead claims handler*, each with different calendars. Activity *Analyze claim* may be assigned to *junior claims handler* and *senior claims handler*, i.e., an instance of *Analyze claim* may be performed by a junior or by a senior claims handler. Meanwhile, activity *Assess claim* may be assigned to *senior claims handler* and *lead claims handler*. Finally, activity *Approve large claim* may be assigned to *lead claims handler*, i.e., only lead claims handlers may perform this activity. Another difference is that in a classic simulation model, each activity is mapped to a distribution of processing times. Meanwhile, in a simulation model with differentiated resources, the distribution of processing times depends not only on the activity, but also on the resource profile. Thus, the distribution of processing times of the activity *Analyze claim* when assigned to a *junior claims handler* is different than when assigned to a *senior claims handler*, e.g., seniors may be faster, on average, than juniors.

Definition 3 (BP simulation model with differentiated resources). *A BP simulation model with differentiated resources DSM is a tuple <E, A, G, F, RProf, BP, AT, AC>, where E, A, G are the sets of events, activities, and gateways of a BPMN model, F is the set of directed flow arcs of a BPMN model, and the remaining elements capture simulation parameters as follows:*

1. *RProf* $= \{r_1, ..., r_n\}$ *is a set of resource profiles, where n is the number of resources in the process, and each resource $r \in R$ is described by:*
   - Alloc $(r) = \{\alpha \mid \alpha \in A\}$ *is the set of activities that r can execute,*
   - Perf $(r, \alpha) = R \times A^m \rightarrow \mathcal{P}^m(\mathbb{R}+)$ *is a mapping from the resource r to a list of density functions over positive real numbers, corresponding to the distribution of processing times of each activity $\alpha \in$ Alloc, with m being the number of activities that r can perform,*
   - Avail(r) *is the calendar (a set of intervals) in which the resource r is available to perform each activity $\alpha \in$ Alloc,*
   - Cost(r) *is the cost of the resource r per time unit (e.g., hour)*
2. *BP, AT, and AC are defined as in Definition 1.*

The key difference between Definition 3 and Definition 1 is that instead of mapping each activity to a pool, Definition 1 maps each resource profile to the set of activities, and for each activity, it captures the corresponding probability density function of processing times. Note that a classic simulation model can be converted into a model with differentiated resources by mapping each resource pool to one resource profile. However, a scenario where an activity is assigned to multiple resource profiles cannot be captured as a classic simulation model. Note also that if every resource profile has a size of one (i.e., one profile per resource), each resource may have different performance and availability. In Sect. 5, we focus on discovering such models with individualized resources.

The operational semantics of simulation models with differentiated resources is captured by Algorithm 1. This algorithm takes as input a simulation model *DSM* according to Definition 3, the number *pCases* of process instances to simulate, and the timestamp *startAt* of the beginning of the simulation. Like in a classic BP simulation engine, the simulation produces a log and the performance

Algorithm 1. Snippet of processes simulation with differentiated resources

```
1:  function SIMULATEPROCESS(DSM, pCases, startAt)
2:      for each resource r ∈ DSM do
3:          readyAt[r] ← minFrom(AVAIL, startAt)
4:      diffResQ ← DIFFRESOURCEQUEUE (ALLOC, AVAIL, SORTINGCRITERIA= min(readyAt))
5:      evtQ ← GENERATEALLARRIVALEVENTS (pCases, DSM, AT, AC)
6:      while evtQ not empty do
7:          e ← POPEVENT(evtQ)
8:          e[r] ← POPRESOURCE(diffResQ, e[α])
9:          e[τ_s] ← max(e[τ_0], readyAt[e[r]])
10:         e[τ_c] ← e[τ_s] + IDLEPROCESSINGTIME (e[τ_s], e[r], e[α], AVAIL, PERF)
11:         readyAt[e[r]] ← e[τ_c] + IdleTime(r, AVAIL, e[τ_c])
12:         UPDATERESOURCEAVAILABILITY(diffResQ, e[r])
13:         UPDATESIMULATEDEVENTLOG(e)
14:         state, enabled ← UPDATEPROCESSSTATE(e[α], e[pState], DSM, BP)
15:         for each α' ∈ enabled do
16:             nE ← Event(α = α', τ_0 = e[τ_c], pState = state)
17:             ENQUEUEENABLEDEVENT(evtQ, nE)
```

indicators in Sect. 3. Due to space limitations, we illustrate steps related to the generation and update of the simulation events, focusing on the functions in Definition 3, but omitting the details of the data structures and algorithms required to handle the event logs, calendars, scheduling, and estimation of performance indicators.

The first issue to handle in models with differentiated resources is that they can be shared among several tasks. Unlike undifferentiated models, which allow only one pool per activity, multiple resource profiles may be allocated to each activity in differentiated scenarios. To address this, we use a multi-queue data structure named DIFFRESOURCEQUEUE, initialized in line 4. The queue groups the resources by activities according to function ALLOC, restricting allocated resources to the remaining shared activities. Besides, resources are sorted in the queue according to a priority function SORTINGCRITERIA given as input. By default, the resource sorting criteria consider the minimum timestamp in which each resource will be ready to perform an activity, i.e., stored in the map readyAt. Thus, the values in the map readyAt (initialized in lines 2–3) are calculated considering the resources working calendars, given by the function AVAIL, and the periods in which resources are busy performing activities during the simulation. The support for multiple sorting criteria in DIFFRESOURCEQUEUE opens many options for prioritizing and sorting resources following different criteria, e.g., allocate resources according to their expertise given some conditions.

Next, function GENERATEALLARRIVALEVENTS in line 5 produces the initial event (see Definition 2) of each process case to simulate, i.e., according to the arrival time distribution AT, in the intervals defined by the arrival calendar AC. The queue evtQ stores and retrieves all the simulated events according to the timestamp in which the corresponding activity α was enabled. Then the simulation proceeds until there is not enabled event in evtQ (line 6). We are using the notation e[r], e[α], e[τ_0], e[τ_s] and e[τ_c] referring respectively to the resource allocated, activity name, enabling, starting and completing times of the event e (see Definition 2). Additionally, e[pState] represents the marking over the flow-arcs of the corresponding process instance at the moment of the event creation.

This marking simulates the token game as specified in the BPMN standard. For each process instance created by the function GENERATEALLARRIVALEVENTS, it generates tokens that traverse the flow-arcs in the model until reaching the end event in the BPMN model. An element in the control flow becomes enabled when one or many tokens arrive at its incoming flow-arcs (i.e., according to the element execution semantics). Similarly, the execution of an enabled element consumes the incoming tokens, generating new ones on its outgoing flow-arcs.

The queue evtQ only stores enabled events. Thus, the attributes $e[r]$, $e[\tau_s]$ and $e[\tau_c]$ are determined and updated once the corresponding event is popped from evtQ, i.e., the event is then executed. In lines 7–8 of Algorithm 1, the event with the lowest enabling timestamp in evQ is allocated to a resource, according to availability and allocation criteria passed to the resources queue diffResQ, i.e., selecting the participant being available the earliest as default criteria.

When the event is enabled, the allocated resource may not be according to their calendar (and vice-versa). Thus, the starting timestamp of the event relies on both task and resource availability (line 9). Next, in line 10, the completion timestamp is calculated by the function IDLEPROCESSINGTIME which adjusts the ideal processing time (if the resource works in the task without interruption according to PERF), plus the time the resource may rest from their calendar in AVAIL. Similarly, function IDLETIME calculates the next timestamp the resource is available after completing the task, updating the resource queue accordingly (lines 11–12). Finally, lines 14–17 update the process state, retrieving the activities enabled after executing the current event, queuing them as events in evtQ with enabling time equal to the completion time of the previous event.

5   Discovering Differentiated Resources Profiles

This section proposes an approach to discover simulation models with differentiated resources described in Sect. 4. Due to space limitations, we focus only on the main steps to discover differentiated resource profiles from event logs, i.e., to model each resource performance and availability independently. Before describing our proposal, Definition 4 formalizes the weekly calendars, followed by Definition 5 introducing some notations we will use across this section.

Definition 4. *A weekly calendar $\widehat{C}$ is binary relation $W \times \Delta$ between the set of weekdays, $W = \{Monday, ..., Sunday\}$, and a set of time granules $\Delta = \{\delta_1, ..., \delta_n\}$ where $\bigcap_{i=1}^{n} \delta_i = \emptyset$. Each time granule $\delta_i \in \Delta$ is a sorted pair of time points $<\tau_s^w, \tau_c^w>$, such that $\tau_s^w, \tau_c^w = <hour, minute, second>$, $hour \in [0, ..., 23]$, $minute, second \in [0, ..., 60]$, and $\tau_s^w \leq \tau_c^w$. A calendar entry κ is a tuple $<\omega, \tau_s^w, \tau_c^w>$ representing a time interval for a given day. For example, $\kappa = <Monday, 08:15:00, 12:00:00>$ describes Monday from 08:15 to 10:30.*

Definition 5 (Notations).

- *Given an event log L: E is the set of all the events in L, R and A are, respectively, the sets of resources and activities in any event $e \in E$. Besides,*

$A_r = \{\alpha \in A \mid \exists\, e \in E, r \in R, \alpha \in A : r, \alpha \in e\}$, and $E_r = \{e \in E \mid r \in R \land r \in e\}$ are the set of activities and events executed by the resource r, respectively. With, $E_\alpha = \{e \in E \mid \alpha \in A \land \alpha \in e\}$ being the set of events, which are instances of the activity α, and $E_{r,\alpha} = \{e \mid e \in E_r \cap E_\alpha\}$ the set of instances of α executed by the resource r.

- Γ is function mapping a timestamp in the event log into a calendar entry $\kappa = <\omega, \tau_s^w, \tau_c^w>$, where $<\tau_s^w, \tau_c^w>$ spans n minutes. Specifically, Γ retrieves an interval of size n containing the timestamp received as input. Note that, Γ retrieves intervals assuming that days are split into intervals of equal size n starting from the $00:00:00$ h, e.g., from $n = 15\,min$ days are split as $[00:00:00 - 00:15:00), [00:15:00 - 00:30:00), ..., [23:45:00 - 00:00:00)$. For example, consider a calendar with time intervals of $15\,min$, for the timestamp $2022 - 01 - 01T08:12$, the function Γ returns the calendar entry candidate $<Saturday, <08,00,00>, <08,15,00>>$.

- $\Omega_r^n = \{\kappa^{m(\kappa)} \mid \forall <\tau_s, \tau_c> \in E_r,\ n > 0,\ \kappa = \Gamma(\tau_s, n) \land \kappa = \Gamma(\tau_c, n)\}$ is a multi-set of calendar entry candidates of duration n mapped from the starting and ending timestamps of each event executed by the resource r, with the supra-index $m(\kappa)$ being the number of calendar entries κ in Ω_r^n.

- $\Omega_{r,\alpha}^n = \{\kappa^m \mid \kappa \in \Omega_r^n \land \alpha \sim \kappa,\ n > 0\}$ is the subset of Ω_r^n containing all the calendar entry candidates that are instances of the activity α, with $\sim$ representing that an instance of α occurred in the calendar entry κ.

To discover resource availability calendars, we take inspiration from the approach in [9], which discovers repetition patterns from a set of time granules with a certain level of confidence and support. The latter approach assumes time intervals that are covered entirely. This condition does not hold when discovering working intervals of a resource, since the event log shows only the start and completion timestamps of each event, and gives no information about what happens in two timestamps. Also, the start of an event is conditioned by the enablement of the related activity, i.e., a resource can be available but still needs to wait to start an activity until it becomes enabled in the process. Thus, we redefined the confidence and support metrics in [9] to discover calendars over time granules not fully described by the input data. Furthermore, we filter the resources with low frequency according to their relative participation, to exclude external resources (i.e., resource who seldom participate in the process), as there is insufficient data to discover availability calendars for such resources individually.

Other approaches such as [12] can be used to discover resource availability calendars. In this latter work, the authors use the activity waiting and processing times of each activity to estimate the intervals resources are available according to an input event log. Thus, [12] assumes that available resources will work as soon as an activity is enabled, and they keep working during the entire activity's execution interval (without any break). In this paper, do not make this assumption. Instead, we only assume that the resource was working when the activity instance starts and when it completes, and in-between, we consider that the resource may or may not be working. In any case, the calendar discovery approach in [12] could be used as an alternative to the approach presented here.

Definitions 6, 7, and 8 describe, respectively, the metrics of confidence, support, and resource participation we use to filter and discover the resource profiles. The metrics retrieve a real number between 0 (worst assessment) and 1, the best possible value. The activity-conditional confidence, given a calendar entry $\kappa = <\omega, \tau_s^w, \tau_c^w>$ related to an activity α, measures the ratio between the number of times α was started or completed on the weekday ω between τ_s^w and τ_c^w, divided by the total of weekdays ω that α occurred. For example, it measures from every Monday a resource was observed executing a given activity, how often it happened between 8:00 AM–8:15 AM. Definition 6 generalizes the metric to a set of tasks executed by a resource in the same time granules as the maximum between the individual value computed for each activity. The support metric computes from all the timestamps a resource was active in the log, what ratio is covered by some calendar entry. Finally, the participation metric estimates the ratio of events performed by a resource compared with the number of events executed by the most frequent resource. The comparison is relative to the activities each resource can perform. For example, resources r_1 and r_2 may execute 10 and 1000 events, respectively. If we compare r_1 and r_2 globally, then r_1 has a participation ratio of 0.01 compared to r_2. However, if r_1 and r_2 execute different activities, and if r_1 is the only executing all the instances of an activity, the relative ratio is 1.0 as r_1 is relevant to the activity r_1 performs alone.

Definition 6. $\text{Confidence}(r, \kappa) = \frac{\max\limits_{\alpha \in A_r, \alpha \sim \kappa} |\Omega_{r,\alpha}^n|}{|\{\omega^m \mid \omega \in W \wedge \omega \in \Omega_{r,\alpha}^n\}|}$ *computes the activity-conditioned confidence of a calendar entry* $\kappa = <\omega, \tau_s^w, \tau_e^w>$. *The multiset in the fraction denominator computes how many times each activity* α *was executed on the weekday* ω.

Definition 7. $\text{Support}(r, \widehat{C}) = \frac{|\{\kappa^m \mid \kappa \in \Omega_r^n \wedge \kappa \in \widehat{C}\}|}{|\Omega_r^n|}$ *computes the support of a given calendar* $\widehat{C}$, *where the multi-set in the fraction numerator computes how many calendar entries* κ *from the multi-set of candidates* Ω_r^n *are covered by* $\widehat{C}$.

Definition 8. $\text{RParticipation}(r) = \frac{\sum_{\alpha \in A_r} |E_{r,\alpha}|}{\sum_{\alpha \in A_r} \max\limits_{r' \in R} |\{E_{r',\alpha}\}|}$ *computes the relative participation of a resource r. The fraction numerator computes the number of events executed by r. The denominator sums up all the events executed by each resource who executed the most events for each activity executed by r.*

Algorithm 2 captures the main steps to calculate differentiated resource profiles. It takes as input an event log, a BPMN model, the size n of the granules in the calendar, the desired support, confidence, and participation values, and the minimum number of data points required to infer the processing-time distributions. Line 2 extracts from the log the sets and multi-sets described in Definition 5, followed by the initialization of the mappings ALLOC, AVAIL and PERF in Definition 3. Lines 4–9 discard the resources with low relative participation (Definition 8), storing (in the mapping AVAIL) the discovered calendars of each resource over the required threshold. Function EXTRACTCALENDARENTRIES, in line 5 transforms the timestamps in which each resource was active into calendar

---

Algorithm 2. Resource Profiles Discovery (from event logs)

```
1:  function DISCOVERRESOURCEPROFILES(L, DSM, n, dSupp, dConf, dPart)
2:      PARSEEVENTLOG(L)                                    ▷ To extract sets and multi-sets in Def. 5
3:      ALLOC, AVAIL, PERF ← ∅, ∅, ∅
4:      for each r ∈ R do
5:          Ω_r^n ← EXTRACTCALENDARENTRIES(E_r, Γ, n)
6:          if RPARTICIPATION(r) ≥ dPart then
7:              AVAIL[r] ← DISCOVERCALENDAR(Ω_r^n, dSupp, dConf)
8:          else
9:              AVAIL[r] ← ∅
10:     for each α ∈ A do
11:         discarded ← ∅
12:         for each r ∈ R : AVAIL[r] = ∅ and E_{r,α} ≠ ∅ do
13:             discarded.ADD(E_{r,α})
14:         jointR ← MAXDISJOINTINTERVALS(discarded)
15:         for each r ∈ jointR do
16:             Ω_r^n ← EXTRACTCALENDARENTRIES(jointR, Γ, n)
17:             Ĉ ← DISCOVERCALENDAR(Ω_r^n, dSupp, dConf)
18:             if Ĉ ≠ ∅ then
19:                 AVAIL[r].ADD(Ĉ)
20:         if ISUNALLOCATED(α) then
21:             BUILDUNRESTRICTEDCALENDAR(jointR, AVAIL)
22:     for each r ∈ AVAIL : AVAIL[r] ≠ ∅ do
23:         ALLOC[r] ← A_r
24:     for each α ∈ A do
25:         PERF[α].ADD(DISCOVERPROCESSINGTIMES(E_α, R))
26:     return ALLOC, AVAIL, PERF
```

Algorithm 3. Calendar Discovery

```
1:  function DISCOVERCALENDAR(Ω_r^n, dSupp, dConf)
2:      Ĉ, discarded ← ∅, ∅
3:      for each <ω, τ_s^w, τ_c^w> ∈ Ω_r^n do
4:          if CONFIDENCE(<ω, τ_s^w, τ_c^w>, Ω_r^n) ≥ dConf then
5:              Ĉ.ADD(<ω, τ_s^w, τ_c^w>)
6:          else
7:              discarded.ADD(<ω, τ̂_s, τ̂_c>)
8:      if SUPPORT(Ĉ, Ω_r^n) < dSupp then
9:          SORTMULTISETBYMULTIPLICITY(discarded[r], order=decreasing)
10:         for <ω, τ_s^w, τ_c^w> ∈ discarded do
11:             Ĉ.ADD(<ω, τ_s^w, τ_c^w>)
12:             discarded.REMOVE(<ω, τ_s^w, τ_c^w>)
13:             if SUPPORT(Ĉ, Ω_r^n) ≥ dSupp then
14:                 break
15:     return Ĉ
```

Algorithm 4. Processing Time Distribution Discovery

```
1:  function DISCOVERPROCESSINGTIMES(E_α, R, binSize = 50)
2:      D̂ ← ∅
3:      pendingResources ← ∅
4:      for each r ∈ R do
5:          if |E_{r,α}| ≥ binSize then
6:              D̂[r] = BESTFITTEDDISTRIBUTION(E_{r,α}, ALLOC, binSize)
7:          else
8:              pendingResources.ADD(r)
9:      jointD ← BESTFITTEDDISTRIBUTION(E_α, binSize)
10:     for each r ∈ pendingResources do
11:         D̂[r] ← jointD
12:     return D̂
```

entries according to Γ (cf. Definition 5). Function DISCOVERCALENDAR in line 7 is described by Algorithm 3.

To discover a calendar, Algorithm 3 receives a multi-set of calendar entry candidates of a given resource r. Then, lines 3–7 iterate over each candidate, adding those with confidence above $dConf$ in the calendar $\widehat{C}$, discarding the remaining ones. Next, line 8 verifies if the calendar achieved the required support $dSupp$. If not, the algorithm adds the most frequent entries until reaching the required support (lines 9–14). Thus, the algorithm relies on confidence only to filter potential outliers among the entry candidates, prioritizing that the calendar always covers the ratio of timestamps described by the support.

Filtering the resource and calendar entries in lines 6–7 of Algorithm 2 may cause the coverage of some tasks to become too low. As a result, an activity that is executed rarely or that is executed by external resources (i.e., resources from outside the organization, who seldom participate in the process) can lose all their resources, if none of them fulfills the participation threshold. This issue is addressed by Algorithm 2 in lines 10–21 by grouping the events of the removed resources related to each activity and assigning them to *aggregated resources*. Function MAXDISJOINTINTERVALS takes those grouped events and: (1) Sort them in ascending order of their start times τ_s, (2) add event e' with the highest τ_s, deleting all events whose time interval intersects e', (3) repeat (1)–(2) until no intervals remain. Next, an aggregated resource is created from each set of events retrieved. The calendar of the aggregated resource is built from the maximal set of mutually disjoint time intervals [4], i.e., by grouping the calendar entries that were discarded due to low confidence. Then, lines 15–19 create a calendar for each aggregated resource. If none fulfills the confidence and support requirement, lines 20–21 retrieve a single calendar as an aggregation of all the discarded events of the related activity without checking for confidence and support values.

Lines 22–23 of Algorithm 2 allocate, to each discovered resource, the activities executed by them in the event log. Then, function DISCOVERPROCESSING-TIMES (line 25) estimates the differentiated resource performance as described in Algorithm 4, which from every pair activity resource (lines 4–5), validates the number of events extracted fulfills a certain level of significance $binSize$ (above 50 by default). Resources below the threshold $binSize$ are grouped, with their performance discovered as an aggregation of all their events (lines 7–11). Function BESTFITTEDDISTRIBUTION adjusts each event duration by the calendar of the corresponding resource. Then, it builds a histogram from the event durations and applies curve-fitting to find a probability distribution, from a library of distributions, that best approximates the histogram (the one with lowest residual sum).

6   Implementation and Evaluation

We implemented the proposed approach as an open-source (Python-based) simulation engine, namely PROSIMOS, available at https://github.com/Automated ProcessImprovement/Prosimos. PROSIMOS supports the simulation of processes with an unpooled allocation model and differentiated availability and performance as per Sect. 4. Besides, it provides a component to automatically discover

Table 1. Characteristics of the business processes used in the experimentation.

	LO-SL/	LO-SH/	LO-ML/	LO-MH/	P-EX/	PRD/	C-DM/	INS/	BPI-12/	BPI-17
Traces	1000	1000	1000	1000	608	225	954	1182	8616	30276
Events	9844	9782	9768	9569	9119	4503	4962	23141	59302	240854
Activities	15	15	15	15	23	23	18	11	8	9
Resources	19	19	34	34	47	54	337	125	68	141
Simulation Time	1.27	1.24	1.25	1.24	1.07	0.72	0.73	1.29	10.32	41.97

a simulation model with differentiated resources from an event log, as described in Sect. 5. PROSIMOS takes as input a BPMN process model with simulation parameters as per Definition 3 (encoded in JSON format). Like other simulation engines, PROSIMOS produces an event log and a set of performance indicators such as waiting, processing, and cycle times, and resource utilization.

Using PROSIMOS, we conducted an empirical evaluation aimed at answering the following sub-questions derived from the question posed in Sect. 1: **EQ1** What impact does unpooled resource allocation have compared to pooled allocation? **EQ2** What impact does differentiated resource performance have compared to undifferentiated performance? **EQ3** What impact does differentiated resource availability have compared to undifferentiated availability?

Datasets. We use five simulated (synthetic) logs and five real-life ones. Since our proposal does not deal with process model discovery, we use the BPMN models generated from the input logs using the Apromore open-source platform,[2], which we manually adjusted to obtain 90% replay-based fitness. Table 1 gives descriptive statistics of the employed logs, including number of traces and events and number of activities and resources. Row "simulation time" shows the average execution times (in seconds) across five simulation runs.

The first four event logs were obtained by simulating a Loan Origination (LO) process model using Apromore. The model contains 15 tasks assigned to 5 resource pools. We first simulated the model by assigning the same calendar to all resource pools. Using this single-calendar (S) model, we generated two logs: one where the resource utilization of each pool is around 50% (Low Utilization – L) and another with a resource utilization of 80% (High Utilization – H). The simulation parameters of the H model were identical to the ones of the L model, except that we adjusted the case arrival rate to obtain higher resource utilization. To test the techniques in the presence of multiple calendars, we simulated the same model after assigning different (overlapping) calendars to each of the five resource pools. We simulated this multi-calendar (M) model twice: once with a low utilization (L) and once with high utilization (H). This procedure led to four simulated logs: LO-SL, LO-SH, LO-ML, LO-MH. The fifth log (*purchasing-example (P-EX)*) is part of the academic material of the Fluxicon Disco tool.[3]

The first real-life log (*PRD*) is a log of a manufacturing process.[4]. The second and third are anonymized real-life logs from private processes. The *C-DM* comes

---

[2] https://apromore.com.

[3] https://fluxicon.com/academic/material/.

[4] https://doi.org/10.4121/uuid:68726926-5ac5-4fab-b873-ee76ea412399.

from an academic recognition process executed at a Colombian University. The *INS* log belongs to an insurance claims process. The fourth real-life log is a subset of the BPIC-2012 log[5] – of a loan application process from a Dutch financial institution. We focused on the subset of this log consisting of activities that have both start and end timestamps. Similarly, we used the equivalent subset of the BPIC-2017 log[6], which is an updated version of the BPI-2012 log (extracted in 2017 instead of 2012). We extracted the subsets of the BPI-2012 and BPI-2017 logs by following the recommendations provided by the winning teams of the BPIC-2017 challenge.[7]

Experiment Setup and Goodness Measures. To address questions **EQ1–EQ3**, we discovered five simulation models from each log using the following approaches:

- **SP-NP-NA** corresponds to an unpooled allocation with undifferentiated performances and availability. We allocate the resources into a single pool, where each resource can execute the same activities as in the log. The resources share an aggregated calendar built from the entire log. The processing time of each activity is discovered by aggregating all its instances without considering the resource who executes them.
- **MP-NP-NA** represents a pooled resource allocation with undifferentiated resource profiles. Resources are grouped into disjoint pools assigned to one or several activities according to [5]. Each resource pool shares a single calendar and shares processing time distribution functions for each related activity, i.e., built by aggregating the events of the resources in the pool.
- **MP-DP-NA** is a pooled resource allocation with differentiated performance and undifferentiated availability. We retain the pools and calendars discovered for **MP-NP-NA**. However, we extract differentiated processing time distributions for each pair activity-resource.
- **MP-NP-DA** is pooled resource allocation with undifferentiated performance and differentiated availability. We retain the pools and processing-time distributions discovered for **MP-NP-NA**. However, we extract a differentiated calendar from the activity instances of each resource in the pool.
- **SP-DP-DA** corresponds to the unpooled resource allocation with differentiated resources and performances proposed in this paper.

We assessed the goodness of the discovered models by simulating them using PROSIMOS and measuring the distance between the simulated logs and the original ones. Camargo et al. [5] propose several measures to assess the goodness of simulation models discovered from data. These measures cover two dimensions: the control-flow and the temporal dimension. The techniques proposed in this paper do not affect the control flow. They only deal with resource performance and availability. Accordingly, we evaluate them using temporal measures. In line with [6], we compare simulated and real logs by extracting temporal

---

[5] https://doi.org/10.4121/uuid:3926db30-f712-4394-aebc-75976070e91f.
[6] https://doi.org/10.4121/uuid:5f3067df-f10b-45da-b98b-86ae4c7a310b.
[7] https://www.win.tue.nl/bpi/doku.php?id=2017:challenge.

histograms from each log and computing the Earth Movers' Distance (EMD) between these histograms. We use two EMD metrics, namely **EMD-CT** and **EMD-WR**. **EMD-CT** compares the distributions of cycle time of the traces in the logs. This metric captures to what extent the total durations produced by the simulation model resemble those in the real log. To calculate the **EMD-CT**, we group the cycle times in the real log into 100 equidistant bins. Then, we discretize the simulated log by grouping the cycle times of its traces into bins of the same width as those of the real log. We then measure the EMD between these histograms. The second metric (**EMD-WR**) compares the distribution of timestamps of the events in the two logs. This measure allows us to assess if the simulated and the real log capture similar work rhythms. To calculate the **EMD-WR**, we transform each log into a histogram by extracting the start and end timestamps of each event in the log, and we group the resulting set of timestamps by hour. We then calculate the EMD between the resulting histograms.

The EMD is defined on an absolute dataset-dependent scale. Thus, EMD distances should not be used to compare the performance of the approach across multiple logs. Below, we use the EMD metrics to assess the relative performance of multiple simulation discovery approaches within a given dataset.

The selection of parameters for simulation model discovery may impact the accuracy. Choosing a small granule size, e.g., $n = 60$ s, may lead to a fragmented calendar with many intervals. Conversely, a large value, e.g., $n = 24$ h, may lead to unrealistic calendars in which resources are always available. With a low support threshold, the algorithm may discard many timestamps in the log, leading to low coverage of the observed events. To mitigate these issues, we run a grid search over a range of parameters to find a configuration with low confidence (to filter outliers), high support (to cover a representative set of events), and mid-to-low resource participation (to discard resources that rarely participate in the process). The grid search returned a granule size of 60 min for all experiments. The confidence values ranged from 0.1 to 0.5, and the support and resource participation ranged between 0.5 and 1.0.

Results. Tables 2 and 3 show the results of the **EMD-CT** and **EMD-WR** metrics, respectively. The results of the **SP-NP-NA** models illustrate that unpooled resource allocations with undifferentiated resource profiles yield, on average, poor results on both metrics. This suggests that undifferentiated availability and performance may lead to less accurate results, especially when resources have considerable differences in availability and performance. Another drawback of this unpooled approach, due to the activities sharing resources, is that resources may become busy executing an activity that they execute rarely. Thus, increasing the waiting times of other shared activities (with higher frequencies) due to the unavailability of the resource. This problem may have more impact on processes with external resources. Still, these unpooled resource allocations with undifferentiated resources may perform well in processes where resources have similar calendars and performance, as shown in the BPI challenge logs.

Comparing the pooled models **MP-NP-NA**, **MP-DP-NA**, and **MP-NP-DA** is not straightforward. On average, they exhibited better results than the

Table 2. Results of the **EMD-CT** metric.

	LO-SL/	LO-SH/	LO-ML/	LO-MH/	P-EX/	PRD/	C-DM/	INS/	BPI-12/	BPI-17	Mean
SP-NP-NA	4.49	3.83	15.77	35.1	17.54	21.73	10.53	11.24	10.04	3.95	13.42
MP-NP-NA	3.77	3.58	4.64	14.44	15.11	17.2	10.53	11.28	9.99	3.94	9.45
MP-DP-NA	3.65	4.15	7.35	17.72	15.54	18.1	10.53	11.23	9.98	3.95	10.22
MP-NP-DA	4.31	6.06	4.64	8.5	10.49	18.32	10.02	11.25	6.55	3.85	8.4
SP-DP-DA	2.19	1.82	2.44	4.9	10.26	7.32	8.83	3.33	3.84	1.32	4.63

Table 3. Results of the **EMD-WR** metric.

	LO-SL/	LO-SH/	LO-ML/	LO-MH/	P-EX/	PRD/	C-DM/	INS/	BPI-12/	BPI-17	Mean
SP-NP-NA	491.4	264.5	341.4	195.1	1728.8	511.9	302.7	9244.1	2510.3	5177.0	2076.7
MP-NP-NA	375.1	276.2	369.5	64.5	1755.7	518.0	254.8	9176.4	2545.8	5141.5	2047.8
MP-DP-NA	507.5	207.6	344.2	64.9	1722.7	447.5	266.9	9178.5	2518.9	5134.5	2039.3
MP-NP-DA	402.5	273.1	388.5	169.5	1807.7	467.1	347.1	9384.5	2638.4	5129.9	2100.8
SP-DP-DA	378.4	273.5	331.3	76.7	1692.2	216.8	238.7	8510.9	2628.9	5277.4	1962.5

unpooled and undifferentiated model **SP-NP-NA**. The latter is a consequence of the pooled models preventing the issue of resources allocated to low-frequency tasks (outliers), but at the cost of not modeling processes with resources shared among tasks. Also, in pooled models, the similarity criteria used to group the resources adjust the data points to discover the aggregated calendars and processing time distributions, leading to more accurate approximations. The experiment shows that, on average, the model **MP-NP-DA** gets better values for the **EMD-CT** metric than the models **MP-NP-NA** and **MP-DP-NA**. Suggesting that a pooled model with differentiated availability and undifferentiated performance approximates trace cycle times better than the baseline of pooled allocation with undifferentiated resources. In contrast, the pooled model with undifferentiated availability and differentiated performance **MP-DP-NA** performs better on the metric **EMD-WR** than **MP-NP-DA** and **MP-NP-NA**.

As highlighted in Table 2, the unpooled model **SP-DP-DA** with fully differentiated performance and availability yields the best results w.r.t. metric **EMD-CT**. On average, the values achieved by **SP-DP-DA** are twice better than **MP-NP-NA** and almost three times better than **SP-NP-NA**. This shows that filtering resources with low resource participation (Definition 8), combined with differentiated modeling of performance and availability, heightens the temporal accuracy of the discovered simulation models. With respect to metric **EMD-WR** (Table 3), the unpooled models with differentiated resource performance and availability exhibited the best average results. Here, differences are not as significant as with the cycle time estimations. However, unlike with the **EMD-CT**, histograms built for the metric **EDM-WR** are also impacted by the inter-arrival times discovered. For example, assume the discovered inter-arrival intervals would produce more dispersed starting events in the simulation than in the actual process. Consequently, it may lead to a shift in the timestamps of the subsequently simulated events. The **EMD-WR** metric compares the exact timestamps in which each event occurs. Then, a shift of those events may have a more significant impact on the metric evaluation than in the **EMD-CT** metric, which compares the trace durations without taking into account the exact timestamps involved. The inter-arrival time discovery is orthogonal to the primary goal of this paper, thus, kept as future work [10].[8]

To summarize, with respect to question **EQ1**, unpooled models offer the best results. However, as expected, these models perform poorly when the process involves homogeneous resource pools. Regarding questions **EQ2–EQ3**, the experiments show that, on average, models with differentiated performance yield better results (w.r.t. replicating the work rhythm) than undifferentiated models. Conversely, models with differentiated availability are able to better replicate the cycle times. If we only take into account one dimension at a time (differentiated performance or availability), we do not observe significant accuracy improvements (w.r.t. to models with undifferentiated resources). Instead, the experiments show that modeling differentiated performance and availability together yield the most visible improvements, both when it comes to replicating the cycle time distribution and the work rhythm.

Threats to Validity. The evaluation reported above is potentially affected by the following threats to validity: (1) *Internal validity*: the experiments rely only on ten events logs. The results could differ on other datasets. To mitigate this limitation, we selected logs with different sizes and characteristics and from different domains. (2) *Construct validity*: we used two measures of goodness based on histogram abstractions. The results could be different if we employed other measures, e.g. similarity measures between time series based on dynamic time warping. (3) *Ecological validity*: the evaluation compares the simulation results against the original log. While this allows us to measure how well the simulation models replicate the as-is process, it does not allow us to assess the accuracy improvements of using differentiated resources in a what-if setting, i.e., predicting the performance of the process after a change.

7 Conclusion

The paper outlined an approach to discover simulation models where each resource may have its own performance profile (differentiated performance) and its own calendar (differentiated availability). The paper empirically shows that models with differentiated performance and availability produce simulation logs that are closer to the actual logs from which the simulation model is discovered.

The proposal has a few limitations that warrant further research. First, to estimate inter-arrival times, it applies curve-fitting to the data series consisting of the start time of the first activity instance of each trace. However, the actual case creation time may be earlier than the start time of the first activity instance. This limitation may be tackled by using specialized approaches such as the one in [10]. Second, the approach to discover availability calendars is designed to discover

---

[8] We estimate the inter-case arrival distribution by applying curve-fitting to the data series consisting of the start time of each trace. Branching probabilities are estimated by replaying the log over the model and counting the conditional flow traversals.

calendars with weekly periodicity. In practice, the availability of a resource may vary across the year (e.g. different availability in summer months than in winter ones), or across a month (e.g., different availability at the start than at the end of a month). Another future work direction is to discover calendars with more complex periodicity. Third, the approach for calendar discovery relies on three parameters: confidence, support, and resource participation. In the current implementation, we apply a grid search over narrow parameter ranges to find an optimal configuration. Another future work direction is to enhance the approach with a hyperparameter tuning algorithm to explore large configuration spaces.

Reproducibility. The experiments on public datasets may be reproduced by cloning the repository https://github.com/AutomatedProcessImprovement/Prosimos (tag bpm2022) and following the instructions given thereon.

Acknowledgment. Work funded by European Research Council (PIX project).

References

1. van der Aalst, W.M.P.: Business process simulation survival guide. In: vom Brocke, J., Rosemann, M. (eds.) Handbook on Business Process Management 1. IHIS, pp. 337–370. Springer, Heidelberg (2015). https://doi.org/10.1007/978-3-642-45100-3_15
2. van der Aalst, W.M.P., Nakatumba-Nabende, J., Rozinat, A., Russell, N.: Business process simulation: how to get it right? In: vom Brocke, J., Rosemann, M. (eds.) Handbook on Business Process Management 1, pp. 313–338. Springer, Heidelberg (2010). https://doi.org/10.1007/978-3-642-00416-2_15
3. Afifi, N., Awad, A., Abdelsalam, H.M.: RBPSim: a resource-aware extension of BPSim using workflow resource patterns. In: ZEUS 2018, pp. 32–39 (2018)
4. Agarwal, P.K., van Kreveld, M.J., Suri, S.: Label placement by maximum independent set in rectangles. Comput. Geom. **11**(3–4), 209–218 (1998)
5. Camargo, M., Dumas, M., González, O.: Automated discovery of business process simulation models from event logs. Decis. Support Syst. **134**, 113284 (2020)
6. Camargo, M., Dumas, M., Rojas, O.G.: Learning accurate business process simulation models from event logs via automated process discovery and deep learning. In: Franch, X., Poels, G., Gailly, F., Snoeck, M. (eds.) CAiSE 2022, pp. 55–71. Springer, Cham (2022). https://doi.org/10.1007/978-3-031-07472-1_4
7. Estrada-Torres, B., Camargo, M., Dumas, M., García-Bañuelos, L., Mahdy, I., Yerokhin, M.: Discovering business process simulation models in the presence of multitasking and availability constraints. Data Knowl. Eng. **134**, 101897 (2021)
8. Freitas, A.P., Pereira, J.L.M.: Process simulation support in BPM tools: the case of BPMN (2015)
9. Li, Y., Wang, X.S., Jajodia, S.: Discovering temporal patterns in multiple granularities. In: TSDM 2000 Workshops, pp. 5–19 (2000)
10. Martin, N., Depaire, B., Caris, A.: Using event logs to model interarrival times in business process simulation. In: BPM 2015 Workshops, pp. 255–267 (2015)
11. Martin, N., Depaire, B., Caris, A.: The use of process mining in business process simulation model construction - structuring the field. Bus. Inf. Syst. Eng. **58**(1), 73–87 (2016)

12. Martin, N., Depaire, B., Caris, A., Schepers, D.: Retrieving the resource availability calendars of a process from an event log. Inf. Syst. **88**, 101463 (2020)
13. Nakatumba, J., Westergaard, M., van der Aalst, W.M.P.: Generating event logs with workload-dependent speeds from simulation models. In: Bajec, M., Eder, J. (eds.) CAiSE 2012. LNBIP, vol. 112, pp. 383–397. Springer, Heidelberg (2012). https://doi.org/10.1007/978-3-642-31069-0_31
14. Rozinat, A., Mans, R.S., Song, M., van der Aalst, W.: Discovering simulation models. Inf. Syst. **34**(3), 305–327 (2009)
15. Russell, N., van der Aalst, W.M.P., ter Hofstede, A.H.M., Edmond, D.: Workflow resource patterns: identification, representation and tool support. In: Pastor, O., Falcão e Cunha, J. (eds.) CAiSE 2005. LNCS, vol. 3520, pp. 216–232. Springer, Heidelberg (2005). https://doi.org/10.1007/11431855_16
16. Workflow Management Coalition: Business process simulation specification, v2.0 (2016). https://www.bpsim.org/specifications/2.0/WFMC-BPSWG-2016-01.pdf

Uncovering Object-Centric Data in Classical Event Logs for the Automated Transformation from XES to OCEL

Adrian Rebmann[1(⊠)], Jana-Rebecca Rehse[2], and Han van der Aa[1]

[1] Data and Web Science Group, University of Mannheim, Mannheim, Germany
{rebmann,han}@informatik.uni-mannheim.de
[2] Management Analytics Center, University of Mannheim, Mannheim, Germany
rehse@uni-mannheim.de

Abstract. Object-centric event logs have recently been introduced as a means to capture event data of processes that handle multiple concurrent object types, with potentially complex interrelations. Such logs allow process mining techniques to handle multi-object processes in an appropriate manner. However, event data is often not yet available in this new format, but is rather captured in the form of classical, "flat" event logs. This flat representation obscures the true interrelations that exist between different objects and associated events, causing issues such as the well-known convergence and divergence of event data. This situation calls for support to transform classical event logs into object-centric counterparts. Such a transformation is far from straightforward, though, given that the information required for object-centric logs, such as explicitly indicated object types, identifiers, and properties, is not readily available in flat logs. In this paper, we propose an approach that automatically uncovers object-related information in flat event data and uses this information to transform the flat data into an object-centric event log according to the OCEL format. We achieve this by combining the semantic analysis of textual attributes with data profiling and control-flow-based relation extraction techniques. We demonstrate our approach's efficacy through evaluation experiments and highlight its usefulness by applying it to real-life event logs in order to mitigate the quality issues caused by their flat representation.

Keywords: Process mining · Object-centric event logs · Semantic analysis

1 Introduction

Process mining focuses on the analysis of event data recorded by information systems in order to gain insights into the true behavior of organizational processes [1]. This behavior is captured in event logs, i.e., sequences of events that denote the execution of activities in the process. Traditional process mining techniques assume that each event in the log refers to exactly one case, represented by a single unambiguous case notion. To define this case notion, researchers commonly choose the main object type that is handled by the process, e.g., a request or an application.

However, defining a single unambiguous case notion becomes problematic if the process handles multiple concurrent object types, with potentially complex interrelations. For example, in an order handling process, multiple items can be part of one

© Springer Nature Switzerland AG 2022
C. Di Ciccio et al. (Eds.): BPM 2022, LNCS 13420, pp. 379–396, 2022.
https://doi.org/10.1007/978-3-031-16103-2_25

order and multiple orders can be shipped in one package. For such a process, there is no main object type to serve as an unambiguous case notion. Instead, one is forced to select an imperfect object type for this purpose, such as an order, and represent the event data accordingly. Once the log is recorded from this perspective, the information related to the other object types is lost, though. It is hence impossible to switch perspectives between object types or to analyze the relations between different objects in the process. Moreover, this "flat" recording leads to spurious behavioral relations in the log [2], which, for example, distort the results of automated process discovery techniques [12].

To overcome these issues, researchers recently introduced object-centric event logs, which can capture multiple types of concurrent objects in the process [2]. These object-centric logs allow process mining techniques to handle multi-object processes in a more appropriate manner. However, there is an abundance of event data captured in the form of classical, "flat" event logs, without access to the original data source from which these logs were extracted (cf. [8,9]). In this case, the only option is to transform the flat event logs into object-centric ones. Such a transformation is far from straightforward, because it requires knowledge about which object types occur in the event log, which object instances exist with which properties, and how these instances relate to the events. This information is to a certain extend contained in flat event logs, but in an unstructured, i.e., hidden way. Uncovering this information manually is a tedious and time-consuming task, considering the complexity of real-life logs, with dozens of attributes and thousands of events. Hence, the transformation from flat into object-centric event logs needs to be supported automatically.

Therefore, we propose an approach that automatically uncovers object-related information in flat event data and uses this information to transform the flat data into an object-centric event log according to the OCEL format [13]. For this purpose, our approach combines the semantic analysis of textual attributes in the flat event log with data profiling and control-flow-based relation extraction. In the following, Sect. 2 first illustrates the challenges that our approach needs to address, before we define preliminaries in Sect. 3. Our approach itself is presented in Sect. 4. Section 5 describes our evaluation, which shows that our approach is able to accurately rediscover flattened OCEL logs and can effectively mitigate quality issues in real-life logs. Section 6 summarizes related work; Sect. 7 discusses limitations and concludes the paper.

2 Problem Illustration

In this section, we illustrate the problems caused by recording object-centric event data in flat event logs and the challenges that must be overcome when transforming these logs into object-centric counterparts. For this, we use an established running example of an order handling process [2], which involves four types of objects: *customers*, *orders*, *items*, and *packages*. As visualized in Fig. 1, a customer can place multiple orders, an item belongs to exactly one order and one package, a package can contain multiple orders, and an order can be split over multiple packages.

Problems of Flat Event Logs. We illustrate the problems of recording a multi-object process in a flat format using the following trace, with an order as the case notion:

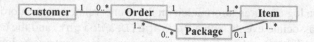

Fig. 1. UML data model of the running example.

t_{order} : ⟨Create order, Reorder item, Pick item, Send package, Pick item, Send package⟩.

The events in t_{order} indicate the picking of two items and the creation of two packages. Although their ordering suggests that these activities occur in an interleaving fashion, there is a clear relation between first picking an item and then sending it in a package. This clear precedence relation on the item level is lost, because there can be several items and packages per order, which we cannot distinguish on the trace level. This phenomenon, called *divergence*, is unavoidable when recording processes with object relations beyond 1:1 in the form of flat event logs [2]. It often occurs together with another unavoidable issue, called *convergence*. Convergence emerges when we use an individual item as the case notion to represent the events from trace t_{order}, which results in the following traces:

$t_{item}(1)$: ⟨Create order, Reorder item, Pick item, Send package⟩,
$t_{item}(2)$: ⟨Create order, Pick item, Send package⟩.

Because both items belong to the same order, the "*Create order*" event is duplicated across the traces. As a result, the information that both items relate to the same order is no longer captured at the trace level. Due to the m:n relations in the process at hand, the impact of this issue is amplified, given that also multiple orders can relate to the same package, and vice versa.

To overcome these issues and their associated information loss, a flat event log needs to be transformed into an object-centric counterpart, as discussed next.

From Flat to Object-Centric Logs. To illustrate the transformation of flat into object-centric event logs, consider the example in Table 1, which provides a flat event log with

Table 1. Flat event log of an order handling process with the *order* as the case notion.

CaseID	Event	Activity	Timestamp	PackageID	Weight	Customer
o1	e1	Create order	05-20 09:07			Pete
o1	e2	Reorder item	05-23 10:40		12.5	Pete
o1	e3	Pick item	05-23 14:20		70.8	Pete
o1	e4	Send package	05-23 17:26	p1	70.8	Pete
o1	e6	Pick item	06-04 15:20		12.5	Pete
o1	e9	Send package	06-06 16:20	p2	20.4	Pete
o2	e5	Create order	06-03 19:17			Pete
o2	e7	Update order	06-04 18:11			Pete
o2	e8	Pick item	06-05 11:48		7.9	Pete
o2	e10	Send package	06-06 16:20	p2	20.4	Pete

two orders. The log captures information on the events related to each order, as well as attributes that associate events with a `PackageID`, a `Weight`, and the `Customer`.

As shown in Table 2 and Table 3, constructing an object-centric version of this event log requires information about: object types (customers, orders, items, and packages), their instances and associated properties (e.g., that package *p1* has a weight of 70.8), and the relations between object instances and events (e.g., that event *e1* creates order *o1*, which relates to items *i1_1* and *i1_2*). However, such crucial information is not explicit in the flat version of the event log, but rather needs to be uncovered in order to transform flat data into an object-centric log. This results in four main transformation tasks:

Table 2. Object-centric event log of the running example.

Event	Activity	Timestamp	Orders	Packages	Items	Customer
e1	Create order	05-20 09:07	{*o1*}	∅	{*i1_1, i1_2*}	{*Pete*}
e2	Reorder item	05-23 10:40	{*o1*}	∅	{*i1_1*}	{*Pete*}
e3	Pick item	05-23 14:20	{*o1*}	∅	{*i1_2*}	{*Pete*}
e4	Send package	05-23 17:26	{*o1*}	{*p1*}	{*i1_2*}	{*Pete*}
e5	Create order	06-03 19:17	{*o2*}	∅	{*i2_1*}	{*Pete*}
e6	Pick item	06-04 15:20	{*o1*}	∅	{*i1_1*}	{*Pete*}
e7	Update order	06-04 18:11	{*o2*}	∅	{*i2_1*}	{*Pete*}
e8	Pick item	06-05 11:48	{*o2*}	∅	{*i2_1*}	{*Pete*}
e9	Send package	06-06 16:20	{*o1,o2*}	{*p2*}	{*i1_1, i2_1*}	{*Pete*}

Table 3. Objects of the object-centric event log.

Type	Instances
Customer	{*Pete* ()}
Order	{*o1* (), *o2* ()}
Package	{*p1* (`Weight`: 70.8), *p2* (`Weight`: 20.4)}
Item	{*i1_1* (`Weight`: 12.5), *i1_2* (`Weight`: 70.8), *i2_1* (`Weight`: 7.9)}

1. **Detect object types.** Object types in a process are not explicitly indicated in flat event logs. Rather, transformation requires these types to be extracted from unstructured activity labels, such as the *order* type in "*Create order*", and from certain event attributes, such as `Customer` in Table 1.
2. **Identify object instances.** Due to divergence and convergence, a transformation approach needs to identify distinct object instances within cases, e.g., that case *o1* deals with two items and two packages, and relate object instances across cases, e.g., that package *p2* appears in both *o1* and *o2*. This involves identifying event attributes that represent identifiers of a specific object, e.g., that `PackageID` defines individual packages. Furthermore, because such identifier attributes may not exist for all object types, it also requires inferring certain object instances from the event log itself, e.g., that events *e3* and *e6* yield two different items (*i1_1* and *i1_2*).

3. **Relate objects to their properties.** Flat event logs do not distinguish between attributes that relate to a specific event, such as a resource performing it, and attributes that provide information about the object handled in the event, such as the `Weight` attribute, which captures information about an individual package or item. When establishing an object-centric log, such relations must thus be derived by separating event attributes from object properties, in order to have a comprehensive view on the instances involved in the process, as captured in Table 3.

4. **Associate object instances with events.** Finally, instead of referring to a specific case, each event in an object-centric log must be mapped to the object instances it relates to. Obtaining a complete mapping requires a thorough analysis of the inter-relations that exist between object instances. For example, this requires the recognition that package $p2$ relates to orders $o1$ and $o2$, as well as items $i1\_1$ and $i2\_1$, and associating all these objects with event $e9$, even though the objects originally stem from a range of different events and cases in the flat log.

In this paper, we propose an approach that tackles these tasks by combining the semantic analysis of the textual attributes of flat events logs with data profiling and control-flow-based relation extraction. In this manner, our approach uncovers object types, their instances and properties, as well as the relations between instances and events.

3 Preliminaries

We define objects, events, flat event logs, and object-centric event logs as follows based on the definitions by Van der Aalst [2].

Objects. An object is a tuple $o = (oi, ot, vmap)$, with oi as its identifier, ot its type, and $vmap$ a value map, which captures the assignment of values to o's properties. For instance, object $i1\_1$ has the identifier $i1\_1.oi =$ "i1_1", the type $i1\_1.ot = item$, and has a value map assigning it a weight, $i1\_1.vmap = \{$Weight : "12.5"$\}$.

Events. An event is a tuple $e = (a, ts, omap, vmap)$, with a its activity label, ts its timestamp, $omap$ the object map, which captures the objects that e relates to, and $vmap$ the value map, which assigns values to e's attributes. For instance, event $e1$ in Table 2 has an activity label $e1.a = Create\ order$, a timestamp $e1.ts = $ 05-20 09:07, an object map $e1.omap = \{o1, i1\_1, i1\_2, Pete\}$, and a value map $e1.vmap = \{$Event : "e1"$\}$.

Flat Event Logs. A flat event log L is a set of events that all have exactly one case identifier in their value map, i.e., $\forall_{e \in L}$CaseID $\in dom(e.vmap) \wedge |e.vmap($CaseID$)| = 1$, whereas their object maps are empty. Events that have the same CaseID are said to be part of the same *case*. Events belonging to the same case are assumed to have a total order, following, e.g., from their timestamps. For instance, event $e1$ in Table 1 has a value map $e1.vmap = \{$CaseID : "o1", Event : "e1", Customer : "Pete"$\}$, whereas, all events in this log have an empty object map, e.g., $e1.omap = \emptyset$. Note that, for instance, $e9$ in Table 2 would be part of two cases of a flat event log, when the *order* type serves as the case notion (cf. $e9$ and $e10$ in Table 1), because each event must have exactly one CaseID, yet, $e9$ refers to two orders, $o1$ and $o2$ resulting in two case identifiers.

Finally, we define $Att = \bigcup_{e \in L} dom(e.vmap)$ as the set of attributes in L. For the log in Table 1, we get: $Att = \{$CaseID, Event, PackageID, Weight, Customer$\}$.

Object-Centric Event Logs. An object-centric event log O is a set of events that have populated object maps. For instance, event $e9$ in Table 2 has an object map $e9.omap = \{o1, o2, p2, i1\_1, i2\_1, Pete\}$ and a value map $e9.vmap = \{$Event : "$e9$"$\}$. The events in O are assumed to have a known partial order, following, e.g., from their timestamps.

4   Approach

As visualized in Fig. 2, our approach for the transformation of a flat into an object-centric event log consists of five main steps. Step 1 extracts the *object types and actions* from the activity label and other textual attributes of events, which yields object types and the applied actions per event. Steps 2 and 3 jointly establish a set of object *instances*: Step 2 first matches extracted object types to attributes that capture identifiers to recognize distinct instances of object types, whereas Step 3 aims to discover instances for object types for which no such identifier attribute was found. Afterwards, Step 4 aims to assign *properties* to object types by identifying attributes that represent object properties. Finally, Step 5 assigns the discovered *object instances to events* by exploiting behavioral relations among object types and instances discovered in previous steps. Based on the result of Step 5, we create an object-centric event log according to the *OCEL* format [13]. In the remainder, we describe each of these five steps in detail.

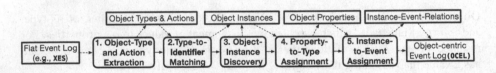

Fig. 2. Overview of the approach.

4.1   Step 1: Object-Type and Action Extraction

The first step of our approach extracts the object types and actions from an event log. As illustrated in Sect. 2, object types need to be derived from unstructured textual attribute values, such as activity labels, and attribute names of a flat event log. An *action* is applied to an object, incurring a change in its state [14]. For instance, the *"Create order"* activity label indicates that a *create* action is applied to an *order*. We extract actions along with the object types since these can contain information about the creation of new object instances, which we will exploit in a later step.

To achieve this, we use a semantic extraction technique from our earlier work [22]. This technique extracts parts of textual attribute values that correspond to different semantic roles, such as *object types* and *actions* in two ways:

1. *Instance-level labeling*: The extraction technique labels parts of unstructured textual attribute values with semantic roles. The parts that correspond to the desired roles are then extracted. For instance, for the *"Send package"* activity label of event *e9*, the technique labels *"package"* as an object type and *"send"* as an action.
2. *Attribute-level classification*: The extraction technique also identifies event attributes that in their entirety correspond to a certain semantic role. It does so based on an attribute's name and its value range. This, e.g., applies to the Customer attribute in Table 1, which allows us to also identify *customer* as an object type contained in the event log, assigning this type to any event that has a value for the attribute.

By taking the output of this extraction technique, Step 1 instantiates a function *extract*, which, given an event $e \in L$, extracts the object types and the actions applied to them (if any) from e. The result maps the object types to a (possibly empty) set of actions, e.g., $extract(e9) = \{package \rightarrow \{send\}, customer \rightarrow \emptyset\}$. Each event's object map is then initialized with its object types, e.g., $e9.omap = \{package \rightarrow \emptyset, customer \rightarrow \emptyset\}$. Finally, we establish a set of identified object types $T = \bigcup_{e \in L} dom(e.omap)$ and move to Step 2, which aims to match these types to identifier attributes.

4.2 Step 2: Type-to-Identifier Matching

In this step, our approach tries to associate identifier attributes with the extracted object types to be able to recognize distinct object instances. For our running example, we can differentiate between the two packages *p1* and *p2* by recognizing that the PackageID is an identifier for *package* objects. Such identifier attributes are not explicitly given, meaning that we need to match object types to attributes. To establish these matches, our approach first identifies a set Att_{ID} of potential object identifiers, by categorizing attributes according to their domain. Then, we match these attributes to object types in T, resulting in a mapping M, consisting of (ot, att) pairs with $ot \in T$ and $att \in Att_{ID}$.

Finding Potential Identifier Attributes. To identify the set $Att_{ID} \subseteq Att$, we recognize that identifiers generally use alphanumeric domains, i.e., string or int, such as PackageID (*"p1"* and *"p2"*) and Customer (*"Pete"*) in Table 1. Therefore, we categorize attributes according to their domain's data type and add those with string and int domains to Att_{ID}. In this manner, we discard attributes corresponding to, e.g., timestamps, boolean values, and floats, such as the Weight attribute.

Matching Identifier Attributes and Object Types. Next, we aim to identify matches between the object types in T and potential identifier attributes in Att_{ID}, resulting in the set M. Some object types and attributes can be directly matched. For others, we first establish candidate matches, and then verify the validity of these candidates.

Direct Matching. Object types in T that stem from the attribute-level classification of Step 1 reflect objects that correspond to the name of a specific event attribute, such as the *customer* object type corresponding to the Customer attribute. Because these types were identified in this manner, we know that their identifiers are captured in the corresponding attributes, if indeed these are part of Att_{ID}. Therefore, we can directly add such pairs, e.g., (*customer*, Customer), to the matches in M.

Establishing Candidate Matches. For object types that cannot be directly matched, such as the *package* type extracted from activity labels, we first establish candidate matches, collected in a set M_C, using two strategies, considering attribute names and values.

First, we establish candidate matches by checking if the name of an unmatched object type encompasses the name of an unmatched attribute, or vice versa. In this manner, we recognize a candidate match between the *package* type and the `PackageID` attribute, or between the *item* type and a, hypothetical, `order_item` attribute.

Then, for attributes in Att_{ID} that are not yet in a candidate match with an object type, we apply a strategy inspired by *data profiling* [4], which checks if an attribute exclusively *co-occurs* with an object type. For the running example, all events associated with the *package* type (*e4, e9, e10*) have values for the `PackageID` attribute, but this attribute does not apply to any other events. Therefore, even if `PackageID` was named `pID` and hence not a name-based candidate match, our approach would still be able to recognize it as a potential identifier for the *package* type and add it to M_C.

Validating Candidate Matches. Although name-based similarity and co-occurrence are useful indicators to identify relations between object types and attributes, there is no guarantee that the candidate matches actually capture proper identifiers. Therefore, we next validate each candidate match $(ot, att) \in M_C$ by determining if each unique value of att is indeed associated with a specific instance of ot, and vice versa.

This validation task is complex, though, given that multiple events in a case can relate to the same object instance (e.g., creating and updating an order) or multiple instances of the same object type (e.g., shipping multiple packages for one order), and that, due to duplication issues, the same event can essentially appear in two cases (cf. *e9* and *e10*). For an object type ot, we deal with these issues by aiming to establish a set of events $E'(ot)$ that should each relate to a different instance of ot. Given $E(ot) \subseteq L$ as the events related to ot (i.e., that have ot in their $omap$ after Step 1), we obtain $E'(ot)$ by avoiding duplicate events and by selecting only a single event per case. Our approach avoids duplicates by only selecting events from $E(ot)$ that have a unique combination of an activity label, timestamp, and event attributes (aside from their `CaseID` and event ID, if available). In this manner, we detect *e9* and *e10* as duplicates. Given the identified duplicates, we select a single event per case related to ot, in a manner that maximizes the size of $E'(ot)$. For instance, given $E(package) = \{e4, e9, e10\}$, we select *e10*, because *e4* and *e9* stem from the same case, and obtain $E'(package) = \{e4, e10\}$.

Finally, if the attribute values of att are unique for the events in $E'(ot)$, we consider att as a valid identifier of ot and add (ot, att) to M. For instance, we consider (*package*, `PackageID`) a valid match, given that the two events in $E'(package)$ have unique values for the attribute, "*p1*" and "*p2*". If there are multiple valid candidates for the same object type, we match the type to the attribute with the largest number of unique values and discard the other candidates.

Object-Instance Creation. For all matches $(ot, att) \in M$, our approach creates an object instance o with its type ot for each unique value of att, i.e., its identifier oi, and adds these instances to the object maps of the events that refer to this instance. For example, we add package *p1* to event *e4* and package *p2* to events *e9* and *e10*.

4.3   Step 3: Object-Instance Discovery

Next, we aim to discover instances for those object types for which no explicit identifier attribute was found in the previous step, such as the *item* type in the running example. For this, we try to find activities that indicate the instantiation of objects, either based on their activity labels or based on the life cycle of an object.

Instance Discovery Based on Creation Actions. We first aim to identify activities whose meaning hints at the instantiation of an object, such as *"Create order"*. To this end, we use the action classification framework of the MIT Process Handbook [19], which defines a set of 15 *creation actions* (see Table 4) describing the creation of some output.

Given an event, we check if any of its actions, extracted in Step 1, corresponds to an action in this set. If so, the occurrence of this event implies the instantiation of a new object. For instance,

Table 4. Creation actions [19] used by Step 3.

build	compute	construct	copy	create
design	develop	document	duplicate	generate
make	manufacture	perform	produce	record

we recognize that events *e1* and *e5*, corresponding to the *"Create order"* activity label, result in two new orders.

Although we here identify creation actions based on the 15 actions from the MIT Process Handbook, our work is independent of this specific resource. It can be replaced or enhanced with alternatives, such as the *build verbs* from the classification framework by Levin [15], multilingual resources, such as *ConceptNet* [23], or a self-defined set.

Instance Discovery Based on Object Life Cycles. Although creation actions are a reliable indicator for the creation of new objects, they are not always available for an object type. Therefore, our approach next analyzes the life cycles of object types in terms of the applied activities per case. To illustrate this, consider the life cycles in Fig. 3.

life cycle 1	⟨Receive request, Update request, Complete request⟩
life cycle 2	⟨Receive request, Receive request, Complete request, Complete request⟩

Indicator activity: *"Receive request"*

Fig. 3. Recognizing activities that indicate new object instances.

For an object type *ot*, we aim to identify an *indicator activity*, which corresponds to a new object instance. We look for such an indicator by checking if there are any activities related to *ot* that occur for every case of this type. For example, assuming only the two depicted cases relate to requests in the process at hand, both *"Receive request"* and *"Complete request"* are candidate activities, since they occur in both life cycles. In case of such a tie, we select the activity that most commonly occurs first among the candidates—*"Receive request"* in the example—as the activity that we use

388   A. Rebmann et al.

to identify new object instances. Therefore, we recognize that the cases in Fig. 3 relate to three distinct *request* objects: one in the first life cycle and two in the second.

Object-Instance Creation. For each event that indicates a new object instance, based on a creation action or indicator activity, our approach establishes an object instance, for which we generate a unique identifier oi, and add it to the event's object map. Duplicate events, as identified in Step 2, form an exception here. Since they correspond to the creation of the same object instance, which is why we assign the same instance to them.

Note that we discard all object types for which neither Step 2 nor Step 3 identified any instances, by removing the type from T as well as from any event's object map.

4.4   Step 4: Property-to-Type Assignment

In this step, our approach tries to associate properties to object types, which are attributes that capture information about an object instance associated with an event, rather than relate to the event itself. For instance, although event $e3$ ("*Pick item*") has a `Weight` attribute with a value of "70.8", it is clear that this refers to the weight of the item, not of the event. Therefore, in this step we aim to establish a mapping between a log's attributes Att and the object types in T.

To establish this mapping, we first select all attributes that were not recognized as object identifiers in Step 2. Then, we consider an attribute att to be a property of an object type ot if (1) events related to ot have a value for att and (2) all events related to the same object instance have the same value for att. The former ensures co-occurrence, ascertaining that att indeed relates to ot, whereas the latter ensures that object properties are immutable per object instance, in line with their definition in the OCEL format. In this manner, we identify that `Weight` is an attribute of both the *item* and *package* types, whereas attributes such as a `timestamp` or `employee` are not identified as properties, because they change across the events related to the same object instance.

Finally, we avoid assigning an attribute att as a property to multiple object types if the attribute name indicates a clear relation to one of the types. For example, we avoid assigning an `item_category` attribute to the *package* type, given that this property clearly relates to items, irrespective of the co-occurrence of the attribute and packages.

4.5   Step 5: Instance-to-Event Assignment

Finally, our approach aims to complete the mapping between events and object instances, which is necessary to account for missing *instance-to-event* and *instance-to-instance* relations. The former involves events that correspond to a particular object type, but for which no particular instance has yet been discovered. For example, event $e2$ ("*Reorder item*") is already recognized as relating to the *item* type, yet we still need to identify that this event refers to the same item that is later handled by event $e6$ ("*Pick item*"). The latter refers to the inter-relations that can exist among object instances, which need to be reflected in the object maps of the corresponding events. For example, since package *p1* relates to order *o1*, event *e4*, which creates this package, should also be associated with that order. We identify these missing relations as follows.

Finding Missing Instance-to-Event Relations. To find missing instance-to-event relations, we first identify the events that are associated with an object type (through Step 1), but for which no instance was discovered in Step 2 or 3. This applies, e.g., to event $e2$ (*"Reorder item"*) and $e7$ (*"Update order"*). Then, given such an event, we search within the case for other events that are associated with an object instance of the same type and verify that the object's properties match across the events. For instance, since event $e2$ has a `Weight` of 12.5, we do not want to associate it with the item of event $e3$, which has a weight of 70.8, but rather with the same item as event $e6$, which also relates to an item weighing 12.5 kg. Should multiple object instances satisfy this requirement, we associate the event to the instance of its nearest predecessor or successor.

Finding Missing Instance-to-Instance Relations. We look for instance-to-instance relations by (1) considering relations between instances and cases, (2) identifying strict orders among object types, and (3) consolidating cross-case relations.

Discovering Case Objects. We first exploit that, commonly, each case in a flat event log corresponds to an instance of a particular object type, such as an *order* in our running example. If such a *case object* can be identified, we know that any other object instance handled in the same case also relates to that object, e.g., that the items and packages handled in the first case all relate to order $o1$ as well.

However, to recognize such inter-relations, we need to identify the case object type, if any, for a particular event log. Given that instances of this object type must, by definition, be in a 1:1 relation with the cases in a log, we first discount all object types for which this does not apply, i.e., which are affected by convergence and divergence issues. Given an object type ot, we thus ensure that (1) no instance of ot is associated with multiple cases in the log, such as the *package* type in the example, and (2) that no case in the log is associated with multiple instances of ot, such as the *item* type.

If these checks yield a single case object type, ot_c, then we add each object instance of that type to the object map of all events e in their case, using `CaseID` as the instance's identifier. For the example, *order* is the only object type that passes the checks, which means that we assign orders $o1$ and $o2$ to all events in their respective cases.

Strict Order Between Object Types. We next identify instance-to-instance relations by looking for the existence of strict orders between object types. Here, we consider an object type ot_1 to be in a strict order with type ot_2 if every time an event related to ot_2 occurs, an event related to ot_1 (directly or indirectly) precedes it. In this manner, we, for example, observe a strict order between items and packages in the running example.[1]

Given such a strict order between ot_1 and ot_2, we relate an instance o_1 of type ot_1 to an instance o_2 of ot_2 if the life cycle of o_1 completes before the life cycle of o_2 begins, i.e., if the last event related to o_1 comes before the first event related to o_2. For example, we relate item $i1\_1$, which last occurs in $e6$, to package $p2$, which first occurs in $e9$.

Consolidating Cross-Case Relations. Last, we consolidate inter-relations across cases by ensuring that duplicate events are associated with the same sets of object instances.

---

[1] Note that these object types can still occur in an interspersed manner, as e.g., seen in case $o1$, where events related to items also occur in between packages.

Given two duplicates, e and e', we achieve this by associating both events with all object instances stemming from the union of their object maps. For example, having recognized that events $e9$ and $e10$ are duplicates, we add all object instances stemming from case $o1$ (associated with $e9$) to the object map of $e10$, and vice versa. In this manner, we, e.g., recognize that package $p2$, which is created by these duplicate events, deals with item $i1\_1$ (stemming from case $o1$) as well as item $i2\_1$ (stemming from $o2$).

Having associated events with all object instances, our approach has uncovered the necessary information to construct its output, an object-centric event log.

4.6 Output

Our approach returns an object-centric event log according to the OCEL format [13], which, at a high level, consists of an *objects* and an *events* map.

The *objects* map relates object identifiers to instances, which are in turn associated with their type and property values. To populate this map, we add all instance identifiers, either detected in Step 2 or generated in Step 3, to the map and associate these instances with their properties identified in Step 4. Simultaneously, we also disassociate any object property from the events that they were associated with in the flat event log, e.g., rather than having `Weight` as an attribute of event $e4$, we represent it as a property of the respective package: $objects[p1] = \{package, \{\texttt{Weight}:\text{"70.8"}\}\}$.

The *events* map associates identifiers with events, which are associated with object instances through their $omap$. It is important to recognize that these events are no longer grouped per case. As a result, we can omit any duplicate event from consideration, e.g., by removing event $e10$ and preserving $e9$. The map is then populated with the remaining events, which are each associated with the identifiers of their respective object instances, as assigned in Steps 2, 3, and 5, e.g., $e1.omap = \{\text{"}o1\text{"}, \text{"}i1\_1\text{"}, \text{"}i1\_2\text{"}\}$.

Based on the established maps, we return the object-centric log, which can directly be used by object-centric process mining techniques [3,17].

5 Evaluation

We implemented our approach as a Python prototype[2], using the PM4Py library [7] for event log handling. Based on this prototype, we perform evaluation experiments to assess our approach's capability to rediscover artificially flattened object-centric logs (Sect. 5.1). Then, we illustrate its practical value by showing that it can resolve divergence and convergence in real-life scenarios (Sect. 5.2). Finally, we discuss the main insights from our evaluation and its limitations (Sect. 5.3).

5.1 Rediscovering Object-Centric Event Logs

We assess whether our approach is able to rediscover an artificially flattened object-centric log by comparing its output with the original OCEL log.

---

[2] https://gitlab.uni-mannheim.de/processanalytics/uncovering-object-centric-data.

Data. For our evaluation experiments, we use a publicly available OCEL log of an order handling process.[3] Currently, this is the only available log suitable for this evaluation. It contains 22,367 events and 11,522 object instances of five object types: 2,000 orders, 8,159 items, 20 products, 17 customers, and 1,325 packages. From this original OCEL log, we create three flattened logs, using the *item*, *order*, or *package* as the case notion. The resulting logs capture 1:n, n:1, and n:m relationships between objects and include object types both in attribute names and activity labels. Thus, all relation types are covered, meaning that all strategies employed by our approach can be assessed.

Setup. To assess the ability of our approach to correctly discover relevant object-centric information in the flat event log, we conduct experiments using two settings:

(1) All attributes. In this setting, we use all information from the flattened event logs as input for the rediscovery task.

(2) Masked ID attributes. To assess the robustness of our approach, we also purposefully reduce the information that is available by masking object ID attributes in the flattened event logs. This increases the dependency of our approach on its instance discovery techniques employed in Step 3. Specifically, we mask each ID attribute once for each of the three flattened logs. Since the *item* and *order* logs include identifiers for all four other types, and the *package* log captures only a customer identifier, we obtain nine *masked* logs, one with *package*, four with *item*, and four with *order* as the case notion.

We measure the performance of our approach in terms of the well-known *precision*, *recall*, and F_1-*score* metrics with respect to the original OCEL log per type of element, i.e., object types, object instances, properties, and instance-to-event assignments. Using A to denote the set of elements uncovered by our approach and G for the set of elements in the ground truth, i.e., the OCEL log, precision is the fraction of elements uncovered by our approach that are actually correct ($|A \cap G|/|A|$), recall is the fraction of elements in the OCEL log that were also correctly uncovered by our approach ($|A \cap G|/|G|$), and the F_1-score is the harmonic mean of precision and recall. Because flattening the log causes a loss of information about entire object types, we only include object types in G that are actually contained in a particular flattened log. To avoid propagating false positives from object-type extraction (Step 1), we only include elements in A that relate to object types actually present in the original OCEL log for the other steps.

Results. Table 5 reports on the results of our rediscovery experiments, micro-averaged over the logs for the respective settings. In the following, we discuss the results for the different tasks that our approach addresses.

Object-Type Extraction. For the extraction of object types, our approach achieves a recall of 1.00 and a precision of 0.71, yielding an F_1-score of 0.82. We thus accurately identify all object types from the original log. The lower precision is caused by the extraction of two additional object types, *payment reminder* and *delivery*. Although not contained in the original OCEL log, their extraction is not problematic and can even enable additional insights, e.g., on the number of payment reminders sent per order.

Object-Instance Identification. Our approach identifies object instances with perfect accuracy in both the regular and masked settings. This highlights its ability to find and

---

[3] http://ocel-standard.org/1.0/running-example.jsonocel.zip.

match ID attributes to object types (Step 2) and the usefulness of our instance-discovery strategies (Step 3), which can identify instances for types with masked ID attributes.

Property-to-Type Assignment. When assigning properties to object types, our approach achieves a perfect recall, but a rather low precision of 0.37. An in-depth look reveals that these different assignments are not problematic, though. For example, the attribute `cost` is assigned to both *product* and *item*, whereas in the original it is only associated with products. However, given that also items have costs, such assignments are redundant, but not wrong. Similarly, our approach associates attributes such as `price` and `weight` with orders, items, packages, and products. While these are realistic assignments, the attributes are not considered as properties in the original OCEL log, but are associated with events. Thus, our approach actually provides a more complete mapping.

Instance-to-Event Assignment. For instance-to-event assignment, we achieve an excellent recall (0.998, rounded in Table 5) and a high precision (0.94) in the all-attributes setting. Thus, our approach assigns relevant object instances to events they relate to. An in-depth look into the constructed OCEL logs reveals that the superfluous assignments of instances to events are mainly assignments of packages to events that relate to items shipped in the respective package. Such assignments are not considered in the original log, but can enable insights into the packaging process in a post-hoc analysis.

When masking identifier attributes, precision and recall decrease slightly, which indicates that our approach occasionally makes incorrect assignments. This is especially the case for 1:n relationships between object types and the case notion. For example, in the *order* event log, where one order may contain many different items, items with the same properties may be assigned incorrectly. However, it is important to recognize that such assignments are simply not possible based on the information in the masked log, whether done by an automated approach or manually.

Table 5. Results of the evaluation experiments averaged over flattened logs.

Element	All attributes				Masked ID attribute			
	Count	Precision	Recall	F$_1$-score	Count	Precision	Recall	F$_1$-score
Object types	12	0.71	1.00	0.82	42	0.71	1.00	0.82
Object instances	24k	1.00	1.00	1.00	94k	1.00	1.00	1.00
Object properties	10	0.37	1.00	0.54	34	0.39	1.00	0.56
Instance-to-event	411k	0.94	1.00	0.97	1,559k	0.93	0.97	0.95

5.2   Real-Life Application Cases

We next demonstrate the practical value of our approach by showing that it is capable of resolving convergence and divergence issues in well-known real-life event logs. The full results and OCEL logs obtained by our approach can be found in our repository (see Page 13). In the following, we use individual cases and events from these logs to illustrate in detail how our approach mitigates divergence and convergence.

Divergence. We use the BPI17 application log [8] to show how our approach mitigates divergence issues. The log captures a loan application process, containing 1,202,267 events, 31,509 cases, and 26 distinct activities. Divergence is particularly frequent, because the log uses the *application* as the case notion and one application can have multiple offers. This means that cases in the log often contain multiple events that denote execution of the same activity for distinct offers (divergence). Applying process discovery to the log leads to loop-backs, as visualized in Fig. 4. This shows the directly-follows graph (DFG) discovered for one case of the log, which is already quite complex.

When applying our approach to mitigate the divergence issue, we discover that 42,995 offers are handled in the 31,509 applications and that offers have several properties, such as an offered amount and a monthly cost. For the particular case in Fig. 4, we find that four distinct offers are handled in this application, that these all have different properties, and that the process is linear with respect to a single offer, e.g., ⟨*Create Offer, O Created, O Sent, O Canceled*⟩. It is important to stress that this information on the sub-case level is not readily available in the flat log and has to be uncovered by identifying the distinct offers handled in a single case. Our approach achieves this by extracting the *offer* type, finding an identifier attribute for it, and assigning it, among others, the MonthlyCost and OfferedAmount properties.

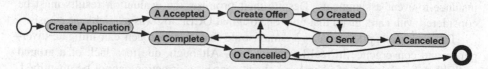

Fig. 4. Directly-follows graph of application 196483749 of the BPI17 log.

Convergence. To illustrate how our approach can mitigate convergence issues, we chose the BPI19 event log [9], which captures data on the purchasing process of a multinational company and contains 1,595,923 events across 251,734 cases with 42 distinct activities. Each event relates to a single *purchase order item* and multiple purchase order items can belong to the same *purchasing document*. Consequently, events on the purchasing-document level are duplicated across cases (convergence). For example, the duplication of "*Vendor creates invoice*" events suggests the creation of invoices per purchase order item, whereas in reality invoices can cover multiple such items.

When applying our approach to mitigate convergence, we discover, among others, 251,734 purchase order items, 76,349 purchasing documents, 86,868 invoices, 1,975 vendors, and 4 companies. The resulting OCEL log reveals the relationships between object types, as shown in Fig. 5: Purchasing documents consist of any number of purchase order items, a vendor creates multiple invoices, and each invoice is associated with one purchasing document. Notably, in contrast to the input log, events related to purchasing documents and invoices, such as "*Vendor creates invoice*" or "*Document created*", are not captured at the level of individual purchase order items, but at the level of purchasing documents, thus eliminating duplicate events. This demonstrates that our approach is able to reveal actual relationships among objects and mitigate the convergence issue present in real-life logs.

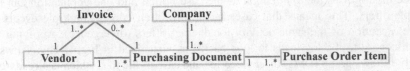

Fig. 5. UML data model with relations between object types found in the BPI19 log.

5.3 Discussion

Our evaluation experiments show that our approach is capable of accurately uncovering object-centric information from an artificially flattened object-centric event log, using different settings and case notions. We also observe that our approach even uncovers more information than originally captured in the OCEL log. This includes additional object types, properties, and relations, which allow for deeper insights into the process. The main difficulty for our approach was the recognition of object inter-relations for objects in a 1:n relation with the case object, which resulted in several incorrect instance-to-event assignments. Despite their promise, the evaluation results must be considered with care, given that only one original OCEL log was available as a basis.

The real-life application cases demonstrate that our approach can mitigate divergence and convergence in real-life event logs. Although, due to a lack of a ground truth, the completeness of uncovered object-centric information cannot be quantified, the results nevertheless show that our approach provides considerable practical value by extending the analysis potential for flat event logs of multi-object processes.

6   Related Work

Our work primarily relates to research on object-centric representations of event data and discovering object-centric information from event logs.

After storing event data in flat formats like XES [5] for many years, the first data format proposed for object-centric event logs was XOC [16], which does not require a case notion and therefore avoids flattening multi-dimensional data. More recently, researchers introduced the OCEL format [13], which allows for more efficient storage and processing than its predecessor. Beyond log formats, another proposed option for storing multi-dimensional object-centric event data are event graphs, which enable the analysis of behavior of different objects handled in a process [11]. For our approach, we adopt OCEL as the output format, which, among others, enables the subsequent application of techniques for discovering object-centric process models, such as object-centric behavioral constraint models [17] and object-centric Petri nets [3].

Approaches for the discovery of object types and their behavioral relations from event data usually require relational data or rich logs that cover multiple perspectives of a process as input. This includes approaches for the discovery of artifact (i.e., object) life cycles from raw logs of artifact-centric systems [10,21] as well as the discovery of behavioral dependencies between object types based on such logs [20] or based on data extracted from ERP systems [18]. Compared to these approaches, our approach takes flat event logs, where no explicit relations between objects are given, and transforms them into object-centric logs. The approach by Bano et al. [6] also uses flat event logs as input data, but their goal is to discover UML models from activity labels and attribute names to provide analysts with domain-specific context information.

7   Conclusion

In this paper, we proposed an approach to uncover object-centric data from flat event logs to automatically transform them into object-centric logs according to the OCEL format. To this end, our approach combines the semantic analysis of textual attributes with data profiling and control-flow-based relation extraction. It extracts object types, discovers object instances and their properties, and assigns these instances to events they relate to. We demonstrated our approach's efficacy in an evaluation by showing that it is able to rediscover an artificially flattened object-centric log and that it can mitigate convergence and divergence issues in real-life event logs.

Our approach is subject to certain limitations. First, object types must at least be mentioned in the flat log for our approach to extract them. However, once an object type is extracted, instances, properties, and relations can be identified through the use of diverse strategies that include and go beyond the semantic analysis of events. Second, to accurately handle n:1 and m:n relations with respect to the case notion, our approach relies on duplicate detection, which requires (non-duplicate) events to have discriminative timestamps or attribute values. Finally, because the assignment of objects to events often depends on domain knowledge about inter-object relations, our approach can currently not handle all scenarios. For example, it is not clear without domain knowledge that items shall relate to packages but not vice versa. Nevertheless, as our evaluation shows, our approach achieves promising results and thus provides an important contribution towards the applicability of object-centric process mining.

In future work, we aim to give domain experts the option to provide input regarding the higher-level relations between different object types in the form of rules. For example, they could state that, if an event results in the creation of an order, it also relates to the items of that order. Then, only assignments that adhere to these rules could be made. Moreover, we want to integrate common-sense knowledge into our approach, which could help to derive relations between object types through their meaning.

Reproducibility: The implementation, employed data, and obtained OCEL logs are available through the repository linked in Sect. 5.

References

1. van der Aalst, W.: Process Mining: Data Science in Action. Springer, Heidelberg (2016). https://doi.org/10.1007/978-3-662-49851-4
2. van der Aalst, W.M.P.: Object-centric process mining: dealing with divergence and convergence in event data. In: Ölveczky, P.C., Salaün, G. (eds.) SEFM 2019. LNCS, vol. 11724, pp. 3–25. Springer, Cham (2019). https://doi.org/10.1007/978-3-030-30446-1_1

3. van der Aalst, W., Berti, A.: Discovering object-centric petri nets. Fundamenta Informaticae **175**(1–4), 1–40 (2020)

4. Abedjan, Z., Golab, L., Naumann, F.: Profiling relational data: a survey. VLDB J. **24**(4), 557–581 (2015). https://doi.org/10.1007/s00778-015-0389-y

5. Acampora, G., Vitiello, A., Di Stefano, B., van der Aalst, W., Günther, C., Verbeek, E.: IEEE 1849tm: the XES standard. IEEE Comput. Intell. Mag. 4–8 (2017)

6. Bano, D., Weske, M.: Discovering data models from event logs. In: Dobbie, G., Frank, U., Kappel, G., Liddle, S.W., Mayr, H.C. (eds.) ER 2020. LNCS, vol. 12400, pp. 62–76. Springer, Cham (2020). https://doi.org/10.1007/978-3-030-62522-1_5

7. Berti, A., van Zelst, S., van der Aalst, W.: Process mining for python (PM4Py): bridging the gap between process-and data science. In: ICPM Demo Track, pp. 13–16. CEUR-WS (2019)

8. van Dongen, B.: BPI Challenge (2017). https://doi.org/10.4121/uuid:5f3067df-f10b-45da-b98b-86ae4c7a310b

9. van Dongen, B.: BPI Challenge (2019). https://doi.org/10.4121/uuid:d06aff4b-79f0-45e6-8ec8-e19730c248f1

10. van Eck, M., Sidorova, N., van der Aalst, W.: Guided interaction exploration in artifact-centric process models. In: Business Informatics, pp. 109–118. IEEE (2017)

11. Esser, S., Fahland, D.: Multi-dimensional event data in graph databases. J. Data Semant. **10**(1), 109–141 (2021)

12. Fahland, D.: Artifact-centric process mining. In: Sakr, S., Zomaya, A.Y. (eds.) Encyclopedia of Big Data Technologies, pp. 108–117. Springer, Cham (2019). https://doi.org/10.1007/978-3-319-77525-8_93

13. Ghahfarokhi, A.F., Park, G., Berti, A., van der Aalst, W.M.P.: OCEL: a standard for object-centric event logs. In: Bellatreche, L., et al. (eds.) ADBIS 2021. CCIS, vol. 1450, pp. 169–175. Springer, Cham (2021). https://doi.org/10.1007/978-3-030-85082-1_16

14. Leopold, H., van der Aa, H., Offenberg, J., Reijers, H.A.: Using hidden Markov models for the accurate linguistic analysis of process model activity labels. Inf. Syst. **83**, 30–39 (2019)

15. Levin, B.: English Verb Classes and Alternations: A Preliminary Investigation. University of Chicago Press, Chicago (1993)

16. Li, G., de Murillas, E.G.L., de Carvalho, R.M., van der Aalst, W.M.P.: Extracting object-centric event logs to support process mining on databases. In: Mendling, J., Mouratidis, H. (eds.) CAiSE 2018. LNBIP, vol. 317, pp. 182–199. Springer, Cham (2018). https://doi.org/10.1007/978-3-319-92901-9_16

17. Li, G., de Carvalho, R.M., van der Aalst, W.M.P.: Automatic discovery of object-centric behavioral constraint models. In: Abramowicz, W. (ed.) BIS 2017. LNBIP, vol. 288, pp. 43–58. Springer, Cham (2017). https://doi.org/10.1007/978-3-319-59336-4_4

18. Lu, X., Nagelkerke, M., Van De Wiel, D., Fahland, D.: Discovering interacting artifacts from ERP systems. IEEE Trans. Serv. Comput. **8**(6), 861–873 (2015)

19. Malone, T., Crowston, K., Herman, G.: Organizing Business Knowledge: The MIT Process Handbook. MIT Press, Cambridge (2003)

20. Popova, V., Dumas, M.: Discovering unbounded synchronization conditions in artifact-centric process models. In: Lohmann, N., Song, M., Wohed, P. (eds.) BPM 2013. LNBIP, vol. 171, pp. 28–40. Springer, Cham (2014). https://doi.org/10.1007/978-3-319-06257-0_3

21. Popova, V., Fahland, D., Dumas, M.: Artifact lifecycle discovery. Int. J. Coop. Inf. Syst. **24**(01), 1550001 (2015)

22. Rebmann, A., van der Aa, H.: Extracting semantic process information from the natural language in event logs. In: Advanced Information Systems Engineering, pp. 57–74 (2021)

23. Speer, R., Chin, J., Havasi, C.: Conceptnet 5.5: an open multilingual graph of general knowledge. In: Thirty-First AAAI Conference on Artificial Intelligence (2017)

Systems

Why Companies Use RPA: A Critical Reflection of Goals

Peter A. François[1](✉) (iD), Vincent Borghoff[1] (iD), Ralf Plattfaut[1] (iD), and Christian Janiesch[2] (iD)

[1] Fachhochschule Südwestfalen, Lübecker Ring 2, 59494 Soest, Germany
{francois.peter,borghoff.vincent,plattfaut.ralf}@fh-swf.de
[2] TU Dortmund University, Otto-Hahn-Straße 14, 44319 Dortmund, Germany
christian.janiesch@tu-dortmund.de

Abstract. Scholarly literature discusses several goals companies pursue with robotic process automation. These goals include both instrumental and humanistic objectives, such as increased process efficiency or higher job satisfaction. However, single case studies often focus only on single goals – not on the full spectrum of goals. An overview and critical reflection of the range of potential goals for employing robotic process automation is still missing. In this article, we review scholarly literature and report on an analysis of multiple expert interviews. In total, we identify 28 goals companies pursue with robotic process automation. Further, we found that the goal dimensions mirror and extend beyond the well-known devil's quadrangle for business process management and form a robotic process automation goal pentagon that also includes people as a goal dimension. Despite the breadth of goals covered in literature, we found that practitioners predominantly focus on financial goals. With our results, we enable an extended discussion on theoretical insights of robotic process automation goals and provide guidance for users and software vendors alike to context-sensitively shape development and implementation projects by focusing on a set of relevant goals.

Keywords: RPA · Robotic Process Automation · Goals · Meta-Synthesis

1 Introduction

Robotic Process Automation (RPA) is an important automation technology for companies today. In a survey of over 400 decision-makers from various industries, 53% of respondents claimed to have started implementing or using RPA [1].

There is a broad spectrum of benefits in the literature that can be put into practice by using RPA (e.g., [2]). The benefits range from the low cost of implementation over the possibility that end users can implement solutions themselves to RPA being non-invasive using the user interface of standard IT systems. In addition, a lot of work on how to introduce RPA into companies exists (e.g., [3, 4]). Yet, in literature, there are only brief mentions of the goals companies pursue with RPA. As RPA is still a practice-driven topic, many insights surface via single case studies. Some of these case studies briefly mention the companies' goals. However, summative syntheses are comparably

© Springer Nature Switzerland AG 2022
C. Di Ciccio et al. (Eds.): BPM 2022, LNCS 13420, pp. 399–417, 2022.
https://doi.org/10.1007/978-3-031-16103-2_26

rare (e.g., [4]), and there has not been a focused study of RPA implementation and usage goals in practice. Hence, the picture of the goals for implementing RPA projects is vague. However, the ability to communicate the motives and targets that drive an RPA implementation is a critical factor for its success [3]. In the same vein, Syed et al. call for research to find ways to help companies to realize RPAs benefits following their idiosyncratic business context [5]. As such, further research on the goals companies pursue with RPA is necessary.

A systematic overview of goals would further the scientific understanding of why companies use RPA. Thus, it would also be helpful for companies thinking about RPA introduction or reflecting on their own use of it. From a research perspective, a deeper understanding of different organizational goals allows researchers to contextualize future studies on RPA. Comparing theory and practice on the importance of the goals complements this picture further. Also, it aids in aligning RPA technology development with customer goals. Thus, with this research effort, we aim to close this gap and answer the following research questions (RQ):

- RQ1: Which goals do companies pursue with RPA?
- RQ2: How can these goals be systematized?

This paper is structured accordingly: Sect. 2 presents a brief theoretical background. Section 3 details our method to answer the research questions. The results are presented in Sect. 4 and discussed in Sect. 5. The paper closes with a brief conclusion.

2   Background

2.1   Goals and Goal Setting

Goals formalize well-considered desired outcomes that entities plan to achieve. They have two attributes: content (the result being sought) and intensity (goal setting) [6]. Organizational goals reflect these on an individual, team, or organizational level and guide a company's action as well as measure its performance [7]. Therefore, organizational goals direct a company's behavior by setting guardrails for further development [8]. As organizational goals are dynamic, these guardrails are formed by and can vary depending on changes within a company's internal and external context [9]. Kotlar et al. distinguish organizational goals by being financial or non-financial [8]. While financial goals can manifest in improving quality or direct cost reduction, non-financial goals can be found in increased flexibility, control, or higher employee wellbeing. Similarly, Sarker et al. differentiate between instrumental and humanistic goals [10].

Goal setting is a thoughtful activity to establish a desired future state [6]. As numerous studies have shown, setting the right goals improves performance in four ways: they serve a directive function, they have an energizing function, they affect persistence, and they arouse problem-solving strategies (for an overview cf. [11]). The process can be guided by criteria or rules such as SMART [12] or advanced goal taxonomies such as the balanced scorecard [13]. The latter proposes measures in the four perspectives financial, customer, internal business processes, as well as learning and growth. Such taxonomies enable a concerted and wary definition of goal systems, which may contain complementary and competing goals.

In this respect, organizational goal objectives and goal setting are highly influenced by the strategic planning and the context a company operates in [14]. Strategic planning falls under the regime of long-term internal corporate vision. In contrast, context is, among other factors, influenced by available tools and mechanisms, which facilitate to achieve or even allow for the setting of specific goals. These goals in turn pose (possibly new) reachable benefits [15, 16]. The availability of certain technological solutions, such as RPA, can be an example of this contextual influence [17].

2.2   Robotic Process Automation

Even though RPA is a relatively new technology, it features a wealth of definitions (see e.g., [18–20]). For the purpose of this paper, we understand RPA as "a technology that allows the development of (multiple) computer programs (i.e., bots) that automate rules-based business processes through the use of graphical user interfaces" [2].

That is, RPA enables the automation of already digitized, yet manually executed business processes. Therefore, it mimics user interactions with other business applications to interact with and manipulate data [21]. In contrast to other automation solutions, RPA is a non-invasive automation technology: underlying systems or infrastructures are not affected by the introduction of RPA and do not need to be adapted [22]. This fundamentally changes the options of how processes can be automated. Consequently, business process management (BPM) practices and corresponding goals need to be revisited, considering this new technology and its versatility.

BPM pursues distinct goal dimensions when (re)designing specific processes. Dumas et al. structure these goal dimensions within the devil's quadrangle [23]. The quadrangle consists of the four dimensions of time, cost, quality, and flexibility. The underlying assumption is that a change in one of the dimensions directly affects the manifestation of the other dimensions. This interconnection between the goal dimensions helps to highlight goal conflicts when applying BPM methods. The four dimensions are operationalized through distinct performance indicators [23].

BPM is context-sensitive, meaning the actual choice of appropriate BPM methods and technologies varies depending on multiple situational factors, including the goal dimension [24]. This implies that applying different automation technologies (such as RPA or BPM) can form distinct goal configurations and vice versa.

3   Method

In our research, we followed a linear three-step approach starting with a structured literature review following vom Brocke et al. and Webster and Watson to cover already published goals in the literature [25, 26]. We chose a broad search approach by using "Robotic Process Automation" itself as the search term. To cover various aspects of the BPM field, we searched multiple databases with application-related research. This included the AIS library, the Business Process Management Journal, and the proceedings of the BPM conference. Further, we included the AIS senior scholar's basket not to miss high-quality articles not available in the AIS library but relevant to the field of information systems. Lastly, we included SSRN to also cover unpublished work. We used the search term in all relevant fields (abstract, keywords, title, body). We have not limited our search to a specific time period, but we included all relevant literature ($n = 227$). We browsed through each article's corresponding references for other potentially relevant articles (backward search) ($n = 26$). All articles were read and classified in terms of their relevance [27]. We excluded duplicates ($n = 4$). Then, we carefully screened the remaining items and further excluded papers that had a focus other than RPA or mention RPA only as a marginal phenomenon (e.g., in the form of an exemplary technology) ($n = 131$). From the initially identified 253 articles, we found 118 to be of relevance for our research.

In a second step, we added ten semi-structured expert interviews (interviews 1–10) together with six additional interviews from an open data dataset (interviews 11–16) to our data base [28]. For the selection of our new interview partners, we applied purposeful sampling [29]. To maximize variation within our ten original interviews, we composed the sample with companies from varying business sectors and sizes in terms of number of employees per company. Besides the variation achieved through purposeful selection on an organizational level, we also headed for a maximum of permeation on an individual level. Therefore, we included different interviewee positions within the companies (consultants, managers, team leads, etc.) as well as the type of access to RPA (users, consultants, educators). Through this selection, we also strive to cover both mainstream and more niche goals within one sample. We continued to conduct interviews until saturation was reached. All interviews were conducted either via video conferencing software or via telephone, later transcribed manually, and (if necessary) translated. For the interviews, we applied a semi-structured approach, focusing on the implementation and usage patterns and structures related to RPA within the respective companies. Therefore, the guideline was loosely oriented on the RPA lifecycle, organizational peculiarities, and the governance of RPA operations. All topics included the goals pursued with the respective structure and phase. While following the interview guideline, we consciously allocated time for the interviewees to address further relevant topics as they emerged. The interviews had a duration of between 40 and 60 min. Table 1 gives an overview of the interviewees.

In a third step, we took the two sets of collected data (relevant scholarly publications and 16 interviews with RPA experts) and extracted the goals companies pursue with RPA (see Sect. 4.2). To this end, we read the texts and used open coding to identify goals [30]. During the process, two researchers individually read the relevant papers and interview transcripts and coded for goals of RPA implementation and use. We then compared the corresponding codes and coded text passages and, in the event of any discrepancies, reached an agreement in scientific discourse – if necessary, with all four authors. Then, we used the initial coding as a starting point for multilevel axial coding [31]. In a first step, the goals found were merged to summarize goals that are equivalent in meaning. This stage was also iterative, and the procedure for disagreement was analog to that described above. In a further stage, the goals found were further examined and abstracted into distinct dimensions (see Sect. 4.1).

Table 1. Overview of interviewees.

No.	Type	Interviewee position	Sector	Employee count
#1	Consultant	Process automation consultant	IT consulting	<500
#2	Consultant	Process automation consultant	IT consulting	<500
#3	User	RPA manager	Chemical industry	~15.000
#4	Consultant	Organizational unit manager, RPA consultant, RPA software developer	Manufacturing	~35.000
#5	User	RPA manager	Civil engineering	~55.000
#6	Consultant	Management consultant	Strategy consulting	~30.000
#7	Consultant	RPA consultant (especially scaling)	Strategy consulting	~30.000
#8	User	RPA manager	Service provider	~35.000
#9	User	Manager of IT division	Health insurance	~4.000
#10	Education provider	Chief executive officer	RPA consulting and training	<100
#11	User	Team leader RPA center of excellence	Finance	<500
#12	User	Division manager projects and processes	Consumer goods	~4.500
#13	User	Sales manager in business customer sales, process analyst, RPA developer	Logistics	~1.000
#14	User	RPA process owner of all marketing processes	System house	<100
#15	Consultant	RPA consultant	IT consulting	<100
#16	User	Department head in a bank	Finance	<500

4   Results

4.1   The RPA Goal Pentagon

We uncovered several dimensions of goals for RPA that emerged from the data using our methodological procedure as outlined above. Upon a more detailed inspection, we found notable similarities between these dimensions and the four performance dimensions of the *devil's quadrangle* of BPM as outlined by Dumas et al. [23]. The devil's quadrangle includes the dimensions of *time, cost, quality*, and *flexibility*. As RPA can be understood as an automation technology alternative to traditional BPM systems, naturally some characteristics are shared.

However, we found that not all goals can be clearly related to the four classic dimensions. In addition, we noticed that the anthropomorphic nature of RPA [32] takes shape in goals as well, distinctively extending beyond the above-mentioned four dimensions.

In the following, we describe the classic four dimensions in their RPA coloring as they emerged from our data. In addition, we introduce a novel fifth dimension present in our data that we refer to as *people* to form the *RPA goal pentagon* (see Fig. 1):

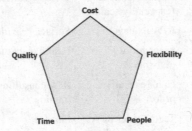

Fig. 1. The RPA goal pentagon

Time. This dimension comprises all goals that we clustered because they affect the acceleration of temporal sequences. According to Dumas et al. [23], goals in this dimension focus on optimizing the cycle or throughput time of a process and in the course improve one or both of the following: processing time and waiting time. The technological distinctiveness of RPA as opposed to BPM (see also Sect. 2) fuels further goals that highlight its lightweight nature and the rapid implementation with RPA.

Cost. This dimension comprises all goals that affect the company's income and expenditure. According to Dumas et al. [23], "Within process redesign efforts, it is widespread to focus on reducing operation cost, particularly labor cost." This coincides, for example, with RPA's ability to automate tasks even more directly as well as RPA's comparatively low license and implementation costs.

Quality. This dimension includes all goals that strive to improve the quality overall, or of specific artifacts required in the company. This includes the quality perceived by the customer and the quality perceived by other business process participants. A significant factor of a process's quality is the "[…] amount, relevance, quality, and timeliness of the information [...]" [23] used in the business process. By using RPA, companies can improve processes through improving data quality (e.g., by reducing human errors) as well as general accuracy in business processes [2].

Flexibility. This dimension encompasses all goals that aim to make processes, the use of IT, operations, or the entire company more flexible and increase the ability to react to changes [23]. It relates, for example, to RPA's benefits in being able to quickly establish and then change digital processes as well as the ability to cope with changing transaction volumes.

People. This dimension constitutes a novel dimension of its own and highlights the importance of humanistic goals for RPA. It comprises any goal that directly benefits a human actor involved in or concerned with the process. Hence, it relates to the anthropomorphic nature of RPA being able to mimic human workers and relieve them directly of arduous tasks or processes without significant intervention in existing work procedures [32]. So far, in our data we see evidence for employee-centered goals as most robots perform back-office tasks. For example, this dimension comprises improved work environments or improved services that mitigate demographical or seasonal changes. Goals of this dimension often impact the individual, but they are not exclusive to this level. Further, it is conceivable to have goals at the team or organization level as well as customer-focused rather than employee-focused goals.

Beyond the five dimensions of RPA goals, we found one distinctive and unique goal relating to innovative technologies and collected it as a further goal without a category (28.). It was noticeable in our interview study, and we felt obliged to highlight its existence and discuss how to approach it in the following section.

While we have not had the opportunity to assess the interplay of the five dimensions of the RPA goal pentagon for goal setting and realization empirically, we assume it to behave similarly to the devil's quadrangle. That is, pursuing a goal along one dimension may weaken the performance along another dimension. As we found some goals to be associated with multiple dimensions (see Sect. 4.2), we assume pursuing a particular goal can strengthen or weaken multiple dimensions at the same time, respectively. Subsequently, we use these dimensions as attributes to group the goals to better structure, assess, and discuss them.

4.2 Goals for Introducing and Using RPA

In total, we found 28 goals for implementing and using RPA. We found 24 goals in literature. We could confirm all but one in our interviews. In addition to these, we found four goals in our interviews that were not mentioned in literature. Table 2 gives an overview of the goals, their origin, and their respective goal dimensions. In the following, we detail the goals with a focus on those not well understood since they were not mentioned in either literature or practice.

As argued above, we could confirm the practical relevance of 23 goals that were mentioned in literature. This includes the goals mentioned in case studies or directly abstracted by us from the benefits and characteristics mentioned in literature. In contrast,

Table 2. List of goals from literature and interviews sorted by dimensions.

Goal	Found in		Dimension					
	Interviews	Literature	Time	Cost	Quality	Flexibility	People	Other
1. Improving process speed/Increasing productivity/efficiency (in general, not further specified)	(#2, #8, #11, #14, #16)	[18, 33–45]	x	x	x			
2. Reducing IT-implementation time	(#1, #6, #8, #9, #15, #16)	[18, 34, 37, 39, 42, 44, 46–48]	x	x		x		
3. Freeing up resources within IT	(#4, #15)	[39, 42]	x	x		x		
4. Mitigating the effect of sporadic high workloads	(#8)	[48, 49]	x	x		x	x	
5. Jumping the IT development queue	(#15)	[37]	x			x		
6. Reducing cost (in general, not further specified)	(#2, #3, #5, #6, #8, #12, #14, #15)	[33, 34, 37, 40–42, 44, 45, 50, 51]		x				
7. Using RPA to automate steps in BPM-projects/automating processes/using RPA as one element in a BPM toolset/using RPA to conduct iterative BPM and automation	(#4, #5, #8, #10, #12)	[35, 46, 52–56]		x				
8. Dealing with high transaction volume	(#9, #10, #16)	[19, 21, 34, 48, 54]		x				
9. Reducing employee count/reducing full-time equivalent (FTE) count/getting more work done with the same employees/obtaining the ability to grow, scale	(#2, #3, #4, #5, #7, #9, #11, #12, #13, #15, #16)	[19, 34, 39, 41, 42, 44–46, 48, 49, 57, 58]		x		x		
10. Filling/mitigating open positions	(#9, #16)	[33]		x				

(*continued*)

Table 2. (*continued*)

Goal	Found in		Dimension					
	Interviews	Literature	Time	Cost	Quality	Flexibility	People	Other
11. Improving quality (in general, not specified)/improving service quality/improving data quality/increasing accuracy/reducing errors/reducing error cost	(#2, #11, #13, #14)	[18, 33–36, 38–42, 44, 45, 54, 56, 58, 59]		x	x			
12. Improving process innovation	(#8, #10, #11)	[36, 42, 49, 60]		x	x			
13. Standardizing activities and processes/improving compliance/improving documentation, transparency, auditability	(#8, #12, #14)	[33, 34, 36, 39, 44, 49, 59]		x	x			
14. Mitigating risk (e.g., uncertainty of cost of IT projects)	(#8)	[35, 39]		x	x			
15. Avoiding shadow IT	(#1, #8)			x	x			
16. Ensuring operations outside of business hours		[34, 39, 58]		x	x	x		
17. Reducing IT implementation effort and cost	(#9, #14, #16)	[2, 34–36, 41, 42, 46, 48, 51, 58, 61]		x		x		
18. Mitigating the impact of demographic change	(#8)	[49, 61]		x			x	
19. Ensuring availability	(#14)	[33, 34, 62]			x			
20. Using RPA to implement substitution rules for employees and ensure fail-safe operation	(#12)				x			
21. Mitigating staffing shortcomings due to previously cut positions	(#3, #5)	[62]			x			

(*continued*)

Table 2. (*continued*)

Goal	Found in		Dimension					
	Interviews	Literature	Time	Cost	Quality	Flexibility	People	Other
22. Relieving employees of boring or unattractive tasks	(#4, #8, #10, #11, #12, #15)	[33, 35, 38–45, 47, 55, 56, 58, 59, 62–64]			x		x	
23. Connecting to systems without APIs/making legacy software with new processes/overcoming data silos/using RPA as a bridging technology	(#1, #4, #6, #8, #9, #10, #11, #14)	[35, 36, 39, 48, 55, 56]				x		
24. Getting the task of process automation closer to the domain knowledge	(#8)	[35, 37, 39, 51]				x		
25. Giving the departments the opportunity to adapt systems to fast-changing processes	(#11)	[38]				x		
26. Raising acceptance of automation in departments due to fast implementation	(#8)	[47, 49]					x	
27. Using RPA as a door opener for BPM with departments	(#8)						x	
28. Using RPA because it is a trend	(#15)							x

four of the goals found in the interviews were not included in the literature before. We could not confirm the relevance of one goal (16.) in practice. In the following, we explain selected goals and also detail the four goals not previously described in literature in more detail.

Many of the goals do not fall into only one goal dimension. Instead, the goals satisfy the intentions of multiple dimensions. For example, we classified the goal of mitigating the effect of sporadic high workloads (4.), in which RPA is used to keep employees from being overworked when periods with uncharacteristically high workloads occur, to belong simultaneously to the dimensions of time, cost, flexibility, and people as it also ensures all tasks being processed in time without a (costly) reduction in service levels.

One of the goals heavily discussed in literature is the goal of relieving employees of boring or unattractive tasks (22.). While we were able to establish the relevance of this goal in practice, only two interviewees of the type "user" (#8, #12) stated this to be a goal of interest to them despite being a goal of positive nature. The goal of reducing employee count (9.) appears inverse in effect: while getting rid of unattractive tasks is positive for employees, reducing their count and thus threatening their jobs will most certainly be perceived negatively by employees.

Another common goal is that of freeing up resources in the IT department (3.). Companies pursuing this goal may have issues finding sufficiently qualified IT personnel. Another reason may be to enable existing IT personnel to focus on more complex IT problems or to have less complex automation problems be solved by "less expensive" employees in the specialist departments. This goes hand in hand with the goal of jumping the IT development queue (5.). This goal may also be pursued by companies whose IT departments are overworked or with the intent of enabling specialty departments to determine and change the development sequence and development speed more flexibly. Interviewee #15 stated: *"If it is possible* [to solve an automation issue] *otherwise, but* [it is] *still done with RPA, then most of the reasons are* [...] *that you have a very low priority with IT.* [...] *it will be implemented in 2023, but they want to have it today."*

Another noteworthy goal is using RPA to conduct iterative BPM and automation (7.). Companies pursuing this goal see RPA as one tool of many in the pursuit of successfully performing an automation project. This can correspond with the goal of using RPA as a door opener for BPM with departments (27.), where companies aim to bypass social barriers to BPM. By following through with RPA implementation, business departments disclose the process flow and, in the course, create the opportunity for process improvement or as interviewee #8 puts it: *"We use RPA to access areas that we would not have access to with other technologies, because then it's a typical IT project again* [which] *takes 2 or 3 years until I see results.* [...] *we have already noticed that so much is already achieved with standardization and optimization alone* [...]*".*

Literature mentions the risk of creating shadow IT when using RPA [65]. In the interviews, RPA was sometimes used contrary to the goal of avoiding shadow IT (15.) by using governance measures to control RPA initiatives [#1, #8], giving the departments an official way to solve their automation needs.

Aside from the goals classifiable with our pentagon, consultant interviewee #15 stated that some companies start RPA initiatives with the goal of using RPA because it is a trend (28.): *"And the other thing that might be interesting is that RPA is simply a trend. People like to jump on trends and have a look at them. A lot of PoCs* [proof of concept] *start when somebody has heard from another company that RPA is all the rage right now, and in the future, it will be even better with artificial intelligence. That is also always a motivation."* This is not a particularly rational goal, but it expresses a lack of digitization and BPM strategy. It highlights that actual goals can be rather implicit and rooted in human deficiencies and should be reviewed for their true rationale. We relate this "goal" to novelty bias and the bandwagon effect, both of which are well-studied and provide advice on how to approach the effects. Due to its obvious practical relevance, we felt obliged to highlight its existence.

The goal of ensuring operations outside of business hours (16.) stated in the literature [34] could not be revalidated in the interviews. Yet, in the interviews, we found the goal of using RPA to implement substitution rules for employees and ensure fail-safe operation (20.) in the interviews, which was not previously covered in the literature, but is closely related to ensuring operations outside of business hours (16.). Here the interviewee intended to implement substitution rules within processes using RPA.

4.3   Discussion of the Prevalence of RPA Goals

The characteristics of RPA that are most frequently mentioned in literature focus on the flexibility of the technology and its ability of relieving employees from performing boring tasks (22.) [2, 34]. In contrast, these goals took a secondary role in the interviews. Cost reduction (6.) and FTE savings (9.) were mentioned the most dominantly. All interviewees mentioned at least one goal of the cost dimension, while fewer mentioned goals in the quality, flexibility, and time dimensions. Goals from the people dimension were mentioned the least often by our interviewees.

Most interviewees did not rank the importance of the pursued goals. Nevertheless, some gave information on the importance of the goals. For example: *"The topics higher productivity, higher volume, thereby indirect cost savings. These are the classic* [goals for using RPA]. *"* (#16). Interviewees #3, #5, #6, #9, and #16 even clearly stated that they would mainly follow financial goals. Interviewee #6 stated: *"I personally see RPA as just a cheap version to automate things"*, hinting at a clear financial focus. Interviewees #12 and #13 pointed out that mainly quality goals where pursued. Yet, #12 stated: *"We have not carried out an overarching business case. Not one where we said how many processes there are in the company, what potential they have, and now we are calculating a business case. Instead, we are convinced that savings can be achieved* [in our RPA project]. *"* In this case, the goal of increased quality was a sub-goal of decreasing cost. Interviewee #13 stated: *"The occurrence of errors is of course also associated with time and cost, especially towards the customer."* Therefore, interviewee #13 did not pursue goals from the quality dimension for their own sake either but to reduce error costs.

Building on the second finding, we found that many of the companies who mentioned multiple goals also stated to have created a business case before starting an RPA project. Therefore, we argue that those companies, in fact, did have cost or profit (directly or indirectly) as at least one of their main goals since their projects would not have started without a positive business case. Some companies carry this focus foreword into choosing which processes are implemented in RPA. One interviewee responsible for managing bot creation projects and choosing processes to automate even stated: *"If you come to me without a business case, I'll kick you out again"* (#14).

Within those financially-driven business cases, FTE reduction or being able to scale without increasing worker count was named as one of the most important levers of RPA. This was also essential to RPA projects in general: *"We are a very cost-driven company, and headcount reduction, or FTE reduction, is definitely on the agenda* [with RPA]. *Even if people sometimes try to sweep it under the carpet, because of course, it is well received by the shareholders but not necessarily by our own employees"* (#3).

In summary, we can conclude that in practice reducing cost is the most important goal for using RPA.

5 Implications and Future Research

5.1 Implications for Theory and Practice

Based on our analysis, we found 28 goals companies pursue with RPA. Of these goals, four have not been discussed in literature before. Thus, we extend the body of knowledge by adding these goals to the body of knowledge. Thereby, we complement the picture of RPA established so far by adding to the overviews of characteristics and benefits [2, 34].

Further, we propose the RPA pentagon as a means to cluster the goals and provide orientation and structure for goal setting. The pentagon shares four dimensions with Dumas et al.'s devil's quadrangle and adds a people dimension due to the unique (and anthropomorphic) characteristics of RPA as opposed to traditional automation with BPM software.

Our RPA pentagon, similar to other goal-setting criteria or taxonomies such as the devil's quadrangle and the balanced scorecard, can help establish a more well-adjusted goal system for RPA endeavors leading to better long-term utility and satisfaction with software robots rather than purely following cost-centered goals. This is especially true for large-scale installations that can have significant social and cultural impact on a company. The goals in the RPA goal pentagon are not only goals in conflict but – as mentioned earlier – can be complementary to each other as well as applicable to multiple dimensions. That is, by following specific goals such as mitigating the impact of demographic change (18.), one will improve both the cost and the people dimension while potentially weakening the other three dimensions.

Typically, the use of RPA is contextualized by the main objectives pursued. Theory on the use of RPA must, therefore, always consider these goals as a setting. Knowledge about RPA gained in cost-focused contexts does not necessarily apply in people-focused contexts. The focus on different goals by different companies may also influence the success factors for RPA implementation. Therefore, current implementation models should be re-evaluated on how they can be adapted to bring out those benefits most aligned to the company's goals. A clear picture of goals allows for a better alignment across users within a company or across companies. For this, new models of setting up governance to propagate goals across the company are necessary.

While it is conceivable that different companies will focus on varied goals depending on their strategic orientation, project characteristics, and other context factors, one may assume that all five dimensions are about equally important in general. However, our findings from the expert interviews diverge starkly from this naïve assumption. Monetary goals take precedence while other goals are pursued as secondary targets or as windfall profit when fulfilling the monetary goals. Be that as it may, it was not our objective to formulate a goal-setting strategy procedure for RPA goals but to focus on compiling and structuring goals relevant and potentially unique to RPA. It remains to be seen if this process differs from traditional goal setting in IT projects.

On a broader scale, we also contribute to the ethical discussion regarding RPA. Sarker et al. [10] stress that information systems research is a socio-technical discipline. According to them, often the technical implications are more pronounced in research, which may lead to "a lack of ethical standing of the discipline in society due to the failure of IS scholars and practitioners to reflect on the consequences of information technology and to critique and actively oppose initiatives where IT might facilitate the development of a dehumanized and dystopian society" [10].

In the light of this call, the current use of RPA and the goals thereof should be questioned. Despite the range of social benefits of RPA discussed in literature, these appear not to be the focus of implementation projects in practice. Using RPA without paying mind to the social aspects bears risks both for technology acceptance and for achieving the full potential of RPA. We aspire that our findings – and especially our fifth dimension people – facilitate and extend the discussion on how to mitigate these risks for humans. Here, additional guidelines may become necessary as robots are commonly employed to perform any or all of the 3D to 5D jobs that are described as dull, dirty, dangerous, dear, or difficult. Not all of the above apply to the software robots of RPA (in particular as knowledge work is rarely dangerous). Yet, performing difficult tasks for humans may evoke other reactions from people that performing dull or (digitally) messy tasks. It remains to be seen how applicable guidelines for robot use (see e.g. [66–68]) are for software robots.

Lastly, the finding that some companies only use RPA because it is a trending topic opens implications for theory. We should be aware of this mimetic isomorphism in relation to RPA and other automation technologies. First studies in this direction have been conducted [69]. However, future research is still needed.

Our holistic overview of goals enables practitioners to be clearer on their own goals for implementing and using RPA and question them and their balancing. Such clarity will enable more successful implementation projects, especially from a socio-technical perspective, by enabling a more targeted consideration of the framework conditions, such as the observance or rejection of certain critical success factors [3]. The occurrence of goals we identified allows companies to question the relevance of different goals for their projects and have a more open discussion. A transparent and aligned set of goals will facilitate the sustainable success of RPA projects. Moreover, companies can question whether they fall into the fallacy of implementing RPA because it is a trend. Lastly, our findings can aid RPA vendors and consultants to improve their solutions and ensure that implementation projects are set up to address the specific desired goals.

5.2 Future Research

Due to the high impact of the goals for theory and practice, we need to better understand the goals companies pursue with RPA as well as their respective nature. Therefore, future research should examine additional case studies using the overview of goals we established to broaden the understanding of current RPA projects. New case studies should include the companies' goals for conducting RPA projects and their importance. This would help assessing the activities within RPA projects from the point of view of their objectives.

There might also be a divergence of the pursued goals or their importance in companies of different sectors, sizes, or with different strategies. This might have further implications for the success of RPA projects. Therefore, further research should systematize the goals companies pursue with RPA according to these environmental characteristics or additional contextual factors. Some interviewees hinted at a change in the importance of goals during the implementation and usage phases, but they did not directly describe such effects. Therefore, future research should examine how and at which points in the implementation circle the goals of using RPA in companies change.

Additionally, we assume that there may be a hierarchy, different priorities, or causal as well as temporal interrelation between different goals. Our data does not allow for insights into such relations. Therefore, future research should examine the nature and extent to which the goals are interrelated. This may also explain why practice focuses rather on monetary goals (potentially short term) while research more broadly considers humanistic goals as well (long term perspective).

Furthermore, it should be investigated how the companies' goals influence the direction and success of RPA implementation and use. This also may include enquiring how the knowledge of goals for using RPA and their occurrence affect previous insights on the characteristics, benefits, and critical success factors of RPA. We also need to question established theory and previous findings in light of the insights gained in this work. Research should be conducted on how the complete set of goals presented reflects on the case studies available in literature. It should be evaluated how goals could have influenced the course of action of these projects and if there are any underlying goals that were not discussed in the interviews that did impact projects in practice.

A further discussion on the social aspects of using RPA would be beneficial to understand the use and success of RPA in companies. Research should assess why the social goals of using RPA do not play a more prominent role in practice yet despite the societal discourse on autonomous IT. Furthermore, it should be evaluated how RPA projects should be set up to reap the full potential of RPA – not only focusing on monetary aspects but also on the social (and potentially the environmental) effects.

In addition, research is needed on a holistic or integrated goal framework covering both heavyweight projects focused on BPM technology and lightweight projects focused on RPA technology as both tend to interact and complement each other. That is, from a practical as well as theoretical point of view goals should intersect and not be disparate to prearrange for BPM technology that may be enhanced with RPA bots and RPA bots that may be retired for comprehensive automation using BPM software [4].

Lastly – since goals are only valuable if they can be realized – further research should examine RPA goal setting and determine under which circumstances and using which procedures the goals can be realized. The influence factors determining the achievement of goals should be evaluated in practice.

6   Conclusion

In this paper, we found and systematized 28 goals companies pursue with RPA in literature and practice. We compared which goals are mentioned in literature and which bear relevance in practice and found four previously not reported goals. Further, we uncovered five distinct goal dimensions, four of which relate to the devil's quadrangle of BPM and one novel dimension, people, which forms an RPA goal pentagon.

The goals in this pentagon include instrumental and humanistic objectives, for example, increased process efficiency or higher job satisfaction. We discussed the prevalence of the goals and the implications of the interplay between the goal dimensions. In particular, our analysis shows that – with few exceptions – practitioners have so far almost exclusively focused on the cost dimension's financial goals.

Despite our careful research design, this paper has some limitations. In our keyword-based literature search, we focused on the benefits and characteristics of RPA. There might be additional work not captured by this search. We tried to mitigate this risk through a backward search. We interviewed primarily Germany-based companies, which may introduce cultural and economic bias. Due to the explorative nature of our research, our findings on the occurrence of goals can only be indicative.

References

1. Watson, J., Wright, D.: The robots are ready. Are you? (2017). https://www2.deloitte.com/content/dam/Deloitte/tr/Documents/technology/deloitte-robots-are-ready.pdf. Accessed 15 June 2022
2. Plattfaut, R., Borghoff, V.: Robotic process automation – a literature-based research agenda. J. Inf. Syst. (2022). https://doi.org/10.2308/ISYS-2020-033
3. Plattfaut, R., Borghoff, V., Godefroid, M., Koch, J., Trampler, M., Coners, A.: The critical success factors for robotic process automation. Comput. Ind. **138**, 103646 (2022). https://doi.org/10.1016/j.compind.2022.103646
4. Herm, L.-V., Janiesch, C., Helm, A., Imgrund, F., Hoffman, A., Winkelmann, A.: A framework for implementing robotic process automation projects. Inf. Syst. e-Bus. Manag. (2022). https://doi.org/10.1007/s10257-022-00553-8
5. Syed, R., et al.: Robotic process automation: contemporary themes and challenges. Comput. Ind. **115**, 103162 (2020)
6. Locke, E.A., Latham, G.P., Smith, K.J., Wood, R.E.: A theory of Goal Setting & Task Performance. Prentice Hall, Englewood Cliffs (1990)
7. Mohr, L.B.: The concept of organizational goal. Am. Polit. Sci. Rev. **67**, 470–481 (1973). https://doi.org/10.2307/1958777
8. Kotlar, J., de Massis, A., Wright, M., Frattini, F.: Organizational goals: antecedents, formation processes and implications for firm behavior and performance. Int. J. Manag. Rev. **20**, 3–18 (2018). https://doi.org/10.1111/ijmr.12170
9. Ansoff, H.I.: Strategic Management. Palgrave Macmillan, Basingstoke (2007)
10. Sarker, S., Chatterjee, S., Xiao, X., Elbanna, A.: The sociotechnical axis of cohesion for the IS discipline: its historical legacy and its continued relevance. MIS Q. **43**, 695–719 (2019). https://doi.org/10.25300/MISQ/2019/13747
11. Locke, E.A., Latham, G.P.: Building a practically useful theory of goal setting and task motivation: a 35-year odyssey. Am. Psychol. **57**, 705–717 (2002)
12. Doran, G.T.: There's a S.M.A.R.T. way to write management's goals and objectives. Manag. Rev. **70**, 35–36 (1981)
13. Kaplan, R.S., Norton, D.P.: The balanced scorecard – measures that drive performance. Harvard Bus. Rev. **83**(7), 172 (1992)
14. Gooding, R.Z., Goel, S., Wiseman, R.M.: Fixed versus variable reference points in the risk-return relationship. J. Econ. Behav. Organ. **29**, 331–350 (1996). https://doi.org/10.1016/0167-2681(95)00067-4

15. de Massis, A., Kotlar, J., Wright, M., Kellermanns, F.W.: Sector-based entrepreneurial capabilities and the promise of sector studies in entrepreneurship. Entrep. Theory Pract. **42**, 9–23 (2018). https://doi.org/10.1177/1042258717740548

16. Decker, C., Mellewigt, T.: Thirty years after michael E. Porter: what do we know about business exit? Acad. Manag. Perspect. **21**, 41–55 (2007). https://doi.org/10.5465/amp.2007.25356511

17. Kotlar, J., de Massis, A.: Goal setting in family firms: goal diversity, social interactions, and collective commitment to family-centered goals. Entrep. Theory Pract. **37**, 1263–1288 (2013). https://doi.org/10.1111/etap.12065

18. Cooper, L.A., Holderness, D.K., Sorensen, T.L., Wood, D.A.: Robotic process automation in public accounting. Account. Horiz. **33**, 15–35 (2019). https://doi.org/10.2308/acch-52466

19. Lacity, M., Willcocks, L.: Robotic process automation at Telefónica O2. MIS Q. Exec. **15**, 21–35 (2016)

20. Willcocks, L.P., Oshri, I., Kotlarsky, J. (eds.): Dynamic Innovation in Outsourcing. Springer, Cham (2018). https://doi.org/10.1007/978-3-319-75352-2

21. Hallikainen, P., Bekkhus, R., Pan, S.L.: How OpusCapita used internal RPA capabilities to offer services to clients. MIS Q. Exec. **17**, 41–52 (2018)

22. Lacity, M., Willcocks, L.: Robotic Process Automation: The Next Transformation Lever for Shared Services. The Outsourcing Unit Working Research Paper Series 15/07 (2015)

23. Dumas, M., La Rosa, M., Mendling, J., Reijers, H.A.: Fundamentals of Business Process Management. Springer, Berlin (2018). https://doi.org/10.1007/978-3-662-56509-4

24. vom Brocke, J., Zelt, S., Schmiedel, T.: On the role of context in business process management. Int. J. Inf. Manag. **36**, 486–495 (2016). https://doi.org/10.1016/j.ijinfomgt.2015.10.002

25. vom Brocke, J., Simons, A., Riemer, K., Niehaves, B., Plattfaut, R., Cleven, A.: Standing on the shoulders of giants: challenges and recommendations of literature search in information systems research. Commun. AIS **37**, 201–225 (2015). https://doi.org/10.17705/1CAIS.03709

26. Webster, J., Watson, R.T.: Analyzing the past to prepare for the future: writing a literature review. Manag. Inf. Syst. Q. **26**, xiii–xxiii (2002)

27. Page, M.J., et al.: The PRISMA 2020 statement: an updated guideline for reporting systematic reviews. BMJ (Clinical research ed.) **372**, 71 (2021). https://doi.org/10.1136/bmj.n71

28. Helm, A., Herm, L.-V., Imgrund, F., Janiesch, C.: Interview Guideline, Transcriptions, and Coding for "A Consolidated Framework for Implementing Robotic Process Automation Projects" (2021). https://b2share.eudat.eu/records/402d2d1544124d24902182652d1bc77a. Accessed 15 June 2022

29. Palinkas, L.A., Horwitz, S.M., Green, C.A., Wisdom, J.P., Duan, N., Hoagwood, K.: Purposeful sampling for qualitative data collection and analysis in mixed method implementation research. Adm. Policy Ment. Health Mental Health Serv. Res. **42**(5), 533–544 (2013). https://doi.org/10.1007/s10488-013-0528-y

30. Strauss, A., Corbin, J. (eds.): Grounded Theory in Practice. Sage Publishing, Thousand Oaks (1997)

31. Gioia, D.A., Corley, K.G., Hamilton, A.L.: Seeking qualitative rigor in inductive research. Organ. Res. Methods **16**, 15–31 (2013). https://doi.org/10.1177/1094428112452151

32. Herm, L.-V., Janiesch, C., Reijers, H.A., Seubert, F.: From symbolic RPA to intelligent RPA: challenges for developing and operating intelligent software robots. In: Polyvyanyy, A., Wynn, M.T., Van Looy, A., Reichert, M. (eds.) BPM 2021. LNCS, vol. 12875, pp. 289–305. Springer, Cham (2021). https://doi.org/10.1007/978-3-030-85469-0_19

33. Hofmann, P., Samp, C., Urbach, N.: Robotic process automation. Electron. Mark. **30**(1), 99–106 (2019). https://doi.org/10.1007/s12525-019-00365-8

34. Meironke, A., Kuehnel, S.: How to measure RPA's benefits? A review of metrics, indicators, and evaluation methods of RPA benefit assessment. In: WI Proceedings. Virtual, 5 (2022)

35. Shome, N.: RPA for Telcos: The Next Wave of BPM Evolution. SSRN, 3712422 (2017)
36. Eulerich, M., Pawlowski, J., Waddoups, N.J., Wood, D.A.: A framework for using robotic process automation for audit tasks. Contemp. Account. Res. **39**, 691–720 (2022). https://doi.org/10.1111/1911-3846.12723
37. Beerbaum, D.O.: Artificial Intelligence Ethics Taxonomy - Robotic Process Automation (RPA) as Business Case. SSRN (2021). https://doi.org/10.2139/ssrn.3834361
38. Zhang, C., Thomas, C., Vasarhelyi, M.: Attended process automation in audit: a framework and a demonstration. J. Inf. Syst. (2021). https://doi.org/10.2308/ISYS-2020-073
39. Chugh, R., Macht, S., Hossain, R.: Robotic process automation: a review of organizational grey literature. Int. J. Inf. Syst. Proj. Manag. **10**, 5–26 (2022)
40. Januszewski, A., Kujawski, J.: Best practices in robotic process automation in global business services. In: AMCIS Proceedings. Virtual, 6 (2021)
41. Söderström, F., Johansson, B., Toll, D.: Automation as migration? – Identifying factors influencing adoption of RPA in local government. In: ECIS Proceedings. Virtual, 38 (2021)
42. Kaniadakis, A., Linturn, L.: Organisational adoption of a hyped technology: the case of robotic process automation. In: ECIS Proceedings. Virtual 46 (2021)
43. Elsayed, N.S.S., Kassem, G.: Assessing process suitability for robotic process automation: a process mining approach. In: WI Proceedings. Virtual, 18 (2022)
44. Staaby, A., Hansen, K., Grønli, T.-M.: Automation of routine work: a case study of employees' experiences of work meaningfulness. In: HICSS Proceedings, pp. 156–165 (2021)
45. Johansson, B., Söderström, F.: IS capabilities for robotic process automation - Feeny-Willcocks framework revisited. In: ECIS Proceedings, Timisoara, 127 (2022)
46. Penttinen, E., Kasslin, H., Asatiani, A.: How to choose between robotic process automation and back-end system automation? In: ECIS Proceedings. Portsmouth, 66 (2018)
47. Lacity, M., Willcocks, L., Gozman, D.: Influencing information systems practice: the action principles approach applied to robotic process and cognitive automation. J. Inf. Technol. **36**, 216–240 (2021). https://doi.org/10.1177/0268396221990778
48. Koch, J., Vollenberg, C., Matthies, B., Coners, A.: Robotic process flexibilization in the term of crisis: a case study of robotic process automation in a public health department. In: ECIS Proceedings. Timisoara, 59 (2022)
49. Zhang, C.A., Issa, H., Rozario, A., Søgaard, J.S.: Robotic process automation (RPA) implementation case studies in accounting: a beginning to end perspective. Account. Horiz. (2022). https://doi.org/10.2308/HORIZONS-2021-084
50. Lacity, M., Willcocks, L.: A new approach to automating services. MIT Sloan Manag. Rev. 41–49 (2016)
51. van der Aalst, W.M.P.: Hybrid Intelligence: to automate or not to automate, that is the question. Int. J. Inf. Syst. Proj. Manag. **9**, 5–20 (2021)
52. Cewe, C., Koch, D., Mertens, R.: Minimal effort requirements engineering for robotic process automation with test driven development and screen recording. In: Teniente, E., Weidlich, M. (eds.) BPM 2017. LNBIP, vol. 308, pp. 642–648. Springer, Cham (2018). https://doi.org/10.1007/978-3-319-74030-0_51
53. Keerthana, Prasannakumar, Abishek, Arul, Vijayalakshmi: Data filtering and visualization for sentiment analysis of ecommerce websites. In: ICICNIS Proceedings, Kottayam (2021)
54. Desai, D., Jain, A., Naik, D., Panchal, N., Sawant, D.: Invoice processing using RPA & AI. In: ICSMDI Proceedings, Tiruchirappalli (2021)
55. Eulerich, M., Waddoups, N., Wagener, M., Wood, D.A.: The Dark Side of Robotic Process Automation. SSRN (2022). https://doi.org/10.2139/ssrn.4026996
56. Magaletti, N., Cosoli, G., Leogrande, A., Massaro, A.: Process Engineering and AI Sales Prediction: The Case Study of an Italian Small Textile Company. SSRN (2022). https://doi.org/10.2139/ssrn.4026183

57. Gerbert, P., et al.: Powering the Service Economy with RPA and AI (2017). https://www.bcg.com/de-de/publications/2017/technology-digital-operations-powering-the-service-economy-with-rpa-ai. Accessed 15 June 2022

58. Patri, P.: Robotic process automation: challenges and solutions for the banking sector. Int. J. Manag. **11**, 322–333 (2021)

59. Denagama Vitharanage, I.M., Bandara, W., Syed, R., Toman, D.: An empirically supported conceptualisation of robotic process automation (RPA) benefits. In: ECIS Proceedings. Virtual, 58 (2020)

60. Bygstad, B.: The coming of lightweight IT. In: ECIS Proceedings. Münster, 22 (2015)

61. Plattfaut, R.: Robotic process automation–process optimization on steroids? In: ICIS Proceedings, Munich, 3 (2019)

62. Willcocks, L.: Robo-Apocalypse cancelled? Reframing the automation and future of work debate. J. IT **35**, 286–302 (2020). https://doi.org/10.1177/0268396220925830

63. Kokina, J., Gilleran, R., Blanchette, S., Stoddard, D.: Accountant as digital innovator: roles and competencies in the age of automation. Account. Horiz. **35**, 153–184 (2021). https://doi.org/10.2308/HORIZONS-19-145

64. Wallace, E., Waizenegger, L., Doolin, B.: Opening the black box: exploring the socio-technical dynamics and key principles of RPA implementation projects. In: ACIS Proceedings, 86 (2021)

65. Schuler, J., Gehring, F.: Implementing Robust and Low-Maintenance Robotic Process Automation (RPA) Solutions in Large Organisations. SSRN (2018). https://doi.org/10.2139/ssrn.3298036

66. Marr, B.: The 4 Ds of Robotization: Dull, Dirty, Dangerous and Dear (2017). https://www.forbes.com/sites/bernardmarr/2017/10/16/the-4-ds-of-robotization-dull-dirty-dangerous-and-dear/?sh=22d89f4c3e0d. Accessed 15 June 2022

67. Takayama, L., Ju, W., Nass, C.: Beyond dirty, dangerous and dull: what everyday people think robots should do. In: ACM/IEEE HRI Proceedings, Amsterdam, pp. 25–32 (2008)

68. Hangar Technology: Robotics (Drones) Do Dull, Dirty, Dangerous & Now Difficult (2018). https://medium.com/hangartech/robotics-drones-do-dull-dirty-dangerous-now-difficult-a860c9c182a4. Accessed 15 June 2022

69. Tingling, P., Parent, M.: Mimetic isomorphism and technology evaluation: does imitation transcend judgment? J. AIS **3**, 113–143 (2002). https://doi.org/10.17705/1jais.00025

A Trustworthy decentralized Change Propagation Mechanism for Declarative Choreographies

Amina Brahem[1,3]([✉]), Tiphaine Henry[2,4], Sami Bhiri[3], Thomas Devogele[1],
Nassim Laga[2], Nizar Messai[1], Yacine Sam[1], Walid Gaaloul[4],
and Boualem Benatallah[5]

[1] LIFAT, University of Tours, Tours, France
`amina.brahem@univ-tours.fr`
[2] Orange Labs, Paris, France
`tiphaine.henry@orange.com`
[3] OASIS, University Tunis El Manar, Tunis, Tunisia
[4] Telecom SudParis, UMR 5157 Samovar, Institut Polytechnique de Paris, Paris, France
[5] Dublin City University, Dublin, Ireland

Abstract. Blockchain technologies have emerged to serve as a trust basis for the monitoring and execution of business processes, particularly business process choreographies. However, dealing with changes in smart contract-enabled business processes remains an open issue. For any required modification to an existing smart contract (SC), a new version of the SC with a new address is deployed on the blockchain and stored in a contract registry. Moreover, in a choreography, a change in a partner process might affect the processes of other partners. Thus, the change effect must be propagated to partners of the choreography affected by the change. In this paper, we propose a new approach overcoming the limitations of SCs and allowing for the change management of blockchain-enabled declarative business process choreographies modeled as DCR graphs. Our approach allows a partner in a running blockchain-based DCR choreography instance to change its private DCR process. A change impacting other partners is propagated to their affected processes using a SC. The change propagation mechanism ensures the compatibility checks between public DCR processes of the partners. We demonstrate the approach's feasibility through an implemented prototype.

Keywords: Process choreography · Change propagation · DCR graph · SC

1 Introduction

Blockchain technologies have emerged to serve as a trust basis for the monitoring and execution of business processes [18], and particularly business process choreographies [2]. This is due to several mechanisms, be it the consensus method

© Springer Nature Switzerland AG 2022
C. Di Ciccio et al. (Eds.): BPM 2022, LNCS 13420, pp. 418–435, 2022.
https://doi.org/10.1007/978-3-031-16103-2_27

applied among the nodes to validate a transaction, the immutable nature of transactions, process automation using SCs and its ability to manage decentralized, peer-to-peer interactions [10].

Both imperative and declarative process modelling paradigms have been used to deal with blockchain-based business process execution. Proposed techniques include translation of BPMN collaboration models into SCs [2] and execution engines of declarative orchestration processes called Dynamic-Condition-Response-(DCR) graphs [3]. However, dealing with changes in blockchain-enabled business processes remains an open research issue [17].

Business processes managed by "static" SCs cannot be upgraded because the SCs are immutable once deployed. Efforts exist to support versioning in SCs [22]. For any required modification on the existing SC, a new version of the SC with new address is deployed on the blockchain and stored in a contract registry. So the process may have new version and in future interactions should be consistent with it. However, with SC having many versions, it is difficult to maintain inter-dependent SCs links and to copy data from old to new version of the contract [22]. Moreover, these are all costly operations. We aim to enable a way to integrate change management into SCs implementing the business logic of process choreographies without deploying them again. This circumvents the aforementioned problems related to versioning.

In a choreography, each partner manages its private process and interacts with other partners via its public process. The model comprising all interactions is called choreography process [7,9]. In a running choreography instance, a change may consist of a simple change operation (ADD/REMOVE/UPDATE) or combination of change operations [4,7]. A change in the instance of a partner process may affect other partners' process instances. Hence, change must be propagated to the affected partners of the choreography instance [7,20]. In a trip e-booking process, for example, a hotel may close its catering facility for reparations and thus DELETE the dining service. A tourist having booked the hotel with dinner included will be unable to reach the hotel service "ProvideDinner". Additionally, a new restaurant may want to establish (ADD) a convention with the hotel. This new relationship will affect the tourist interested in trying the restaurant. Thus, the ADD change must be propagated to the tourist process instance.

One has also to ensure that neither the structural nor behavioral compatibility of partners processes are violated after a change [1,7,11,12]. Structural compatibility checks consist of ensuring that there is at least one potential *send* message assigned to a partner with a corresponding *receive* message assigned to another partner [12]. Behavioral compatibility refers to ensuring that the choreography process after the change is safe and terminates in acceptable state. In other words, no deadlocks should occur between partners public processes during the choreography execution after change [7,11]. For example, in the trip e-booking process, the task "HaveDinner" is a public task. It is composed of two messages, namely the send and receive messages, that are respectively assigned to Tourist and NewRestaurant. When NewRestaurant DELETES the receive message "HaveDinner", a structural incompatibility occurs as the corresponding send message is still present in the Tourist process.

To the best of our knowledge, the integration of change management, and especially change propagation, in blockchain-based declarative choreographies management systems has not been studied. In this paper we focus on the following research question: *(RQ) How to guarantee correct change propagation in declarative blockchain-based DCR choreography process instances?*

We propose a change mechanism to bring adaptiveness to the trustworthy execution of declarative choreographies. We adopt the three levels of granularity: (i) choreography, (ii) public and (iii) private processes used in imperative languages such as BPMN and apply it to the declarative language called DCR graphs [7]. With DCR, processes are modelled as a set of events linked together with relations (a kind of temporal dependencies) [5,6]. In [14], authors propose an approach for a trustworthy deployment and execution of DCR choreographies. Choreography participants build incrementally the choreography process managed by a SC. Meanwhile, participants execute their private events off-chain in their local process execution engine. We build on and extend this work with the change management mechanism, focusing on the introduction, negotiation and propagation phases at the process instance level. Similarly to declarative languages, a DCR graph is specified as a set of rules. These rules are interpreted at runtime. As they represent business requirements, it is easier to add or update constraints if a requirement changes [5].

Our approach allows a partner in a running DCR choreography instance to change its private process. Changes affecting private activities are applied off-chain while changes impacting interactions with other partners are managed on-chain through the SC [14]. Changes are mainly ADD/REMOVE and UPDATE operations applied to the DCR choreography events and relations [4,7]. We only focus on these change operations as that they are challenging by themselves and that any change to a process can be written as a combination of these operations [13]. The on/off-chain separation ensures (i) the privacy of the partners as private information in private processes is not shared and kept off-chain and (ii) trust as the blockchain provides an immutable history of execution logs attesting the enforcement of correct execution of the choreography interactions [14]. Besides, SC transactions act like "approval check points" during change negotiation and propagation. Hence, claim resolution is eased between partners in case of a misbehavior as the blockchain stores the negotiation and propagation history on-chain. For example, when a partner wrongfully projects the change and creates a behavioral incompatibility after the change, execution logs can be used to check who is the source of and what is the erroneous behavior. To summarize, we complement the work in [14] with the following contributions:

- We augment the SC managing the choreography process with change management techniques to enable change operations on a deployed instance of a choreography.
- We propose a protocol that allows (i) partners to first negotiate the change on-chain, (ii) then to dynamically update the choreography process instance managed by the SC with the new process change information, and (iii) finally propagate this information across partners processes affected by the change.

– We leverage the platform in [14] to integrate change to running DCR choreography instances.

The remainder of this paper is organized as follows. Section 2 presents fundamental definitions used in the approach and introduces an illustrating example. Section 3 reviews the main known related work. Section 4 details our approach. Section 5 presents an implemented prototype and some evaluation tests. Finally, Sect. 6 concludes the paper and gives insights into future work.

2   Basic Concepts and Illustrating Example

DCR graphs are one of many declarative business process modeling languages whose formalism is presented in [6]. A DCR graph G is represented by a triplet (E, M, Rel). E is a set of labelled *events*. M denotes the *marking* of the graph and is represented by the triplet (currently included events In, currently pending responses Pe, previously executed events Ex). Finally, Rel is the set of *relations* of the graph. Relations are of five types: condition$\longrightarrow \bullet$, response$\bullet \longrightarrow$, milestone$\longrightarrow \diamond$, include$\longrightarrow +$ and exclude$\longrightarrow \%$.

A **DCR choreography** models interactions between partners. Its execution is done in a distributed way. A DCR choreography is defined as follows [14]:

Definition 1. A DCR choreography C is a triple (G, I, R) where G is a DCR graph, I is a set of *interactions* and R is a set of *roles*. An interaction i is a triple (e, r, r') in which the event e is initiated by the role r and received by the roles $r' \subset R \backslash \{r\}$.

Which leads us to the definitions of one partner's **public** and **private** DCR processes. A **public DCR process** of one partner represents the **projection** of the DCR choreography over this partner (see definition 4 in [14] for more details). The **private DCR process** of one partner is a kind of a refinement of the public DCR process, i.e., it comprises the public interactions in addition to the internal events related to this partner.

Figure 1 presents the trip e-booking scenario that was initially presented in the introduction (c.f. Sect. 1) translated into DCR: Fig. 1(a) presents the DCR choreography and Fig. 1(b) presents the tourist private DCR process. The choreography process is managed in the different partners processes, namely Tourist, TouristOfficer, Hotel, and CastleAdmin, to ensure a separation of concerns. *Pay-Pass (p3)* is an internal event of the role Tourist, managed off-chain to preserve the privacy of Tourist. *PurchasePass (e1)* is a choreography interaction sent by Tourist and received by TouristOfficer. It is managed on-chain (c.f. [14]). To execute the send event, Tourist triggers the SC from its private DCR process. Table 1 shows the choreography markings of Fig. 1 during a run. Each column stands for the events of the choreography. Rows indicate markings' changes as events on the left are triggered. For example, initially no event is executed nor pending and the event $e1$ is included. Thus, its marking is (1, 0, 0). Once Tourist executes $e1$, the marking becomes (1, 0, 1). Partners have control over the set of

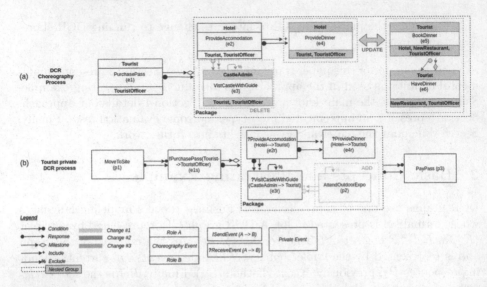

Fig. 1. DCR choreography process and tourist private DCR process of the trip e-booking process (Color figure online)

Table 1. Evolution of the markings (included, pending, executed) of the DCR choreography process in Fig. 1 (before changes)

	Markings			
	e1	e2	e3	e4
(init)	(1, 0, 0)	(0, 0, 0)	(0, 0, 0)	(0, 0, 0)
e1	(1, 0, 1)	(1, 1, 0)	(0, 0, 0)	(0, 0, 0)
e2	(1, 0, 1)	(1, 0, 1)	(1, 1, 0)	(1, 1, 0)

internal and choreography interactions they are involved in. This set of events, mentionned hereinafter as a partner private DCR process, is illustrated by the tourist's one in Fig. 1(b).

Both choreography and private DCR processes in Fig. 1 are susceptible to changes. A change is composed of a set of change elements and a combination of change operations. We define in the following these two concepts.

Definition 2. Change element. Let $C = (G, I, R)$ be a DCR choreography. Let ϵ be the set of *internal events* in G, i.e., events having one initiator $r \in R$. I is the set of *choreography interactions*. G_{Ref} is a change element (also called refinement element) iff one of the following conditions are met:

1. $G_{Ref} \in \{\epsilon \cup I \cup \longrightarrow \bullet \cup \bullet \longrightarrow \cup \longrightarrow \diamond \cup \longrightarrow + \cup \longrightarrow \%\}$
2. $G_{Ref} = (e \in \{\epsilon \cup I\}, m_e(in, pe, ex), \{\longrightarrow \bullet \cup \bullet \longrightarrow \cup \longrightarrow \diamond \cup \longrightarrow + \cup \longrightarrow \%\})$

This means that, G_{Ref} is a refinement element if it is either (1) an atomic element, i.e., (i) an internal event in the set of internal events ϵ such as $p2$ in

Fig. 1(b), or (ii) an interaction such as $e3$, or (iii) one of the five relations such as the condition relation linking $e3$ and $e2$. (2) a DCR fragment, i.e., a subgraph with a minimal configuration: {one event, initial marking of the event, one relation}. For example in Fig. 1, change #3 consists into adding the DCR fragment $(p2: (1, 0, 0), p2 \longrightarrow \bullet e3, p2 \longrightarrow +e3, p2 \longrightarrow \%p2)$.

Definition 3. Change operation. To define the change operations, we refer to [4] where authors propose three change operations on DCR orchestration processes. We re-adapt these operations to be used in the context of DCR choreographies and where the change element can be one of the three types defined in Definition 2. Change operations are of three major types[1]:

- $C \oplus G_{Ref}$ to *ADD* the refinement element G_{Ref} to the original DCR choreography C. To apply the change, one has to compose the refinement element with the original graph, i.e., one has to take the union of events, labels, relations and markings of the two parts of the composition.
- $C \ominus G_{Ref}$ to *REMOVE* a change element G_{Ref} from C. For example, to remove an interaction, one has to remove it from the set of interactions I, its marking from the marking (In, Pe, Ex) of the graph as well as the incoming and outgoing constraints coming to/going from this interaction.
- $C[G_{Ref} \mapsto G'_{Ref}]$ to *UPDATE* a change element. For the case of an event (internal or interaction): the *UPDATE* operation is used for replacing one event by another or re-labelling it. To replace, for example, one interaction with another, one has to update the set of interactions I with the new interaction, the marking of the graph and the set of incoming and outgoing constraints.

An example of an UPDATE operation is change #1 in red in Fig. 1. Here, the pass purchased by the Tourist in the event $e1$ undergoes a change: it will let the Tourist to have a dinner in NewRestaurant instead of having it in the Hotel. Hence, the TouristOfficer, who manages the pass and can add new participants, establishes a new convention with NewRestaurant. Consequently, the change operations to make are: (i) add the partner NewRestaurant, (ii) an UPDATE operation where the interaction $e4$ is replaced with the DCR fragment{$e5$, $e6$}. These changes are called *public changes*.

To proceed with such a change: (i) the change should be negotiated (agreed on or not) by the involved partners, (ii) the change proposition should be examined by all involved partners, (iii) the negotiation outcome should be tamper-proof to avoid that someone diverges from the common understanding, and (iv) the change should be correctly propagated [5,6].

In the following section, we present the related work regarding change mechanisms in cross-organizational business processes.

3   Related Work

Change management at runtime in procedural processes has been studied in [7] where change propagation algorithms ensure behavioral and structural soundness

---
[1] We use the same notation of the operations defined in [4].

of choreography partners private processes after the change. In [8, 15], authors consider the change negotiation phase but no mechanism is proposed to ensure that all partners have trustfully applied the change, and no blockchain is used to deal with this problem.

Change management has also been studied in DCR processes, mainly through runtime changes. The first efforts appear with the notion of DCR fragments where simple change/add/remove operations are implemented [4]. Authors follow the *build-and-verify* approach to apply incremental changes to the fragments. This approach consists of the continuous iterations of (i) modeling, (ii) deadlocks and livelocks freedom verification, and (iii) executing until a further adaptation is required. Nonetheless, partner trust into change propagation of DCR choreographies is not addressed in this work. In [19], authors use a *correct-by-construction* approach on running instances of DCR graphs. The structure underlying a DCR is a labelled transition system. Starting from a user-defined change, authors define a reconfiguration workflow. During the transition period, old requirements are disabled and verified subpaths of activity executions are enabled. This setting holds until new requirements are verified. However, not every reconfiguration problem has a solution and for every change, one has to build a new reconfiguration workflow. It requires heavy calculations to discover the verified subpaths, which is not easy for large models. Finally, in [5], authors use a set of rules ensuring the correctness of new instances of DCR graphs by design. New change operations must respect these rules to prevent a misbehavior.

Regarding change management in blockchain-enabled processes, in [17], authors propose an approach that allows collaborative decisions about (1) late binding and un-binding of actors to roles in blockchain-based collaborative processes, (2) late binding of subprocesses, and (3) choosing a path after a complex gateway. A policy language enables the description of policy enforcement rules such as who can be a change initiator and who can endorse a change. However, authors do not consider ADD/REMOVE/UPDATE change operations like we do. Additionally, the private processes of roles are not considered and neither is the propagation of the effect of the new decisions over partners.

To summarize, most related work consider change in process orchestrations only [4, 5]. Additionally, approaches binding actors to roles in a process collaboration [17] currently push the burden of checking the transitive effect of new changes onto the new parties. This checking, likely done in a manual way, which can lead to errors. Finally, even when the change propagation soundness is dealt with, the proposed approach does not provide a mechanism that ensures choreography partners project the change and propagate it trustfully.

4   Proposed Approach

DCR business processes monitored in the blockchain are represented into SCs as follows (c.f. [14]): the SC holds a set of activities, each assigned to an actor, and linked to an execution state. A relation matrix which summarises the execution constraints is used to update activity states based on smart-contract based execution requests.

Table 2. Proposed allowed and denied changes for a DCR process

Type	Rule
AR1	Change condition/response/milestone relations
DR1	Inclusion of an excluded event
DR2	Exclusion of an included event
AR2	Block temporarily/permanently an included event

Partners coordinate their own processes connected to the blockchain, and propose/receive changes to/from other partners (step 1 in Sect. 4.1). Our goal is to make it possible for each partner to (i) modify its private DCR process, and (ii) suggest a change to the DCR choreography monitored in the blockchain. If the change request is fully private, for e.g., it concerns an internal event or a relation linking two internal events, (private-to-private relation) or a relation linking an interaction to an internal event (public-to-private relation), then the private process of the partner updates accordingly. If the change is public, it is managed onchain (step 2 in Sect. 4.2). Public changes concern an interaction or a relation linking two interactions (public-to-public relation) or a relation linking an internal event to an interaction (private-to-public relation). Then, a negotiation stage starts (step 3 in Sect. 4.3), followed by a propagation stage (step 4 in Sect. 4.4.)

4.1 Step 1: Change Proposal

The role initiator defines the change of its private DCR process off-chain. She may modify its internal events and interactions, as well as relations linking events. The introduction of a change is called refinement (cf. Definition 2). It is done before submitting it to other partners for examination.

A set of integrity rules need to be defined to ensure the correctness of the updated graph. A DCR graph is correct iff it is safe, i.e., free of deadlocks and live, i.e., free of livelocks. A DCR graph is deadlock free if for any reachable marking, there is either an enabled event or no included required responses. Whereas liveness describes the ability of the DCR graph to completion by continued execution of pending response events or their exclusion. To do so, we leverage non-invasive adaptation rules, originally introduced in the context of DCR orchestrations, to DCR choreographies [5]. We divide these rules in rules describing (i) allowed change rule (AR) and (ii) denied change rule (DR) presented in Table 2[2].

One can ADD/REMOVE/UPDATE condition, response and milestone relations (AR1). The only restriction is not to have cycles of condition/response relations to avoid deadlocks. However, one cannot include an already excluded

---

[2] The reader can check [5] for more details about denied and allowed change operations in DCR orchestrations that inspired the proposed changes.

event (DR1) neither can she exclude an already included event (DR2). One alternative to this is to block temporarily or permanently an event (AR2). We suppose that we want to block permanently a DCR graph G of executing an event e. We refine with the fragment Q: $Q = \{e: (0, 1, 0), g: (0, 1, 0)), g \longrightarrow \bullet$ $g, g \longrightarrow \bullet e\}$. Here, e can never fire (again) because it depends on g. Moreover, by excluding and including g, one can selectively enable and disable e.

In our example, the change proposal #1 consists into replacing $e4$ by the fragment composed of the events $\{e5, Pe6\}$. Here, one did not add an exclude relation to the already included event $e4$, i.e., (DR2) evaluates to *false*. Consequently, one is not concerned by blocking temporarily/permanently an event, i.e., (AR2) is also verified. Only milestone and response relation are added to the graph and thus (AR1) evaluates to *true*. Moreover, change # 1 does not contain an include relation to an already excluded event and so (DR1) is also respected. Hence, the change proposal evaluates to *true* because $\forall i$, (ARi) evaluates to *true* and $\forall j$, (DRj) evaluates to *false*.

4.2   Step 2: Change Request for Public-Related Changes

The SC stores the list of change requests assigned to process instances as a hashmap. Ongoing process instance changes are recorded with the identification hash of the current process instance h_{curr}. The identification hash corresponds to the IPFS hash of the process instance description[3]. This hash is generated by the change initiator upon a change request, before the SC call. During the change request lifecycle, the request is assigned to a status belonging to $\{Init,$ *BeingProcessed*, *Approved*, *Declined*$\}$. Status is set to *Init* if no change request is ongoing, to *BeingProcessed* during the negotiation stage, to *Approved* or *Declined* once the change request is processed by all endorsers.

Algorithm 1 presents the SC function registering a change request. The identity of the change initiator is checked: it should belong to the list of partner addresses (line 2). Then, the change request is created for the current process instance (line 3–8). The hash of the redesigned workflow is stored in h_{req} (line 4). This identification hash corresponds to the IPFS description of the requested redesigned public workflow h_{req}. The status of the change request is set to *BeingProcessed* (line 5). The addresses of the change initiator and endorsers E are attached to the request (line 6–7). Endorsing partners are, for example, in the case of adding a choreography interaction i (i) the sender and the receiver(s) of the event and, (ii) partners connected directly with the choreography interaction. The change initiator also sets two response deadlines t1 for change endorsement and t2 for change propagation to be checked by the SC (line 8–9). Finally, the SC emits a change request notification to all partners listening to the SC (line 10). If one of the change endorsers does not reply before deadline t1 during endorsement or t2 during propagation, an alarm clock triggers a SC function cancelling the change request. If one of the change endorsers does not reply

---

[3] InterPlanetary File System (IPFS) is a peer-to-peer protocol that uses content addressing for storing and sharing files on the blockchain (https://docs.ipfs.io/).

Algorithm 1: Request change smart contract function

Data: *changeRequests* the list of change requests, E the list of endorser addresses, h_{curr} the current ipfs workflow hash, h_{req} the ipfs hash of requested change description, $t1$ the deadline timestamp for change endorsement, and $t2$ the deadline timestamp for change propagation

Result: emits change request notifications to endorsers

1 **Function** requestChange(h_{curr}, h_{req}, E, $t1$, $t2$):
2   require $msg.sender$ belongs to the list of business partners;
3   **if** *changeRequests[h_{curr}].status == Init* **then**
4   set *changeRequests*[h_{curr}].h_{req} ← h_{req};
5   set *changeRequests*[h_{curr}].status ← "*BeingProcessed*";
6   set *changeRequests*[h_{curr}].initiator ← $msg.sender$;
7   set *changeRequests*[h_{curr}].endorsers ← E;
8   set *changeRequests*[h_{curr}].t1 ← $t1$;
9   set *changeRequests*[h_{curr}].t2 ← $t2$;
10   emit RequestChange(h_{curr}, h_{req}, E, $msg.sender$);
11   **else**
12   emit Error; // an ongoing change request is being processed
13 **End Function**

---

before deadline t1 during endorsement or t2 during propagation, an alarm clock triggers a SC function cancelling the change request at a specified block in the future corresponding to t1 or t2. It consists into a SC function being called by incentivized users triggering the SC at the desired timestamp [16]. Upon trigger, the SC function sets the change request status to cancelled and emits an event notifying partners that the change has been cancelled. By so doing, we prevent any deadlock that could occur due to one of the partners not responding.

In Fig. 1, Change #1 is public as it concerns three partners, namely Tourist, Hotel, and NewRestaurant. Hence a negotiation must occur between the partners to reach a consensus on the proposed change before propagating it. NewRestaurant launches the change negotiation by triggering the SC. The SC updates the change requests list linked to h_{curr} with the following information: [(1) h_{req} the IPFS hash of the updated process description which comprises the operation UPDATE(e4) with (e5 + e6), (2) the list of endorsers: $\{address_{Hotel}, address_{Tourist}\}$, (3) Change negotiation deadline t1 = 72 h, (4) Change propagation deadline t2 = 120 h]

4.3 Step 3: Change Negotiation for Public-Related Changes

All partners subscribe to the change request events emitted by the SC. Endorsing partners must send their decision request to the SC based on the rules in Table 2. If the change once computed on the endorser's process respects all ARi and DRj rules, then the endorser approves the request. It is otherwise rejected. The rules checks are manual and can be automated in the future work. The SC collects the

---

Algorithm 2: Endorser decision management smart contract function

Data: *changeRequests* the list of change requests, e_s the endorser address, E the list of registered endorsers, h_{curr} the hash of the current workflow, h_{req} the hash of the desired workflow, *rsp* the endorser response $\in \{0,1\}$

1 **Function** endorserRSP(h_{curr}, e_s, rsp):
2   require($block.timestamp <= changeRequests[h_{curr}].t1$);
3   require($e_s \in E$);
4   require($changeRequests[h_{curr}].changeEndorsement[e_s] \; != \; 1$);
5   require($changeRequests[h_{curr}].status == ("BeingProcessed")$);
6   **if** $rsp == 1$ **then**
7   set $changeRequests[h_{curr}].changeEndorsement[e_s] \longleftarrow 1$;
8   emit AcceptChange(h_{req}, e_s);
9   lockInstanceChecker(h_{curr})
10   **else if** $rsp == 0$ **then**
        // declineapprovalOutcomes
11   set $changeRequests[h_{curr}]$.status $\leftarrow$ "*Declined*";
12   emit DeclineChange(h_{req}, e_s);
13   **else**
14   emit Error(h_{req}, e_s);
15 **End Function**

---

different decisions from the endorsers to lock (or not) the choreography instance and proceed (or not) with the change. We detail both stages hereinafter.

Algorithm 2 presents the SC function receiving one endorser's decision. The alarm clock should not have been raised (line 2), the endorser address e_s should belong to the list of registered addresses (line 3), and not having answered to the change request already (line 4). The change request should also be processable, i.e., its status should be set to *BeingProcessed* (line 5). If all conditions are met, the endorser response *rsp* is processed. If *rsp* equals 1 (line 6), the endorser has accepted the change. Its response is saved into the change endorsement list (line 7), the notification of acceptance is sent to all endorsers as well as the change initiator (line 8), and the SC checks whether the instance needs to be locked (line 9). The *lockInstanceChecker* function assesses whether all endorsers have accepted the change: the *changeEndorsement* list should be filled with ones. At this stage, no further execution of included events is allowed and the mechanism waits for pending events to terminate. The change status is then updated to *Approved*.

In our example, we suppose that both endorsers confirmed the change request ($rsp_{Hotel} = 1$ and $rsp_{Tourist} = 1$) while respecting t1. The SC locks the instance for change propagation. As it manages the negotiation process, a tamper-proof record of the negotiation is accessible by all partners. This prevents conflicts and eases potential claim resolutions.

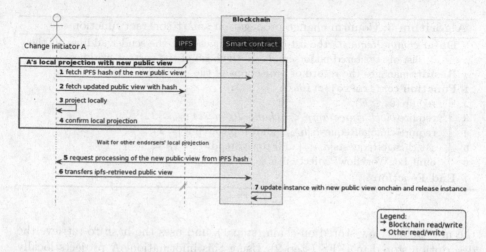

Fig. 2. Sequence diagram of the propagation stage illustrating the interactions between partners and the SC with A being the change initiator

4.4 Step 4: Change Propagation

Change propagation is to apply the change effect after the negotiation phase succeeds to (i) the affected partners DCR public processes, (ii) each partner propagates the change effect to its private DCR process. To ensure the correctness of the change propagation, we introduce the following property where c is a change, r, E are respectively c initiator and endorsers, G_r, $G_{r'}$ are respectively the public DCR process of r and r' where $r' \in E$:

Property 1. *If* $effect(c)_{\|G_r}: \Rightarrow G_r$ *is correct-by-construction and* $\forall r \in E$, $effect(c)_{\|G_{r'}}: \Rightarrow G_{r'}$ *is correct-by-construction then* G_r *and* $G_{r'}$ *are compatible* $\forall r \in E$.

Property 1 states that if G_r is correct-by-construction and if $\forall r' \in E$, $G_{r'}$ are also correct-by-construction, then compatibility is verified. In fact, a public DCR process G_r is correct-by-construction means that computing the effect of a change c over r introduces no deadlocks in G_r (see Sect. 4.1). Thus, if the DCR public models of the change initiator and endorsers are safe, i.e., no deadlocks can occur, then they are able to communicate in a proper way after the change and are consequently compatible with each other. The SC enforces propagation correctness as it maintains the tamper-proof record for the endorsement and application of the change effect across partners. Another correctness criterion is checking the consistency between one partner's private and public DCR processes. This is out of the scope of the present work and will be done in future work.

Indeed, Fig. 2 depicts the sequence diagram of the change propagation interactions taking place between partners and the SC. Each partner projects locally the DCR choreography in its projection using the process description given in the IPFS hash (Fig. 2 step 1–4). Participant A first fetches the IPFS hash of

---

Algorithm 3: Confirm change propagation smart contract function

---

Data: *changeRequests* the list of change requests, e_s the sender address, E the list of registered endorsers, h_{curr} the hash of the current workflow

Result: manages the record of projections of the new public view

1 **Function** confirmProjection(h_{curr}, e_s):
2   require($e_s \in E$);
3   require($block.timestamp <= changeRequests[h_{curr}].t2$);
4   require(changeRequests[h_{curr}].didPropagate[id] != 1);
5   set changeRequests[h_{curr}].didPropagate[id] $\longleftarrow$ 1;
6   emit LogWorkflowProjection(h_{curr});
7 **End Function**

---

the new public view stored on-chain (step 1), and uses the hash to retrieve the description stored in IPFS (step 2). Using this information, A projects locally the new version of the process, merging its private process with the updated public activities. Once completed, A notifies the SC to confirm the projection (step 4). Algorithm 3 presents the function triggered by partners to confirm the projection to the SC. A list *didPropagate* keeps track of the propagation status, i.e., it records the private projection of each partner. The function checks that the partner belongs to the list of endorsers (line 2), that the alarm clock has not been raised (line 3), and that the endorser has not projected locally yet (line 4). Each participant must proceed before the change propagation deadline $t2$. Else, the propagation is cancelled, and the instance returns to its initial state before the change request. Other endorsers follow the same steps.

The SC detects all local projections once *didPropagate* is filled with ones and notifies the change initiator. The change initiator then retrieves the new DCR choreography that was saved into IPFS using h_{req} (Fig. 2 step 5) and forwards it to the SC (Fig. 2 step 6). The SC updates the relations and markings stored into the process instance and resets the change status of the workflow instance: a new change request can be processed (Fig. 2 step 7). In total, all participants must complete two transactions with the SC and one transaction with IPFS. The change initiator must also complete two additional transactions with the SC to update the view stored on-chain and unlock the instance.

In our motivating example, the propagation of change #1 occurs with all endorsers Tourist, Hotel and Restaurant updating their private DCR process with the approved change. To do so, they retrieve the change description stored in IPFS under h_{req}. They project the updated public change description on their role following the same approach as in [14]. For example, Tourist will retrieve the events $\{e5, e6\}$. Tourist then combines this projection with its private events $\{p1, p2, p3\}$. Once all projections have been done and notified to the SC, the change initiator NewRestaurant finally triggers the SC to update the DCR choreography of the running instance with the updated process description e.g., the updated relation matrices, event markings and access controls (c.f. [14] for a more detailed description).

5 Implementation and Evaluation

5.1 Implementation

In [14], authors presented a solution aiming at executing DCR choreographies in a hybrid on/off-chain fashion. The DCR choreography description comprises events descriptions (i.e., labels, roles, markings), relation matrices, and actors linked to events. Here, we leverage this platform to integrate change at the process instance level[4].

We use a Ganache testnet to deploy a public SC S which manages each process. S comprises (1) execution constraint rules, (2) a list of workflows initially empty, and (3) the list of change requests linked to the list of workflows. At the time of writing, 1ETH = 2581,86\$. The initial cost of deployment of S is 0.10667554 ETH (439.21\$) for a gas usage of 3,555,855. Additionally, a SC manages roles authentication and access control rules. Its deployment takes 1,953,149 gas (0.05859442 ETH or 241.25\$). The SC is deployed in Ropsten at: 0x523939C53843AD3A0284a20569D0CDf600bF811b[5]. For each workflow, RoleAdmin (1) generates the DCR choreography bitvector representation, (2) saves the textual DCR choreography input to IPFS, and (3) registers the new workflow on-chain (cf. [14]). The workflow is identified by the IPFS unique hash.

Each partner can edit the running instance. Editing is done using the panel manager, a tool to update DCR graph descriptions. Users can add private and choreography interactions, as well as condition, response, include, exclude, and milestone relations. They can also use the panel manager to remove and update events and relations. The panel manager implements integrity rules presented in Subsect. 4.1: the panel verifies the soundness of a desired change operation. Hence, we obtain a redesigned DCR graph that is correct-by-construction. After edition, the panel manager triggers the SC if it detects a public change. The SC registers the request and forwards it to the identified partners. Each partner accesses the change request and answers back to the SC. If the change request is accepted by all, change propagation starts.

5.2 SC Evaluation Costs

The initial cost for deploying the motivating example instance is 0,00933308 ETH (24,097\$) for a gas usage of 311,103. Indeed, the consensus algorithm used in the Ethereum blockchain is a proof of work [10], hence each SC transaction excepting read transactions are payable to compensate miners from computation costs. We evaluate the transaction costs to assess the computation costs related to the change negotiation and propagation functionalities.

In our motivating example, three changes occur. Change#1, initiated by NewRestaurant, is fully public: e5 and e6 replace e4, and two public-to-public relations (response and milestone) are added. In the following, we investigate the public negotiation and propagation SC costs for this change.

---

[4] Code of the implemented prototype augmented with change management is accessible at https://github.com/tiphainehenry/adaptiveChangeDCR/.

[5] This address can be used with Etherscan to access the record of transactions.

Table 3. SC change propagation gas costs and gas fees

Stage	Step	Partner	Gas	Cost (ETH)	Cost ($)
Nego.	LaunchNego	NewRestaurant	213194	0,00639582	16,513
	Case decline	TouristOfficer	46773	0,00140318	3,623
	Case accept	TouristOfficer	78999	0,00157998	4,079
		Tourist	86428	0,00181256	4,68
Propag.	Upd. projection	TouristOfficer	96448	0,00201296	5,197
	Upd. projection	Tourist	96375	0,0020115	5,193
	Upd. projection	NewRestaurant	87648	0,00175296	4,526
	Upd. SC instance	NewRestaurant	1321496	0,02642992	68,238

Table 3 presents the gas usage induced by the execution of the SC during the negotiation and propagation stages. It is used to compensate miners for their computation power in the blockchain cryptocurrency. The table also presents the transaction costs in ETH and USD.

Regarding the negotiation stage, Tourist first launches the change request for the replacement of one public task by a new fragment of two public tasks. The transaction fees for the request are 0,00639582 ETH, and are the highest fees of the negotiation stage. Indeed, the fee to be paid to decline or accept a role is worth around 0.0015 ETH. Nonetheless, all fees are of the same order of magnitude (0.001 ETH). *Regarding the propagation stage*, the transaction fees of the SC correspond to two stages. First, the change endorsers apply the change effect to their private processes. No transaction fee is requested to fetch the IPFS hash of the new DCR choreography. However, a transaction fee is necessary to update the SC list *didPropagate* recording the projections. The SC notification of the local update is worth 0,00201296 ETH and 0,0020115 ETH for both endorsers (around 5$ per local projection). The change initiator finally updates its projection. The cost to switch the workflow locally is 0,00175296 ETH. NewRestaurant sends a transaction to update the DCR choreography on-chain using the same tool used to deploy a new instance on-chain. The cost for switching the DCR choreography on-chain is 0,02642992 ETH. It is one order of magnitude higher compared to other transaction fees, but close to the cost of instantiating a new instance on-chain, due to the update of relation matrices and markings. Hence, propagation transaction fees are higher than the negotiation ones. Additionally, the cost of the propagation mainly comprises the cost of the DCR choreography update. *Execution times*, represent the results obtained after the enactment of one trace. The reported execution time factors the transaction confirmation time obtained on the test network. In average, the execution time of on-chain interactions is 14.8 s. Additionally, the average time for IPFS transactions is 7.6 ms. The change initiator NewRestaurant needs to process four on-chain transactions and two off-chain transactions with IPFS. Both endorsers TouristOfficer and Tourist must process three on-chain transactions and two IPFS transactions. Hence, in total, the whole cycle of change management takes 152.6 s if all participants launch their transactions on trigger (4 + 3 + 3 on-chain transactions requiring 14.8 s in average, and 2 + 2 + 2 IPFS transactions requiring 7.6 ms in average).

6 Discussion and Conclusion

In this paper, we propose a change propagation mechanism to bring adaptiveness to trustworthy execution of declarative choreography instances. Our approach comprises three main steps. First, the change introduction, where a partner in a running DCR choreography instance wants to change its private process. Here, we declare rules that specify the allowed and prohibited changes. These rules provide a correct-by-construction DCR choreography after the change and ensure that no deadlocks nor livelocks occur. All changes are applied at the process instance level. Local changes are managed off-chain. Changes impacting an interaction start on on-chain negotiation phase. If the negotiation succeeds, the change effect is propagated to the partners affected directly by the change. We suppose that all partners project trustfully the updated DCR choreography. The SC records partners' involvement in a tamper-proof fashion during the change negotiation and propagation stages. If a misbehavior occurs, the blockchain logs can be used as a shared source of truth.

We present a prototype implementation as a proof-of-concept to evaluate the technical feasibility of the approach, and evaluate it experimentally by looking at transaction fees for a typical change. We leverage IPFS temporarily during the negotiation and propagation stages to process the updated DCR choreography for cost optimization considerations. Only a hash of the DCR process is stored into the SC. Our experiments show that the transaction fees required for the propagation are one order of magnitude higher than the ones for the negotiation, as the cost of the propagation mainly comprises the cost of the DCR choreography update. We appraise these costs to have more significance for heavy negotiation scenarios, and to be proportional to the number of participants involved in a public change, as the more participants, the more interactions are necessary with the SC.

In this paper, we only consider the compatibility checks between public DCR processes of partners as a correctness criterion. This ensures that the DCR choreography is safe and terminates in acceptable state, i.e., is deadlock free after the change. We are currently working on proving the consistency checks between one partner's private and public DCR processes. In the present work, we focus on the current instance of the process. Nonetheless, it is also interesting to consider the change at the process model level and that after change is validated, all future instances follow the change.

A limitation of the approach is the fact that the change initiator specifies the endorsers. This can be handled differently by considering a pre-specified list of the endorsers and the choreography participants agree on this list before starting the process instance [17]. In this way, the agreement on change negotiation and propagation can be placed off-chain. An on-chain transaction saying that the agreement is reached is stored in a multi-signed document in IPFS (this might require the use of a different blockchain platform). However, even with multi-sig mechanisms, the risk of private key loss remains and recovery schemes such as using secured wallets should be investigated [21]. Finally, governance should also be considered carefully when choosing the access control setup. For public blockchains, not every endorser should necessarily run their own full node to preserve the consensus. For permissioned blockchains, governance should be well shared between change endorsers to avoid any tampering or transaction misuse.

References

1. van der Aalst, W.M.P., Weske, M.: The P2P approach to interorganizational work-flows. In: Dittrich, K.R., Geppert, A., Norrie, M.C. (eds.) CAiSE 2001. LNCS, vol. 2068, pp. 140–156. Springer, Heidelberg (2001). https://doi.org/10.1007/3-540-45341-5_10
2. Weber, I., Xu, X., Riveret, R., Governatori, G., Ponomarev, A., Mendling, J.: Untrusted business process monitoring and execution using blockchain. In: La Rosa, M., Loos, P., Pastor, O. (eds.) BPM 2016. LNCS, vol. 9850, pp. 329–347. Springer, Cham (2016). https://doi.org/10.1007/978-3-319-45348-4_19
3. Madsen, M.F., et al.: Collaboration among adversaries: distributed workflow execution on a blockchain. In: FAB, p. 8 (2018)
4. Mukkamala, R.R., et al.: Towards trustworthy adaptive case management with dynamic condition response graphs. In: EDOC, pp. 127–136. IEEE (2013)
5. Debois, S., et al.: Replication, refinement & reachability: complexity in dynamic condition-response graphs. Acta Informatica **55**(6), 489–520 (2018). https://doi.org/10.1007/s00236-017-0303-8
6. Hildebrandt, T.T., Slaats, T., López, H.A., Debois, S., Carbone, M.: Declarative choreographies and liveness. In: Pérez, J.A., Yoshida, N. (eds.) FORTE 2019. LNCS, vol. 11535, pp. 129–147. Springer, Cham (2019). https://doi.org/10.1007/978-3-030-21759-4_8
7. Fdhila, W., et al.: Dealing with change in process choreographies: design and implementation of propagation algorithms. Inf. Syst. **49**, 1–24 (2015)
8. Fdhila, W., et al.: Multi-criteria decision analysis for change negotiation in process collaborations. In: EDOC, pp. 175–183. IEEE (2017)
9. van der Aalst, W.M.P., Lohmann, N., Massuthe, P., Stahl, C., Wolf, K.: From public views to private views – correctness-by-design for services. In: Dumas, M., Heckel, R. (eds.) WS-FM 2007. LNCS, vol. 4937, pp. 139–153. Springer, Heidelberg (2008). https://doi.org/10.1007/978-3-540-79230-7_10
10. Buterin, V., et al.: A next-generation smart contract and decentralized application platform. White paper **3**(37) (2014)
11. Debois, S., Hildebrandt, T., Slaats, T.: Safety, liveness and run-time refinement for modular process-aware information systems with dynamic sub processes. In: Bjørner, N., de Boer, F. (eds.) FM 2015. LNCS, vol. 9109, pp. 143–160. Springer, Cham (2015). https://doi.org/10.1007/978-3-319-19249-9_10
12. Decker, G., Weske, M.: Behavioral consistency for B2B process integration. In: Krogstie, J., Opdahl, A., Sindre, G. (eds.) CAiSE 2007. LNCS, vol. 4495, pp. 81–95. Springer, Heidelberg (2007). https://doi.org/10.1007/978-3-540-72988-4_7

13. Hallerbach, A., Bauer, T., Reichert, M.: Capturing variability in business process models: the Provop approach. J. Softw. Maint. Evol. Res. Pract. **22**(6–7), 519–546 (2010)
14. Henry, T., Brahem, A., Laga, N., Hatin, J., et al.: Trustworthy decentralized execution of declarative business process choreographies. In: ICSOC. Springer (2021)
15. Indiono, C., Rinderle-Ma, S.: Dynamic change propagation for process choreography instances. In: Panetto, H., et al. (eds.) OTM 2017. LNCS, vol. 10573, pp. 334–352. Springer, Cham (2017). https://doi.org/10.1007/978-3-319-69462-7_22
16. Li, C., Palanisamy, B.: Decentralized privacy-preserving timed execution in blockchain-based smart contract platforms. In: HiPC, pp. 265–274. IEEE (2018)
17. López-Pintado, O., Dumas, M., García-Bañuelos, L., et al.: Controlled flexibility in blockchain-based collaborative business processes. Inf. Syst. **104**, 101622 (2020)
18. Mendling, J., Weber, I., Van Der Aalst, W., et al.: Blockchains for business process management-challenges and opportunities. ACM TMIS **9**(1), 1–16 (2018)
19. Nahabedian, L., Braberman, V., D'ippolito, N., Kramer, J., Uchitel, S.: Dynamic reconfiguration of business processes. In: Hildebrandt, T., van Dongen, B.F., Röglinger, M., Mendling, J. (eds.) BPM 2019. LNCS, vol. 11675, pp. 35–51. Springer, Cham (2019). https://doi.org/10.1007/978-3-030-26619-6_5
20. Ryu, S.H., Casati, F., Skogsrud, H., Benatallah, B., Saint-Paul, R.: Supporting the dynamic evolution of web service protocols in SOA. TWEB **2**(2), 1–46 (2008)
21. Xiong, F., Xiao, R., et al.: A key protection scheme based on secret sharing for blockchain-based construction supply chain system. IEEE Access **7**, 126773–126786 (2019)
22. Xu, X., Weber, I., Staples, M.: Blockchain patterns. In: Xu, X., Weber, I., Staples, M. (eds.) Architecture for Blockchain Applications, pp. 113–148. Springer, Cham (2019). https://doi.org/10.1007/978-3-030-03035-3_7

Architecture of decentralized Process Management Systems

Kai Grunert[1,2]([mail]) [ID], Janis Joderi Shoferi[1,2], Kai Rohwer[1,2],
Elitsa Pankovska[1,2], and Lucas Gold[1,2]

[1] Technische Universität Berlin, Straße des 17. Juni 135, 10623 Berlin, Germany
[2] Service-centric Networking, Ernst-Reuter-Platz 7, 10587 Berlin, Germany
kai.grunert@tu-berlin.de
https://www.tu.berlin/snet/

Abstract. In the progressing digitalization, automated processes increasingly integrate IoT devices. However, incorporating field devices located in unstable environments can easily lead to situations in which central automation systems are no longer connected to the devices and their functionality. A temporary disconnection is especially likely for location-independent mobile systems such as smartphones, drones, and mobile robots. This situation can result in the suspension or interruption of the process execution. As a solution, we propose and analyze an old architectural approach: the decentralized execution of one process over multiple, collaborating process engines placed directly on the devices. We name the overall system a decentralized Process Management System (dPMS) and explain the architecture, the interfaces, and several aspects of the process deployment. For the latter, one of the most interesting procedures is Dynamic Deployment: it allows portable and, to some extent, self-organizing processes.

Keywords: decentralized Process Management System · Distributed and heterogeneous workflow enactment service · Business process management system · Process automation · Process execution environment · Process engine

1 Introduction

The digitization of work steps and process flows are omnipresent in today's companies. IT systems enable and support the automation of processes, especially of repeating business processes. In order to increase efficiency, this topic affects all sectors: business, public and private. Under consideration of the economic value, attempts are made to exploit the automation potential as far as possible [13].

From a technical point of view, automating process flows consists of two parts: the preparation/implementation of every process step and the connection of each step in the correct sequence. The former can be achieved, for example, by building and running a new software application, executing a business rule system, calling

© Springer Nature Switzerland AG 2022
C. Di Ciccio et al. (Eds.): BPM 2022, LNCS 13420, pp. 436–452, 2022.
https://doi.org/10.1007/978-3-031-16103-2_28

the interface of an existing IT system, or creating work instructions for a process participant. The latter connects and controls the data flow in the proper order to realize the entire process. This connection logic could be hard-coded within every application that realizes a process step. However, it would mean that any process change requires an application change. So, for a more flexible realization, central systems, such as Business Process Management Systems (BPMS) [11], usually manage the process and data flow based on a process description.

Conventional process automation systems, especially in the business domain, interact with systems running on a stable backend with a performant network connection. However, the growing number of integrations of all kinds of IT systems in a digitalized process affects automation systems with the classic problems of distributed computing [18]. This becomes increasingly problematic in areas where IoT devices are becoming more ubiquitous, such as in Smart Homes, Industry 4.0, Smart City, or logistics. Automating processes that include field systems like robots, smartphones, smart devices, and drones do not provide the same reliable connectivity. Some of these machines even have non-static locations, meaning they can move from one place to another. Thereby, the network connection quality usually changes, which can result in a complete loss of connectivity.

Depending on the use case, a disconnection of minutes, hours, or days between an IT system and the central controller can have critical consequences for the business operation. Although most systems can work autonomously on a process step, not being able to transmit the results will block the whole process flow. Trying to avoid this situation brings challenges for a process automation system. It requires deep technical knowledge, which delays the implementation and therefore weakens the business-IT alignment.

The presented problem has been addressed in research in the past, particularly under the notion of distributed, heterogeneous, or decentralized Workflow Enactment Services. In addition, the use of multiple collaborating agents has also been investigated in the context of process management systems (see Related Work in Sect. 5). However, a detailed explanation of the overall concept, the general architecture, the components, and the patterns of a direct and decentralized process execution system is still missing. In summary, we make the following contributions in this paper:

- We explain and define the concept of, what we call, a decentralized Process Management System (dPMS)[1] with its components, such as the Distributed Process Engines (DPE).
- We abstractly define the interfaces of the components and their modules.
- We examine the options and patterns for process deployment.
- We introduce an open-source implementation of a dPMS called PROCEED.

In the field of process systems, different terms have been used in recent years for the same or similar things. In the context of this paper, we also use some terms interchangeably:

---

[1] We do not extend the term Business Process Management System (BPMS) because the word "Business" is not inclusive enough for the private and public sectors.

* Workflow Management System (WfMS), (Business) Process Management System (BPMS, PMS), and automation system
* Process/Workflow/Execution Engine and Workflow Enactment Service
* Process, process definition, process specification, process description, and process model
* Work item, process step, task, and activity
* Tasklist and worklist

2   Example Use Case and Problem Definition

In the next decade, IoT devices will be ubiquitously deployed [6]. There will be an extreme federation between network-enabled machines by their connection with the Internet and the surrounding. Moreover, every machine will be an active element of many automated processes. These processes need to be managed and run reliably, but this is not always easy, as we will demonstrate in the following example (extended from [10]).

Figure 1 illustrates the communication architecture of an automated process in the Smart Home. The home network consists of a motion sensor, a camera, a door opener, and a loudspeaker, all positioned next to the door. A smartphone for displaying tasks and information is also connected to the network. To realize individual Smart Home processes, the homeowner needs an automation system (Process Management System, PMS) which nowadays is mostly a centralized service in the cloud. The orange lines depict the logical data connections from the automation system through the device manufacturer's cloud services to the devices. In contrast, the blue lines indicate that the physical data connections pass through more systems like multiple routers and networks.

The PMS enacts the process shown in BPMN at the bottom of the diagram: the motion sensor triggers a camera if it detects movement in front of the door. Then, the camera forwards the picture to the owner's smartphone, who needs to decide to lock or open the door. In case of a negative decision, the homeowner informs the visitor via the loudspeakers.

For the process execution, the automation system must continuously monitor the states of the devices and, depending on the process definition, start the subsequent step by contacting the involved devices. Unfortunately, the existing architecture introduces several possible sources of infrastructure errors (illustrated with red flashes) that can occur regardless of the correct functionality of the devices used to realize a process step. It may happen that the central systems, i.e., the PMS or the manufacturer's cloud services, crash or malfunction. Also, the underlying network infrastructure for data transmission may drop out, become congested, or become misconfigured.

Problem 1. Executing processes in a centralized Process Management System may be interrupted by many unrelated infrastructure errors, although the required systems may be fully functional.

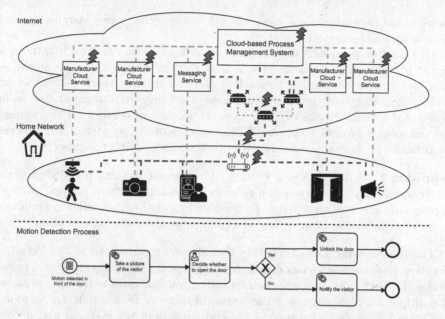

Fig. 1. The diagram shows the components and data flows for realizing an automated Smart Home process, which is depicted at the bottom in BPMN notation. The orange lines indicate the logical connections, whereas the blue line depicts the physical data flow. The process execution in a centralized Process Management System (PMS) usually requires connecting to multiple services, devices, and networks. The PMS, the middleware services, or the connections can fail (red flashes), leading to situations in which processes are interrupted, although the actual end devices are fully functional. (Color figure online)

These errors can completely stop the execution and observations of all processes, make the execution get stuck halfway through, or just be suspended. For example, a simple Internet outage of the Smart Home can cause the camera to no longer take the photo, although all systems in the home network work correctly and reach each other. Applying such infrastructure issues to the business, safety, and health domain, the situation can end critical if processes are not executed any further. However, ensuring that everything is correct, consistent, durable, and recoverable for process systems can become a challenge [1].

3   A decentralized Process Management System

For the problems shown, the fundamental reason for the interruption of the process flow is the centralized nature of the (Business) Process Management System. It coordinates the process flow from a server, and for every process step, it interacts with systems and machines located elsewhere. Hence, it can become challenging to ensure a correct, consistent, durable, and recoverable process execution [1]. Solutions like compensational or ACID transactions can sometimes resolve some issues for some use cases. However, other cases require a more continuous, uninterrupted execution.

Therefore we propose a decentralized process execution approach without a central coordinator. The basic idea is that the devices, systems, or machines know the process description themselves and can directly reach each other at runtime for subsequent process steps. Hence, we suggest running the steps in *Distributed Process Engines (DPE)* that are located next to or on the system that executes a process step. One DPE executes only one part of the process definition and cooperates directly with other DPEs to fulfill the process.

Definition 1. *A* decentralized Process Management System (dPMS) *defines, manages, and enacts processes in a decentralized manner, where multiple parts of a process are executed on different Distributed Process Engines, allowing multiple engines to control the process flow of a process instance during execution.*

The fundamental idea of a dPMS was already summarized in the WfMC's Workflow Reference Model as *distributed* and *heterogeneous workflow enactment service* [11, Sect. 3.3]. Processes can be split, grouped, executed, and monitored on multiple workflow engines, either from the same or from a different vendor.

Figure 2 displays and compares the components of the classical and decentralized approach. Essentially, the figure shows the same architecture: a process can be managed and executed inside multiple process engines with the same Interfaces[2]. In addition, there are tools for defining process models and monitoring the execution. However, the decentralized Process Management System focuses on enacting one process within multiple process engines, with no single engine having the overall responsibility for the process flow.

For successfully enacting a process, a dPMS uses multiple Distributed Process Engines – a variant of a standard process engine. Its main tasks are the execution of process steps and the coordination of the execution location of the following process step. Each DPE can support different functionalities internally (Sect. 3.3 describes a DPE in further detail). This configurability allows creating different Distributed Process Engines for different system requirements of the various Machine types. A *Machine* can be every system in an environment with a processing unit and a network adapter, from low-power, constrained devices to high-performance computers. Examples are every kind of IoT device, usual server systems, laptops, smartphones, drones, or mobile robots.

Machines may have *Capabilities*, i.e., functional abilities that relate directly to a Machine and that can be performed on or by the Machine, usually because of a physical property or direct connection. For example, these are sensing/acting activities on an IoT device, the possibility to take a photo with a connected camera, or the ability to control a drone. A running process instance within a Distributed Process Engine can use such Capabilities via Interface 6 (described in Sect. 3.3).

---

[2] In the course of the shift from Workflow to BPM systems, the naming has changed a bit: Workflow Engine = Process Engine, Workflow Client Applications = Worklist/Tasklist.

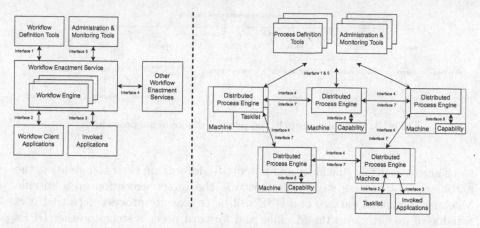

Fig. 2. The components of a Workflow System (left) defined by the WfMC [11, Sect. 3.2] and the components of a decentralized Process Management System (dPMS) (right). Although essentially the same, a dPMS emphasizes the decentralized coordination of the process flow by using multiple Distributed Process Engines (DPE) located on various Machines. The interfaces used are mostly identical. In addition, a dPMS has Interface 6 for direct use of the Capabilities of a Machine and Interface 7 for decentralized coordination of the DPEs.

3.1   The Execution of Decentralized Processes

In order to realize a decentralized process, the process designer starts by creating an executable process description in a Process Definition Tool. Executable means that the tool generates a machine-readable serialization of the description, that each element of the process notation is defined semantically unambiguously, and that the specification is self-contained by including all execution instructions.

The Process Definition Tool is also responsible for initially deploying the process or its parts to the Machines. There are several deployment mechanisms for transferring a process description to multiple DPEs. For example, many process steps require a specific Capability and therefore need to be transferred to different Machines. The various mechanisms and the initial deployment are detailed in Sect. 3.2.

At runtime, the DPE of a Machine creates a process instance and executes the first process step by interpreting the process description and running the execution instructions. After finishing the first step, the DPE searches for a suitable execution location for the next process step with the surrounding DPEs/Machines by evaluating the next steps' requirements. This results in allocating the step to either an external or the initiating DPE. Depending on the deployment pattern, it may or may not happen that the process description is deployed to the next engine during this allocation (see Sect. 3.2). Next, the initiating DPE transfers the instance data and process-related data like process variables to the next executing DPE. This procedure repeats after each completed process step until the entire process is finished. By distributing process

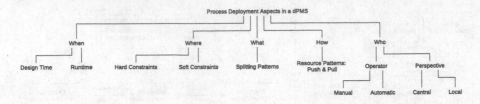

Fig. 3. Several aspects of the process deployment procedure.

steps across multiple Machines and potentially leveraging their Capabilities, the automation of process steps shifts towards the direct execution on a specific Machine. A typical pattern of a DPE will be to execute process steps that it is capable of executing on the Machine and forward process steps to other DPEs that it is not capable of executing.

Once the process execution has started, the process owner can monitor, disable, or modify the process at runtime using the Administration & Monitoring Tools. The tools can communicate with the involved DPEs by requesting data or by receiving data (DPE-initiated transmission). The challenge with decentralized process execution is that the execution can continue on local Machines while a central administration system loses the connection to the Machines. This can sometimes be handled by indirect data transmission, e.g., via a message queue, if it is guaranteed that all involved systems have access to it. Nonetheless, there are situations when Machines have constrained connectivity where centralized administration is not possible anymore. In these cases, it is usually still possible to monitor the execution in the local environment. For example, when the administration software is run directly in the Smart Home.

3.2 Deployment Patterns

For enacting one process on multiple engines, the deployment procedure of the executable process definition becomes essential. *Process deployment* is the act of finding and transferring the process description with its execution specification to the correct Distributed Process Engines. It does not include the transfer of the instance data at runtime. Figure 3 shows various aspects of the delivery: it must be decided when, where, and how are which process parts deployed and by whom. (In the following paragraphs, we explain the aspects. Because the paper's approach assumes that there is a DPE installed on each Machine, we reference it with the abbreviation "D/M" or only use the terms DPE or Machine.)

An important aspect of a dPMS process deployment is which parts of a process definition need to be delivered to a D/M for enacting one step (*what*). I.e., how the process description is split into one or more *process fragments*. A fragment contains one or more process steps. A *process step* is an atomic unit that a process engine can execute, such as a task, event, or gateway. A Machine must receive at least the process steps that it is intended to execute itself. However, a fragment can contain more steps than necessary for execution.

This can be relevant for administrative tasks or other deployment aspects, such as runtime re-deployment.

When defines the time when the deployment happens. It can be at design time or runtime. In fact, deployments cannot be exclusively at runtime because the process would never start without prior delivery to a process engine. Thus, deployment at runtime is always preceded by a one-time deployment at design time, but afterwards it allows process parts to be forwarded during runtime.

The disadvantage of deploying all process steps solely at design time is that the predefined D/Ms could no longer be reachable in the environment at runtime, which can happen with location-independent Machines. So, it is possible that the next process step cannot be triggered, which would lead to a deadlock. In contrast, a DPE can select a currently available Machine if it further deploys the process definition at runtime.

Each process step can have requirements that determine *where* or by whom it is processed, i.e., on which D/M or by which human process participant. Requirements are either hard or soft constraints. The former are conditions that must be satisfied before a D/M is allowed to execute a step. The "Creation" resource patterns in [19] define some possible hard constraints and can be used for assignments in a dPMS environment. However, since these patterns mainly focus on human agents and the physical Machines also play an essential role in a dPMS, hard constraints can refer to all possible properties of a Machine. Examples include the required Capabilities, a specific Machine address, a physical location, or – human-wise – a person/role that must work on a task.

Soft constraints are optimization functions to select the best D/M for a criterion, e.g., the one with the highest battery status or the lowest workload.

All requirements are evaluated at the time of deployment resulting in a pre-selection of concrete, suitable D/Ms for a process step. A Decider component and the new Interface 7 are responsible for this (see Sect. 3.3). The evaluation includes the discovery of Machines and matching the Machine's properties against the next step's requirements. The mechanisms are fundamentally the same as in centralized workflow systems, and the resource patterns description [19] discussed the aspects in detail. Nevertheless, the involvement of multiple process engines in one process flow is different. Thus, resource discovery includes searching for D/Ms in a dynamic network environment. Furthermore, the location of the requirement matching can be distributed: all properties can either be collected and matched at the deployment initiator, or all D/Ms receive the requirements and decide for themselves whether they satisfy them.

How describes the actual transmission to the resource, meaning the offering of a process step to the matching D/Ms and the final allocation. The pull and push resource patterns can be used for this purpose [19]. However, challenges occur with the transmission of a process step to human participants: unlike Machines, where a task can be transferred directly, deployment can only be done indirectly for humans via a Tasklist. In a centralized architecture, this is trivial because the responsible Tasklist that the user is working with is known. In a decentralized approach, it is much more difficult because a process participant can access and work on multiple Machines simultaneously. For example, the display of a user task on a smartphone, a laptop, or a smart TV may depend on the process participant's current working location. Therefore, the deployment must typically transfer the process step to each suitable D/M where the user is logged in (Pattern: distribution by offer – multiple resources [19]). The user selects the task in one of her Tasklists, which requires either a deletion on the other Tasklists or a continuous status synchronization. This distributed coordination without a central system gets even more complicated for indirect assignment patterns, such as role-based allocation.

Another aspect of deployment is *who* triggers it and what kind of view they have on the dPMS environment. It can be a manual (human) or an automatic operator. The operator's perspective influences to which D/Ms a process fragment can be delivered. The centralized view of the process owner can probably see many engines but cannot tell whether the local machines can reach each other. Therefore, centralized deployment can lead to deadlocks at runtime. On the other hand, the local perspective is the view of a D/M, which is particularly interesting for runtime deployments.

The shown aspects can be combined to deployment methods. In the following, we define two methods likely to be used in a dPMS.

Static Deployment deploys the process fragments at design time with direct allocation to a DPE from a central perspective. Therefore, by evaluating all the requirements, the fragments of a process are created. Each fragment is *statically* assigned and delivered to a DPE before runtime. As a result, at runtime, the selected DPEs only need to exchange the instance and process-related data for executing a process.
When: Design Time, Where: any, What: any, How: "Distribution by allocation" Push resource pattern, Who – Operator: any, Who – Perspective: Central

Dynamic Deployment refers to the transmission of the process description from DPE to DPE at runtime[3]. Typically, a DPE executes the process step that it is able to run and afterwards evaluates the next step's requirements. As a result, it selects an appropriate DPE[4] and sends the process description to it (in addition to the instance and process-related data).
When: Runtime, Where: any, What: constrained, How: any, Who – Operator: any, Who – Perspective: Local

With deployment at runtime from the local Machines, a special type of process emerges: a **Portable Process**. It is a self-contained, transportable package,

---

[3] As mentioned before, the process description must be transmitted once at design time for the process's inception.

[4] If the DPE finds no suitable execution engine, it may wait until one is available. This is desirable because an environment contains Machines that are added or removed over time.

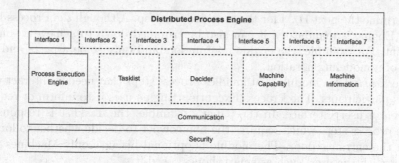

Fig. 4. The modules of a Distributed Process Engine, which implement Interface 1 to 7. The modules with a dashed border are optional.

similar to a transportable agent in [5]. It contains the complete process description, every process step specifications (e.g., source code), and metadata (e.g., requirements) for enacting a process step. Based on this data, a D/M searches for a suitable or optimal engine for executing the next process step. Then, it passes on the process package together with all runtime data. This way, the process moves from one D/M to another and from one network to another. It is even possible that the Machine itself, which stores the Portable Process, is transported to another environment, where it can search for a suitable execution location.

3.3 Building Blocks of a Distributed Process Engine

In a decentralized Process Management System, Distributed Process Engines (DPE) run on multiple Machines, executing parts of a process instance and coordinating the transfer of the runtime data to other DPEs. They can interact with process participants, the Machine's Capabilities, and external applications to fulfill the execution instructions.

Definition 2. *A Distributed Process Engine (DPE) is a workflow engine with the primary purpose to create, manage and enact process instances partially. It coordinates the separated process execution with other DPEs. A Distributed Process Engine runs on a Machine whose Capabilities it can make available to the process instance.*

A DPE has several modules, depicted in Fig. 4. They implement Interfaces 1 to 7. Some modules are optional to configure a DPE for a particular purpose, such as a specialized engine with a small footprint that runs on constrained devices.

The *Process Execution Engine* is a mandatory module. It allows storing, updating, deleting, and executing process specifications and fragments. The module understands the process notation to create and manage process instances. During execution, it offers the ability to start/stop/continue the process at any time. After each executed step, it usually interacts with the Decider to determine the next D/M for the next process step. Although the Process Execution Engine is mandatory, it can also be trimmed down to achieve a smaller footprint. For example, the engine may understand only some aspects and elements of a modeling notation.

The *Tasklist* is a module that handles process steps that need to interact with process participants. The Process Execution Engine uses it to inform or request input via a user interface. In the paper's example, the Tasklist is responsible for the process step, which requires the homeowner to decide whether to lock or unlock the entrance door. The module is optional because only Machines that interact with the user, such as smartphones, need it.

To execute a process in a decentralized manner, different Machines execute parts of the process. The *Decider* determines where to deploy the next process step for execution. To find a suitable location, the Decider compares the requirements of a process step (e.g., the demanded Capabilities) against the actual properties of all available Machines in the current environment. Communication-wise, there are two ways to achieve this. The first option is that the Decider itself searches for a suitable Machine in the network. Therefore, every other Machine must publish its properties, which is, for example, common in existing Smart Home protocols. The second option is that the Decider sends the information about the requirements to all available Machines, and they answer whether they can execute the upcoming process step.

The *Machine Capability* module provides a standardized interface for accessing the Machine's local functional abilities within the DPE. Its primary uses are within the Process Execution Engine to activate some Capability and within the Decider to validate the needed requirements. The module is optional because not every Machine has Capabilities, and not each process step requires Capabilities. Instead, there could be Distributed Process Engines that purely execute user interactions or computation tasks.

The optional *Machine Information* module collects and returns data about the Machine the DPE is running on, such as the current workload. Besides administering the systems, this can also be useful for the Decider to find an optimal next Machine by considering various other properties than the Capabilities. The *Security* module helps to secure decentralized network communication, including encryption, signing, authentication, and authorization.

The network connection is a prerequisite for collaborating with other D/Ms. For a mutual understanding of two network nodes, the communication must use the same protocol suite. The *Communication* module is responsible for negotiating the utilized protocol stack. It allows a DPE to connect with different network technologies such as Wifi, Ethernet, or Bluetooth, enabling various routing scenarios. Therefore, it abstracts the network interface for the DPE's other modules so that multiple protocols can be used in the background. Another task of the Communication module is to monitor the environment continuously, i.e., announce the D/M and discover other D/Ms.

The modules of a Distributed Process Engine realize Interface 1 to 7. The Workflow Reference Model [11] abstractly defines Interface 1 (Process Specification), Interface 2 (Tasklist), Interface 3 (Invoked Applications), Interface 4

(Process Engine Interoperability), and Interface 5 (Administration & Monitoring). The following briefly defines the newly introduced Interfaces 6 and 7.

Interface 6 specifies the interactions with the Machine's Capabilities. It is similar to Interface 3 for calling external applications, but without session establishment:

- *Retrieve* capability descriptions
- *Start/Suspend/Resume/Abort* a capability and *Query* execution status
- *Give/Retrieve* process relevant/application data
- *Signal* events, *Notify* completion

Especially for the Decider, it is essential to retrieve a capability description, which it can match against the process requirements. Furthermore, a process instance uses Interface 6 to activate a Machine's Capabilities during process execution.

Interface 7 specifies the interactions with the Decider component:

- *Retrieve* properties of the DPE and Machine (external)
- *Check* requirements for local execution (external)
- *Find* an optimal D/M for the given requirements (internal and external)

Internally, the Decider module implements Interface 7, and the Process Execution Engine uses it to find the optimal D/M for the next process step. Therefore, Interface 7 allows requesting the properties of other D/Ms for assessing this data against the requirements. However, to protect the process participant's privacy, for example, if the Machine is the user's smartphone, the interface should restrict the queried data. Instead, the Decider should send the requirements to all external D/Ms so they can check for feasibility themselves.

4 Evaluation and Discussion

To validate our approach, we have created an open-source decentralized Process Management System called PROCEED[5]. We used it to evaluate the decentralized execution of the paper's example and other processes.

4.1 The Proceed dPMS

PROCEED is an open-source dPMS that supports the Static and the Dynamic Deployment pattern. It has two main components: a Distributed Process Engine and a Management System. The latter contains the process definition, monitoring, and administration tools within one software application.

The DPE aims to run on various kinds of Machines, from server systems to microcontrollers. We chose JavaScript (JS) as the primary implementation

---

[5] https://gitlab.com/dBPMS-PROCEED/proceed.

language because of its popularity and because many runtimes are available for all types of platforms, even for microcontrollers [9]. However, it has the disadvantage that system functionalities such as networking and storage are not standardized. Instead, the JS runtime has to provide them, making it challenging to create platform-independent applications like the DPE that need to run in multiple runtimes. Therefore, we created a platform-independent abstraction layer for the system functionalities. In this way, the PROCEED DPE can run on all Node.js platforms (such as Windows, Linux, macOS), in the web browser, on Android, on iOS, and on selected microcontrollers.

Processes in PROCEED are specified in BPMN. So the DPE needed a BPMN-based Process Execution Engine written in pure JavaScript that can start/stop/continue at any step in the process description and be interruptible after each executed step. Since this did not exist, we created a new one[6].

The Management System is a complete development environment for processes, allowing process deployment in and monitoring of the local environment. Besides many graphical helper elements, automated process steps are written in JavaScript, and user tasks are created with an editor in HTML/CSS. It is available as a desktop (Windows, Linux, macOS) and a server application.

PROCEED is one of the first systems to enable decentralized process execution. Furthermore, since the code and concepts are publicly documented, it could help pave the way towards standardization of the interfaces or serve as a testbed for new process automation approaches.

4.2   Evaluation of the the Scenario

Figure 5 shows the decentralized execution of the Smart Home example process. Distributed Process Engines are installed on all devices except the loudspeaker (to demonstrate that external applications can still be called traditionally via remote calls). First, the homeowner (in the role of the process designer) creates the process specification in the Definition Tool. Then, she has to choose a deployment method.

The blue lines depict the Static Deployment. They indicate that the process definition can be split into multiple fragments. The homeowner can perform the splitting manually, or the Definition Tool can do this automatically by evaluating the requirements of the process steps. Next, the Definition Tool pushes all fragments to the suitable Machines at design time. With this method, the process specification has been entirely distributed to multiple D/Ms and will need no re-deployment of any fragment at runtime.

In contrast, the green lines show the Dynamic Deployment pattern. There is no process splitting at design time, and the Definition Tool deploys the full process specification only once to a suitable Machine for starting and executing the first process step. This method will require a re-deployment at runtime of the remaining process fragments, which the D/M is not capable of executing anymore.

---

[6] https://gitlab.com/dBPMS-PROCEED/neo-bpmn-engine.

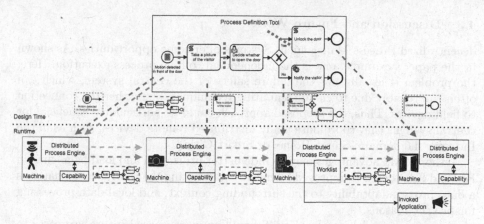

Fig. 5. The deployment and execution of a process in a decentralized Process Management System: the diagram depicts the deployment with the Static (blue arrows) and the Dynamic method (green arrows). At runtime, the Distributed Process Engines exchange the instance and process-related data (orange arrows). (Color figure online)

After the initial deployment, at runtime, the first DPE on the motion sensor starts a new process instance if some movement is detected. The DPE then detects that the next process step is either performed somewhere else (Static Deployment) or requires taking a photo of the visitor, which the motion sensor is not capable of doing (Dynamic Deployment). For the latter, it searches for an appropriate Machine in the environment by evaluating the requirement of finding a camera near the door. If successful, the D/M sends the process definition to a suitable Machine. This re-deployment is unnecessary for Static Deployment since the camera already has its process specification.

For starting the next step on the next DPE, the motion sensor sends the instance data containing the token information to the camera (orange arrow). The camera executes the "taking picture" step, analyzes the next step, and realizes that it is not able to show the picture to the homeowner. So, it passes the instance, the process-related data (the photo), and, if necessary, the process description to the smartphone. Then, via the Tasklist, the homeowner decides whether to open the door. The following process element, the exclusive split, is also run on the smartphone because it has no special requirements. In case of a positive decision, the smartphone forwards the process instance to the electronic door opener. In case of a negative outcome, the smartphone executes the last step by calling the loudspeaker via its API.

The presented way to control the process execution no longer requires a central process engine as a coordinator at runtime. Instead, multiple DPEs interpret parts of one process. This approach results in fewer intermediate infrastructure nodes. Therefore, in the event of connectivity problems or the outage of the centralized Process Management System, the processes can still run on the involved Machines if they can reach each other, which increases reliability.

4.3   Discussion and Future Work

decentralized Process Management Systems offer some opportunities. As shown in the paper's example, one is the independent, offline process execution. Here, the problem is usually not the failure safety of the central system, which can often run reliably due to cloud computing techniques, but rather the connection to field devices. Thus, the proposed approach could be particularly beneficial for central systems that coordinate many distributed remote devices, such as in the Industry 4.0 domain, Smart Home, or Smart Cities.

Other dPMS advantages may include better utilization and load balancing on Fog and Edge devices, improved latency by eliminating the central coordinator as a data proxy, adaptability to the surrounding context, and local data processing for privacy reasons.

As a result, a decentralized PMS could support several future use cases to coordinate the interactions between cloud, edge, and fog to form autonomous applications with intelligent placement of process steps. It supports the self-organized collaboration of systems within a process, which is a future research and development direction [6] to tackle some of the upcoming megatrends.

However, decentralized process execution also brings some challenges. For example, the reliability of Fog devices is usually lower than that of cloud systems, and process recovery is generally easier to manage from a central perspective than in the proposed decentralized approach. In addition, process deadlocks can occur more often in a dPMS than in a BPMS since the coordination with the network infrastructure introduces new failure points. For instance, if a DPE has received the process description at design time but, at runtime, the Machine is no longer available. Or if a process split results in parallel executions that need to be rejoined on one Machine that may not be reached by all incoming instance tokens. Recovering from such failures may require complex operations that are not resolvable in every case, e.g., when a token has already left the current environment.

Other challenges include ensuring security, monitoring, finding the optimal deployment location, and guaranteeing process semantics such as performing transactional operations.

We plan to provide a detailed analysis of the advantages and disadvantages of decentralized Process Management Systems and their potential for different application areas as future work.

In reality, a hybrid architecture is also possible: the centralized system would mainly control the process execution, and the decentralized approach would serve as a backup. This means that all involved Machines must store the process description and the last state of the process instance on their local DPE. Then, in the event of a connection error, the D/Ms can continue the process execution without connecting to the central system.

For the decentralized process architecture to be realized, widely adopted, and interoperable between multiple Machine vendors, it would be necessary for the presented Interfaces 1 to 7 in Sect. 3.3 to be precisely specified and standardized. Manufacturers would then have to implement or install a DPE on their Machines that comply with these standards. In multiple domains, the Machine owners and users would significantly benefit from such a development since it increases the interoperability of different Machines from multiple vendors, as the Smart Home example shows. However, manufacturers would need sufficient incentives and a good tool ecosystem to implement the specifications, which is not easy to achieve, as can be observed, for example, in the Smart Home sector.

5 Related Work

As mentioned earlier, the concept of a dPMS is not new and has been addressed under the notion of distributed, heterogeneous, or decentralized Workflow Enactment Services, e.g., in [2–4,7,11,15,17,20]. It can be noted that very different views on the topic are represented: from data and event distribution to the evaluation of optimal process definitions to the implementation using concrete technologies such as BPEL. Despite the large amount of related work, an overview of the general architecture with an abstract interface definition and the possible deployment mechanisms was still missing.

Moreover, many works focus on the combination with multi agent and grid systems such as [5,8,12,14,16,21]. Even though a dPMS is mainly based on the transfer of process descriptions and instance information, Distributed Process Engines can be interpreted as cooperating agents.

In general, many concepts and technologies have emerged in the last decade that advocate moving the computing parts of an application from centralized systems to a more decentralized approach. Fog Computing, Edge Computing, Content Distribution Networks, and Distributed Ledger Technologies are just some examples. These are not counter-developments to the centralized computing paradigm but a complement to solve some of the deficiencies. The same applies to process execution systems, and a shift towards more decentralized processing can be helpful for some use cases.

6 Conclusion

In this paper, we introduced the concept of a decentralized Process Management System. We explained the Distributed Process Engine, which is installed on multiple Machines and locally coordinates the execution of process steps. As a result, process execution does not require a connection to a central process coordinator, which can improve the continuous process execution. We also analyzed the deployment procedure and defined two different methods: Static and Dynamic Deployment. The latter leads to a Portable Process that can move autonomously over Machines and through environments.

References

1. van der Aalst, W.M.P.: Business process management: a comprehensive survey. Int. Schol. Res. Notices. **2013**, 1–36 (2013)
2. Bauer, T., Dadam, P.: Efficient distributed control of enterprise-wide and cross-enterprise workflows. In: Proceedings of Workshop Informatik 1999, pp. 25–32 (1999)

3. Bauer, T., Dadam, P.: Efficient distributed workflow management based on variable server assignments. In: Proceedings of CAiSE 2000, pp. 94–109 (2000)
4. Bauer, T., Reichert, M., Dadam, P.: Intra-subnet load balancing in distributed workflow management systems. Int. J. Cooper. Inform. Syst. **12**, 295–323 (2003)
5. Cai, T., Gloor, P.A., Nog, S.: DartFlow: A Workflow Management System on the Web using Transportable Agents. Computer Science Technical report, PCS-TR96-283 (1996)
6. European Technology Platform NetWorld2020: Smart Networks in the context of NGI - Strategic Research and Innovation Agenda 2021–2027, September 2020
7. Gokkoca, E., Altinel, M., Cingil, R., Tatbul, E.N., Koksal, P., Dogac, A.: Design and implementation of a distributed workflow enactment service. In: Proceedings of CoopIS 1997: 2nd IFCIS Conference on Cooperative Information Systems, pp. 89–98 (1997)
8. Grundspenkis, J., Pozdnyakov, D.: An overview of the agent based systems for the business process management. In: Proceedings of International Conference on Computer Systems and Technologies (2006)
9. Grunert, K.: Overview of JavaScript engines for resource-constrained microcontrollers. In: 5th International Conference on Smart and Sustainable Technologies 2020, pp. 1–7 (2020)
10. Grunert, K.: Towards uninterruptible smart home processes. In: 2022 IEEE 24th Conference on Business Informatics (CBI), pp. 80–87 (2022)
11. Hollingsworth, D.: Workflow Management Coalition: The Workflow Reference Model. Document Number TC00-1003 (1995)
12. Jennings, N.R., Norman, T.J., Faratin, P., O'Brien, P., Odgers, B.: Autonomous agents for business process management. Appl. Artif. Intell. **14**, 145–189 (2000)
13. Jiménez-Ramírez, A.: Humans, processes and robots: a journey to hyperautomation. In: González Enríquez, J., Debois, S., Fettke, P., Plebani, P., van de Weerd, I., Weber, I. (eds.) BPM 2021. LNBIP, vol. 428, pp. 3–6. Springer, Cham (2021). https://doi.org/10.1007/978-3-030-85867-4_1
14. Joeris, G.: Decentralized and flexible workflow enactment based on task coordination agents. In: 2nd International Bi-Conference Workshop on Agent-Oriented Information Systems, pp. 41–62 (2000)
15. Martin, D., Wutke, D., Leymann, F.: A novel approach to decentralized workflow enactment. In: 2008 12th International IEEE Enterprise Distributed Object Computing Conference, pp. 127–136 (2008)
16. O'Brien, P.D., Wiegand, M.E.: Agent based process management: applying intelligent agents to workflow. Knowl. Eng. Rev. **13**, 161–174 (1998)
17. Reichert, M., Bauer, T., Dadam, P.: Enterprise-wide and cross-enterprise workflow management: challenges and research issues for adaptive workflows. In: Proceedings of Workshop Informatik 1999, pp. 56–64 (1999)
18. Rotem-Gal-Oz, A.: Fallacies of distributed computing explained. In: Whitepaper (2006)
19. Russell, N., ter Hofstede, A.H.M., Edmond, D., van der Aalst, W.M.P.: Workflow Resource Patterns. In: BETA Working Paper Series, WP 127, Eindhoven University of Technology (2004)
20. Singh, M.P.: Distributed Scheduling of Workflow Computations. Technical report, North Carolina State University, Department of Computer Science (1996)
21. Yu, J., Buyya, R.: A taxonomy of workflow management systems for grid computing. J. Grid Comput. **3**, 171–200 (2005)

Author Index

Printed in the United States
by Baker & Taylor Publisher Services